Ecosystem Engineering

Ecosystem Engineering

Edited by Isaac Hughes

SYRAWOOD
PUBLISHING HOUSE
New York

Published by Syrawood Publishing House,
750 Third Avenue, 9th Floor,
New York, NY 10017, USA
www.syrawoodpublishinghouse.com

Ecosystem Engineering
Edited by Isaac Hughes

International Standard Book Number: 978-1-68286-424-1 (Hardback)

Cataloging-in-publication Data

Ecosystem engineering / edited by Isaac Hughes.
p. cm.
Includes bibliographical references and index.
ISBN 978-1-68286-424-1
1. Biotic communities. 2. Ecological engineering. 3. Biodiversity. I. Hughes, Isaac.

QH541 .E26 2017
628--dc23

Printed in China.

TABLE OF CONTENTS

PREFACE

Ecosystem engineering can be defined as any intervention in nature by living organisms to modify and alter a habitat. This book on ecosystem engineering discusses the various methods of interaction between humans and nature and its relation to restoration ecology, preservation and habitat management. The various advancements in ecosystem engineering are glanced at and their applications as well as ramifications are looked at on detail in this text. It aims to shed light on some of the unexplored aspects and recent researches in this field. Students and researchers in the field of biodiversity, wildlife preservation, microbiology and natural habitat management will find this book very useful. This book attempts to assist those with a goal of delving into the field of ecosystem engineering.

In my initial years as a student, I used to run to the library at every possible instance to grab a book and learn something new. Books were my primary source of knowledge and I would not have come such a long way without all that I learnt from them. Thus, when I was approached to edit this book; I became understandably nostalgic. It was an absolute honor to be considered worthy of guiding the current generation as well as those to come. I put all my knowledge and hard work into making this book most beneficial for its readers.

I wish to thank my publisher for supporting me at every step. I would also like to thank all the authors who have contributed their researches in this book. I hope this book will be a valuable contribution to the progress of the field.

Editor

1

Evidence of Season-Dependency in Vegetation Effects on Macrofauna in Temperate Seagrass Meadows (Baltic Sea)

Maria Włodarska-Kowalczuk*, Emilia Jankowska, Lech Kotwicki, Piotr Balazy

Department of Marine Ecology, Institute of Oceanology Polish Academy of Sciences, Sopot, Poland

Abstract

Seagrasses and associated macrophytes are important components of coastal systems as ecosystem engineers, habitat formers, and providers of food and shelter for other organisms. The positive impacts of seagrass vegetation on zoobenthic abundance and diversity (as compared to bare sands) are well documented, but only in surveys performed in summer, which is the season of maximum canopy development. Here we present the results of the first study of the relationship between the seasonal variability of seagrass vegetation and persistence and magnitude of contrasts in faunal communities between vegetated and bare sediments. The composition, abundance, biomass, and diversity of macrozoobenthos in both habitats were compared five times throughout the year in temperate eelgrass meadows in the southern Baltic Sea. Significant positive effects of macrophyte cover on invertebrate density and biomass were recorded only in June, July, and October when the seagrass canopy was relatively well developed. The effects of vegetation cover on faunal species richness, diversity, and composition persisted throughout the year, but the magnitude of these effects varied seasonally and followed changes in macrophyte biomass. The strongest effects were observed in July and coincided with maximums in seagrass biomass and the diversity and biomass of other macrophytes. These observations indicate that in temperate, clearly seasonal systems the assessment of macrophyte impact cannot be based solely on observations performed in just one season, especially when that season is the one in which macrophyte growth is at its maximum. The widely held belief that macrophyte cover strongly influences benthic fauna in marine coastal habitats, which is based on summer surveys, should be revisited and complemented with information obtained in other seasons.

Editor: Richard K. F. Unsworth, Seagrass Ecosystem Research Group, Swansea University, United Kingdom

Funding: The research leading to these results has received funding from the European Regional Development Fund ("ZOSTERA. Restitution of key elements of the inner Puck Bay ecosystem.", no. POIS.05.01.00-00-205/09-00) and European Community's Seventh Framework Programme (FP7/2007-2013) within the Ocean of Tomorrow call under Grant Agreement No. 266445 for the project Vectors of Change in Oceans and Seas Marine Life, Impact on Economic Sectors (VECTORS). MWK and LK were supported by the Institute ofOceanology Polish Academy of Sciences. The funders had no role in study design, data collection and analysis, decision to publish, or preparation of the manuscript.

Competing Interests: The authors have declared that no competing interests exist.

* Email: maria@iopan.gda.pl

Introduction

Seagrasses and associated macrophytes are important structural and functional components of coastal ecosystems in temperate and tropical seas. They are classic examples of marine ecosystem engineers [1], as they can modify water flow regimes and sedimentation rates, and, hence, modulate the availability of resources to other species [2]. They play the important role of habitat-forming species both under ground, with root systems, and above ground, with vegetative organs, which contribute three-dimensional structures to the sea bed architecture and provide shelter and numerous niches for other organisms, mostly benthic invertebrate and fish species [3]. Seagrasses and associated macrophytes, especially epiphytes, can be direct food sources and sustain a number of grazing invertebrates; thus, they shape carbon flow pathways in shallow-water food webs [4]. The effects of seagrass vegetation on the other components of the benthic systems can be best perceived when seagrass meadows and other habitats are compared [5]. Several studies compare the zoobenthic communities in vegetated and bare sediments and document the positive impact of seagrass vegetation in terms of density, biomass, and diversity [6-8]. However, the vast majority of reports on the positive impacts of seagrasses on benthic fauna are based on results of surveys conducted only in summer when seagrass meadows are logistically most accessible but also when seagrass canopies are at their maximum of annual development. To our knowledge, there have been almost no attempts to explore the persistence of these effects in other seasons. The only report of seasonal contrasts in zoobenthic response to vegetation is Gambi et al. [9], who compare polychaete assemblages in *Posidonia oceanica* and *Zostera noltii* vegetated sediments and on bare sand in the Mediterranean Sea based on a study performed in two seasons - summer and winter. Some indication of seasonality in the importance of seagrass vegetation for associated fauna is reported by Berkenbusch et al. [10]. They identify the biomass of *Zostera novazelandica* as the best explaining factor for macrofaunal community structure in summer at the seagrass growing optimum, while it is much less important in fall. The lack of the seasonal aspect in the rich literature on seagrass influences on zoobenthic communities is a significant gap in understanding the functioning of these systems.

Ecosystem engineer effects can depend on the quantitative characteristics of engineering species populations [1,11,12]. Thus, it can be expected that strong seasonality in these populations can influence their role in local systems. Fortino [13] explores the effects of benthivorous fish on sediment accumulation in a North

American river, but, upon finding no effect in winter, linked it to temperature-driven seasonal reductions in fish activity levels. Berkenbusch et al. [10] report that the bioturbation effects of the ghost shrimp *Callianassa filholi* on intertidal flats persisted throughout the year, but the magnitude of impact varied seasonally correspondingly to changes in bioturbator density. Regarding seagrasses, Hasegawa et al. [14] show that the occurrence and magnitude of the engineering effects of *Zostera marina* (e.g., reduction of current velocity, buffering of sediment resuspension) was season dependent and followed temporal changes in seagrass canopy development. These reports support Fortino's [13] claim that the quantification of engineering organism effects during a single season may exaggerate, or downplay, the role of these processes throughout the year, and a comprehensive assessment of the significance of a given species must be based on information obtained in all seasons.

Seagrass and associated macrophyte canopies can vary substantially in annual cycles [15]. Seasonal variability is generally greater at higher latitudes [16], but clear seasonal trends can be also observed at lower latitudes (e.g. [17]). Duarte [16] analyzed annual biomass variability of 11 macrophyte species, and reports that the coefficient of variation (i.e., variability above the annual mean) ranged from 17% to 120% and was dependent on latitude, but it could also vary locally and with depth and exposure to disturbance. Seasonal dynamics are also species-specific with stronger interannual variability in smaller species (e.g., it is more pronounced in *Z. marina* compared to *P. oceanica*; [18]). Several populations of *Z. marina* in northwest Europe and North America were even reported to employ annual or semi-annual life histories; when no or only a small percentage of vegetative shoots survive winter, and the persistence of a population relies on extensive seed production and new recruitment each spring [19], [20]. In some regions, knowledge of the seasonal variability of macrophyte vegetation in seagrass meadows is lacking or very poor. For example, in the Baltic Sea there have been no reports on the status of seagrass vegetation in different seasons, except for a recent study by Jankowska et al. [21] that was performed in the Puck Bay, which is located in Polish coastal waters.

Duarte [16] suggests that assessing the importance of seagrass at its maximum biomass may not paint a complete picture of its role in systems. Therefore, quantifying biomass variability is important if we are to evaluate the role of seagrasses in coastal ecosystems. Temporal variability in seagrass biomass is likely to determine temporal variability in habitat structure or food availability for the associated biota. Duarte [16] also predicted that latitude-related change in degrees of seasonality can have important ecosystem implications. At low latitudes, benthic communities associated with seagrass beds can benefit from stable levels of macrophyte nutritional and structural support, while in temperate waters, fauna experience considerable seasonal habitat change that can influence its functional and structural balance throughout the year. However, this hypothesis has never been tested.

Jankowska et al. [21] offers the first report on the seasonality of macrophyte cover in Baltic Sea eelgrass meadows. These authors demonstrate that seagrass vegetation persists throughout the year, but both shoot density and the aboveground biomass of the plant as well as the abundance and diversity of associated macrophytes undergo substantial seasonal changes with clear declines in winter. It is intriguing to learn if the seasonality in macrophyte vegetation shapes the patterns of differences in other components of benthic ecosystems between vegetated and bare sediments. Based on the results of other studies performed in Baltic Sea eelgrass meadows [6], [22], we can anticipate that a strong positive impact of macrophyte vegetation on zoobenthic composition, abundance, and diversity occurs in summer, but we do not know if these effects persist in seasons when the macrophyte canopy is not at its maximum. In the present study we aimed to test: 1) if differences in composition, abundance, biomass, and diversity of macrozoobenthic communities between vegetated sediments and bare sand in southern Baltic *Z. marina* meadows are consistent across different seasons; 2) if the persistence and magnitude of these differences mirror seasonal changes in macrophyte cover quantitative and qualitative characteristics. In this paper, we present the first analyses of the relationship between the temporal variability of macrophyte vegetation in a seagrass meadow and its impacts on associated fauna.

Material and Methods

Ethics Statement

The collection of macrophytes, sediment and macrofauna samples undertaken for this research was approved by collection permits issued to a "ZOSTERA. Restitution of key elements of the inner Puck Bay ecosystem," project (no. POIS.05.01.00-00-205/09-00) coordinated by Center of Coordination of Environmental Projects and Regional Directorate of Environmental Protection and Nature Conservation at the Pomeranian Voivodship.

Sampling and laboratory analyses

The samples were collected in the Puck Bay, which is a shallow embayment in the southwestern part of the Gulf of Gdańsk (Baltic Sea) separated from the open sea by the Hel Peninsula. The eight-kilometer Ryf Mew sandbank divides the bay into two parts: the inner Puck Lagoon and the outer Puck Bay. The depths and bottom morphology of these two parts of the bay differ. The outer Puck Bay is notably deeper at an average depth of 20.5 m, while the Puck Lagoon is considerably shallower with an average depth of 3.1 m [23]. Extensive *Zostera marina* meadows were present in the Puck Lagoon and the outer Puck Bay in the 1950s, but from the late 1950s to the 1980s degradation of the meadows and the gradual replacement of eelgrass by filamentous algae and *Zanichella palustris* was observed [24]. A recent inventory of the Polish Exclusive Economic Zone sea bed habitats documented that areas covered by *Z. marina* meadows are growing in size [25].

Two localities were selected for sampling: 1) RM – located in Puck Lagoon close to the shallows of the Ryf Mew sandbank at a depth of 3 m (54°42,7 N 18°33,7 E); 2) JAS – located in the outer part of the Puck Bay east of Jastarnia at a depth of 1.5 m (N 54°41,4 E 18°40,9; Figure 1). Samples were collected five times throughout the year from October 2010 to October 2011. Sampling started in fall 2010 (10–13 October 2010, OCT'). Sampling representative of winter conditions was performed on 30–31 March 2011 (MAR) just after the ice cover disappeared. The winter of 2010/2011 was very cold, and the Puck Bay was covered with ice until 20 March (http://www.smhi.se/oceanografi/iceservice/is_prod_en.php). However, high concentrations of chlorophyll a in the surface sediments at this sampling time [21] suggested that the diatom bloom could have already occurred; thus, based on the pelagic system phenology, this sampling should rather be defined as having been performed in early spring. Regarding the macrophytes, the values of species richness, density and biomass were very low – indicating that vegetation still remained in the winter phase of development. The samples were then collected in late spring (1 June 2011, JUN), summer (19 July 2011, JUL), and fall (8 and 15 October 2011, OCT).

At each sampling location five replicate sets of samples were collected from the vegetated seabed and bare sand. Each set of

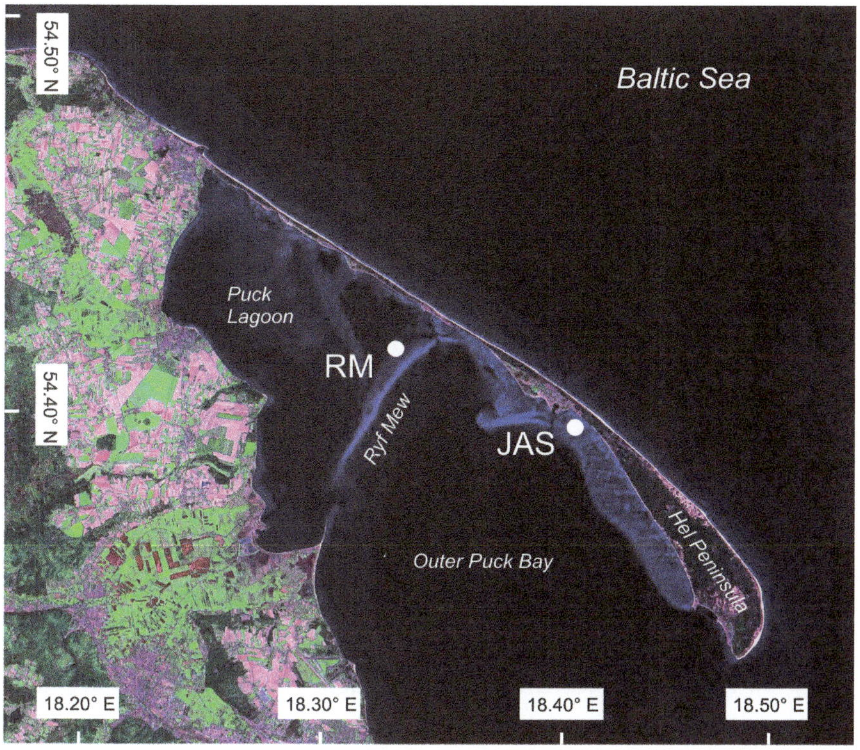

Figure 1. Location of sampling sites in the Puck Bay. The map is based on Landsat picture taken in 2000, publicized by NASA and located in public domain of Wikipedia (http://en.wikipedia.org/wiki/File:Mierzeja_Helska.jpg).

samples included one collected with a 15 cm diameter large core for macrozoobenthos and three sediment samples collected with a 2 cm diameter small core. Sediment samples were collected to analyze grain size (upper 10 cm), POC and TN concentrations, $\delta^{13}C$, $\delta^{15}N$ (upper 2 cm), and photosynthetic pigment concentrations (upper 2 cm). The samples for macrofauna were sieved through 0.5 mm mesh and then fixed in 4% formaldehyde. The sediment samples for photosynthetic pigment analyses were frozen at $-80°C$.

In the laboratory, chlorophyll a and pheopigment concentrations were assessed with the spectrophotometric method [26]. POC, TN, $\delta^{15}N$, and $\delta\ ^{13}C$ in sediments were analyzed using an Elemental Analyzer Flash EA 1112 Series combined with an Isotopic Ratio Mass Spectrometer IRMS Delta V Advantage (Thermo Electron Corp., Germany). To determine the grain-size distribution, sediment samples were dried (48 h, 60°C) and sieved through mesh sizes at thirteen 0.5 phi intervals [27]. In the laboratory all macrofaunal individuals were identified to the level of species or the lowest possible taxonomic level, counted, and weighed. Macrophytes were identified to the lowest possible taxonomic level. Algae and plants were dried in 60°C for 48h and weighed. Seagrass shoots were counted, leaf length was measured, and the dry weight of the aboveground and the belowground parts was determined.

Data analyses

Species richness, defined as the number of taxa in a sample (S), species diversity measured with the Shannon-Wiener diversity loge based index (H), and evenness of distribution of individuals among taxa expressed by the Pielou index (J), were calculated for all macrobenthic samples. Differences in macrozoobenthic univariate characteristics (density, biomass, S, J, H) among the five months (Mt) and between the two stations (St), and two bottom types (Bt, vegetated bottom and bare sand) were tested using the three-way PERMANOVA model based on a similarity matrix created from the Euclidean distances among samples. The PERMANOVA routine does not rely on the assumption of normally distributed data and uses permutations to test null hypotheses. Unbiased estimates of each of the components of variation (CV) were calculated from mean squares to compare the amount of variation that is attributable to different terms in the model. When a significant effect of Bt and significant interaction of BtxMt were detected, pairwise tests for differences between bottom types within the five months were performed. The magnitude of the effect of vegetation on macrofauna univariate characteristics in different months was estimated using of the veg/bare ratio, which is the ratio of the mean recorded on vegetated bottom to that on bare sand.

Bray-Curtis similarities were calculated for presence/absence data (pr/ab), double root transformed, and untransformed data (nonTR) of macrobenthic species abundances in the samples. The three-way PERMANOVA model, with the three factors of Mt, St, and Bt, was applied to the similarity matrices. The patterns of macrobenthic composition were visualized with PCO plot. In addition, CAP (constrained ordination, discriminating among *a priori* groups) was used to visualize the variability along the two axes that best discriminated groups of samples defined by bottom type and month. When the significant effect of Bt and the significant interaction of BtxMt was detected by the main PERMANOVA test, pairwise tests for differences between bottom types in the five months were performed. The magnitude of the effects of vegetation on the macrobenthic community composition in different months was estimated with the distance (Bray Curtis index) between centroids representing the fauna collected from two bottom types. An nMDS plot of Bray-Curtis similarities between averaged species abundances in groups of samples defined by two sites, two bottom types, and five months was constructed to visualize multivariate variability among these groups of samples.

Relationships between environmental variables and macrobenthic community composition were investigated using the Distance-based Linear Models (DISTLM) procedure in PERMANOVA+ [28]. The original data set of environmental variables included 13 quantitative and three categorical (nominal) variables (Mt, St, Bt, as defined above). Quantitative variables included descriptors of organic matter content ($\delta^{13}C$, $\delta^{15}N$, POC, TN, POC/TN), photosynthetic pigments (chla - concentration of chlorophyll a, CPE – chloroplastic pigment equivalents, i.e., the concentration of all photosynthetic pigments, including pheopigments, %chla – percentage of chla in CPE, chla/POC), and the granulometric characteristics of sediments (mean grain size, sorting, fine sand and coarse sand fractions). These variables were tested preliminarily for collinearity using Draftsman plot and the Spearman correlation matrix. Based on the results, only six quantitative variables were left for further analysis (POC, POC/TN, chla, $\delta^{13}C$, mean grain size, fine sand). The forward selection procedure was used to determine the best combination of predictor variables [28].

For samples collected on vegetated bottom the DISTLM analyses was used to explore the relationships between macrophyte biometrics and macrozoobenthic community characteristics. The original data set included 6 quantitative variables: total macrophyte biomass, seagrass above ground biomass, total seagrass biomass, seagrass shoot density, algal biomass and algal species richness. These variables were tested preliminarily for collinearity using Draftsman plot and the Spearman correlation matrix. Based on the results, seagrass total biomass was excluded from further analysis. Similarity matrix created from the Euclidean distances among samples was used for macrozoobenthic univariate characteristics (density, biomass, species richness) while Bray-Curtis similarities of double-root transformed data of species abundances in samples were used for the analyses focused on faunal species composition. The forward selection procedure was used to determine the best combination of predictor variables.

The SIMPER procedure was applied to identify species that were responsible for differences in macrobenthic communities between bare sand and vegetated bottom. SIMPER analysis was performed on Bray-Curtis similarities of both double-root transformed and presence absence data. The SIMPER procedure is based on the breakdown of average dissimilarity between groups into separate contributions from each species [29]. The average contribution to the overall dissimilarity divided by standard deviation (diss/SD) and the percentage of this contribution to total dissimilarity (contr%) were calculated and used to identify discriminating species.

Results

Similar seasonal patterns were observed for all measured seagrass vegetation characteristics (seagrass density, aboveground and belowground biomass, leaf length) with minimum values recorded in MAR and maximum in JUL (Figure 2). The biomass and diversity of other macrophytes also differed among the months studied. In MAR, no macrophyte species other than *Z. marina* was recorded. The highest biomass and species diversity (six species) was recorded in JUL. Brown filamentous algae *Pilayella litoralis* comprised from 30% to 80% of the total biomass of macrophytes accompanying eelgrass in this meadow. Further details of macrophyte vegetation trait variability at the studied stations are reported by Jankowska et al. [21].

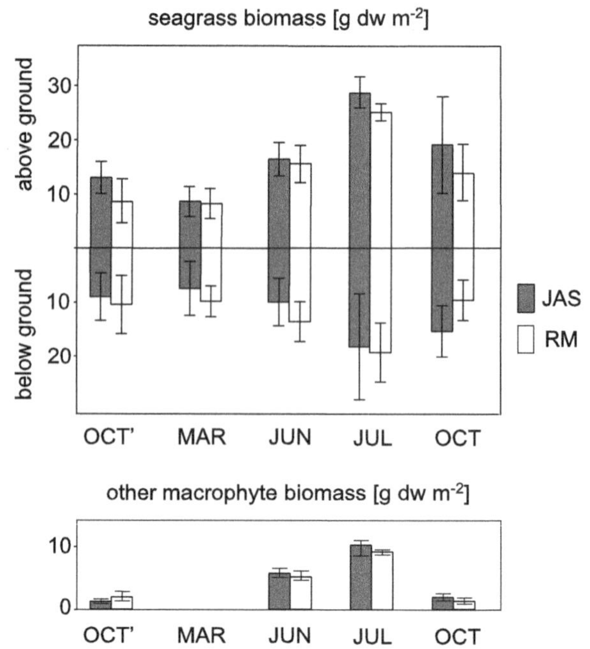

Figure 2. Seasonal variability in macrophyte vegetation characteristics. Seagrass aboveground and underground biomass and other macrophyte biomass [g dw m^{-2}] recorded in 5 months (OCT' - October 2010, MAR - March 2011, JUN - June 2011, JUL - July 2011, OCT - October 2011) at two sampling stations (JAS - gray bars, RM - white bars). Mean and 0.95 CI values are presented. Based on data published by Jankowska et al. [20].

There were significant differences in macrobenthic density among months, stations, and bottom types (PERMANOVA, $P<0.05$, Table 1). The lowest mean density was recorded in OCT on bare sand at the JAS site (1002.2±150.1 SD ind. 0.1 m^{-2}), the highest mean density and high variability was observed in JUL on the vegetated bottom at the RM site (11003.5±7144.6 SD ind. 0.1 m^{-2}, Figure 3). There was significant interaction between month and bottom type, and pairwise tests identified significant effects of vegetation on macrofauna density only in the three months of JUN, JUL, and OCT (Table 2). The strongest effect was documented in JUL, when the mean density on the vegetated bottom was almost three times higher than that recorded on bare sand (Figure 4). Macrobenthic biomass differed significantly between the two stations and two bottom types (PERMANOVA, $P<0.05$, Table 1), while it did not differ among the sampling months. The lowest mean biomass was recorded in OCT on bare sand at the RM site (5.8±1.1 g 0.1 m^{-2}). The very high mean biomass and highest variability in values were recorded in JUN and JUL in vegetated sediment at the JAS site (46.5±35.4 and 57.2±36.5 g 0.1 m^{-2}, respectively). The factor of bottom type had the strongest effect on macrobenthic biomass (the highest CV value, Table 1), but this influence was dependent on month (significant MtxBt interaction, Table 1). The positive effect of vegetation was observed only in JUN, JUL, and OCT, with more than a fivefold increase in faunal biomass on the vegetated bottom documented in JUL (Figure 4).

Forty six taxa were identified in samples (Table S1). Species richness per sample (*S*) differed significantly among months, bottom types, and stations (PERMANOVA, $P<0.05$, Table 1). The three-way PERMANOVA pseudoF values indicated that

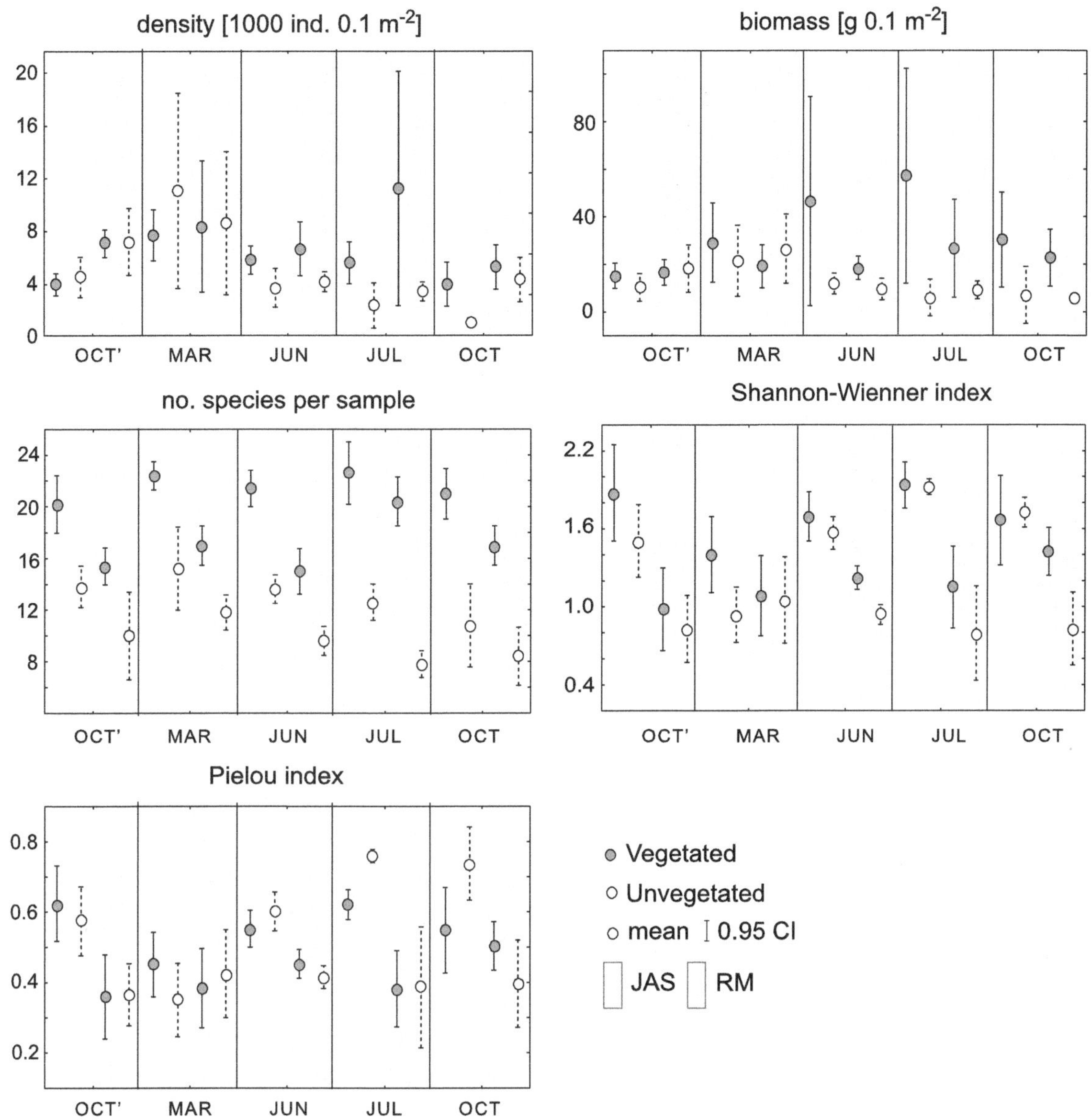

Figure 3. Seasonal variability in macrozoobenthic community characteristics. Density [1000 ind. 0.1 m^{-2}], biomass [g 0.1 m^{-2}], species richness (number of species per sample), diversity (Shannon-Wiener index), and evenness (Pielou index) recorded in 5 months (OCT'- October 2010, MAR - March 2011, JUN - June 2011, JUL - July 2011, OCT - October 2011) at two stations (JAS - white background, RM - gray background) and two bottom types (vegetated, unvegetated). Mean and 0.95 CI values are presented.

bottom type and station had a much stronger effect on species richness than did month (Table 1). Throughout the year *S* was higher at the JAS site (mean 17.4±4.6 SD) than at the RM site (mean 13.2±4.3, Figure 3). *S* was also higher in vegetated bottoms than on bare sand for all combinations of months and sampling sites. The effect of vegetation on *S* was significant in all months, but the ratio of the mean recorded in vegetated sediments to that on bare sand increased from 1.5 in OCT' and MAR and 1.6 in JUN to 1.9 in OCT and 2.1 in JUL (Figure 4). Evenness (*J*) was higher in samples collected at the JAS site than at RM, but this difference was not consistent throughout the sampling period as was indicated by the significant effects of months and significant MtxSt interaction (Table 1). The lowest and highest mean values of *J* were recorded on bare sand at the JAS site, with the lowest in March (0.351±0.084 SD) and the highest in July (0.760±0.015, Figure 3). PERMANOVA did not detect any significant effect of bottom type on *J* (Table 1). Station, month, and bottom type were sources of significant difference in species diversity (*H*), and station had the strongest effect (as indicated by the highest CV). *H* at the JAS site was 1.62 on average (±0.33 SD), while at RM it was 1.03 (±0.27 SD). *H* was significantly higher on vegetated bottoms than on bare sand, but this effect did not interact with the effect of month (Table 1).

PERMANOVA identified significant effects (at $P<0.001$) of Bt, St, and Mt on macrobenthic community composition (Table 2). Bottom type had the strongest effect when pr/ab or double root transformed data were analyzed, as indicated by the highest CV values and illustrated on the MDS plot of group averages (Figure 5). Groups defined by bottom type and site could be easily discriminated on the PCO plot (Figure 6). Samples collected

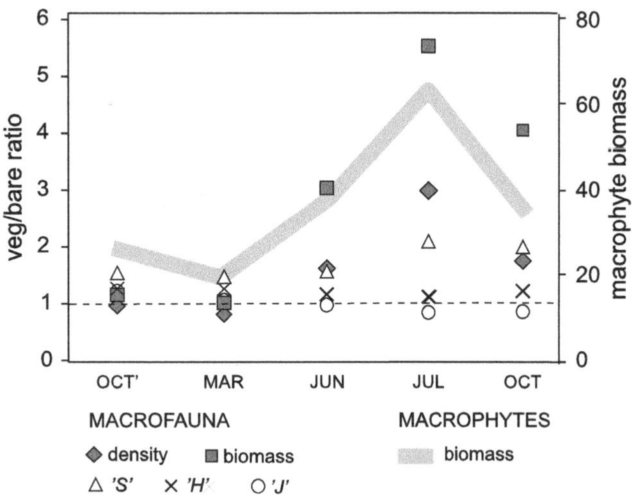

Figure 4. Magnitude of difference in macrofauna univariate characteristics between two bottom types plotted against total macrophyte biomass [g dw m^{-2}]. Veg/bare ratio - ratio of mean recorded on vegetated bottom to mean on bare sand, for samples collected in 5 months (OCT'- October 2010, MAR - March 2011, JUN - June 2011, JUL - July 2011, OCT - October 2011). Veg/bare ratio presented for density, biomass, number of species per sample (*'S'*), Shannon-Wiener index (*'H'*), Pielou index (*'J'*).

on vegetated bottoms and on bare sand were also clearly separated on the (St, Mt) CAP plot (Figure 6). On this plot, points representing different months largely overlapped, especially in the case of samples collected from vegetated sediments. Points representing samples collected from bare sand were much more dispersed, and the different months were better separated.

The significant interaction of BtxMt (Table 2) suggested that the effects of vegetation on macrobenthic composition might vary across the five months studied. Pairwise comparisons of samples collected from two bottom types performed separately for each month documented significant effects in all months for double root and nonTR data, while for pr/ab data, significant effects were observed for all months except March. In all cases (pr/ab, double root transformed, nonTR data), the distance between the samples collected from the two bottom types changed seasonally, and this temporal pattern was similar to that of macrophyte vegetation characteristics in that the minimum was in March and the maximum was in July (Figure 7).

DISTLM analyses confirmed significant effects of all seven environmental variables on macrobenthic community structure (Table 3). DISTLM results indicated that bottom type explained almost 40% of the variation observed in macrofauna communities. Seven variables (Bt, Mt, St, POC, mean grain size, fine sand, chla) were included in the best fitting model, and together explained 61% of total variation. However, only four variables were statistically significant in the model, namely the three categorical variables and POC.

In the vegetated bottom, macrophyte biometrics explained from 10 (density) to 27% (species richness) of variation in macrozoobenthic community characteristics (Table 4). None of the macrophyte biometrics was identified as significant in the model for macrofauna density. Two variables were included in the best fitting model for macrofauna biomass (explaining together 18%) with algal biomass as the only significant variable. Seagrass above ground biomass was the only significant variable among the four variables identified by the model as best fitting to macrofauna

Table 1. Three-way PERMANOVA tests for differences in macrofauna univariate characteristics among stations (St), months (Mt) and bottom types (Bt).

		density		**biomass**		***S***		***H***		***J***	
source	**df**	**PsF**	**CV**	**PsF**	**CV**	**PsF**	**CV**	**PsF**	**CV**	**PsF**	**CV**
St	4	9.4*	359.9	4.8*	3.9	166.7**	2.9	208.1**	0.4	125.6**	0.1
Mt	1	10.2**	595.0	1.9	3.1	6.6**	0.8	7.9**	0.1	10.9**	0.0
Bt	**1**	**7.9***	**326.0**	**30.2****	**10.5**	**607.0****	**5.6**	**32.1****	**0.2**	**0.7**	**0.0**
StxMt	4	2.1*	293.8	1.7	3.7	1.2	0.2	12.2**	0.2	11.1**	0.1
StxBt	4	1.5	122.7	9.5*	8.3	1.9	0.3	1.5	0.0	4.2*	0.0
MtxBt	**1**	**5.4****	**579.8**	**5.2****	**9.3**	**11.0****	**1.6**	**0.2**	**0.0**	**1.5**	**0.0**
MtxStxBt	4	1.2	141.3	0.9	−2.0	2.0	0.7	5.6**	0.2	5.9**	0.1
Res	80		876.1		14.2		1.6		0.2		0.1
Bt (MtxBt)		Jun, Jul, Oct		Jun, Jul, Oct		all		all		—	

Results of tests for density, biomass, number of taxa per sample (*S*), Shannon-Wienner index (*H*), and Pielou index (*J*) are presented. Effects of bottom type and MtxBt interaction printed in bold. PsF - PERMANOVA pseudoF, CV - component of variation. Significant effects: * $P<0.05$, ** $P<0.001$. Bt (MtxBt) - significant effects in pairwise tests for differences between vegetated and unvegetated bottoms performed separately for five months.

Table 2. Three-way PERMANOVA tests for differences in macrobenthic community among stations (St), month (Mt) and bottom type (Bt).

		pr/ab		dbrt		nonTr	
source	**df**	**PsF**	**CV**	**PsF**	**CV**	**PsF**	**CV**
St	4	39.1**	12.1	45.4**	13.0	34.1**	17.8
Mt	1	12.0**	10.4	13.7**	11.0	10.1**	14.8
Bt	**1**	**39.1****	**20.9**	**92.9****	**18.7**	**17.1****	**12.4**
StxMt	4	4.0**	7.7	4.9**	8.6	6.6**	16.3
StxBt	4	4.0*	4.8	5.1**	5.6	7.8**	11.4
MtxBt	**1**	**2.9***	**6.1**	**5.8****	**9.5**	**5.7****	**15.1**
Mt x St x Bt	4	1.6	4.9	2.8**	8.2	3.9**	16.6
Res	80		13.9		13.8		21.8
Bt (MtxBt)		all		all		Oct', Jun, Jul, Oct	

Tests performed on Bray-Curtis similarity matrices for: presence/absence data (pr/ab), double root transformed (dbrt), and untransformed data (nonTr). Effects of bottom type and MtxBt interaction printed in bold. PsF - PERMANOVA pseudoF, CV - component of variation. Significant effects: * P<0.05, ** P<0.001. Bt (MtxBt) - significant effects in pairwise tests for differences between vegetated and unvegetated bottoms performed separately for five months.

species richness variability. Four macrophyte vegetation biometrics were included in the model developed for macrofauna composition with two variables (seagrass above ground biomass and algal biomass) significant at P<0.001.

SIMPER analyses of double root transformed data identified nine species of cont% equal to or higher than 4%. These species altogether contributed to about 45% of the total dissimilarity between vegetated and bare sediments (Table 5). Ten species contributed at least 4% to total dissimilarity between the two bottom types (53% in total) when pr/ab data were analyzed (Table 6). *Theodoxus fluviatilis*, *Idotea balthica*, *Idotea chelipes*, *Mytilus edulis*, and Chironomidae larvae were identified by SIMPER as discriminating the two bottom types in both analyses; they occurred both with higher frequencies and higher densities on vegetated bottoms. None of the discriminating species was found only in samples collected from one bottom type. For most species, higher average densities were recorded on vegetated bottoms, with only *Macoma balthica* occurring with higher numbers on bare sand.

Discussion

This is the first study of natural seagrass and bare sediment communities that provides evidence that the persistence and magnitude of the impact of macrophyte cover on zoobenthos can be seasonally dependent. The significant positive effects of macrophyte vegetation on invertebrate density and biomass were recorded only in June, July, and October, which are the months when the seagrass canopy is relatively well developed. The strongest contrasts in macrofaunal characteristics between vegetated and unvegetated sediments were observed in July and coincided with the maximum peak in all quantitative characteristics of macrophyte vegetation – seagrass biomass, shoot density, and the biomass and diversity of other associated macrophytes. The vegetation effects on faunal species richness, diversity, and composition persisted throughout the year, but the magnitude of these effects varied seasonally in parallel with the pattern of macrophyte cover seasonal development. Again, the strongest effects (i.e. largest contrasts between the two habitats) were documented in July.

Season-related differences in the effects of seagrass cover on fauna are suggested by Pranovi et al. [30], who compared faunal density, biomass, and diversity between bare sand and a plot with experimental transplants of *Cymodocea nodosa* in the Lagoon of Venice. They found no significant difference between these two treatments in March, while significantly higher standing stocks and diversity of the fauna in the vegetated plot were noted in September, when seagrass shoot density was two times and leaf biomass was seven times higher than in early spring. A few studies conducted in natural seagrass beds also showed that seasonal changes in the quantitative characteristics of faunal assemblages mirrored seasonal variability in seagrass vegetation. For example, maximum values of mollusc species richness and abundance in Mediterranean *Posidonia oceanica* meadows occurred in summer months and coincided with the peaks in seagrass shoot density and leaf length [31,32,33], similar observations were reported for mollusc communities inhabiting the green algae *Caulerpa prolifera* meadow [34]. On the other hand, Vonk et al [35] shows that in tropical subtidal seagrass meadows, where the plant biomass does not vary seasonally, the seasonal variation of the fauna is relatively small, much less pronounced that the spatial variability related to the differences in the vegetation biometrics within the meadow. Ecosystem engineering impact depends on the density of the engineering organism, the relationship of which can be non-linear. Moreover, engineering effects can be only detectable above a

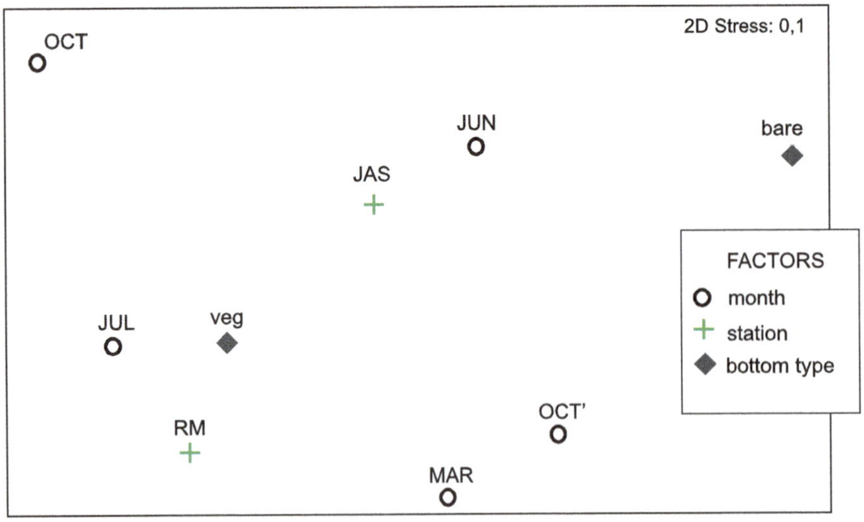

Figure 5. nMDS plots of Bray-Curtis similarities of average species abundances computed for groups of samples representing months, stations, bottom types. Data were double root transformed.

given threshold level of the engineer's density [12]. In the present study no effect of vegetation on zoobenthic density or biomass could actually be detected in months when macrophyte biomass was equal to or lower than 20 g dw m^{-2}. Several studies report that the engineering effects of seagrass are density dependent. Gacia et al. [36] show that sea grass canopies slowed current velocities with intensities proportional to the canopy height of the plants. Shoot density influenced the amount of silt and organic matter in sediments in *Z. marina* beds off the Isles of Scilly [37]. The dependence of invertebrate density and diversity on seagrass shoot density and/or biomass is documented, for example, in *Z. marina* meadows off southwest England for both infaunal [38] and epifaunal communities [5]. In the present study, when only vegetated bottom was considered, faunal community characteristics were also significantly related to different biometrics of macrophyte vegetation (describing both plant and algal components of the meadow).

Bostrom & Bonsdorff [6] list a number of factors that can differentiate the faunal communities associated with seagrass beds and bare sand. They stated that *Zostera* beds had higher habitat complexity, higher food availability, higher organic carbon content, lower flow velocity, finer sediments, higher shelter, lower predation, lower competition, enhanced deposition, and higher sediment stability. It is usually hard to separate the effects of these possible mechanisms and factors when identifying differences between vegetated and bare sands [39]. Seasonal change in the magnitude of differences in faunal communities between sea grass beds and bare sediments in the Puck Bay cannot be explained by the modification of sediment or organic content by plant vegetation. The mean grain size, POC concentration, and $\delta^{13}C$ differed significantly between the bare sand and vegetated bottom at both stations, but the magnitude of these differences remained stable throughout the year and did not follow seasonal variability in seagrass biomass [21]. The DISTLM procedure identified a number of sediment characteristics (e.g., POC content, fine sediments) as significant drivers of zoobenthic community variability, but R^2 values indicated their contributions were rather small. Herkul & Kotta [22] studied the effects of the experimental

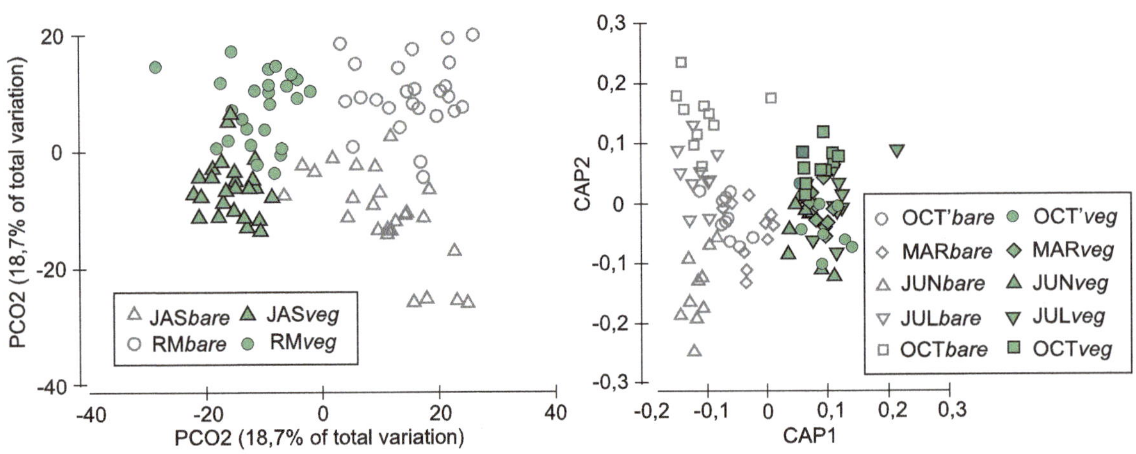

Figure 6. PCO of Bray-Curtis similarities of macrobenthic species abundances in samples (left) and CAP - ordination best discriminating groups of samples defined by bottom types and months (right). Data were double root transformed. Symbols represent sites/bottom types (left) and months/bottom types (right).

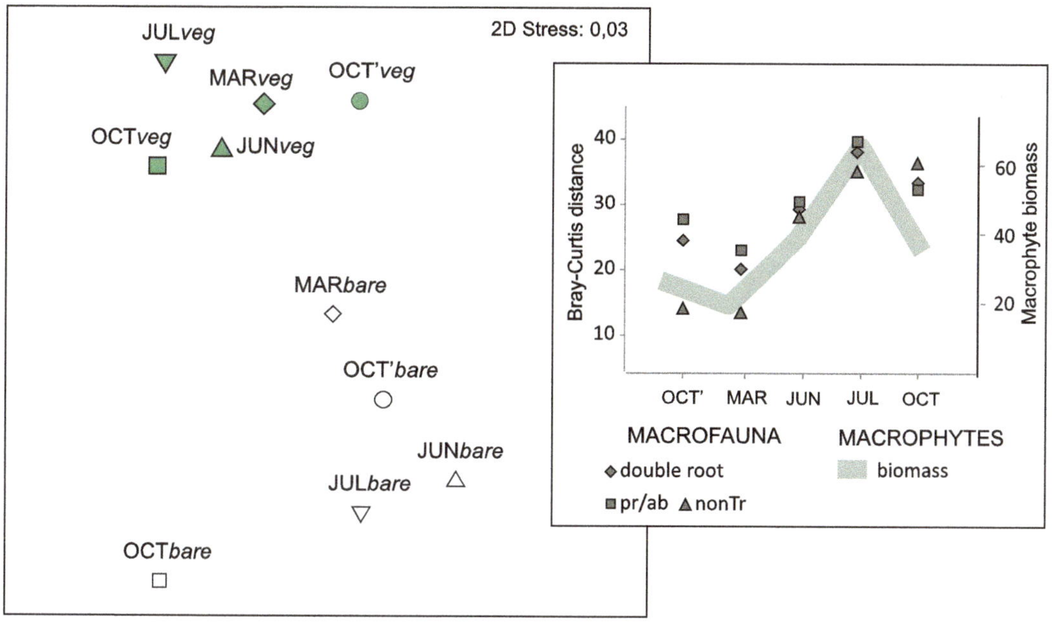

Figure 7. Magnitude of difference in macrofauna community composition between two bottom types in different seasons. nMDS of Bray-Curtis similarities among centroids for groups of samples collected in different months/bottom types (*veg* - vegetated bottom, *bare* - bare sand). Right plot presents Bray-Curtis distances between centroids representing the fauna collected from two bottom types plotted against total macrophyte biomass [g dw m^{-2}] in 5 months. Macrobenthic dissimilarity calculated based on: presence/absence data (pr/ab), double root transformed, and untransformed data (nonTr).

removal of *Z. marina* on sediment characteristics and faunal communities in the northern Baltic Sea. They found sea grass removal had a moderate effect on sediment granulometry (decreased fine grain size fraction), organic matter content, and macrofaunal communities (decreased density and diversity). However, their experiments also included the artificial addition of sand to the seagrass meadow sediments, and this manipulation had no significant effect on the associated macrofauna. These results corroborate our notion that the modification of sediment characteristics is of lesser importance for seagrass meadow macrofauna than other factors, most probably the habitat complexity provided by the vegetation.

Seasonal changes in *Zostera marina* were the most pronounced, but seagrass is not the only phytal element that changed seasonally in the studied meadows. Both the diversity and biomass of other macrophytes changed alongside eelgrass cover seasonal development. In winter, *Z. marina* was the only component of the vegetation, while in summer it was accompanied by six other macrophyte species representing a variety of morphological forms, which together comprised a considerable percentage of the total macrophyte biomass. Habitat structural complexity, which is a key characteristic of seagrass meadows, is produced by a three-dimensional net of interlacing stems, leaves, roots, and rhizomes with a number of gaps and crevices providing a variety of microhabitats to marine infaunal and epifaunal invertebrates [3]. In the Baltic Sea in summer, habitat structural complexity increases largely from increased seagrass biomass (quantitative change), but it also increases from the presence of additional

Table 3. Results of DISTLM procedure for fitting environmental variables to macrobenthic composition.

Marginal tests			Sequential tests			
variable	**psF**	**R^2**	**variable**	**psF**	**R^2**	**cumR^2**
month	4.6**	0.16	bottom type	38.0**	0.28	0.28
site	15.5**	0.14	month	7.0**	0.16	0.44
bottom type	38.0**	0.28	site	30.3**	0.14	0.58
POC/N	2.3	0.02	POC	2.4*	0.01	0.59
POC	8.1**	0.08	mean grain size	1.9	0.01	0.60
Chla	9.8**	0.09	fine sand	1.7	0.01	0.61
mean grain size	4.8**	0.05	Chla	1.0	0.00	0.61
fine sand	5.2**	0.05				
δ13C	5.2**	0.05				

Analyses based on double root transformed macrofaunal data of species abundances in samples. Significant effects: * P<0.05; ** P<0.001.

Table 4. Results of DISTLM procedure for fitting macrophyte cover biometrics to macrobenthic community characteristics for samples collected in vegetated sediments.

Marginal tests			Sequential tests			
variable	**psF**	**R^2**	**variable**	**psF**	**R^2**	**cumR^2**
MACROFAUNA DENSITY						
algal biomass	1.6	0.03	algal biomass	1.6	0.03	0.03
seagr. shoot density	0.1	0.00	seagr. shoot density	3.5	0.07	0.10
seagr. ab. ground biomass	0.0	0.02				
total macrophyte biomass	0.1	0.00				
algal species richness	1.0	0.02				
MACROFAUNA BIOMASS						
algal biomass	8.8*	0.16	algal biomass	8.8*	0.16	0.16
seagr. shoot density	3.5	0.07	seagr. shoot density	1.4	0.02	0.18
seagr. ab. ground biomass	5.5*	0.10				
total macrophyte biomass	5.5*	0.10				
algal species richness	6.7*	0.12				
MACROFAUNA SPECIES RICHNESS						
algal biomass	3.4	0.07	seagr. ab. ground biomass	10.5*	0.18	0.18
seagr. shoot density	3.8	0.07	algal species richness	2.9	0.05	0.23
seagr. ab. ground biomass	10.5*	0.18	total macrophyte biomass	1.3	0.02	0.25
total macrophyte biomass	5.9*	0.11	algal biomass	1.5	0.02	0.27
algal species richness	1.2	0.02				
MACROFAUNA COMPOSITION						
algal biomass	5.0**	0.09	seagr. ab. ground biomass	5.1**	0.10	0.10
seagr. shoot density	4.5**	0.08	algal biomass	4.3**	0.07	0.17
seagr. ab. ground biomass	5.1**	0.10	algal species richness	1.3	0.02	0.19
total macrophyte biomass	4.5**	0.09	total macrophyte biomass	1.2	0.03	0.22
algal species richness	4.6**	0.09				

Macrofauna analysed in terms of density, biomass, species richness and species composition (based on double root transformed species abundances in samples data). Macrophyte cover biometrics include total macrophyte biomass, algal biomass, algal species richness, seagrass shoot density (seagr. shoot density) and seagrass above ground biomass (seagr. ab. ground biomass). Significant effects: * $P<0.05$, **$P<0.001$.

Table 5. Species responsible for discrimination of macrofauna between two bottom types, as indicated by the SIMPER procedure based on double-root transformed data.

	SIMPER		JAS		RM	
	Diss/SD	**Cont%**	**bare**	**veg**	**bare**	**veg**
Theodoxus fluviatillis	1.8	6.2	2.4	17.5	0.8	12.7
Idotea chelipes	2.2	6.0	0.1	11.3		3.6
Mytilus edulis	2.0	5.0	0.6	5.6	0.2	3.4
Hydrobia sp.	1.2	5.0	384.4	410.2	609.6	807.5
Idotea balthica	1.8	5.0	0.1	6.3		3.4
Chironomidae larvae	1.5	4.8	3.8	10.5	0.6	10.6
Cyathura carinata	1.3	4.5	18.8	13.4	0.5	1.8
Gammarus sp. juv	1.4	4.0	4.9	16.2	6.2	10.8
Pygospio elegans	1.2	4.0	56.3	27.7	8.4	15.1

Diss/SD – average contribution to overall dissimilarity divided by standard deviation, Cont% - percentage contribution to total dissimilarity. Mean densities [ind. 0.1 m^{-2}] in groups of samples defined by bottom type (bare, veg) and station (RM, JAS) are presented. Only species of cont% equal or higher than 4 are listed.

Table 6. Species responsible for discrimination of macrofauna between two bottom types, as indicated by the SIMPER procedure, based on presence/absence data.

	SIMPER		JAS		RM	
	Diss/SD	Cont%	bare	veg	bare	veg
Idotea chelipes	2.25	6.98	8	100	0	76
Idotea balthica	2.00	6.90	12	76	0	96
Mytilus edulis	1.73	6.49	24	96	16	92
Bryozoa nd	1.42	5.79	4	60	4	80
Theodoxus fluviatillis	1.23	5.19	40	100	32	80
Chironomidae larvae	1.11	4.78	44	88	40	92
Gammarus oceanicus	1.10	4.45	8	68	0	44
Corophium sp. juv.	0.98	4.18	72	72	36	28
Marenzelleria neglecta	0.98	4.15	84	64	16	28
Lymnaea peregra	0.96	4.00	32	76	16	20

Diss/SD - average contribution to overall dissimilarity divided by standard deviation, Cont% - percentage contribution to total dissimilarity. Mean frequencies of occurrence [%] in groups of samples defined by bottom type (bare, veg) and station (JAS, RM) are presented. Only species of cont% equal or higher than 4 are listed.

structural components provided by other macrophytes (qualitative change). Increased complexity in seagrass meadow vegetation has been shown to positively influence faunal density and diversity in a number of *in situ* observations and experimental surveys [3,40,41]. The seasonal variability in epifaunal density and diversity in seagrass meadows off Florida was related to both increased seagrass biomass and to the occurrence of drift algae, an additional component of habitat complexity, in the summer season [42]. The high total architectural complexity of seagrass habitats, which are produced by the mosaic formed of all structural components, is beneficial to macrofauna as it provides both additional microhabitat niches and protection from predation. Based on their research in Swedish and Finish seagrass beds, Baden et al. [43] report that predation pressure in the Baltic Sea was very low and had no effects on the biomass of mesograzers associated with seagrass. However, the biomass and trophic impact of small benthivorous fish (*Platichthys flesus, Anguilla anguilla, Pomatoschistus minutus, P. microps, Neogobius melanostomus, Zoarces viviparous, Rutilus rutilus*) in the Puck Bay is much higher than in the northern Baltic Sea coastal areas [44] and the protective effects of macrophyte cover could also be more important in this area.

Zostera is not an important direct provider of organic carbon to macrobenthic invertebrates in the Baltic Sea benthic system. In Swedish Baltic Sea coastal waters, Jephson et al. [4] identified only two species that consume seagrass tissues – *Rissoa* sp. and *Theodoxus fluviatilis*, and only the former occurs in the Puck Bay. On the other hand, they did identify a number of grazers that depend on the epiphyte algae that accompany *Z. marina* in seagrass meadows. Seagrasses are rarely treated as preferred food, mostly because of their low nitrogen and high cellulose content as well as the presence of condensed tannins and phenolic acids that can be toxic to herbivores [45]. Epiphytes are identified as the most important food source for fauna associated with seagrass in a number of studies [45,46,47]. Summer increases in seagrass leaf length and shoot density, which increases the overall surface area of aboveground seagrass organs in the Puck Bay, increases the substrate available for epiphyte colonization. We can expect that epiphyte biomass follows this trend, and that the increased amounts of organic carbon stored in the epiphytes support higher densities of invertebrate grazers in the summer season.

The differences in macrobenthic community composition between vegetated and unvegetated sediments in the Puck Bay were produced by shifts in the dominance of co-occurring species rather than a complete change in species composition. Bivalve molluscs that are important components of the Puck Bay macrozoobenthos in terms of both abundance and biomass occurred at different abundances in seagrass meadows and on bare sand. *Macoma balthica* was the only dominant species that occurred with higher densities on bare sand. In a settlement experiment in the Archipelago Sea, Bostrom et al. [48] show that *Z. marina* vegetation can actually inhibit the settlement of the larvae of the bivalves *M. balthica* and *Cerastoderma glaucum*. They also reported that post-settlement transportation of bivalves by resuspension events and secondary settlements of larger individuals were much more common on bare sands and seagrass patches than in continuous *Z. marina* vegetation. These effects were particularly strong for *M. balthica* [48]. In contrast to *Macoma*, the mussel *Mytilus edulis* occurs almost exclusively in seagrass beds, and seagrass vegetation has been shown to collect drifting clumps of mussels, and, thus, facilitate the post-recruitment settlement of *Mytilus* populations [49]. On the other hand, mussels can fertilize seagrass shoot growth through the biodeposition of organic material via feces and pseudofeces [50]. The gastropod *Theodoxus fluviatilis* was among the species that most strongly discriminated the two communities, both with regard to abundances (analyses of double root transformed data) and frequencies of occurrence (presence/absence data). The strong preference of *Th. fluviatilis* for seagrass beds is not surprising, as this species can use seagrass tissue as a direct food source [4]. A number of isopod (*Idotea chelipes, Idotea balthica*) and amphipod (*Gammarus oceanicus, Gammarus salinus, Gammarus zaddachi*) species were important components of the seagrass communities in the Puck Bay. These are small mesograzers that feed on seagrass epiphytes [4,51] and good swimmers that can actively disperse in search of vegetated substrates [52].

The temporal and spatial variability of macrozoobenthos was higher on bare sand than on vegetated bottoms. This was indicated by the larger dispersion of bare sand samples both on PCO ordination, and on CAP ordination plotted along the strongest gradients produced by months and stations (Figure 6). This pattern could have been produced by the protective and stabilizing effects of macrophyte cover. Seagrass canopies slow down near-bottom water velocities with intensities proportional to the canopy height of the plants [33]. Thus, seagrass vegetation can mitigate the effects of extreme events such as storms and strong winds that are very common in this part of the Baltic Sea. Bostrom & Bonsdorf [53], in an experimental study of a succession in Baltic Sea sediments, observe that a single event of strong wind could destroy communities on bare sand, while it had little impact on a parallel community developing on a vegetated bottom. Reusch & Chapman [49] also reported that the storm-driven dislodgement of mussels was much more severe on bare sands than in seagrass beds; this is further support for the notion that macrophyte cover provides natural protection against environmental disturbance.

Conclusions

In this study we have demonstrated that the effects of macrophytes on associated benthic communities in the southern Baltic Sea are seasonally depended. The positive effects of macrophyte cover on the univariate and multivariate characteristics of macrobenthic fauna were either weaker or could not be detected in late fall and early spring when macrophyte vegetation was least developed in terms of both biomass and taxonomic diversity. These findings indicate that in temperate, clearly seasonal systems the assessment of the ecosystem engineer impact cannot be based solely on observations performed in one season only. The common notion that macrophyte cover has a strong influence on benthic fauna in marine coastal habitats is actually mostly based on surveys performed in summer when macrophyte cover is best developed. Thus, the effects of macrophytes can be overestimated to some extent. More studies of seagrass systems performed in other seasons, especially in winter, are strongly recommended if we are to better understand the actual role of macrophyte vegetation in structuring macrozoobenthic communities.

Supporting Information

Table S1 Taxonomic composition and densities of taxa in samples collected at two stations, on two bottom types and in five months. Average values for replicate samples collected at the same stations, bottom types (veg - vegetated bottom, unveg - bare sands) and months are presented.

Acknowledgments

We wish to thank Wojtek Moskal, Marek Zajączkowski, and Kajetan Deja for their help and good company during fieldwork. We also wish to thank Dr Karol Kuliński for his help with chemical analyses of sediment samples.

Author Contributions

Conceived and designed the experiments: MWK EJ LK. Performed the experiments: MWK EJ LK PB. Analyzed the data: MWK EJ. Contributed reagents/materials/analysis tools: MWK. Wrote the paper: MWK EJ LK PB.

References

1. Jones CG, Lawton JH, Shachak M (1994) Organisms as ecosystem engineers. Oikos 69: 373–386.
2. Hemminga MA, Duarte CM (2000) Seagrass Ecology. Cambridge: Cambridge University Press. 289 p.
3. Gartner A, Tuya F, Lavery PS, McMahon K (2013) Habitat preferences of macroinvertebrate fauna among seagrasses with varying structural forms. J Exp Mar Biol Ecol 439: 143–151.
4. Jephson T, Nystrom P, Moksnes P-O, Baden SP (2008) Trophic interactions in *Zostera marina* beds along the Swedish coast. Mar Ecol Prog Ser 369: 63–76.
5. Attrill MJ, Strong JA, Rowden AA (2000) Are macroinvertebrate communities influenced by seagrass structural complexity? Ecography 23: 144–121.
6. Bostrom C, Bonsdorff E (1997) Community structure and spatial variation of benthic invertebrates associated with *Zostera marina* (L.) beds in the northern Baltic Sea. J Sea Res 37: 153–166.
7. Bearley A, Wells FE (2000) Invertebrate fauna in seagrasses on Success Bank, Western Australia. Biol Mar Medit 7: 199–202.
8. Fredriksen S, Backer A, Bostrom C, Christie H (2009) Infauna from *Zostera marina* L. meadows in Norway. Differences in vegetated and unvegetated areas. Mar Biol Res 6: 189–200.
9. Gambi MC, Conti G, Bremec CS (1998) Polychaete distribution, diversity and seasonality related to seagrass cover in shallow soft bottoms of the Tyrrhenian Sea (Italy). Sci Mar 62: 1–17.
10. Berkenbusch K, Rowden AA, Probert PK (2000) Temporal and spatial variation in macrofauna community composition imposed by ghost shrimp *Callianassa filholi* bioturbation. Mar Ecol Prog Ser 192: 249–257.
11. Jones CG, Lawton JH, Shachak M (1997) Positive and negative effects of organisms as physical ecosystem engineers. Ecology 78: 1946–1957.
12. Harley CDG, O'Riley JL (2011) Non-linear density-dependent effects of an intertidal ecosystem engineer. Oecologia 166: 531–541.
13. Fortino K (2006) Effect of season on the impact of ecosystem engineers in the New River, NC. Hydrobiologia 559: 463–466.
14. Hasegawa N, Hori M, Mukai H (2008) Seasonal changes in eelgrass functions: current velocity reduction, prevention of sediment resuspension, and control of sediment-water column nutrient flux in relation to eelgrass dynamics. Hydrobiologia 596: 387–399.
15. Duarte CM, Fourqurean JW, Krause-Jensen D, Olesen B (2006) Dynamics of seagrasses stability and change. In: Larkum AWD, editor. Seagrasses: Biology, Ecology and Conservation. Dordercht: Springer. pp. 271–294.
16. Duarte C (1989) Temporal biomass variability and production/biomass relationships of seagrass communities. Mar Ecol Prog Ser 51: 269–276.
17. Fourqurean JW, Willsie A, Rose CD, Rutten LM (2001) Spatial and temporal pattern in seagrass community composition and productivity in South Florida. Mar Biol 138: 341–354.
18. Guidetti P, Lorenti M, Buia MC, Mazzella L (2002) Temporal dynamics and biomass partitioning in three Adratic seagrass species: *Posidonia oceanica, Cymodocea nodosa, Zostera marina.* Mar Ecol 23: 51–67.
19. Van Katwijk MM, Bos AR, Kennis P, de Vries R (2010) Vulnerability to eutrophication of a semi-annual life history: a lesson learnt from an extinct eelgrass (*Zostera marina*) population. Biol Conserv 143: 248–254.
20. Van Lent F, Vershuure JM (1994) Intraspecific variability of Zostera marina L., (eelgrass) in the estuaries and lagoons of the southwestern Netherlands, I. Population dynamics. Aquat Bot 48: 31–58.
21. Jankowska E, Włodarska-Kowalczuk M, Kotwicki L, Balazy P, Kuliński K (2014) Seasonality in vegetation biometrics and its effects on sediment characteristics and meiofauna in Baltic seagrass meadows, Estuar Coast Shelf S 139: 159–170.
22. Herkul K, Kotta J (2009) Effects of eelgrass (*Zostera marina*) canopy removal and sediment addition on sediment characteristics and benthic communities in the Northern Baltic Sea. Mar Ecol 30: 74–82.
23. Nowacki J (1993) The morphometry of the Gulf. In: Korzeniewski K, editor. Puck Bay. Gdańsk: Fundacja Rozwoju Uniwersytetu Gdańskiego. pp. 71–79.
24. Kruk-Dowgiałło L (1998) Phytobenthos as an indicator of the state of environment of the Gulf of Gdansk. Oceanological Studies 4: 105–121.
25. Gic-Grusza G, Kryla-Staszewska L, Urbanski J, Warzocha J, Weslawski JM (2009) Atlas of Polish marine area bottom habitats: environmental valorization of marine habitats. Gdynia: Broker-Inowacji. 179 p.
26. Dalsgaard T, Nielsen LP, Brotas V, Viaroli P, Underwood G et al. (2000) Protocol handbook for NICE - Nitrogen Cycling in Estuaries: a project under the EU research programme: MAST III. Silkeborg: National Environmental Research Institute. 62 p.
27. Folk RL, Ward WC (1957) Brazos River bar: a study in the significance of grain size parameters. J Sediment Petrol 27: 3–26.
28. Anderson MJ, Gorley RN, Clarke KR (2008) PERMANOVA for PRIMER: guide to software and statistical methods. Plymouth: PRIMER-E Ltd. 214 p.
29. Clarke KR (1993) Non-parametric multivariate analyses of changes in community structure. Aust J Ecol 18: 117–143.
30. Pranovi F, Curiel D, Rismondo A, Marzocchi M, Scattolin M (2000) Variations of the macrobenthic community in a seagrass transplanted are of the Lagoon of Venice. Sci Mar 64: 303–310.
31. Carmen Arroyo MdSC, Rueda JL, Gofas S (2006) Temporal changes of mollusc populations from a *Zostera marina* bed in southern Spain (Alboran Sea) with biogeographic considerations. Mar Ecol 27: 417–430.
32. Urra J, Ramirez AM, Marina P, Salas C, Gofas S et al. (2013) Highly diverse molluscan assemblages of *Posidonia oceanica* meadows in northwestern Alboran Sea (W Mediterranean): seasonal dynamics and environmental drivers. Estuar Coast Shelf S 117: 136–147.

33. Rueda JL, Salas C (2008) Molluscs associated with a subtidal *Zostera marina* L. bed in southern Spain: linking seasonal changes of fauna and environmental variables. Estuar Coast Shelf S 79: 157–167
34. Rueda JL, Salas C (2003) Seasonal variation of a molluscan assemblage living in a *Caulerpa prolifera* meadow within the inner Bay of Cadiz (SW Spain). Estuar Coast Shelf S 57: 909–918.
35. Vonk JA, Christianen MJA, Stapel J (2010) Abundance, edge effect, and seasonality of fauna in mixed-species seagrass meadows in southwest Sulawesi, Indonesia. Mar Biol Res 6: 282–291.
36. Gacia E, Granata TC, Duarte CM (1999) An approach to measurement of particle flux and sediment retention within seagrass (*Posidonia oceanica*) meadows. Aquat Bot 65: 255–268.
37. Bowden DA, Rowden AA, Attrill MJ (2001) Effect of patch size and in-patch location on the infaunal macroinvertebrate assemblages of *Zostera marina* seagrass beds. J Exp Mar Biol Ecol 259: 133–154.
38. Webster PJ, Rowden AA, Attrill MJ (1998) Effect of shoot density on the infaunal macro-invertebrate community within a *Zostera marina* seagrass bed. Estuar Coast Shelf S 47: 351–357.
39. Bouma TJ, Ortells V, Ysebaert T (2009) Comparing biodiversity effects among ecosystem engineers of contrasting strength: macrofauna diversity in *Zostera noltii* and *Spartina anglica* vegetations, Helgol Mar Res 63: 3–18.
40. Stoner AW, Lewis FG (1985) The influence of quantitative and qualitative aspects of habitat complexity in tropical seagrass meadows. J Exp Mar Biol Ecol 94: 19–40.
41. Sirota L, Hovel KA (2006) Simulated eelgrass *Zostera marina* structural complexity: effects of shoot length, shoot density, and surface area on the epifaunal community of San Diego Bay, California, USA. Mar Ecol Prog Ser 326: 115–131.
42. Gore RH, Gallaher EE, Scotto LE, Wilson KA (1981) Studies on decapod Crustacea from the Indian River region of Florida. Estuar Coast Shelf S 12: 485–508.
43. Baden S, Bostrom C, Tobiasson S, Arponen H, Moksnes P-O (2010) Relative importance of trophic interactions and nutrient enrichment in seagrass ecosystems: a broad-scale field experiment in the Baltic-Skagerrak area. Limnol Oceanogr 55: 1435–1448.
44. Tomczak MT, Muller-Karulis B, Jarv L, Kotta J, Martin G et al. (2009) Analyses of trophic networks and carbon flows in south-eastern Baltic Coastal ecosystems. Prog Oceanogr 81: 111–131.
45. Bologna PAX, Heck KL (1999) Macrofaunal associations with seagrass epiphytes. Relative importance of trophic and structural aspects. J Exp Mar Biol Ecol 242: 21–39.
46. Kitting CL, Fry B, Morgan MD (1984) Detection of inconspicuous epiphytic algae supporting food webs in seagrass meadows. Oecologia 62: 145–149.
47. Heck KL, Valentine JF (2006) Pant-herbivore interactions in seagrass meadows. J Exp Mar Biol Ecol 330: 420–436.
48. Bostrom C, Tornroos A, Bondsdorff E (2010) Invertebrate dispersal and habitat heterogeneity: expression of biological traits in a seagrass landscape. J Exp Mar Biol Ecol 390: 106–117.
49. Reusch TBH, Chapman ARO (1995) Storm effects on eelgrass (*Zostera marina* L.) and blue mussel (*Mytilus edulis* L.) beds. J Exp Mar Biol Ecol 192: 257–271.
50. Reusch TBH, Chapman ARO, Groger JP (1994) Blue mussel *Mytilus edulis* do not interfere with eelgrass *Zostera marina* but fertilize shoot growth through biodeposition. Mar Ecol Prog Ser 108: 265–282.
51. Leidenberger S, Harding K, Jonsson PR (2012) Ecology and distribution of the isopod genus *Idothea* in the Baltic Sea: key species in a changing environment. J Crustacean Biol 32: 359–381.
52. Bostrom C, Mattila J (1999) The relative importance of food and shelter for seagrass associated invertebrates- a latitudinal comparison of habitat choice by isopod grazers. Oecologia 120: 162–170.
53. Bostrom C, Bonsdorf E (2000) Zoobenthic community establishment and habitat complexity - the importance of seagrass shoot-density, morphology and physical disturbance for faunal recruitment. Mar Ecol Prog Ser 205: 123–138.

2

Complex Effects of Ecosystem Engineer Loss on Benthic Ecosystem Response to Detrital Macroalgae

Francesca Rossi[1]*, Britta Gribsholt[2], Frederic Gazeau[3,4], Valentina Di Santo[5], Jack J. Middelburg[2,6]

1 Laboratoire Ecologie des Systèmes marins côtiers, Université Montpellier 2, Montpellier, France, **2** Department of Ecosystems, Royal Netherlands Institute for Sea Research, Yerseke, the Netherlands, **3** Centre National de la Recherche Scientifique-Institut National des Sciences de l'Univers, Laboratoire d' Oceanographie de Villefranche, Villefranche-sur-mer, France, **4** Université Pierre et Marie Curie, Observatoire Océanologique de Villefranche, Villefranche-sur-mer, France, **5** Department of Biology, Boston University, Boston, Massachusetts, United States of America, **6** Department of Earth Sciences – Geochemistry, Faculty of Geosciences, Utrecht University, Utrecht, The Netherlands

Abstract

Ecosystem engineers change abiotic conditions, community assembly and ecosystem functioning. Consequently, their loss may modify thresholds of ecosystem response to disturbance and undermine ecosystem stability. This study investigates how loss of the bioturbating lugworm *Arenicola marina* modifies the response to macroalgal detrital enrichment of sediment biogeochemical properties, microphytobenthos and macrofauna assemblages. A field manipulative experiment was done on an intertidal sandflat (Oosterschelde estuary, The Netherlands). Lugworms were deliberately excluded from 1× m sediment plots and different amounts of detrital *Ulva* (0, 200 or 600 g Wet Weight) were added twice. Sediment biogeochemistry changes were evaluated through benthic respiration, sediment organic carbon content and porewater inorganic carbon as well as detrital macroalgae remaining in the sediment one month after enrichment. Microalgal biomass and macrofauna composition were measured at the same time. Macroalgal carbon mineralization and transfer to the benthic consumers were also investigated during decomposition at low enrichment level (200 g WW). The interaction between lugworm exclusion and detrital enrichment did not modify sediment organic carbon or benthic respiration. Weak but significant changes were instead found for porewater inorganic carbon and microalgal biomass. Lugworm exclusion caused an increase of porewater carbon and a decrease of microalgal biomass, while detrital enrichment drove these values back to values typical of lugworm-dominated sediments. Lugworm exclusion also decreased the amount of macroalgae remaining into the sediment and accelerated detrital carbon mineralization and CO_2 release to the water column. Eventually, the interaction between lugworm exclusion and detrital enrichment affected macrofauna abundance and diversity, which collapsed at high level of enrichment only when the lugworms were present. This study reveals that in nature the role of this ecosystem engineer may be variable and sometimes have no or even negative effects on stability, conversely to what it should be expected based on current research knowledge.

Editor: Andrew Davies, Bangor University, United Kingdom

Funding: This work was funded by a PIONIER grant of the Netherlands Organization for Scientific Research to JJM. The funders had no role in study design, data analysis, decision to publish, or preparation of the manuscript.

Competing Interests: The authors have declared that no competing interest exist.

* E-mail: francesca.rossi@univ-montp2

Introduction

It is widely recognized that species loss may affect ecosystem stability and, in turn, have important social and ecological consequences [1–3]. Overall, when more species are present in a community, ecosystem functions are more stable in time and increase their resistance or resilience against disturbance, due to an increased variety of life strategies, functional traits and responses to environmental disturbance [4–6]. Different species contributing to similar ecosystem functions can occur under different environmental conditions and ensure that functions are performed even when some other species are displaced [4,5,7].

Empirical evidence has shown that sometimes species loss can have complex, often idiosyncratic effects on the stability of ecosystem functions and on both their resistance and resilience against disturbance [8,9]. This is especially true for the marine coastal ecosystem, where species identity, functional traits and environmental context may rule ecosystem functions and their response to disturbances [10–12]. One or few species can be key contributors to ecosystem functions to the extent that their loss can outweigh the effect of species richness on ecosystem functions, resilience or resistance [10,13,14]. Such overwhelming effect can, however, vary with the environmental context and change with increasing ecological complexity [15]. It seems thus evident that understanding the role of species loss in marine systems might require a large body of knowledge of the functional role of individual species and of the consequences of losing particular species for ecosystem functions and for their response to disturbance.

Ecosystem engineer species strongly modify resource availability and environmental conditions, which in turn can alter communities and ecosystem functions [16,17]. These species are thus ideal candidates to test hypotheses on how local extinction of key contributors to ecosystem functions can modulate the response of ecosystems to disturbance. In marine sediments and terrestrial soils, ecosystem engineers are often species that contribute to particle mixing and solute penetration into the sediment through the process of bioturbation, which includes sediment reworking and burrow ventilation [18]. These bioturbating engineer species can be great contributors to diagenetic reactions, to the recycling of sedimentary organic matter, and to benthic community structure and biodiversity [18–21]. They also have the potential to strongly modify the response to disturbance of benthic communities and functions [22]. For instance bioturbation by *Marenzellaria viridis* has been found to decrease hypoxic crises in the Baltic Sea [23]. It has also been found that in soil, reworking activity of dominant contributors to bioturbation, such as earthworms, modulates the response of plant communities to invasions [24]. Surprisingly, however, ecosystem engineer loss has been poorly considered in studies investigating ecosystem response to disturbance.

In marine coastal systems, a large portion of seaweeds and seagrasses is not consumed by herbivores but returns to the environment as decaying organic matter [25]. In shallow-water coastal habitats, detrital seaweeds and seagrasses can supply important amounts of organic matter to benthic ecosystems. They represent an important source of disturbance because during decomposition they can alter sediment biogeochemistry and, in turn, benthic communities of consumers. Despite detrital seaweed and seagrass supply is often a natural event, proliferation and blooms of seaweeds as a consequence of eutrophication can dramatically increase detrital seaweed biomass and exacerbate the effects on benthic ecosystems [26–29]. Biomass decomposition can induce hypoxia followed by the release of reduced toxic compounds such as sulphides, which are toxic for several benthic species and can therefore strongly reduce species abundance and diversity [26,27,29–33]. Detrital decomposition can also provide food, directly for detritivores or indirectly, by stimulating bacterial metabolism, thereby altering the benthic food web [28,34]. These effects strongly depend on the amount of detrital inputs and on its decomposition patterns, which may strongly vary as a consequence of benthic invertebrate composition and identity [35].

Bioturbating fauna can modulate the recycling of detrital macroalgae enhancing organic matter mineralization and microbial activity [36–38]. Some burrowing bioturbating species, such as certain crabs or polychaetes, can also incorporate detritus into their burrow wall lining and represent important sinks for detrital macroalgae [39]. In addition, often bioturbating infauna are also detritivores that can consume directly detritus and contribute to its recycling in the food web. In a laboratory experiment, for instance, it was found that nitrogen and carbon fluxes at the sediment-water interface changed following both the addition of the detrital green macroalgae *Chaetomorpha linum* and the presence of the bioturbating polychaete *Nereis diversicolor*. This polychaete ingested macroalgal fragments, enhanced microbial metabolism and contributed to diffuse solutes in the interstitial sediment and at the sediment-water interface [36].

One common well-recognized bioturbating species typical of temperate waters is the burrowing lugworm *Arenicola marina*. This worm lives head down in J-shaped burrows where it ingests sediment and defecates on the surface. Its activity has been recognized to destabilize sediment and counteract the effect of microphytobenthos, increase sediment microhabitats and alter sediment biogeochemistry, by enhancing solute penetration into sediment depth and re-distributing organic material along the vertical profile [40]. Consequently, this species may alter other fauna and vegetation [36,37,41]. This study reports on how the loss of this bioturbating lugworm may alter the response to detrital macroalgae of intertidal benthic macrofauna and sediment biogeochemistry. It was hypothesized that the deliberate exclusion of lugworms would increase the magnitude of changes caused by the disturbance of macroalgal detritus for (i) carbon availability and benthic respiration; (ii) microalgal biomass and (iii) macrofauna assemblages. Recycling of the organic carbon derived from detrital decomposition was also investigated to explore some of the mechanisms that might regulate the effect of *A. marina* on benthic responses to macroalgal detritus.

Materials and Methods

Study Site and Experimental Design

Two manipulative experiments were done on an intertidal flat of the Oosterschelde estuary (The Netherlands) dominated by the bioturbating lugworm *Arenicola marina*, during summer 2006. The study intertidal flat, named Slikken van Viane (51° 37′ 00″ N, 4° 01′ 00″ E), was a sheltered area situated on the upper intertidal close to a salt marsh. Sediment grain-size was 65% fine sand and 35% very fine sand (Rossi F., unpublished data). The Oosterschelde estuary is a national park designated as zone VI by the International Union for Conservation of Nature (IUCN). This IUCN classification has among its objectives the facilitation of scientific research and environmental monitoring, mainly related to the conservation and sustainable use of natural resources (http://www.iucn.org). No specific permits were therefore required for performing these scientific experiments, which did not involve endangered or protected species. The first experiment tested the effect of *A. marina* exclusion on the response of macrofauna assemblages and sediment biogeochemistry to different amounts of detrital macroalgae. The second experiment measured the changes in detrital recycling following *A. marina* exclusion. For the first experiment (Fig. 1), all lugworms were excluded from 15 randomly chosen sediment plots of 1×1 m, some meters apart, by burying mosquito nets to the depth of approximately 7 cm (hereafter exclusion A− treatment), following the method proposed by [41] in a more extensive area. A procedural control for the effect of the net and of the burial was done for other 15 sediment plots of 1×1 m by burying mosquito nets with 10×10 cm holes (hereafter procedural control A+ treatment).

One month after lugworm exclusion, the sediment plots were organically-enriched by addition of detrital *Ulva* spp. Fresh *Ulva* thalli were previously collected in the Oosterschelde estuary and washed to remove animals and sediment. Thalli were then weighted and frozen in separate plastic bags. Freezing is

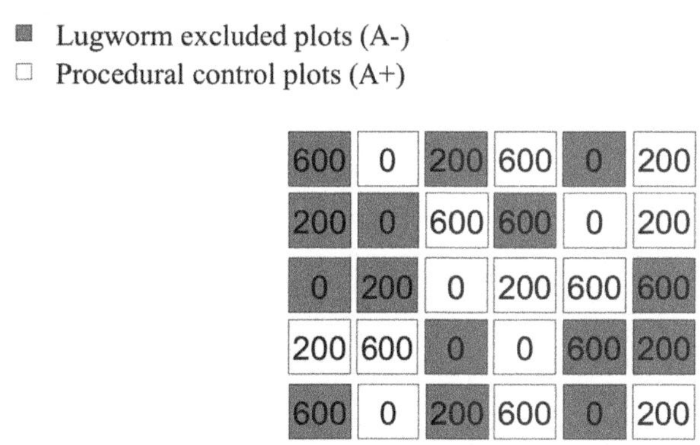

Figure 1. Schematic experimental design for experiment 1. Each rectangle corresponds to 1 m^2 plot. Numbers refers to the detrital enrichment. (0: no addition; 200:200 g WW were added twice at a 1 week interval; 600:600 g WW were added twice with a 1 week interval). Distance among plots are not in scale.

commonly used in experiments that aims at testing effects of detrital macroalgae and decomposition patterns. This treatment is necessary in order to create detritus and ensure that *Ulva* is dead when buried since this seaweed can survive for very long periods in the dark [30]. The thalli were left defrosting at air temperature and then added to the top 1 cm of the sediment by gently hand-churning the detritus with the sediment. Enrichment was done twice, on July 3 and July 10, 2006. Each time either 200 or 600 g of wet weighted *Ulva* (hereafter treatments 200 and 600, respectively) were added to 10 exclusion (A−) and 10 procedural control (A−) plots. These amounts corresponded to 40 and 120 g dry weight. Other 10 plots (5 A− and 5 A+) were not enriched (hereafter treatment 0), but the sediment was hand-churned to control for the manipulation. There was no procedural control for the physical presence of detritus as previous experiments had demonstrated no effects [28–30]. The amount of *Ulva* spp. was based on the available literature to cause medium and high levels of disturbance (e.g. [27,30,32,33]).

The second experiment was run at the same time of experiment 1. Two additional procedural control plots for burial and 2 exclusion plots were enriched with 200 g ^{13}C-labelled *Ulva* twice, resembling the treatment "200". These four plots were used to trace the fate of the carbon released from detrital *Ulva*. The labeling of *Ulva* was done in the laboratory. Leaves were washed and carefully cleaned to eliminate epiphytes and animals. Then, thalli were transferred to aquaria containing filtered seawater supplemented with $NaH^{13}CO_3$ (99% ^{13}C) to the concentration of 0.2 g L^{-1}. The aquaria were oxygenated and covered with transparent PVC to limit ^{13}C bacterial respiration.

Field Sampling

For the first experiment, sampling was done on July 30 and August 1, 2006, one month after detrital enrichment. This timing was chosen because previous studies suggested that effects of algal mats on macrofauna occur within 2 to 10 weeks from the deposition [31–33,42] and that decomposition of *Ulva* occurs within a month (e.g. [43,44]).

Sediment samples were taken for organic carbon (hereafter OC), porewater total inorganic carbon ($DIC = H2CO_3 + HCO_3^{-} + CO_3^{2-}$), microalgal biomass (chlorophyll *a*) and macrofauna species composition, abundance and diversity. The sediment for chlorophyll *a* and OC was collected using a cut syringe of 3 mm inner diameter (i.d.). For chlorophyll *a*, the sediment was collected to 1 cm depth, kept in the dark and preserved at −80°C. The sediment for OC was collected to the depth of 6 cm, sectioned into 0–1, 1–2 and 2–6 cm depths and preserved freeze-dried. *Ulva* detritus was removed from this sediment under the stereo-microscope. The sediment for porewater DIC extraction was collected with cores of 5 cm i.d. and sectioned into 0–1, 1–2 and 2–6 cm depth intervals at the lowest and highest levels of organic enrichment (0 and 600) in both the control and the exclusion treatments. Two cores were collected from each plot and then pooled to reach the amount of water needed for the analyses. Porewater was extracted by centrifugation (1344 g, 15 min) and the supernatant was filtered on Millipore 45 μm. The extracted water was stored in glass vials, sealed, immediately poisoned with saturated $HgCl_2$ to stop bacterial activity and analyzed within 3 d. Only 3 out of 5 randomly-chosen replicate plots were sampled for OC and TCO_2.

The sediment for macrofauna was collected using 11 cm i.d. PVC core. This sediment was sieved (0.5 mm mesh) and macrofauna fixed in 4% buffered formol. Before sampling, all plots were photographed to estimate the abundance of *A. marina* using visible mounds. In the laboratory, all plots were independently examined by four researchers to reduce estimate bias. Five randomly-chosen natural plots were also photographed and then sampled using 11 cm i.d. cores to test for the effect of the procedural control A+ on the natural abundance of *A. marina* and of the other benthic macrofauna. No differences were detected between natural and procedural control treatments (see Results section) and this latter was used for evaluating the effect of detrital enrichment and lugworm exclusion.

Sediment cores were also used to measure fluxes of inorganic carbon (TCO_2) and oxygen (O_2) at the sediment-water interface. Intact sediment cores were collected by pushing Plexiglas cores (11 cm i.d.) into the sediment to a depth of 20 cm at each of three replicate plots for the lowest and the highest levels of organic enrichment (0 and 600) in both the control and the exclusion treatments. The net was carefully cut around the cores to allow the core penetration into the sediment. Cores were sealed with rubber stoppers and dug out of the sediment. All cores were brought to a climate controlled room within 2 h, and kept in the dark at *in situ* temperature (10°C) before further handling. The same day, the cores were incubated in the dark for 12 h. Incubation in the dark

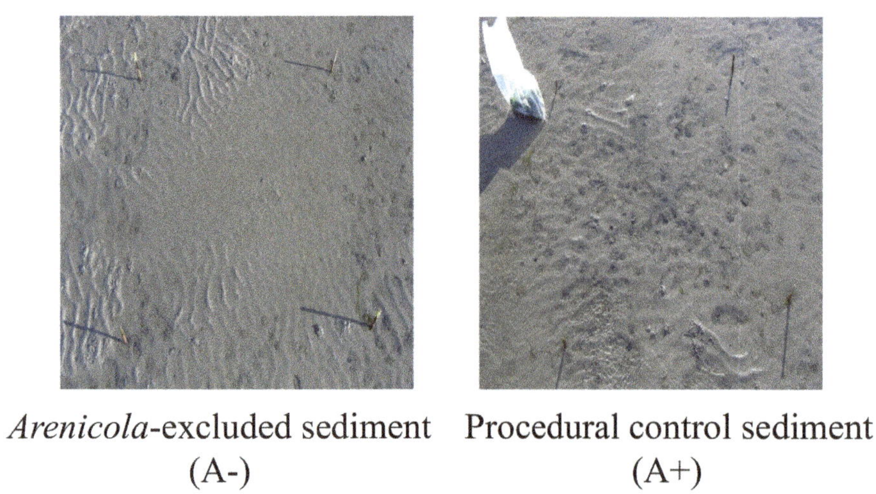

Figure 2. Comparison of sediment surface appearance after *Arenicola marina* exclusion at low tide. A− represent the sediment after 1 month from the burial of net to exclude burrowing species. A+ is the procedural control treatment (see Materials and Methods for details).

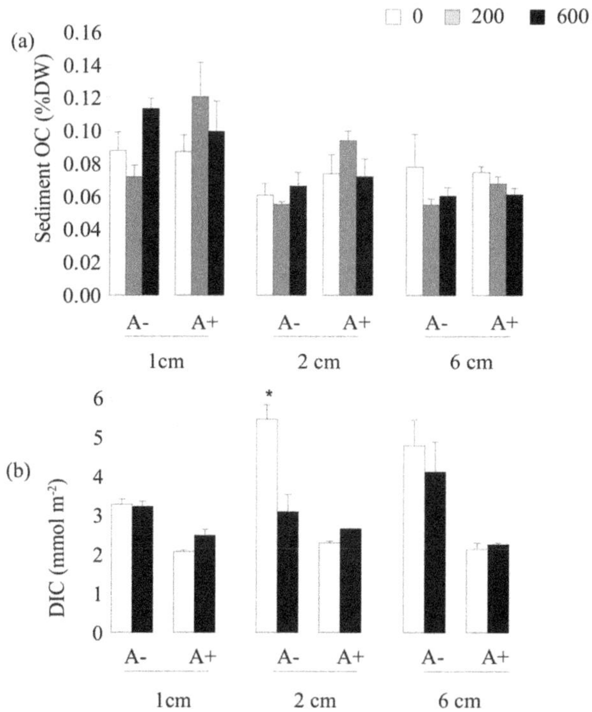

Figure 3. Mean (+1SE) for (a) bulk sediment organic carbon (OC) and (b) porewater dissolved inorganic carbon (DIC) at different sediment depths (*1 cm*: surface sediment to 1 cm depth; *2 cm*: from the depth of 1 to 2 cm; *6 cm:* from the depth of 2 to 6 cm). In the graphs, A−: *A. marina* excluded; A+: procedural control for burial; the legend indicate the levels of detrital addition (0: no addition; 200:200 g WW were added twice at a 1 week interval; 600:600 g WW were added twice with a 1 week interval). Data were sampled one month after detrital addition. Asterisk indicates significant differences.

did not allow to gather additional information concerning the (auto)trophic state of the sediment, which would need a light incubations of other sediment cores. Our system was however limited to carry 12 cores at a time and no further analyses could be done. Prior to incubations, cores were inundated with well-oxygenated Oosterschelde water collected the same day ($[NH_4^+] = 20$ µmol L$^-$, $[NO_3^-] = 110$ µmol L^{-1}, salinity = 22) and left to acclimatize in the dark for 6–8 h. During incubation, the overlying water-column (25–30 cm) was continuously stirred with a magnetic stirrer, maintaining continuous water circulation at a rate well below the resuspension limit. Dissolved O_2 never decreased below 65% of saturation. Water samples for TCO_2 and O_2 were collected in glass vials and sealed avoiding air bubbles. All samples were immediately poisoned with saturated $HgCl_2$ to stop bacterial activity and then analyzed within 3 d.

For experiment 2, sediment samples were taken from the four plots with labelled *Ulva*. Sampling was done on four occasions, 6, 8, 15 and 17 days after detrital enrichment. Each time, one plexiglass intact sediment core (11 cm i.d.) was collected to 20 cm depth, following the modality used to sample the cores for estimating benthic respiration. To reduce plot damage during core collection, the central part of the plot (20 cm from the edges) was divided into 4 quadrants of 30×30 cm. Each sampling occasion, a sediment core was taken from a different quadrant. An area of 15×15 cm was thus damages each time, corresponding to 2.25% of the entire plot area. The sampled sediment was replaced with plexiglass tubes to avoid any collapse and to leave the remaining

Table 1. Two-way analyses of variance for sediment organic carbon (OC) and porewater DIC at different sediment depths one month after detrital enrichment.

		0–1 cm		1–2 cm		2–6 cm	
Sediment OC	**df**	**MS**	**F**	**MS**	**F**	**MS**	**F**
Exclusion = E	1	0.67 10^{-3}	0.40	1.60 10^{-3}	3.80	0.09 10^{-3}	0.52
Detrital enrichment = D	2	0.61 10^{-3}	0.96	0.09 10^{-3}	0.40	0.62 10^{-3}	2.22
E×D	2	1.7 10^{-3}	2.64	0.42 10^{-3}	1.90	0.17 10^{-3}	0.62
Residual	12	0.63 10^{-3}		0.22 10^{-3}		0.28 10^{-3}	
Porewater DIC							
Exclusion = E	1	1.89	17.50	5.06	1.03	10.15	4.80
Detrital enrichment = D	1	0.06	2.00	1.28	5.12	0.15	0.28
E×D	1	0.11	3.43	4.93	**19.71**	0.30	0.57*
Residual	8	0.03		1.00		2.11	

Significant values ($P<0.05$) are in bold. Data were not-transformed ($P_{C\text{-cochran}} > 0.05$). Asterisk indicates pooling of interaction when $P>0.25$.

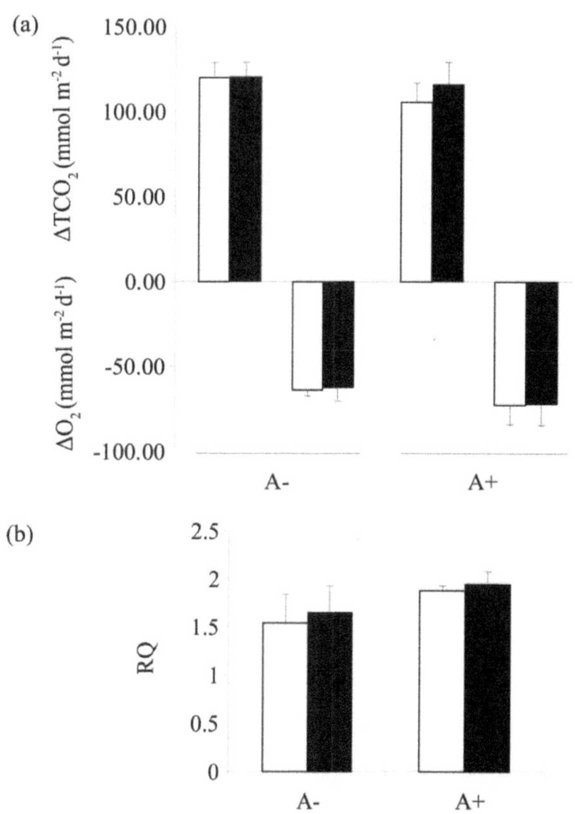

Figure 4. Mean (+1SE) for (a) benthic oxygen (O_2) consumption and total carbon dioxide (TCO_2) released during 12 hours dark incubation; (b) Respiratory quotient (RQ). In the graphs, A−: *A. marina* excluded; A+: procedural control for burial; the legend indicate the levels of detrital addition (0: no addition; 200:200 g WW were added twice at a 1 week interval; 600:600 g WW were added twice with a 1 week interval). Data were sampled one month after detrital addition.

Table 2. Two-way analyses of variance for sediment-water exchanges of O_2 and TCO_2 and for the benthic respiratory coefficient RQ.

		O_2		TCO_2		RQ	
	df	**MS**	**F**	**MS**	**F**	**MS**	**F**
Exclusion = E	1	8.8 10^{-6}	2.38	252.08	0.78	0.29	2.39
Detrital enrichment = D	1	8.0 10^{-9}	0.00	102.08	0.31	0.02	0.18
E×D	1	4.38 1^{-7}	0.11*	70.08	0.22*	0.00	0.01*
Residual	8	4.10 1^{-6}		325.08		0.14	

Significant values ($P<0.05$) are in bold. Data were not-transformed ($P_{C\text{-cochran}}>0.05$), except for O_2, which was arc-tangent transformed to homogenise the residual variances. Asterisk indicates pooling of interaction when $P>0.25$.

part of the plots undisturbed. The cores were handled like the cores used for evaluating oxygen and carbon exchange at the sediment-water interface for the experiment 1. However, before incubating for wet respiration and determining release of $T^{13}CO_2$ in the water column, the labelled cores were incubated for dry fluxes during 6 h. At the beginning and at the end of the incubation, air samples were collected using syringes and pumped into tubes containing soda lime to trap CO_2 and measure the $^{13}CO_2$ derived from *Ulva* decomposition. A LICOR automated CO_2 analyzer was used to quantify CO_2 release every hour. Sediment-water exchange rate and macrofauna data were generated as for experiment 1 (see "Analytical measures" section).

Following flux incubations for both experiments, the cores were sieved at 0.5 mm to collect the macrofauna. No lugworm was included in the cores, meaning that any observed pattern represented the result of sediment and macrofauna alteration due to *Arenicola*, but not the direct activity of the lugworm during incubation.

Analytical Measures

In the laboratory, macrofauna were identified to the species level, counted and dried at 60°C for 48 h to obtain dry weight biomass to be used for stable isotope analyses relative to the experiment 2. The detrital *Ulva* remaining in the samples was also collected, dried and weighed. Chlorophyll *a* was extracted in darkness for 24 h at 0–4°C using a 90% acetone solution. The sediment was homogenized and sub-samples of about 1 g dry weight were taken. Pigments were measured spectrophotometrically before and after 0.1 N HCl acidification. Pre-combusted (450°C for 3 h) sediment was used as a blank. Measurements for bulk sediment OC were made using a Perkin-Elmer CHN element analyzer on freeze-dried material, after carbonate removal with HCl.

Flux rates of TCO_2 and O_2 between the sediment and the overlying water were estimated as difference between the initial and the final TCO_2 or O_2 concentrations during dark incubation, assuming constant solute exchange with time. The oxygen dissolved into the water samples was analyzed using the standard method of Winkler titration, while water TCO_2 and porewater DIC were determined by the flow injection/diffusion cell technique, which is based on the diffusion of CO_2 through a hydrophobic membrane into a flow of deionized water, thus generating a gradient of conductivity proportional to the concentration of CO_2.

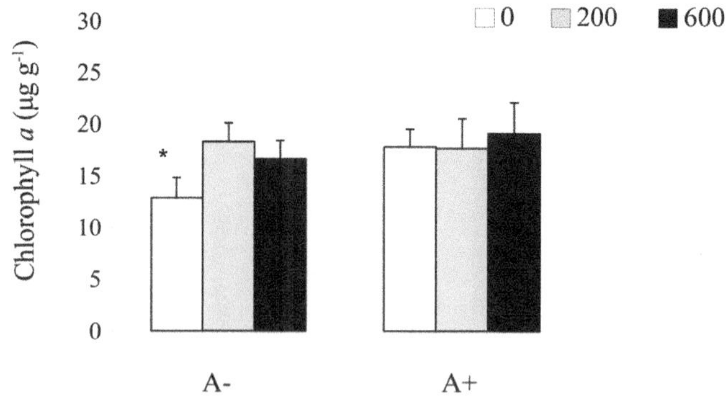

Figure 5. Mean (+1SE) for chlorophyll *a* concentration at the surface sediment (1 cm depth). In the graphs, A−: *A. marina* excluded; A+: procedural control for burial; the legend indicate the levels of detrital addition (0: no addition; 200:200 g WW were added twice at a 1 week interval; 600:600 g WW were added twice with a 1 week interval). Data were sampled one month after detrital addition. Asterisk indicates significant differences.

Water TCO_2 (wet fluxes), gaseous CO_2 (dry flux), bulk sediment and macrofauna collected in the labelled plots were analyzed for carbon isotope composition ($\delta^{13}C$). The gaseous $^{13}CO_2$ (dry flux) captured using soda lime was analyzed after acidification. The ^{13}C isotope signature for TCO_2 dry and wet fluxes was analyzed using a GC column coupled to a Finningan delta S mass spectrometer IRMS. The carbon isotopic composition for sediment, residual detritus and animals was determined using a Fisons elemental analyzer coupled to a Finningan delta S mass spectrometer. ^{13}C to ^{12}C ratio was referred to Vienna PDB and expressed in units of ‰. Reproducibility of the measurements is better than 0.2‰ [45]. The excess of the heavy isotope of carbon (above background) was expressed as total uptake (I) in milligrams of ^{13}C. I was calculated as the product of excess ^{13}C (E) and carbon or biomass (C):

$$I = E \times C$$

$$E = F_{control} - F_{sample}$$

where F indicates the fraction of isotope added and it is calculated as:

$$F = R/R + 1$$

where R can be calculated from the measured $\delta^{13}C$ using $R_{reference}$ of 0.0112372:

$$\delta^{13}C = \left[R_{sample}/R_{reference} - 1\right] \times 10^3$$

Statistical Analyses

For experiment 1, data were analyzed with two-way orthogonal mixed model of analysis of variance (ANOVA), with exclusion (2 levels: procedural control A+ and exclusion of *A. marina*, A−) and detrital enrichment (3 or 2 levels: 0, 200 and 600) as fixed and random factors, respectively. Detrital enrichment was considered a random factor because the amount of detritus to add was chosen within a range of values to simulate conditions of low and high

Table 3. Two-way analyses of variance for chlorophyll *a*, macrofauna abundance and number (N) of species.

		Chlorophyll a		Abundance		N of species	
	df	MS	F	MS	F	MS	F
Exclusion = E	1	72.23	1.27	0.48	0.69	0.03	0.00
Detrital enrichment = D	2	53.74	**4.02**	0.29	0.56	30.00	**7.11**
E×D	2	56.80	**4.25**	2.34	**4.81**	17.73	**4.21**
Residual	24	13.37		0.49		4.22	

Significant values ($P<0.05$) are in bold. Data were sqrt-transformed for macrofauna abundance.

enrichment, but no specific hypothesis concerned the chosen amounts. A significant interaction between exclusion and detrital enrichment was expected if the exclusion of *A. marina* changed the response to detrital enrichment. These analyses were done for the total abundance of macrofauna individuals, number of macrofauna species, chlorophyll *a*, sediment OC, porewater DIC, water O_2 and TCO_2 exchange rate. *C-cochran* test was used to test for residual variance homogeneity before running the analyses. When this test was significant ($P<0.05$), data were transformed. Pooling was done when $P>0.25$. *A posteriori* multiple comparison SNK test was run when interaction term was significant.

The variables collected for evaluating the recycling of detrital ^{13}C (experiment 2) were analyzed with repeated measure ANOVA, with time (6, 8, 15 and 17 sampling days from the end of detrital addition) as the repeated measure random factor and *A. marina* exclusion as fixed factor. Plots (2 levels) was nested in exclusion. The interaction of time with the exclusion treatment was measured over the interaction time×plot. When significant interaction was found, the main effect exclusion was not interpreted. Assumption about the correlation among the different observations of the repeated measure factor (time) was tested with Mauchley's sphericity test.

Decomposition rate of macroalgal detritus was estimated by fitting the first-order model:

$$B_t = B_0 e^{kt}$$

Where B_t is the g of detrital biomass at time t (days), B_0 is the initial biomass, at time 0, at the moment of the second detrital addition and k is the specific decomposition constant. The two detrital additions were done at a distance of one week, which may greatly affect the estimates of B_0. We therefore considered the estimates of residual *Ulva* as the sum of two decomposition curves, with the same constant k but different decomposing time:

$$B_t = B_{(0-7)} e^{k(t+7)} + B_{(0)} e^{kt}$$

The curves of ^{13}C benthic incorporation (including loss due to respiration) were fitted to the logistic curve:

$$I_t = Y/(1 + qe^{-rt})$$

$$q = (Y - I_0)/I_0$$

Where Y is the ^{13}C loss from macroalgal decomposition and r is the capacity of benthos to incorporate ^{13}C.

Results

After the exclusion of *A. marina*, no burrows were visible in the excluded plots (Fig. 2) and the number of burrows in the procedural control (treatment A+) was comparable to that of natural sediment (mean±SE, n = 5:8.72±1.77 and 10.33±1.17 burrows m^{-2} for the natural and the procedural control, respectively). Overall, there were no differences in macrofauna species composition between the natural sediment and the procedural control plots. Macrofauna abundance (mean±SE, n = 5) was 294±42 and 220±53 individual m^2, whereas the number of species was 10±0.2 and 10±0.5 for A+ and natural sediment, respectively. Therefore, A+ could be used as a control for the effect of lugworm exclusion and detrital macroalgae.

Overall, in the natural sediments and in the A+ plots the gastropod *Hydrobia ulvae* was numerically dominant, representing 90% of macrofauna abundance. The remaining 10% was made up by oligochaetes (3%), deposit feeder polychaetes (3%) such as spionids, nereids, capitellids and cirratulids and bivalves (3%), mainly *Macoma balthica* and *Cerastoderma edule*.

Carbon Availability and Benthic Respiration

At the time of sampling, sediment organic carbon (OC) did not vary following detrital enrichment, the exclusion of *A. marina* or their interactions (Table 1, Fig. 3a; ANOVA, $P>0.05$). Macroalgal debris was still present in the sediment where detrital macroalgae were added at the highest enrichment level (treatment 600). The biomass of this macroalgal debris was smaller in the lugworm-excluded sediment than in procedural controls (mean±SE: lugworm exclusion, treatment A−: 2.3±1.4 gDW m^{-2}; A+: 12.0±6.6 gDW m^{-2}; ANOVA: $F_{1,\,8} = 6.19$, $P = 0.038$).

There was a significant effect of the interaction between lugworm exclusion and detrital enrichment on porewater DIC to the depth of 1–2 cm (Table 1; Fig. 3b). In the non-enriched sediment, porewater DIC concentration was higher in the lugworm exclusion than in the control plots (SNK test at $P = 0.05$: non-enriched plots, treatment 0: A−>A+). Moreover, in the lugworm-exclusion plots, porewater DIC was lower in the enriched than in the non-enriched plots (SNK test at $P = 0.05$: *Arenicola* exclusion, A−: 0>600). Values in detrital-enriched, lugworm-excluded sediment were similar to those measured for the A+ treatment (Fig. 3b). A similar pattern was observed for superficial (0–1 cm) and deeper sediment (2–6 cm in Fig. 3b), but no significant changes were measured.

Both *A. marina* exclusion and detrital enrichment or their interaction did not significantly change sediment-water exchanges of TCO_2 and O_2 nor the RQ respiration coefficient (TCO_2:O_2 ratio) (Table 2; Fig. 4).

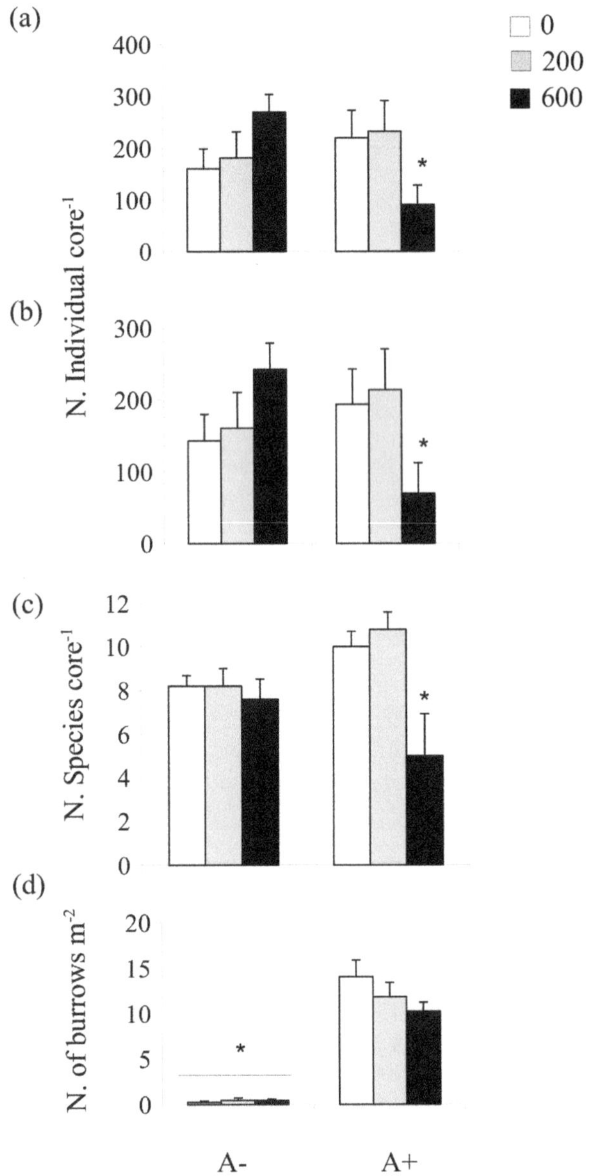

Figure 6. Mean (+1SE) for (a) total number of macrofauna individuals, (b) abundance of *Hydrobia ulvae*, (c) number of macrofauna species and (d) numbers of *Arenicola marina* burrows in the plots where the mosquito net was buried (A−) and in the procedural control plots (A+) at the end of the experiment, e.g; 1 month from the second addition of detrital *Ulva* (0: no addition; 200:200 g WW were added twice at a 1 week interval; 600:600 g WW were added twice with a 1 week interval). Data were sampled one month after detrital addition. Asterisk indicates significant differences.

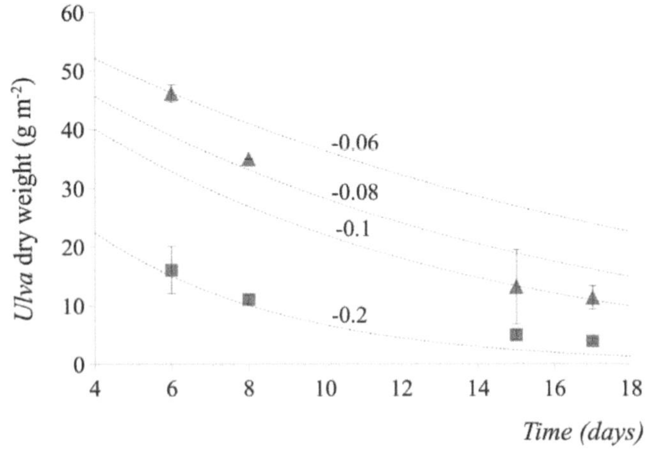

Figure 7. Decomposing curves for *Ulva* detritus in lugworm excluded (treatment A−; square symbols in the graphs) or procedural control sediments (treatment A+; triangle symbols in the graph). Symbols indicate mean (± SE) for macroalgal biomass and are relative to experiments 2.

Microalgal Biomass and Macrofauna

There was a significant interaction between *A. marina* exclusion and detrital enrichment for microalgal biomass (Table 3; Fig. 5). Lugworm exclusion had a negative effect on microalgal biomass in the non-enriched plots (SNK test: 0: A+>A−). In the lugworm-excluded sediment, detrital enrichment enhanced microalgal biomass (SNK test: A−: 0<200 = 600). Values were similar to those measured for the A+ control plots (treatments 200 and 600 in Fig. 5).

A significant interaction between *A. marina* exclusion and detrital enrichment was also found for macrofauna abundance and number of species (Table 3). Abundance was lowered by the highest level of detrital enrichment in the plots where *A. marina* was present (Fig. 6a; SNK test: A+: 0 = 200>600). Overall, both the dominant gastropod *Hydrobia ulvae* and all other remaining macrofauna decreased in abundance, especially nereids, spionids and bivalves (ANOVA: $F_{2,\ 24} = 4.26$, $P = 0.026$ for *H. ulvae*, Fig. 6b; $F_{2,\ 24} = 4.65$, $P = 0.019$ for remaining taxa, data were square-root transformed; SNK tests for both: A+: 0 = 200>600). Number of species showed a similar pattern, decreasing in response to the highest level of enrichment in the A+ plots (Fi. 6C; SNK test: A+: 0 = 200>600). Detrital enrichment seemed to not affect the abundance of the lugworms as the number of *Arenicola* burrows remained relatively stable at different levels of detrital enrichment (2 way ANOVA: $F_{2,\ 12} = 0.88$, $P = 0.34$, Fig. 6d).

The Fate of Detrital Carbon

There were differences in the rate of *Ulva* decomposition following *A. marina* exclusion. The decomposition constant k was −0.2 for the detritus added to the *A. marina* excluded sediment (A− plots in Fig. 7). Interestingly, the A+ plots did not fit well the exponential curve using the same constant k through the decomposition. Rather, at the beginning of decomposition, they fitted k values between −0.06 and −0.08, whereas at the end they fitted a K constant of −0.1 (Fig. 7). There was a significant interaction between exclusion treatment and time after detrital enrichment for the flux rate of $T^{13}CO_2$ from the sediment to the water column (Table 4). More $T^{13}CO_2$ was released by the sediment in the exclusion treatment within 6 days from burial (Fig. 8a). Respiration rates were (mean ± SE) 0.84±0.25 and 0.12±0.03 mg $^{13}C\ m^{-2}\ day^{-1}$ for dry and wet fluxes, respectively.

Part of the ^{13}C remained in the benthos and it was incorporated in the bulk sediment and in the macrofauna. The amount incorporated was variable among plots, but it did not significantly vary between exclusion treatments or time (Table 4; Fig. 8 b–c). *H. ulvae* incorporated the largest amount of ^{13}C among all macrofauna taxa (mean ± SE: 3.2±0.8 and 1.2±0.4 mg $^{13}C\ m^{-2}$, for *H. ulvae* and remaining fauna, respectively).

Table 4. Repeated measure analyses of variance and sphericity tests for the ^{13}C incorporation in macrofauna and bulk sediment, and for the $^{13}CO_2$ released from the benthos (benthic respiration) during detrital decomposition at an intermediate level of detrital enrichment (200 gWW plot^{-1}added twice).

		Macrofauna		Bulk sediment		Benthic respiration	
	df	MS	F	MS	F	MS	F
Exclusion = E	1	5.97 10^{-7}	0.12	4.27 10^{-4}	0.22	1.52 10^{-5}	**9.73**
Time = T	3	3.13 10^{-7}	0.60	8.78 10^{-4}	1.02	1.84 10^{-5}	**14.45**
Residual (Plots)	2	4.96 10^{-6}		1.96 10^{-3}		1.56 10^{-6}	
T×E	3	6.21 10^{-7}	0.12	8.79 10^{-4}	1.02	8.98 10^{-6}	**7.07**
Residual (T×Plots)	6	5.21 10^{-6}		8.59 10^{-4}		1.27 10^{-6}	
Sphericity test (χ^2) n = 6		7.80		4.83		8.06	

Significant values ($P<0.05$) are in bold. Sphericity test was not significant.

Discussion

Bioturbation can enhance solute penetration in deeper sediments, their exchange between the sediment and the overlying water and re-distribute organic material along the vertical profile [36,37,41]. As such, the loss of species that are important contributors to bioturbation may greatly change the response to detrital decomposition of sediment biogeochemistry. In laboratory experiments, for instance, *A. marina* and the polychaete *Nereis diversicolor* were found to increase turnover of carbon and nutrients during detrital macroalgal decomposition [19,36]. The exclusion of *A. marina* from intertidal sediments was thus expected to modify the biogeochemical response to macroalgal detrital enrichment. Conversely, at the end of decomposition (1 month after the second enrichment) benthic respiration (O_2 and TCO_2 sediment-water exchanges) and sediment organic carbon content did not change as an effect of lugworm exclusion.

Ulva amount and time of sampling (one month after enrichment) allowed decomposition and release of a large quantity of organic matter, which could be mineralized or accumulated in the sediment, thereby modifying sediment organic carbon, inorganic carbon dissolved in the interstitial water and fluxes at the sediment-water interface. *Ulva* halftime decomposition rate is within 15 days [30]. By considering the range of k constant calculated for this study (between −0.06 and −0.2), at least 85% *Ulva* would be decomposed at the time of sampling for both enrichment levels. If one considers that dried sediment weighs about 1600 g l^{-1} and *Ulva* organic carbon content is around 40% of dried tissues [34], this decomposed detritus corresponds to organic carbon concentrations ranging between 0.15 and 1.2% sediment dry weight for the top first cm, roughly 2–10 times the concentration of organic carbon found in natural sediments for low and high enrichment, respectively (see the non-enriched sediment in Fig. 2a). The lack of sediment carbon accumulation and changes in TCO_2 fluxes at the sediment-water interface indicated that the benthic system was able to remove the carbon derived from *in situ* decomposition and that bioturbation did not play a central role in detrital carbon recycling.

It is important to re-iterate that fluxes were measured during 12 h dark incubation of sediment cores without lugworms, which indicated the overall indirect effect of lugworms on detrital benthic mineralization, but did not include any direct effects of lugworms due to burrow ventilation during the period of incubation. The importance of ventilation was instead assessed in the field by measuring porewater DIC, which increased after lugworm removal, indicating a decreased solute exchange due to *A. marina* loss. Interestingly, these changes were variable along the vertical sediment profile and were offset by the addition of detrital macroalgae, which supported the idea that fast mineralization of detrital carbon in these sediments is only in part due to bioturbation. The complexity of the relationships between species, functions and response to disturbance can increase with increasing ecological realism [15,46]. Our experiments were done *in situ* and they were designed to test effects of *A. marina* under natural environmental conditions. Detrital macroalgae were deliberately added to the sediment, without using artificial nets that could limit detrital loss. This method exposed decomposing detritus to tides and waves, especially when detritus remained at the sediment surface. The role of bioturbation on sediment biogeochemistry has been mainly assessed through laboratory studies. The few field studies on the role of *Arenicola* on porewater, fluxes and organic matter decomposition have revealed that its role may vary and decrease considerably in nature [19,37,47]. By comparing field to laboratory experiments, it was found, for instance, that solute exchange was more variable in the field because it was not only ruled by bioturbation but also by advection related to waves and tides and by differences in the reactivity of sediment organic matter related to the complexity of biological communities of primary producers and consumers [47]. This is particularly true when experiments are done in permeable sediment, where *A. marina* is often a dominant species [48]. In these sediments, hydrodynamics can enhance advective flow from the porewater to the overlying water column, rule the exchange of solutes and organic matter at the sediment-water interface and overwhelm the biological effect of bioturbation [49]. Although our experiments were done in a relatively sheltered sand flat, as indicated by the dominance of fine-grained sand, numerous ripple marks were visible at low tide (Fig. 2) and they probably indicated strong hydrodynamics generated by waves and tides. In these sediments, hydrodynamics could accelerate the exchange of solutes produced during decomposition, the removal of macroalgal detritus from the sediment and override the effect of bioturbation on the response of sediment biogeochemistry to detrital macroalgae.

The decomposition rate of *Ulva* in these sediments was fast, as indicated by the estimated k constants (k = −0.2 and −0.1 for A− and A+, respectively; Fig. 7). These values were at the limit or even exceeded the range of values typical of *Ulva* detritus, which varies between −0.04 and −0.1 [34]. More importantly, the estimated k

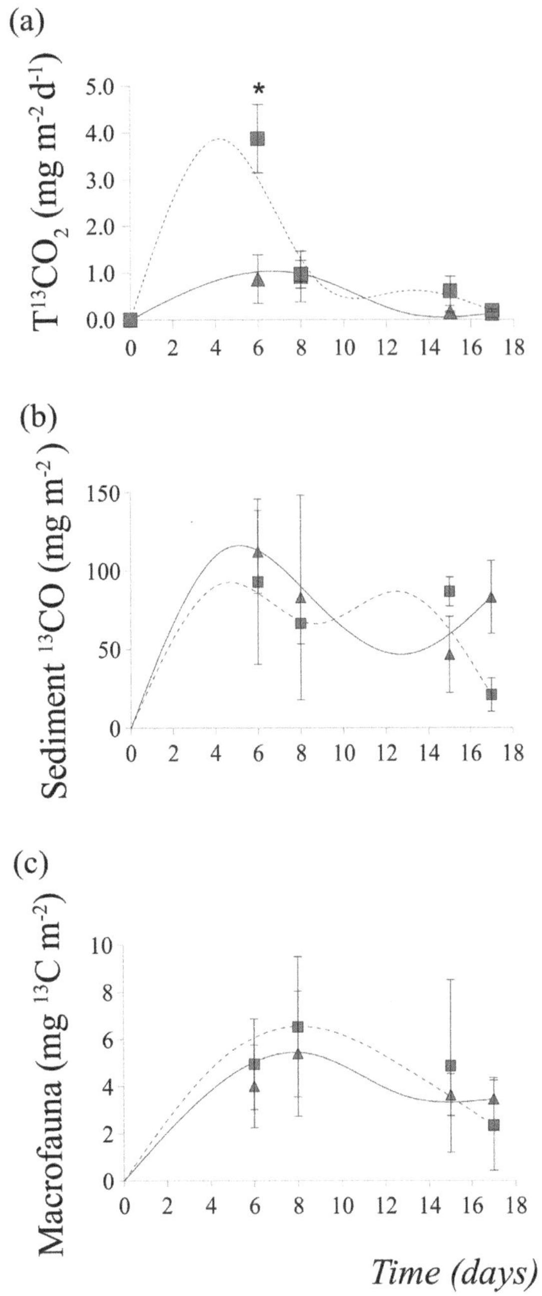

Figure 8. Mean ± SE for (a) $T^{13}CO_2$ released from the sediment to the water column (sum of dry and wet fluxes); (b) amount of ^{13}C organic carbon in the bulk sediment (averaged over depth profile) and (c) amount of ^{13}C carbon incorporated by macrofauna. Data are fitted to the LOESS smoothing with an interval of 4 and 1 polynomial degree. Values are averaged over the two plots where *A. marina* was excluded (treatment A−; square symbols in the graphs) and the two procedural control plots (treatment A+; triangle symbols in the graph). Data refers to the experiment 2. Asterisk indicates significant differences.

constant for the lugworm-exclusion sediment was twice that of the control plots. In addition, the biomass of macroalgal debris was, on average, 6 times larger in controls than in lugworm-excluded sediments (2 and 12 g DW per plot^{-1} for A+ and A−, respectively). This may bring to the conclusion that despite hydrodynamics could play an important role in determining benthic response to detrital macroalgae, *A. marina* could represent an important sink for detritus and it could modify *in situ* decomposition rate, as found for other burrowing bioturbators [39,50]. These species can mechanically bury macroalgae in their funnels, thereby decreasing detritus availability to surface sediment and slowing down decomposition rate, while maintaining a pool of organic matter in the sediment. Buried detritus and slowed decomposition rate could, in turn, modify mineralization processes. In this study, a peak in $T^{13}CO_2$ release to the water column was identified at an early stage of decomposition in *A. marina*-excluded sediment plots, suggesting that the loss of *A. marina* accelerated the mineralization of detrital *Ulva* at an early stage of decomposition. This explanation of increased mineralization following *A. marina* loss was supported by the finding that porewater DIC decreased in response to detrital addition in *A. marina* excluded sediments. In addition, simultaneously to this decrease, macroalgal biomass increased. Probably, without the mechanical action of burying macroalgal fragments by *Arenicola*, detrital macroalgae remaining on sediment surface and decomposing faster released nutrients that enhanced bacterial metabolism and their carbon consumption. Evidence for increased carbon consumption during biomass decomposition has been found for plant litter in freshwater ecosystems [51,52]. In addition, benthic microalgae are often regulated by bottom-up processes [53] and they could greatly benefit from nutrient release, thereby contributing to recycling the mineralized detrital carbon. Interestingly, detrital enrichment and lugworms had similar effects on microalgal biomass. Probably, both lugworm bioirrigation and detrital decomposition can provide nutrients necessary to regulate microalgal growth [19,28,37,47].

A. marina is believed to influence the assembly rules and the distribution of macrofauna and, as such, their patterns of response to disturbances by altering the sedimentary habitat (e. g. [37,41]). Our results showed a negative effect of *A. marina* on macrofauna response to detrital addition since both macrofauna abundance and diversity decreased following the highest level of enrichment in presence of lugworms. The increased detrital burial by *A. marina* could play a central role in determining this response. Patches of detrital macroalgae in sediments can govern the spatial and temporal variability of macrofauna patterns of distribution. The presence of detritus may decrease numbers of individuals and diversity because of hypoxia during decomposition. Once detritus has decomposed, fast recolonizing taxa can promptly occupy the previously disturbed patches and benefit from increased food supply [29]. Detrital burial due to lugworm activity probably prolonged hypoxic conditions and slowed down macrofauna recovery. The decrease in abundance common to all taxa and the loss of diversity could corroborate this explanation as anoxic conditions related to detrital decomposition deplete most taxa abundance and diversity [29]. Moreover, burial could decrease detrital availability to surface consumers and therefore prevent their colonization, once hypoxia ceases. Macroalgal supply to sediment surface can represent an important source of food for surface grazers and detritivores, such as the gastropod *Hydrobia ulvae* [34,54]. Part of detrital carbon was transferred to macrofauna, especially to *H. ulvae* (experiment 2, Fig. 7c), which was the numerically dominant species in these area. Its important role for detrital carbon transport to the food web was found previously in the surrounding area [34]. This species is highly mobile and it crawls on sediment surface for feeding. Its local abundance might thus vary rapidly following food availability on surface sediment.

In summary, our field experiment showed that *A. marina* can be an important sink for detrital *Ulva*. However, conversely to what it should be expected based on current research knowledge, this species has moderate, sometimes negative effects on the response to detrital enrichment of sediment biogeochemistry and macrofauna. *A. marina* loss causes the sedimentary ecosystem to become a faster recycler of detrital macroalgae, enhancing detrital loss, carbon consumption and mineralization. This may bring to the conclusion that the loss of this species might be unimportant or somewhat beneficial for certain intertidal ecosystems. However, fast recycling and detrital dispersal could undermine carbon storage capacity of benthos and, in turn, the stability of whole estuarine ecosystem facing organic enrichment and eutrophication.

These findings clearly demonstrate that under natural conditions, the local extinction of an ecosystem engineer may have complex effects on the ecological stability of marine sediment biogeochemistry and benthic communities of consumers. They also highlight the importance of performing experiments under natural conditions if we are to understand and predict the response of ecosystems to perturbation and species loss.

Acknowledgments

The authors wish to thank the technicians for their invaluable help in the chemical analyses and E. Weerman, JJ Jansen and F Montserrat and numerous other students for their help in the field. S. Thrush is kindly thanked for his critical revision of an early version of this MS. This study is part of the BIOFUSE project, within the MARBEF network of excellence.

Author Contributions

Conceived and designed the experiments: FR BG. Performed the experiments: FR BG FG VDS. Analyzed the data: FR BG FG VDS. Contributed reagents/materials/analysis tools: JJM. Wrote the paper: FR BG FG VDS JJM.

References

1. Grime JP (1997) Ecology - Biodiversity and ecosystem function: The debate deepens. Science 277: 1260–1261.
2. McCann KS (2000) The diversity-stability debate. Nature 405: 228–233.
3. Worm B, Barbier EB, Beaumont N, Duffy JE, Folke C, et al. (2006) Impacts of biodiversity loss on ocean ecosystem services. Science 314: 787–790.
4. Tilman D, Downing AL (1994) Biodiversity and stability in grasslands. Nature 367: 363–365.
5. Yachi S, Loreau M (1999) Biodiversity and ecosystem productivity in a fluctating environment: The insurance hypothesis. Proceedings of the National Academy of Sciences of the United States of America 96: 1463–1468.
6. Downing A, Leibold M (2010) Species richness facilitates ecosystem resilience in aquatic food webs. Freshwater Biology 55: 2123–2137.
7. Ives A, Carpenter S (2007) Stability and diversity of ecosystems. Science: 58–62.
8. Pfisterer A, Schmid B (2002) Diversity-dependent production can decrease the stability of ecosystem functioning. Nature 416: 84–86.
9. Allison G (2004) The influence of species diversity and stress intensity on community resistance and resilience. Ecological Monographs 74: 117–134.
10. Emmerson MC, Solan M, Emes C, Paterson DM, Raffaelli D (2001) Consistent patterns and the idiosyncratic effects of biodiversity in marine ecosystems. Nature 411: 73–77.
11. Goodsell PJ, Underwood AJ (2008) Complexity and idiosyncrasy in the responses of algae to disturbance in mono- and multi-species assemblages. Oecologia 157: 509–519.
12. Rossi F, Vos M, Middelburg J (2009) Species identity, diversity and microbial carbon flow in reassembling macrobenthic communities. Oikos: 503–512.
13. Solan M, Cardinale BJ, Downing AL, Engelhardt KAM, Ruesink JL, et al. (2004) Extinction and ecosystem function in the marine benthos. Science 306: 1177–1180.
14. Bolam SG, Fernandes TF, Huxham M (2002) Diversity, biomass, and ecosystem processes in the marine benthos. Ecological Monographs 72: 599–615.
15. Romanuk T, Vogt R, Kolasa J (2009) Ecological realism and mechanisms by which diversity begets stability. Oikos 118: 819–828.
16. Jones CG, Lawton JH, Shachak M (1994) Organisms as ecosystem engineers. Oikos 69: 373–386.
17. Berke S (2010) Functional Groups of Ecosystem Engineers: A Proposed Classification with Comments on Current Issues. Integrative and Comparative Biology 50: 147–157.
18. Kristensen E, Penha-Lopes G, Delefosse M, Valdemarsen T, Quintana CO, et al. (2012) What is bioturbation? The need for a precise definition for fauna in aquatic sciences. Marine Ecology Progress Series 446: 285–302.
19. Kristensen E (2001) Impact of polychaetes (Nereis and Arenicola) on sediment biogeochemistry in coastal areas: Past, present, and future developments. Abstracts of Papers of the American Chemical Society 221: U538–U538.
20. Braeckman U, Provoost P, Moens T, Soetaert K, Middelburg J, et al. (2011) Biological vs. Physical Mixing Effects on Benthic Food Web Dynamics. Plos One 6: :-e18078. doi:10.1371/journal.pone.0018078.
21. Lohrer AM, Thrush SF, Gibbs MM (2004) Bioturbators enhance ecosystem function through complex biogeochemical interactions. Nature 431: 1092–1095.
22. Eklof J, van der Heide T, Donadi S, van der Zee E, O'Hara R, et al. (2011) Habitat-Mediated Facilitation and Counteracting Ecosystem Engineering Interactively Influence Ecosystem Responses to Disturbance. Plos One 6: e23229.-doi:10.1371/journal.pone.0023229.
23. Norkko J, Reed DC, Timmermann K, Norkko A, Gustafsson BG, et al. (2012) A welcome can of worms? Hypoxia mitigation by an invasive species. Global Change Biology 18: 422–434. doi:10.1111/j.1365-2486.2011.02513.x.
24. Eisenhauer N, Milcu A, Sabais A, Scheu S (2008) Animal Ecosystem Engineers Modulate the Diversity-Invasibility Relationship. Plos One 3.
25. Cebrian J (2004) Role of first-order consumers in ecosystem carbon flow. Ecology Letters 7: 232–240.
26. Raffaelli DG, Raven JA, Poole LJ (1998) Ecological impact of green macroalgal blooms. Oceanography and Marine Biology, an Annual Review 36: 97–125.
27. Ford RB, Thrush SF, Probert PK (1999) Macrobenthic colonisation of disturbances on an intertidal sandflat: the influence of season and buried algae. Marine Ecology Progress Series 191: 163–174.
28. Rossi F, Underwood AJ (2002) Small-scale disturbance and increased nutrients as influences on intertidal macrobenthic assemblages: experimental burial of wrack in different intertidal environments. Marine Ecology Progress Series 241: 29–39.
29. Kelaher BP, Levinton JS (2003) Variation in detrital enrichment causes spatio-temporal variation in soft-sediment assemblages. Marine Ecology Progress Series 261: 85–97.
30. Rossi F (2006) Small-scale burial of macroalgal detritus in marine sediments: Effects of Ulva spp. on the spatial distribution of macrofauna assemblages. Journal of Experimental Marine Biology and Ecology 332: 84–95.
31. Bolam SG, Fernandez TF, Read P, Raffaelli D (2000) Effects of macroalgal mats on intertidal sandflats: an experimental study. Journal of Experimental Marine Biology and Ecology 249: 123–137.
32. Hull SC (1987) Macroalgal mats and species abundance: a field experiment. Estuarine Coastal and Shelf Science 25: 519–532.
33. Thrush SF (1986) The sublittoral macrobenthic communtiy structure of an irish sea-lough: effect of decomposing accumulatoins of seaweed. Journal of Experimental Marine Biology and Ecology 96: 199–212.
34. Rossi F (2007) Recycle of buried macroalgal detritus in sediments: use of dual-labelling experiments in the field. Marine Biology 150: 1073–1081.
35. Godbold JA, Solan M, Killham K (2009) Consumer and resource diversity effects on marine macroalgal decomposition. Oikos 118: 77–86.
36. Hansen K, Kristensen E (1998) The impact of the polychaete Nereis diversicolor and enrichment with macroalgal (Chaetomorpha linum) detritus on benthic metabolism and nutrient dynamics in organic-poor and organic-rich sediment. Journal of Experimental Marine Biology and Ecology 231: 201–223.
37. O'Brien A, Volkenborn N, van Beusekom J, Morris L, Keough M (2009) Interactive effects of porewater nutrient enrichment, bioturbation and sediment characteristics on benthic assemblages in sandy sediments. Journal of Experimental Marine Biology and Ecology: 51–59.
38. D'Andrea A, DeWitt T (2009) Geochemical ecosystem engineering by the mud shrimp Upogebia pugettensis (Crustacea: Thalassinidae) in Yaquina Bay, Oregon: Density-dependent effects on organic matter remineralization and nutrient cycling. Limnology and Oceanography 54: 1911–1932.
39. Vonk J, Kneer D, Stapel J, Asmus H (2008) Shrimp burrow in tropical seagrass meadows: An important sink for litter. Estuarine Coastal and Shelf Science 79: 79–85.
40. Meysman FJR, Galaktionov ES, Gribsholt B, Middelburg JJ (2006) Bioirrigation in permeable sediments: Advective pore-water transport induced by burrow ventilation. Limnology and Oceanography 51: 142–156.
41. Volkenborn N, Reise K (2006) Lugworm exclusion experiment: Responses by deposit feeding worms to biogenic habitat transformations. Journal Of Experimental Marine Biology And Ecology 330: 169–179.
42. Norkko A, Bonsdorff E (1996) Population responses of coastal zoobenthos to stress induced by drifting algal mats. Marine Ecology Progress Series 140: 141–151.
43. Buchsbaum R, Valiela I, Swain T, Dzierzeski M, Allen S (1991) Available And Refractory Nitrogen In Detritus Of Coastal Vascular Plants And Macroalgae. Marine Ecology-Progress Series 72: 131–143.
44. Nedergaard RI, Risgaard-Peterson N, Finster K (2002) The importance of sulfate reduction associated with Ulva lactuca thalli during decomposition: a mesocosm experiment. Journal of Experimental Marine Biology and Ecology 275: 15–29.

45. Herman PMJ, Middelburg JJ, Widdows J, Lucas CH, Heip CHR (2000) Stable isotopes as trophic tracers: combining field sampling and manipulative labelling of food resources for macrobenthos. Marine Ecology Progress Series 204: 79–92.
46. Mckie B, Schindler M, Gessner M, Malmqvist B (2009) Placing biodiversity and ecosystem functioning in context: environmental perturbations and the effects of species richness in a stream field experiment. Oecologia: 757–770.
47. Papaspyrou S, Kristensen E, Christensen B (2007) Arenicola marina (Polychaeta) and organic matter mineralisation in sandy marine sediments: In situ and microcosm comparison. Estuarine Coastal and Shelf Science 72: 213–222.
48. Needham H, Pilditch C, Lohrer A, Thrush S (2011) Context-Specific Bioturbation Mediates Changes to Ecosystem Functioning. Ecosystems 14: 1096–1109.
49. Huettel M, Roy H, Precht E, Ehrenhauss S (2003) Hydrodynamical impact on biogeochemical processes in aquatic sediments. Hydrobiologia 494: 231–236.
50. Retraubun A, Dawson M, Evans S (1996) The role of the burrow funnel in feeding processes in the lugworm Arenicola marina (L). Journal of Experimental Marine Biology and Ecology 202: 107–118.
51. Fontaine S, Bardoux G, Abbadie L, Mariotti A (2004) Carbon input to soil may decrease soil carbon content. Ecology Letters 7: 314–320.
52. Lennon JT (2004) Experimental evidence that terrestrial carbon subsidies increase CO2 flux from lake ecosystems. Oecologia 138: 584–591.
53. Hillebrand H, Worm B, Lotze H (2000) Marine microbenthic community structure regulated by nitrogen loading and grazing pressure. Marine Ecology Progress Series 204: 27–38.
54. Riera P (2010) Trophic plasticity of the gastropod Hydrobia ulvae within an intertidal bay (Roscoff, France): A stable isotope evidence. Journal of Sea Research 63: 78–83.

3

Patchiness and Co-Existence of Indigenous and Invasive Mussels at Small Spatial Scales: The Interaction of Facilitation and Competition

Johan Erlandsson[1]*, Christopher D. McQuaid[2], Martin Sköld[3]

1 ARONIA Coastal Zone Research Team, Åbo Akademi University/Novia University of Applied Sciences, Ekenäs, Finland, **2** Coastal Research Group, Department of Zoology and Entomology, Rhodes University, Grahamstown, South Africa, **3** Department of Mathematics, Stockholm University, Stockholm, Sweden

Abstract

Ecological theory predicts that two species with similar requirements will fail to show long-term co-existence in situations where shared resources are limiting, especially at spatial scales that are small relative to the size of the organisms. Two species of intertidal mussels, the indigenous *Perna perna* and the invasive *Mytilus galloprovincialis*, form mixed beds on the south coast of South Africa in a situation that has been stable for several generations of these species, even though these populations are often limited by the availability of space. We examined the spatial structure of these species where they co-exist at small spatial scales in the absence of apparent environmental heterogeneity at two sites, testing: whether conspecific aggregation of mussels can occur (using spatial Monte-Carlo tests); the degree of patchiness (using Korcak B patchiness exponent), and whether there was a relationship between percent cover and patchiness. We found that under certain circumstances there is non-random conspecific aggregation, but that in other circumstances there may be random distribution (i.e. the two species are mixed), so that spatial patterns are context-dependent. The relative cover of the species differed between sites, and within each site, the species with higher cover showed low Korcak B values (indicating low patchiness, i.e. the existence of fewer, larger patches), while the less abundant species showed the reverse, i.e. high patchiness. This relationship did not hold for either species within sites. We conclude that co-existence between these mussels is possible, even at small spatial scales because each species is an ecological engineer and, while they have been shown to compete for space, this is preceded by initial facilitation. We suggest that a patchy pattern of co-existence is possible because of a balance between direct (competitive) and indirect (facilitative) interactions.

Editor: Simon Thrush, National Institute of Water & Atmospheric Research, New Zealand

Funding: This study was funded by the Hellmuth Hertz Foundation, The Royal Swedish Academy of Sciences (www.kva.se), Stint Foundation, Sweden, and Rhodes University (www.ru.ac.za), South Africa (to JE). This work is also based upon research supported by the South African Research Chairs Initiative of the Department of Science and Technology and the National Research Foundation (to CM). The funders had no role in study design, data collection and analysis, decision to publish, or preparation of the manuscript.

Competing Interests: The authors have declared that no competing interests exist.

* E-mail: jerlands@abo.fi

Introduction

General ecological theory predicts that stable coexistence of two or more similar species will be unusual and dependent on resource availability or predation [1,2,3]. Coexistence of competing species at larger local scales, that is scales beyond those at which species directly interact, is often explained by environmental heterogeneity, or by species interactions generating spatially segregated distributions even in the absence of environmental heterogeneity [1,3,4,5,6,7]. Thus, at spatial scales that are small relative to body size, it is considered unlikely that two competing species can coexist, unless the inferior competitor is a superior colonizer of empty patches or there is spatial variability in their competitive abilities so that the superior competitor is at a disadvantage in at least some part of its distribution [2,6,8]. It is even less likely that two related species are able to coexist if one is an introduced species and the other indigenous, since the introduced species is often capable of becoming invasive precisely because it is a superior competitor [2].

Exploitative competition results in indirect negative interactions, while interference competition involves direct negative interactions, e.g. territoriality, overgrowth or chemical competition. In classical competition theory, there are no benefits, only costs, for species engaged in interference competition. In nature, however, interference between interacting species can be both costly and beneficial for an inferior resource exploiter, e.g. the production of chemicals as defence involves a cost, while predation on the eggs and larvae of the competitor is beneficial in terms of growth etc. [2].

Competition will occur only if the resource is limiting, and resources can be either renewable or non-renewable. In the case of exploitation competition for a non-renewable resource, this can lead to systems being controlled by lotteries, the classic example being coral reef fishes competing for habitat [9]. In benthic marine systems competition is primarily for either food, which is renewable, or for space, especially primary space, which is often non-renewable [10]. Competition for space in these systems involves both exploitation, with some species being faster at colonising space than others [11], and interference competition with overgrowth being a common mode of interaction [12,13]. Because space is an absolute requirement for sessile or sedentary species, it is more likely to lead to competitive exclusion [10]. It has been suggested that introduced alien species are more likely to be successful if they are dissimilar from the dominant indigenous species, for example if they belong to different genera [14,15], precisely because they will have less similar resource requirements. However, in the case of competition for primary space (as opposed to secondary space) it is difficult to see how the resource can be differentiated and it is unclear that taxonomic similarity will have a strong effect on invasive success. For example, the introduced Mediterranean mussel *Mytilus galloprovincialis* (*Lamarck 1819*) has replaced the indigenous mussel *Aulacomya ater* (Chemn.) on the west coast of South Africa [16] but competes for space equally aggressively with congeneric *M. trossulus* (*Gould 1850*), on the west coast of the USA, excluding it from the southern end of its distribution [17,18].

In South Africa *M. galloprovincialis* now poses a threat on the south coast to the indigenous mussel *Perna perna* (*Linnaeus 1758*) [19], which has generally lower fecundity, recruitment and growth rates than other mytilid mussels [20,21]. At present, the abundance of *M. galloprovincialis* on the south coast is highly variable and site-specific [22]. At sites where *M. galloprovincialis* is abundant, it dominates the upper mussel zone, and *P. perna* the lower zone, with an overlap in distribution (co-existence at cm-m scales) in the mid zone [19]. Being different species, these mussels have different tolerances to wave action and desiccation, and in the upper zone they exhibit different post-settlement mortality of recent recruits, factors that partially explain these patterns of zonation [23,24,25,26]. Thus, the two species exhibit partial habitat partitioning across a gradient of environmental conditions, exhibited as differences in zonation. Although habitat partitioning through differences in physiological tolerances can be interpreted as a consequence of past competition, this does not apply to this situation where one of the species has been recently introduced to the system, rather habitat segregation reflects both direct and indirect interactions between the two. Direct interference competition between *P. perna* and *M. galloprovincialis* occurs through different growth rates, so that they can usurp each other's space [24,27]. Exploitative competition between mussels includes different settlement and recruitment intensity [25] that results in different recolonization rates after disturbance. *M. galloprovincialis* re-colonizes free space faster than *P. perna* in the sympatric mid zone after disturbance by severe storm waves [11]. This accords with the suggestion of Amarasekare [2] that disturbance or human exploitation can help invasive species to dominate, at least temporarily.

In addition to competition for space, there is indirect evidence that these mussels are likely to compete for food. Depletion of food in the water column seems unlikely, but mussel growth rates can vary in response to very small scale (10 s cm) changes in food availability [28], while van de Koppel et al. [3] have shown that self-organized spatial structure in *M. edulis* is driven partly by food depletion, and Bertness and Grosholz [29] identify one of the disadvantages for bivalves of living in aggregations as reduced growth through intraspecific competition for food. As we can imagine no mechanism for partitioning food between our two species, we assume they exhibit interspecific competition for this resource. Certainly, they have indistinguishable stable isotope (δC^{13} and δ^{15}) signatures [30], while upwelling, which increases local food availability, strongly affects their abundances and size distribution [31].

Although *M. galloprovincialis* and *P. perna* exhibit partial habitat segregation, and have similar requirements for food and space, they also exhibit co-existence in the mid-mussel zone that seems to be stable in the mid-term i.e. over decades or multiple generations [19]. Facilitation or positive interactions between species [32] is believed to be less common between closely related invasive and indigenous species than between taxonomically distinct organisms [14], yet *P. perna* and *M. galloprovincialis* exhibit not just competitive interactions, they also facilitate each other's survival in at least three ways. This is essentially because they are both ecosystem engineers so that each species can moderate abiotic conditions enough to allow survival of the other in areas where the second species would otherwise be excluded. Firstly, *P. perna* rarely settles on bare rock and its recruitment rates are facilitated by the much faster recolonisation of bare rock surfaces of *M. galloprovincialis* in the mid zone following disturbances such as storms [11]. Secondly, *M. galloprovincialis* survives wave action much better on the low shore when it is intermingled experimentally with *P. perna* than in monospecific plots [23,24]. Thirdly, *P. perna* survives desiccation in the mid-mussel zone better when mingled experimentally with the more tolerant *M. galloprovincialis* [27]. Thus, both competition and facilitation occur between these two mussel species.

Consequently, this system allows us to examine co-existence at small spatial scales between two species competing for the same resources (food and space) in the absence of apparent environmental heterogeneity. If interspecific competition is of over-riding importance, we would expect some degree of conspecific aggregation due to small-scale competitive exclusion. However, because the two species can facilitate each other's survival on parts of the shore where they are otherwise vulnerable to either desiccation (*P. perna*) or wave action (*M. galloprovincialis*), we anticipate that individuals of each species would be likely to exist close to individuals of the other species, i.e. being more mixed (randomly dispersed), in the mid zone. Here we examine the small-scale spatial structure (at cm scales) of these two mussels where they co-exist in the mid-mussel zone. In particular, we examine the link between abundance (percent cover) and the degree of patchiness of each species to estimate whether numerical dominance of a species is an important predictor of patchiness, and also ask whether mussels in mixed beds are significantly aggregated with conspecifics.

Methods

Study area and sampling design

We studied two sites where *M. galloprovincialis* occurs at high densities on the south coast of South Africa [22]. The sites, Look-Out Beach and Robberg, are rocky shores separated by ca 5 km in Plettenberg Bay (34°05′S; 23°20′E). Intertidal mussel beds in this region are wave exposed and can be divided into low, mid- and upper mussel zones. The width of each mussel zone depends on the slope of the shore and the degree of wave action, but generally each zone is 2–3 metres wide. Mussel beds are usually monolayered rather than multilayered, with the byssus threads of adults attached directly to the rock surface [33]. The two species are interspersed in the mid-mussel zone to form a mosaic within patches of 100% cover, although percent cover in the overall zone may be <100% as patches can be separated by unoccupied space.

Two sub-sites, separated by ca 25 m, were sampled at each site during consecutive spring low tides. Within each sub-site, non-overlapping photographs (real surface area 30×20 cm^2) were taken haphazardly within a 5 m stretch in the mid-mussel zone. We took ca 30 photographs per sub site to ensure sufficient good quality images and analysed a sub-set of these. In each of 10 randomly selected photographs, the Korcak B patchiness exponent was estimated for the small-scale distribution of one of the two mussel species (*M. galloprovincialis* or *P. perna*) and 10 other photos were selected for the other species so that the degree of patchiness for each species could be estimated independently (see below).

Using 3 photographs selected randomly from the 30 taken at one of the sub-sites from each site, we tested the hypothesis of non-random small-scale aggregation, with the null hypothesis being random distribution (see below). Discrimination between the two mussel species in the photographs was based mainly on shell colour, *M. galloprovincialis* having darker and blue colours (sometimes black) and *P. perna* having brown (usually light brown) and more purple colours [34]. Where there was uncertainty, we used more subtle shell colour differences (e.g. narrow blue lines indicate *M. galloprovincialis*, while reddish tints characterise *P. perna*) described by Bownes [35] and Bownes et al. [36]. In addition, the percent cover of each mussel species was estimated in a 10×10 cm^2 quadrat in each photograph to examine the relationship between cover and patchiness. Cover was estimated visually using a transparent frame of 16 small quadrats (each constituting 6.25%) put over the larger quadrat within each photo.

Analyses and modelling

Different methods and techniques of estimating spatial heterogeneity and patchiness exist, including geostatistical techniques such as spatial autocorrelation, variograms/semivariograms and fractal analysis/dimension, that have also been used in intertidal systems [37,38,39,40]. However, in the present study we needed an index that would measure patch sizes of each species directly while also estimating the size distribution of patches over different continuous small scales (in contrast to e.g. variance/mean ratio). The Korcak B patchiness exponent is such an index and therefore related to the concept of fractal dimension [41,42]. B is an estimate of the slope of the logarithmic relationship between the number of patches ≥ specific threshold sizes and the different threshold sizes of those patches (here estimated as the number of individuals in a patch; see below). Thus, the number of patches ≥ the specific predetermined sizes is plotted on a log-linear graph [41,42,43]. The Korcak B patchiness exponent was originally used for describing the distribution/numbers of different sized islands (different areas) in an archipelago, but has since been used for different purposes although this has been rare in ecology [41]. It is also a useful index for the estimation of habitat fragmentation [42,43] since it takes into account two of the three criteria for this process, i.e. the higher number of patches produced and the division into smaller patches [44]. The greater isolation/separation between patches observed during the habitat fragmentation process is not estimated by this technique as it is with, for example, the nearest-neighbour analysis, but the nearest-neighbour analysis does not estimate or take into account the higher number of smaller patches produced. Our second technique, Monte-Carlo testing, used to assess deviation from random patterns is, together with other random permutation tests, commonly used in ecological studies [45]. What is special about the technique used in the present study is the application of a spatially explicit Monte-Carlo technique in a user-friendly way.

Deviation from small-scale random distributions. Here, our aim was to test the hypothesis that individuals of the two mussel species were non-randomly aggregated with their conspecifics at small scales in the 10×10 cm^2 quadrats in each of the 3 photographs at one sub-site from each site. As a measure of the degree of conspecific aggregation, we counted the total number of pairs of neighbouring mussels that belonged to the same species within each quadrat of the photos (the total density of mussels in these 6 photos ranged from 40 to 122). We also took into account the mussels just outside each quadrat, so that all neighbours to mussels inside the quadrat were detected, in order to avoid edge effects. Two mussels were judged to form a neighbouring pair if they were connected by shell contact or separated by a distance equal to or less than half of the width of a mussel individual without their boundaries being intersected by another mussel. This allowed for possible shell contact when the shell valves were open and possible byssus contact. The tests were performed conditionally on the observed number of each species and their spatial configuration with species labels blinded. Under these circumstances, random samples of configurations can easily be generated by randomly assigning species to the mussels while keeping their relative numbers fixed (see Fig. 1). For each photo, a Monte-Carlo based p-value for the hypothesis under study was computed as the proportion of a large number of generated random species configurations with the degree of conspecific aggregation greater than the observed. For the Monte-Carlo tests,

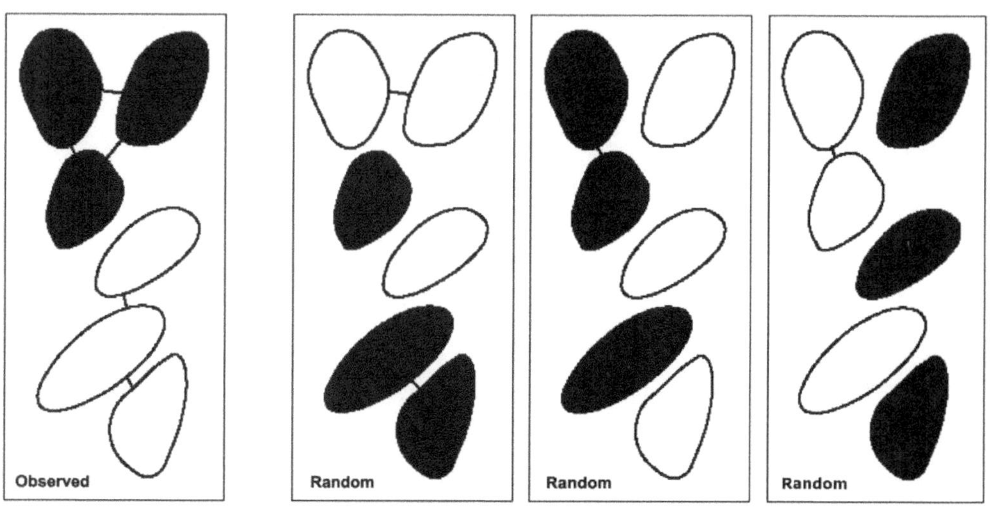

Figure 1. Simplified illustration of Monte-Carlo approach. Hypothetical configuration (observed; randomized) showing aggregations of two species (black/white). Neighbouring conspecifics are joined by arcs. Our measure of the degree of conspecific aggregation equals the number of arcs in the frame (i.e. 5 in frame 1). Frames 2-4: Sample random reconfigurations of the species labels, the number of arcs being 2, 1 and 1.

the sample unit was the individual mussel, so that sample size was the number of mussels in each quadrat. To allow a more robust test, we subsequently combined the probabilities derived from the 3 photographs for each site using Fisher's method [46], providing meta-analyses of the results.

Patchiness structure of aggregations. To describe the spatial structure of the aggregations (degree of patchiness) of the two mussel species in the mid zone, we used the Korcak patchiness exponent B [41]. In each of 10 photographs at each sub-site we counted the number of patches or clumps (N) with the number of mussel individuals (10 different photos for each species) larger or equal (≥) to i where i is equal to 1, 2, 4, 8, and 16 individuals. An aggregation was defined as mussels of each species that were connected by shell contact or a distance equal to or less than half of the width of a mussel individual between mussels. The Korcak patchiness exponent B is estimated from:

$$N(i) = constant \times i^{-B} \quad (1)$$

or

$$\log N(i) = \log(constant) - B \log i \quad (2)$$

B is estimated from the slope of the log-log relationship between N and i. Higher B estimates indicate smaller aggregations of individuals and more isolated mussels (i.e. higher patchiness), while lower B estimates mean that mussels are distributed in larger clumps with more individuals in each clump (i.e. lower patchiness).

Three-factor ANOVA (species as a fixed factor; site as a random factor; subsite as a random factor nested in site) was done for B aggregation estimates and for percent cover. Correlation analysis between percent cover and B of each species and site (and pooled data) was also conducted.

Results

Deviation from random distributions

The Monte-Carlo test of random distribution of species gave p-values of 0.00001, 0.013, and 0.11 respectively for the three photos at Look-out Beach and values of 0.013, 0.053 and 0.22 at Robberg. While not all of these p-values showed significant deviation (i.e. $p<0.05$) from randomness, they may be combined site-wise using Fisher's method [46], which gives a p-value of 0.0001 for Look-out Beach and 0.006 at Robberg. Hence, our test supports the interpretation that non-random aggregations can occur at both sites. However, this can be seen as a pilot study as we can neither accept nor reject the possibility that non-random aggregations exist over the whole spatial scale of these sites, since only a very small subset of this was examined. Thus, under certain circumstances there is non-random conspecific aggregation of *P. perna* and *M. galloprovincialis* (i.e. context-dependent pattern), but random distribution (i.e. the two species are mixed at small scales) also occurs, meaning that neither conspecific or random aggregation is the rule at these sites. Since we did not design the study to analyse the circumstances or contexts that promote either random or aggregated patterns, there was no reason to analyse further samples. Instead we focus on describing the degree of patchiness of each species separately at each site, and whether there was any relationship between patchiness and percent cover of each species.

Patchiness structure of aggregations

At both sub-sites in Look-out Beach there was significantly higher cover of *P. perna* than *M. galloprovincialis*, while patchiness was significantly lower for *P. perna* (Korcak B estimates closer to 0) than for *M. galloprovincialis* (B closer to 1; Table 1 & 2; Fig. 2; SNK-post hoc test, $P<0.05$). In contrast, at both sub-sites in Robberg, there was significantly higher cover of *M. galloprovincialis* than *P. perna*, with patchiness significantly lower for *M. galloprovincialis* than for *P. perna* at this site (see Table 1, 2; Fig. 2; SNK-post hoc test, $P<0.05$). The two 3-factor nested ANOVAs of Korcak B index and percent cover of mussels respectively showed slightly heterogeneous variances (Cochran's test: SQRT (Korcak B+1) transformed data, $C=0.3016$, $C_{crit}=0.2926$, $p<0.05$; percent cover untransformed data, $C=0.3004$, $C_{crit}=0.2926$, $p<0.05$). However, slightly heterogeneous variances are not a problem when samples are balanced and when the numbers of different treatment groups and samples are relatively large, usually more than 5 treatments and n larger than 6, as ANOVA is robust enough for this [45,47]. In the present study the numbers of treatment groups and samples were 8 and 10 respectively.

Thus, at each site the species with the highest cover had the lowest B estimates (patchiness low), indicating that it showed fewer, larger patches, while the species with the lowest cover had high B estimates (patchiness high) indicating that its distribution consisted of more, smaller patches. However, the correspondence between percent cover and patchiness of mussels between sites was not maintained within sites for either species, since there was no significant negative correlation between percent cover and Korcak B estimates of each species at each site. (such a correlation would be very hard to detect since the range of observed cover was very

Table 1. Three-factor nested ANOVA (Sub-site nested in Site; Species a fixed factor) of Korcak B patchiness exponent of mussels in the mid zone.

Source of variation	df	MS	F	p	Error term
Species	1	0.1600	0.12	0.7839	Spec.×Site
Site	1	0.0055	0.15	0.7377	Subsite(Site)
Subsite(Site)	2	0.0373	9.32	0.0003	Residual
Spec.×Site	1	1.2823	49.73	0.0195	Spec.×Subs.(Site)
Spec.×Subs.(Site)	2	0.0258	6.43	0.0027	Residual
Residual	72	0.0040			

Analysis done of square root transformed (SQRT (B+1)) data, which made variances more homogeneous, although still slightly heterogeneous.

Table 2. Three-factor nested ANOVA (Sub-site nested in Site; Species a fixed factor) of percent cover of mussels in the mid zone.

Source of variation	df	MS	F	p	Error term
Species	1	3328.20	0.13	0.7818	Spec.×Site
Site	1	1080.45	13.67	0.0660	Subsite(Site)
Subsite(Site)	2	79.02	0.81	0.4499	Residual
Spec.×Site	1	26136.45	127.73	0.0077	Spec.×Subs.(Site)
Spec.×Subs.(Site)	2	204.62	2.09	0.1310	Residual
Residual	72	97.84			

Analysis of non-transformed data as transformed data did not make variances more homogeneous.

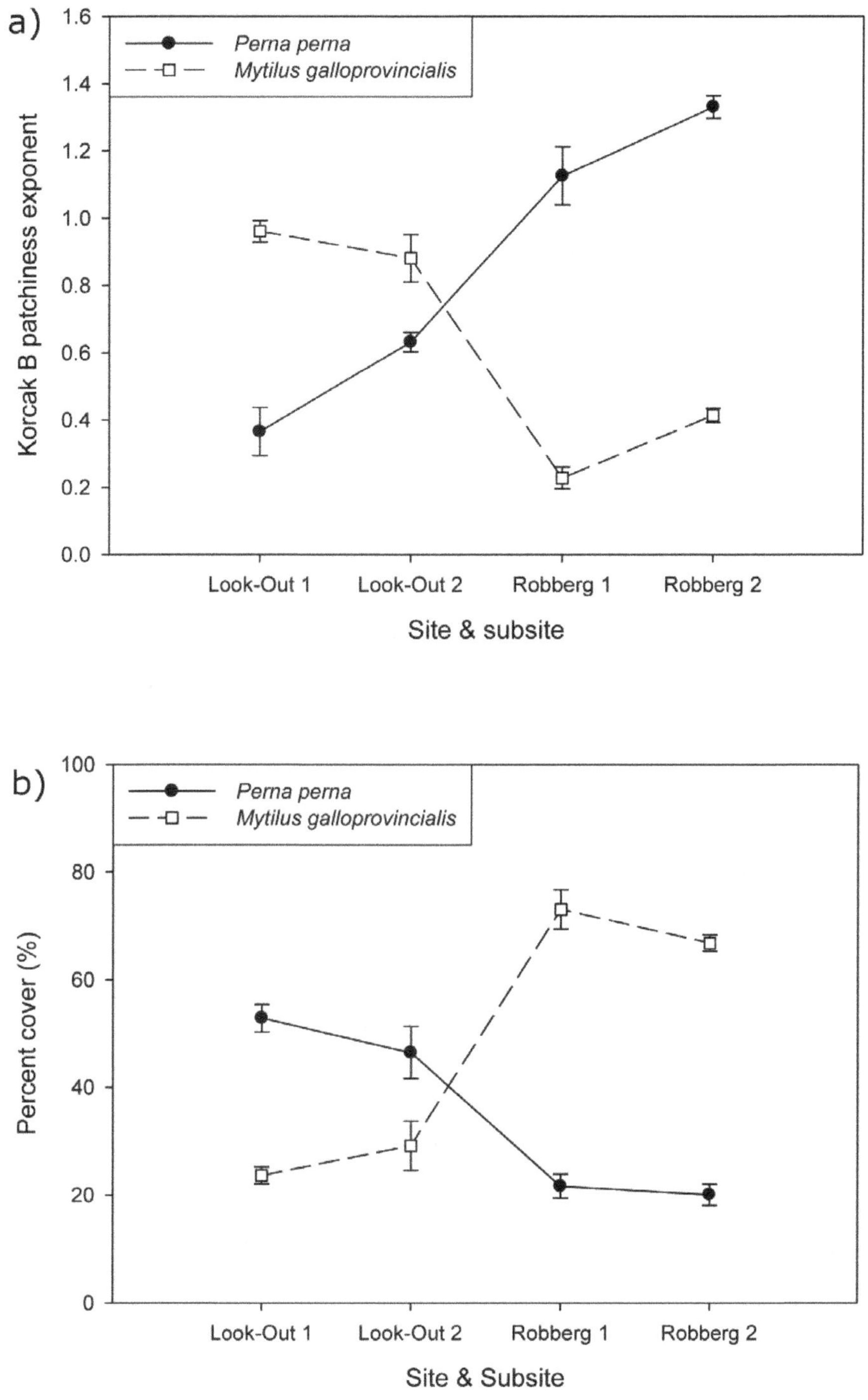

Figure 2. Differences in a) patchiness and b) mussel cover between the sites (Look-Out; Robberg) and sub-sites.

narrow within subsites), although correlations were significant if species and/or sites were pooled (Fig. 3; both species combined at: Look-Out 1: $t = -5.91$, $p = 0.00001$, $r = -0.81$; Look-Out 2: $t = -1.96$, $p = 0.06$, $r = -0.42$; Robberg 1: $t = -7.63$, $p = 0.0000001$, $r = -0.87$; Robberg 2: $t = -15.48$, $p = 0.0000001$, $r = -0.96$; Look-Out 1+2: $t = -5.03$, $p = 0.00001$, $r = -0.63$; Robberg 1+2: $t = -13.42$, $p = 0.0000001$, $r = -0.91$). All Korcak B estimates (i.e. the slope of the log-log regression between N and i) showed r^2-values ranging between 0.64 and 0.99 (mean = 0.89, SD = 0.08) indicating that log-log linear regressions between N and i (the number of patches with i individuals in) were very good.

Discussion

Monospecific populations of soft-bottom blue mussels are known to show strong self-organisation in both the laboratory and the field [7]. Here we show that both non-random and random distribution of two mussel species (i.e. conspecific or mixed aggregation) can occur at scales that are small relative to body size under field conditions on rocky shores, and that patchiness is related to the numerical dominance of each species. This is important, as it is linked to the long-term co-existence of two ecologically similar species in situations where

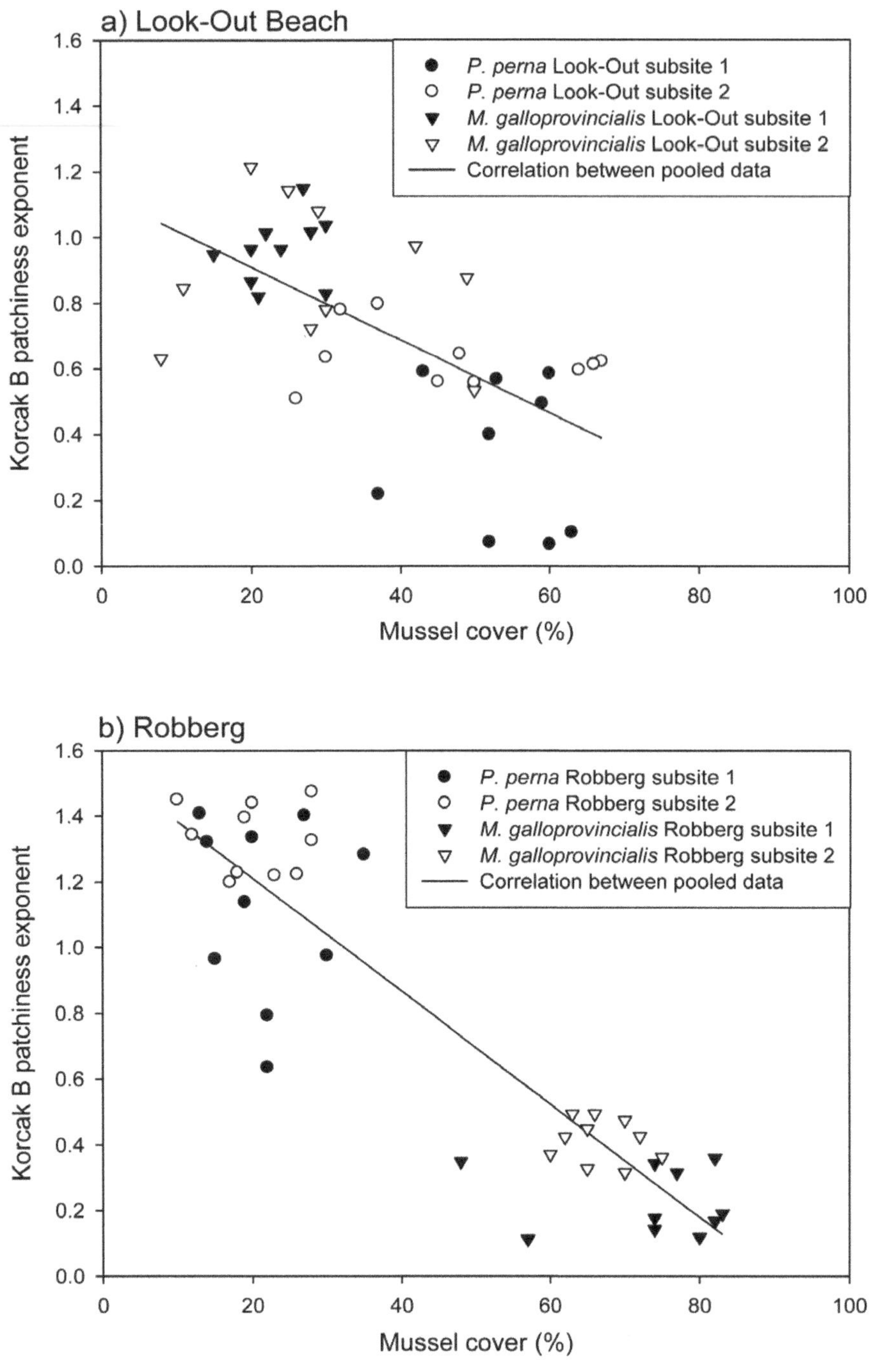

Figure 3. Linear regression between patchiness index and mussel cover at a) Look-Out and b) Robberg.

at least one resource (space) is limiting. In this system co-existence could be explained by the balance between facilitation and competition between the two species. If populations of these species are shaped primarily by interspecific competition, then we would expect conspecific aggregation, while facilitation would promote mixed (random) aggregation. In the event we observed both mixed and conspecific aggregation, which presumably reflects that in the mid-mussel zone both competition and facilitation occur [27].

There is overlapping habitat use by our two study species, i.e. they co-exist in the mid-mussel zone with each being excluded from one part of the shore by environmental stress or a combination of competition and environmental stress. *M. galloprovincialis* has weaker byssal attachment [48] and is excluded from the low shore by its susceptibility to wave action as well as competitive exclusion by *P. perna* [24]. *P.perna* is excluded from the upper mussel zone by recruitment failure [25] and its vulnerability to desiccation [27]. Thus *P.perna* is excluded from the high mussel zone and *M. galloprovincialis* from the low mussel zone largely by pre- and post-recruitment processes respectively [49].

Here, we showed that the two species can exhibit both conspecific and mixed aggregation (i.e. context-dependent aggregation) in the mid-mussel zone. Conspecific aggregation can be viewed as a consequence of interspecific competition (individuals are eliminated by competitors) or as active aggregation that could be established at different ontogenetic stages. Mussels have external fertilisation with dispersive planktonic larvae, and there is evidence for differences between the two mussels in spawning seasonality [50]. When coupled with seasonality in the ocean currents that disperse larvae, this can affect dispersal and population connectivity [51], and could result in conspecific aggregation as conspecific larvae will tend to arrive on the shore together. Secondly, larvae of these species show changes in their responsiveness to the presence of other recruits as they age. Field experiments indicate that larvae are attracted to biofilm as they settle out of the plankton, while older recruits (late plantigrades) are attracted to the combination of biofilm and other recruits [52]. This indicates that settler behaviour may be flexible so that conspecific attraction at the settlement stage is a possibility. Recruits (i.e. individuals that have settled and undergone metamorphosis) tend to move and clump together faster than adults (Porri et al. unpubl. data), while among adults, *M. galloprovincialis* is significantly more mobile than *P. perna* [53]. In the laboratory, medium-large sized (4–10 mm) recruits of *P. perna* show preferential movement towards conspecific adults (unpubl. data), but we have no information on how individual movement could allow conspecific aggregation under field conditions. A possibility unrelated to mussel behaviour is predator preference for one of the two species, although there is evidence that top-down control can eliminate rather than promote spatial structure [54]. Predators have significant effects on recruits of these species under experimental field conditions [55], but there is no evidence that they exhibit preferences between the two species at such sizes. The situation is different for adults as oystercatchers (*Haematopus moquini*) selectively feed their chicks with *M. galloprovincialis* where both mussel species are available [56], but it seems unlikely that this could produce the observed patterns. Although secondary settlers of mussels show attraction to other mussel recruits [52], we have no evidence that this attraction is species-specific. Overall it seems likely that conspecific aggregation results from a combination of such attraction plus synchronised settlement of monospecific clouds of larvae (although more extensive studies need to confirm this pattern).

Analyses of patchiness in the mid zone revealed that patchiness and numerical dominance can be linked depending on scale. The two are related among sites and species, so that the numerically dominant species at a given site had lower patchiness, but this effect did not occur within species and site (i.e. no significant correlation between percent cover and Korcak B estimates of each species at each site). This may simply reflect the probability of coalescence of small patches into fewer larger patches as abundance increases. In a similar way the probability of mussel removal by storms in the mid zone at the same sites is related to the numerical dominance of each species [11] conforming to the compensatory mortality model described by Connell [57]. We speculate that the negative relationship between patchiness and percent mussel cover would differ depending on the balance between competition and facilitation. With facilitation alone, there would be no competitive exclusion and a weak effect of cover on B (i.e. no significant relationship), while, with no facilitation, competition would result in exclusion of the weaker competitor and a strong negative effect of total mussel cover on B for that species as its population becomes more fragmented by more intense competition. Indeed, there is a trend (Fig. 3) that the slope of the curve at Robberg (total mussel cover ca 85–100%) is steeper than at Look-Out Beach where total mussel cover was lower (ca 70–85%), and competition for space presumably weaker.

There are a number of interactions between the two species that can be considered as facilitation, as defined by Bruno et al. [32]. For example, *P. perna* is extremely slow to recolonise free primary space after experimental removal [58] and its settlement is facilitated by the presence of *M. galloprovincialis*, which recolonises free space 2–3 times faster following disturbance in the mid zone [11]. Furthermore, field experiments have shown that in the mid- mussel zone *P. perna* is weakly facilitated by *M. galloprovincialis* as adult *P. perna* survive better when mixed with *M. galloprovincialis*, indicating that *P. perna* is to some extent also protected from desiccation by the physical matrix provided by the more desiccation-tolerant *M. galloprovincialis* [27]. This is presumably related to the greater susceptibility of *P. perna* to hypoxia; during aerial exposure it is obliged to gape the shell to allow oxygen uptake, making it more vulnerable to desiccation [26].

Similarly, on the low shore the survival of *M. galloprovincialis* is initially facilitated by *P. perna*. *M. galloprovincialis* in experimental monospecific patches on wave-exposed shores show much higher mortality than when mixed with *P. perna*, although it is ultimately ousted by interference competition for space [24]. Thus each species is partially facilitated by the other in zones where it is physically challenged through amelioration of abiotic stress, though this does not prevent partial habitat segregation. *M. galloprovincialis* appears to have two competitive advantages in the mid zone: firstly, faster recolonisation of free space [11], and secondly *P.perna* exhibits intraspecific competition in this zone [27]. However, *M. galloprovincialis* also ameliorates desiccation of *P.perna*, allowing it to survive there, so that the patchy pattern observed in the mid zone may be explained by a combination of different factors.

The observation that co-existence, even at small spatial scales (cm-m scales), is possible between two species that compete for at least one and probably two resources (space and food) challenges existing theory, especially as one is indigenous and the other invasive. Importantly, this is possible because the two species both compete and facilitate each other's survival [27]. This occurs because both species are ecosystem engineers that modify the physical environment, ameliorating conditions for their competitor, and is in contrast to Branch's [10] suggestion that

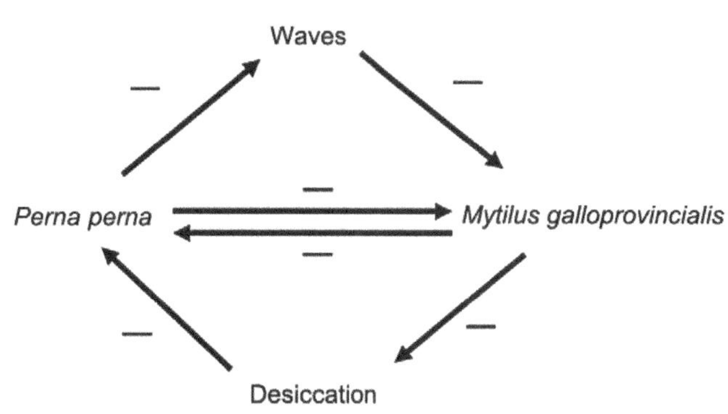

Figure 4. Summary of suggested explanations of how two negative effects interact to produce a positive effect. The amelioration of abiotic stresses by the more tolerant species constitutes an indirect positive effect (two negative effects making a positive one) on the less tolerant competitor. In the case of wave action, this is an indirect effect of *P. perna* on *M. galloprovincialis*. In the case of desiccation stress, the reverse is true. Direct interactions between the two are negative.

competitive exclusion on rocky shores is likely when two species compete for space. Co-existence at these small spatial scales is possible through the balance between direct (competitive) negative interactions and indirect (facilitative) interactions that are positive in sum, because they combine two negative effects (Fig. 4).

Acknowledgments

We thank Sarah Bownes and Purba Pal for help in the field. We are also grateful to two anonymous reviewers for improving the manuscript.

Author Contributions

Conceived and designed the experiments: JE CM. Performed the experiments: JE CM. Analyzed the data: JE MS. Contributed reagents/materials/analysis tools: JE MS. Wrote the paper: JE CM MS. Analysis/modeling using Monte-Carlo tests: MS. Analyses using Korcak B exponent: JE.

References

1. Holt RD (1984) Spatial heterogeneity, indirect interactions, and the coexistence of prey species. Am Nat 124: 377–406.
2. Amarasekare P (2002) Interference competition and species coexistence. Proc R Soc Lond B 269: 2541–2550.
3. van de Koppel J, Rietkerk M, Dankers N, Herman PMJ (2005) Scale-dependent feedback and regular spatial patterns in young mussel beds. Am Nat 165(3): E66–E77.
4. Chesson PL, Warner RR (1981) Environmental variability promotes coexistence in lottery competitive systems. Am Nat 117(6): 923–943.
5. Hassell MP, Comins HN, May RM (1994) Species coexistence and self organizing spatial dynamics. Nature 370: 290–292.
6. Amarasekare P, Nisbet R (2001) Spatial heterogeneity, source-sink dynamics and the local coexistence of competing species. Am Nat 158: 572–584.
7. van de Koppel J, Gascoigne JC, Theraulaz G, Rietkerk M, Mooij WM, et al. (2008) Experimental evidence for spatial self-organization and its emergent effects in mussel bed ecosystems. Science 322: 739–742.
8. Chesson P (2000) Mechanisms of maintenance of species diversity. Ann Rev Ecol Syst 31: 343–366.
9. Geange SW, Stier AC (2009) Order of arrival affects competition in two reef fishes. Ecology 90: 2868–2878.
10. Branch GM (1984) Competition between marine organisms: ecological and evolutionary implications. Oceanogr Mar Biol Ann Rev 22: 429–593.
11. Erlandsson J, Pal P, McQuaid CD (2006) Re-colonisation rate differs between co-existing indigenous and invasive intertidal mussels following major disturbance. Mar Ecol Prog Ser 320: 169–176.
12. Quinn JF (1982) Competitive hierarchies in marine benthic communities. Oecologia 54: 129–135.
13. Bell JJ (2008) Sponges as agents of biological disturbance. Mar Ecol Prog Ser 368: 127–135.
14. Ricciardi A, Atkinson SK (2004) Distinctiveness magnifies the impact of biological invaders in aquatic ecosystems. Ecol Lett 7: 781–784.
15. Stachowicz JJ, Tilman D (2005) Species invasions and the relationships between species diversity, community saturation, and ecosystem functioning. In: Sax DF, Stachowicz JJ, Gaines SD, eds. Species Invasions: Insights into Ecology, Evolution and Biogeography Sinauer Associates, Inc., Massachusetts. pp 41–64.
16. Robinson TB, Griffiths CL, McQuaid CD, Rius M (2005) Marine alien species of South Africa – status and impacts. Afr J Mar Sci 27: 297–306.
17. McDonald J, Koehn R (1988) The mussels *Mytilus galloprovincialis* and *M. trossulus* on the Pacific coast of North America. Mar Biol 99: 111–118.
18. Rawson P, Agrawal V, Hilbish T (1999) Hybridization between the blue mussels *Mytilus galloprovincialis* and *M. trossulus* along the Pacific coast of North America: evidence for limited introgression. Mar Biol 134: 201–211.
19. Bownes SJ, McQuaid CD (2006) Will the invasive mussel *Mytilus galloprovincialis* Lamarck replace the indigenous *Perna perna* L. on the south coast of South Africa? J Exp Mar Biol Ecol 338: 140–151.
20. Lasiak TA, Barnard TCE (1995) Recruitment of the brown mussel *Perna perna* onto natural substrata: a refutation of the primary/secondary settlement hypothesis. Mar Ecol Prog Ser 120: 147–153.
21. Harris JM, Branch GM, Elliott BL, Currie B, Dye AH, et al. (1998) Spatial and temporal variability in recruitment of intertidal mussels around the coast of southern Africa. S Afr J Zool 33: 1–11.
22. von der Meden C, Porri F, Erlandsson J, McQuaid CD (2008) Coastline topography affects the distribution of indigenous and invasive mussels. Mar Ecol Prog Ser 372: 135–145.
23. Zardi GI, Nicastro KR, McQuaid CD, Rius M, Porri F (2006) Hydrodynamic stress as a determinant factor in habitat segregation between the indigenous mussel *Perna perna* and the invasive *Mytilus galloprovincialis* in South Africa. Mar Biol 150: 79–88.
24. Rius M, McQuaid CD (2006) Wave action and competitive interaction between the invasive mussel *Mytilus galloprovincialis* and the indigenous *Perna perna* in South Africa. Mar Biol 150: 69–78.
25. Bownes SJ, McQuaid CD (2009) Mechanisms of habitat segregation between an invasive and an indigenous mussel: settlement, post-settlement mortality and recruitment. Mar Biol 156: 991–1006.
26. Nicastro KR, Zardi GI, McQuaid CD, Stephens L, Radloff S, et al. (2010) The role of gaping behaviour in habitat partitioning between coexisting intertidal mussels. BMC Ecology 10: 17–27.
27. Rius M, McQuaid CD (2009) Facilitation and competition between invasive and indigenous mussels over a gradient of physical stress. Basic Appl Ecol 10: 607–613.
28. McQuaid CD, Mostert BP (2010) The effects of within-shore water movement on growth of the intertidal mussel *Perna perna*: an experimental field test of bottom-up control at centimetre scales. J Exp Mar Biol Ecol 384: 119–123.
29. Bertness MD, Grosholz E (1985) Population dynamics of the ribbed mussel, *Geukensia demissa*: the costs and benefits of an aggregated distribution. Oecologia 67: 192–204.
30. Hill JM, McQuaid CD, Kaehler S (2006) Biogeographic and nearshore-offshore trends in isotope ratios of intertidal mussels and their food sources around the coast of southern Africa. Mar Ecol Prog Ser 318: 63–73.
31. Cole VJ, McQuaid CD (2010) Bioengineers and their associated fauna respond differently to the effects of biogeography and upwelling. Ecology 91: 3549–3562.
32. Bruno JF, Stachowicz JJ, Bertness MD (2003) Inclusion of facilitation into ecological theory. Trends Ecol & Evol 18(3): 119–125.
33. McQuaid CD, Lindsay JR, Lindsay TL (2000) Interactive effects of wave exposure and tidal height on population structure of the mussel *Perna perna*. Mar Biol 137: 925–932.
34. Branch GM, Griffiths CL, Branch ML, Beckley LE (1994) Two oceans: a guide to the marine life of southern Africa David Philip Publishers, Cape Town.
35. Bownes SJ (2005) Habitat segregation in competing species of intertidal mussels in South Africa. PhD thesis, Rhodes University, Grahamstown.
36. Bownes SJ, Barker NP, McQuaid CD (2008) Morphological identification of primary settlers and post-larvae of three mussel species from the coast of South Africa. Afr J Mar Sci 30: 233–240.
37. Legendre P (1993) Spatial autocorrelation: trouble or new paradigm? Ecology 74: 1659–1673.
38. Hewitt JE, Thrush SF, Cummings VJ, Pridmore RD (1996) Matching patterns with processes: predicting the effect of size and mobility on the spatial distributions of the bivalves *Macomona liliana* and *Austrovenus stutchburyi*. Mar Ecol Prog Ser 135: 57–67.
39. Dale MRT (2000) Spatial pattern analysis in plant ecology Cambridge University Press, Cambridge.
40. Kostylev V, Erlandsson J (2001) A fractal approach for detecting spatial hierarchy and structure on mussel beds. Mar Biol 139: 497–506.
41. Hastings HA, Sugihara G (1993) Fractals: A User's Guide for the Natural Sciences Oxford University Press, Oxford.
42. Imre AR, Cseh D, Neteler M, Rocchini D (2011) Korcak dimension as a novel indicator of landscape fragmentation and re-forestation. Ecol Indicat 11(5): 1134–1138.
43. Peralta P, Mather P (2000) An analysis of deforestation patterns in the extractive reserves of Acre, Amazonia from satellite imagery: a landscape ecological approach. Int J Remote Sensing 21(13–14): 2555–2570.
44. Fahrig L (2003) Effects of habitat fragmentation on biodiversity. Ann Rev Ecol Evol Syst 34: 487–515.
45. Quinn GP, Keough MJ (2002) Experimental design and data analysis for biologists Cambridge University Press, Cambridge.
46. Fisher RA (1932) Statistical Methods for Research Workers. 4th edn Oliver and Boyd: London.
47. Underwood AJ (1997) Experiments in ecology: their logical design and interpretation using analysis of variance Cambridge University Press, Cambridge.
48. Nicastro KR, Zardi GI, McQuaid CD (2010) Differential reproductive investment, attachment strength and mortality of invasive and indigenous mussels across heterogeneous environments. Biol Inv 12: 2165–2177.
49. Bownes SJ, McQuaid CD (2010) Mechanisms of habitat segregation between an invasive (*Mytilus galloprovincialis*) and an indigenous (*Perna perna*) mussel: adult growth and mortality. Mar Biol 157: 1799–1810.
50. Zardi GI, McQuaid CD, Nicastro KR (2007) Balancing survival and reproduction: seasonality of wave action, attachment strength and reproductive output in indigenous *Perna perna* and invasive *Mytilus galloprovincialis* mussels. Mar Ecol Prog Ser 334: 155–163.

51. Carson HS, Lopez-Duarte MP, Rasmussen L, Wang D, Levin LA (2010) Reproductive timing alters population connectivity in marine metapopulations. Curr Biol 20: 1926–1931.
52. von der Meden CEO, Porri F, McQuaid CD, Faulkner K, Robey J (2010) Fine-scale ontogenetic shifts in settlement behaviour of mussels: changing responses to biofilm and conspecific settler presence in *Mytilus galloprovincialis* and *Perna perna*. Mar Ecol Prog Ser 411: 161–171.
53. Nicastro KR, Zardi GI, McQuaid CD (2008) Movement behaviour and mortality in invasive and indigenous mussels: resilience and resistance strategies at different spatial scales. Mar Ecol Prog Ser 372: 119–126.
54. Weerman EJ, Herman PMJ, van de Koppel J (2011) Top-down control inhibits spatial self-organization of a patterned landscape. Ecology;doi: 10.1890/10-0270.1.
55. Plass-Johnson JG, McQuaid CD, Porri F (2010) Top-down effects on intertidal mussel populations: assessing two predator guilds in a South African marine protected area. Mar Ecol Prog Ser 411: 149–159.
56. Kohler S, Bonnevie B, McQuaid C, Jaquemet S (2009) Foraging ecology of an endemic shorebird, the African Black Oystercatcher (*Haematopus moquini*) on the south–east coast of South Africa. Estuar Coast Shelf Sci 84: 361–366.
57. Connell JH (1978) Diversity in tropical rain forests and coral reefs. Science 199: 1302–1310.
58. Dye AH (1992) Experimental studies of succession and stability in rocky intertidal communities subject to artisanal shellfish gathering. Neth J Sea Res 30: 209–217.

4

Before the Endless Forms: Embodied Model of Transition from Single Cells to Aggregates to Ecosystem Engineering

Ricard V. Solé[1,2,3,4]*, Sergi Valverde[1,3,4]

1 ICREA-Complex Systems Lab, Universitat Pompeu Fabra, Barcelona, Spain, 2 Santa Fe Institute, Santa Fe, New Mexico, United States of America, 3 Institut de Biologia Evolutiva, UPF-CSIC, Barcelona, Spain, 4 European Centre for Living Technology, C Foscari University of Venice, Venice, Italy

Abstract

The emergence of complex multicellular systems and their associated developmental programs is one of the major problems of evolutionary biology. The advantages of cooperation over individuality seem well known but it is not clear yet how such increase of complexity emerged from unicellular life forms. Current multicellular systems display a complex cell-cell communication machinery, often tied to large-scale controls of body size or tissue homeostasis. Some unicellular life forms are simpler and involve groups of cells cooperating in a tissue-like fashion, as it occurs with biofilms. However, before true gene regulatory interactions were widespread and allowed for controlled changes in cell phenotypes, simple cellular colonies displaying adhesion and interacting with their environments were in place. In this context, models often ignore the physical embedding of evolving cells, thus leaving aside a key component. The potential for evolving pre-developmental patterns is a relevant issue: how far a colony of evolving cells can go? Here we study these pre-conditions for morphogenesis by using CHIMERA, a physically embodied computational model of evolving virtual organisms in a pre-Mendelian world. Starting from a population of identical, independent cells moving in a fluid, the system undergoes a series of changes, from spatial segregation, increased adhesion and the development of generalism. Eventually, a major transition occurs where a change in the flow of nutrients is triggered by a sub-population. This ecosystem engineering phenomenon leads to a subsequent separation of the ecological network into two well defined compartments. The relevance of these results for evodevo and its potential ecological triggers is discussed.

Editor: Johannes Jaeger, Centre for Genomic Regulation (CRG), Universitat Pompeu Fabra, Spain

Funding: This work has been supported by grants of the James McDonnell Foundation (RVS, SV), Fundación Marcelino Botín (RVS, SV), FIS2004-05422 (RVS) and by the Santa Fe Institute. The funders had no role in study design, data collection and analysis, decision to publish, or preparation of the manuscript.

Competing Interests: The authors have declared that no competing interests exist.

* E-mail: ricard.sole@upf.edu

Introduction

A key problem in evolutionary biology is the emergence of complex life forms under the cooperation of several interacting cells [1,2]. Multicellularity emerged through evolution several times (at least 25) and has been a prerequisite for the generation of complex types of development [3–6]. This major transition brought division of labour and opened the door for the emergence of development and body plans [7–9]. But for many reasons, and in spite of its obvious importance, the evolution of multicellularity is not yet well understood. The fossil traces of the transition are still incomplete, although rapidly improving. However, dedicated efforts to unravel the phylogeny of multicellular living forms, the analysis of special model organisms and the cues provided by the presence of potential genetic toolkits predating the emergence of complex metazoans are defining the potential minimal requirements for the transition towards complex multicellular life forms.

This transition is particularly relevant for the critical changes that took place around 560 Myr ago, associated to the so called Cambrian event [1,10] but its roots predate a much earlier time window, as indicated by the analysis of ancestral genomes. Moreover, the picture gets more complicated as we consider additional components related to the physical environment and the constraints and opportunities posed by ecological interactions. Actually, the multiple facets of the debate on the origins of multicellular organisms have to do with the role played by the different potential shapers of the event. These multiple factors are not independent, and are likely to have interacted in complex ways.

In general terms, the multicellular state is characterized by the existence of cell-cell interactions of some sort that provide a source for collective adaptation to energy limitations, physical fluctuations and eventually division of labor. In multicellular organisms, lower-level entities (cells) have relinquished their ability to reproduce as independent units and instead replicate exclusively as part of the larger whole. But long before a developmental body plan was even defined, in what has been dubbed the "pre-Mendelian world" [11], several layers of complexity were required. This as a particularly relevant problem deeply tied with the problem of hierarchies in evolution [12–13].

Before developmental programs allowed true multicellular organisms to emerge, single cells developed into monomorphic aggregates and later on into differentiated aggregates [13]. Moreover, cell adhesion mechanisms required for the emergence of multicellularity have a much early origin [14]. In this context, long before complex metazoans appeared, some key components

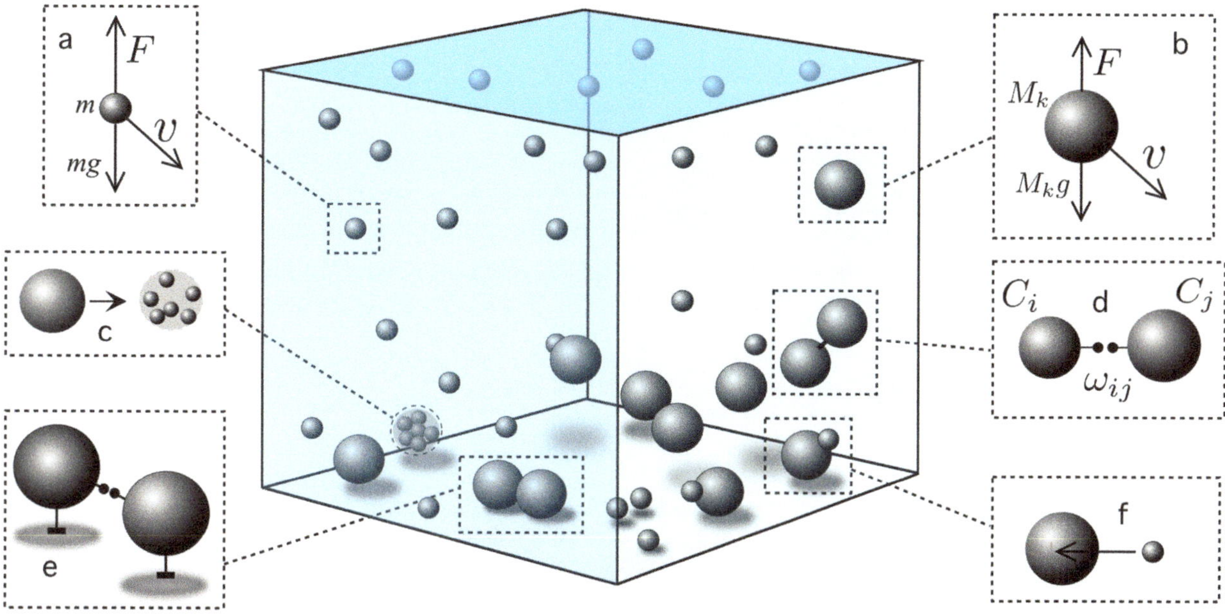

Figure 1. Basic scheme of the components of the CHIMERA model. The system is confined within a rigid cube with a floor where particles fall and to which cells can attach. Five additional square boundaries are also present which cannot be crossed. The upper boundary allows a flow of nutrient particles (here small spheres) at a constant rate. Particles fall under gravity (a,b) and experience local turbulence as a random velocity field. As they reach the floor, they can keep moving under the same flow and also disappear as they degrade (c) into detritus particles. Cells can evolve adhesion among them (d) as well as with the substrate (e). Finally, cells and particles (f) ineract through collisions. If the cell is able to exploit that particular type of energy, the particle involved disappears and is transformed into cell's biomass (see text).

of the toolkit were already in place. How did these components affected the transition to multicellularity is an open question, and theoretical models can help to address it.

Most mathematical and computational models dealing with early evolution of development assume that either genetic networks or even body plans are already in place or instead deal with pattern-forming colonies and their potential to form structures under given spatial and nutrient constraints [15–17]. However, less attention has been given to the physics associated to these processes, particularly in relation with early scenarios lacking fine-tuned genetic regulation of development. By physics we refer to two different levels. One includes diffusion, excitability, oscillations or even cellular interaction forces, which can be captured by cell sorting models based on energy minimization functionals [11,18,19,20]. In this context, it is possible to evolve morphologies and observe the interplay between cell differentiation, growth and communication [21–23] with a properly defined optimization algorithm. The other level deals with the embedding of cells and organisms within a physical medium. Here, forces are closer to standard physics, i.e., the way cells might displace in the three-dimensional environment, how cells interact with the substrate and how cells find nutrients in a fluctuating medium. Previous work on physically-embedded artificial systems was pioneered by a number of researchers, who evolved artificial "organisms" [24,25]. Related work has considered the interplay between genetic networks and morphogenesis [26–30].

In this paper we would like to address, under a well defined framework, some questions related to the pre-multicellular world where cell aggregates (but not true organisms or body plans) could develop, providing some key preconditions for multicellularity to emerge. In particular, we would like to understand how the physical context, cell-cell adhesion properties, ecological and epigenetic factors concur to favor the emergence of cell aggregates. Moreover, we would like to understand how the environment influences (and can be influenced by) the evolution of cell diversity and cell-cell interactions. These are, we believe, important pieces in the extended evolutionary synthesis provided by evo-devo [31].

We also try to see how this explicit three-dimensional embodiment can play a role in favoring the emergence of innovations. We explore such a pre-body plan scenario, and its potential for generating complexity, by allowing a physically-embedded model of a cellular community to freely evolve. The interaction between our simple evolved aggregates and the environment is shown to trigger the emergence of *ecosystem engineering* (EEN) an important component in macroevolutionary patterns [32]. EEN can be defined as a modification of the abiotic environment by a species that affects resource availability for other species [33–37]. Because it involves a persistent ecological

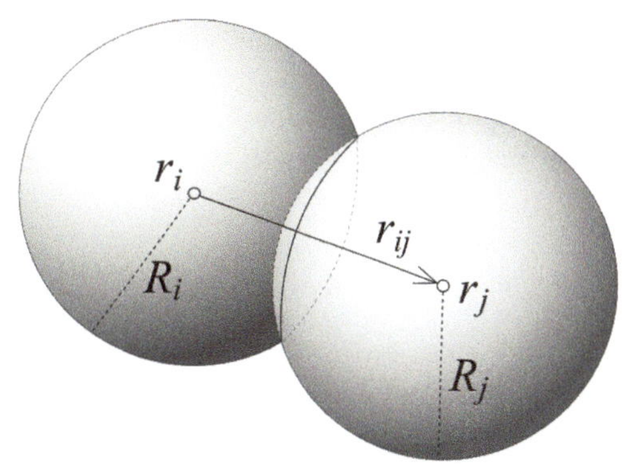

Figure 2. Simplified geometry of cell-cell collisions. Here, $\mathbf{r}_i$ and $\mathbf{r}_j$ are the cell positions, $\mathbf{r}_{ij}$ is the offset vector, R_i and R_j are the cell radius, and $R_i+R_j-|\mathbf{r}_{ij}|$ is the interpenetration depth.

modification, its presence implies the existence of ecological inheritance.

As we will shown below, a simple model of physically interacting cells with adhesion properties starting from a set of independent, genetically identical cells exploiting a single energy resource (from a given repertoire) evolves in time towards a spatially segregated community involving a trophic chain. The ecological network includes both a population of generalists feeding on all available food sources along with a population of specialized detritivores. The transition from the original monomorphic population to the spatially organized aggregate with ecological structure takes place through the emergence of an innovation grounded in evolving adhesion between cells and walls as well as cell-cell adhesion. In spite of its simplicity, it fairly well illustrates the potential of this type of model to explore the emergence of major transitions in pre-Mendelian scenarios.

Environment, Cells and Physical Simulation

In our model, evolution takes place within a spatially confined environment (the spatial domain is a cube with floor and walls) where physical forces play a role as external constraints. Organisms are spatially embedded structures and their embodiment is relevant as it provides the proper link with the external world and the biotic scenario where other organisms inhabit.

Physical models of cell interactions have been developed for a broad range of problems involving multicellular assemblies [38–41] and our study follows some of the standard methods of computational physics [42]. Figure 1 displays a basic scheme of the system considered here along with the different components of the physical interactions that will be taken into account.

1.1 Cells and particles

Our starting point is a population of single-cell organisms, where each cell in the initial population is identical. Cells and particles are simulated with rigid bodies moving within a fluid-like environment. A cell (particle) has spherical geometry with radius R_i, mass M_i, spatial position r_i and velocity $\mathbf{v}_i = d\mathbf{r}_i/dt$. The motion of a cell is described by the standard second law:

$$M_i \frac{d\mathbf{v}_i}{dt} = \mathbf{F}_i \tag{1}$$

Numerical integration gives cell velocity at time $t + \Delta t$:

$$\mathbf{v}_i(t+\Delta t) = \mathbf{v}_i(t) + \frac{\mathbf{F}_i}{M_i}\Delta t \tag{2}$$

where Δt is the size of the integration step, and the total force acting on M_i will be:

$$\mathbf{F}_i = \mathbf{F}_e + \mathbf{F}_i^c + \mathbf{F}_i^n + \mathbf{F}_i^l - k_d\mathbf{v}_i + M_i\mathbf{g} \tag{3}$$

applied to any cell is the sum of environmental forces $\mathbf{F}_e$, the gravitational field $\mathbf{g}$, the collision force $\mathbf{F}_i^c$, the cell-wall adhesion $\mathbf{F}_i^n$ and the cell-cell adhesion $\mathbf{F}_i^l$ term.

1.2 Environmental forces

For simplicity, we asume that the external environment exerts the same force to every cell or particle, i. e.:

$$\mathbf{F}_e = k_e\mathbf{n} \tag{4}$$

where n is a random vector with $|\mathbf{n}| = 1$ and $k_e > 0$ is a constant parameter that indicates the strength of the external field. This choice implies that external fluctuations in the fluid medium are homogeneous, thus affecting all parts of the system uniformly. This approximation thus neglects potential effects played by small-scale eddies, which might actually play a role in adaptational changes.

Movement of particles in a fluid is subject to dissipation, as defined by a viscous drag $(-k_d\mathbf{v}_i)$ where k_d is the drag coefficient associated to the surrounding fluid. The effect of drag is to resist motion, making the particle gradually come to rest in the absence of other influences.

1.3 Cell-cell collisions

We apply the discrete element method [43] to the computation of collision forces between the $i-th$ cell and all its interpenetrating cells, that is, cells located at a distance below a given threshold (see figure 2):

$$\mathbf{F}_i^c = \sum_{i \neq j} \nu \left[\mathbf{v}_{ij} - \frac{\mathbf{r}_{ij}}{||\mathbf{r}_{ij}||} \left(\mathbf{v}_{ij} \frac{\mathbf{r}_{ij}}{||\mathbf{r}_{ij}||} \right) \right] + \xi \mathbf{v}_{ij} - \nabla \Phi_{ij}(\mathbf{r}_{ij}; R_i, R_j) \tag{5}$$

where ν is the shear coefficient, ξ is the damping coefficient, k is the spring coefficient, and $\mathbf{r}_{ij} = \mathbf{r}_j - \mathbf{r}_i$ is the offset vector between the $\mathbf{r}_i$ and $\mathbf{r}_j$ cell positions, and $\mathbf{v}_{ij} = \mathbf{v}_j - \mathbf{v}_i$ is the difference between cell velocities. The last term in the right-hand side gives the force resulting from a potential function associated to a soft-core interaction, namely.

$$\Phi_{ij}(\mathbf{r}_{ij}; R_i, R_j) = \frac{1}{2} k \left(R_i + R_j - ||\mathbf{r}_{ij}|| \right)^2 \tag{6}$$

Computation of spatial interaction forces requires $N(N-1)/2$ possible collision checks for N bodies in the worst-case scenario. In order to reduce this computational load we will use a spatial partitioning scheme. Here, a 3-D uniform grid subdivides the entire container volume in equally-sized square voxels [42]. All the bodies (cells and particles) are sorted by a mapping function $\mathbf{h}(\mathbf{r})$ that computes their voxel index:

$$\mathbf{h}(\mathbf{r}) = \frac{1}{d}(\mathbf{r} - \mathbf{o}) \tag{7}$$

where $\mathbf{o}$ is the grid origin and d is the voxel size. For each body, we only check their collisions with different particles and cells belonging to a neighboring voxel, i.e., having indexes $\mathbf{h}(\mathbf{r})$ or $\mathbf{h}(\mathbf{r}) \pm 1$. By adjusting the size of the voxel partition, we can minimize the total number of collision tests and enable the real-time simulation of large numbers of physically interacting bodies.

1.4 Cell-substrate adhesion

Attachment of cells to surfaces may provide a favorable micro-environment (e.g., biofilms). Cell-wall distance $D(\mathbf{r}_i)$ is:

$$D(\mathbf{r}_i) = \frac{ax_i + by_i + cz_i + d}{\sqrt{a^2 + b^2 + c^2}} \tag{8}$$

where $ax + by + cz + d = 0$ is the plane equation for the closest wall to the cell located at $\mathbf{r}_i = (x_i, y_i, z_i)$. When a cell with adhesion strength to the substrate $J_i^f > 0$ is closer than the adhesion range $\delta_w > 0$, i.e., $D(\mathbf{x}_i) < \delta_w$, we attach a spring connecting the cell $\mathbf{x}_i$ with its projection on the wall $\mathbf{x}_i^w$ (see figure 3). Now, the wall spring exerts the following attraction force:

$$\mathbf{F}_i^n = -k_s(||\mathbf{r}_i - \mathbf{r}_i^w|| - d_s)\frac{\mathbf{r}_i - \mathbf{r}_i^w}{||\mathbf{r}_i - \mathbf{r}_i^w||} \quad (9)$$

where d_s is the spring equilibrium distance, k_s is the spring constant and $\mathbf{F}_n^i = 0$ when the cell is not attached to any spring. Existing cell-wall springs can be removed with certain probability $1 - q(i)$ or when the spring length is above the maximal length, i.e., $||\mathbf{r}_i - \mathbf{r}_i^w|| > d_s^m$. As we will see, cells can evolve cell-wall adhesion J_i^f in order to maximize the intake of nutrient particles.

1.5 Cell-cell adhesion

Cells can form aggregates by attaching to other cells. Each cell has an intrinsic probability J_i^c to create a new adhesion link. Given two close cells located at $\mathbf{r}_i$ and $\mathbf{r}_j$, we will set an adhesion string connecting them with probability $(J_i^c + J_j^c)/2$. The adhesion force to any cell is the sum of forces contributed by all the active cell-cell adhesion springs:

$$\mathbf{F}_i^l = -\sum_j k_l(||\mathbf{r}_i - \mathbf{r}_j|| - d_l)\frac{\mathbf{r}_i - \mathbf{r}_j}{||\mathbf{r}_i - \mathbf{r}_j||} \quad (10)$$

where d_l is the spring equilibrium distance, and k_l is the spring constant. Adhesion springs break spontaneously with rate $\delta \approx 0.001$ or when the spring is very large, i.e., $||\mathbf{r}_i - \mathbf{r}_j|| > d_l$.

On the other hand, interpenetration collisions between cells and walls are not resolved with explicit forces (see figure 3A). When a body interpenetrates a wall, it bounces along its negative velocity direction, i.e., $\mathbf{v}_i(t+\Delta t) = \mu\mathbf{v}_i(t)$ by a constant factor $\mu < 0$. In addition, we relocate the body outside the wall (see below). This approximation ensures that particles will never move beyond container limits and does not change the main results presented here. The final cell (or particle) position is obtained from:

$$\mathbf{r}_i(t+\Delta t) = \mathbf{r}_i(t) - \delta\mathbf{n} + \mathbf{v}_i(t+\Delta t)\Delta t \quad (11)$$

where $\delta > 0$ is the cell-wall interpenetration distance and n is the unit normal vector of the wall at the collision point with $|\mathbf{n}| = 1$ (see figure 3A).

Finally, the computation of particle force $\mathbf{f}_i$ is more simple than in the case of cells. Food particles can collide with boundary walls, other particles and with cells (see below). But particles cannot adhere to the walls, or to any other body. The total force exerted on a particle is

$$\mathbf{f_i} = \mathbf{F}_e + \mathbf{f}_i^p + \mathbf{f}_i^c - k_d\mathbf{v}_i + \mathbf{g} \quad (12)$$

where $\mathbf{F}_e$ is the environmental force (as defined for cells, see above), $\mathbf{v}_i$ is the particle velocity and g is the gravity field. Particle collision force $\mathbf{f}_i^p$ is similar to the equation used to compute cell collision responses (5) but using particle mass (M) and radius (r). Similarly, the term $\mathbf{f}_i^c$ accounting for the collision response between any pair of interpenetrating particles mirrors equation (5).

We have calibrated the parameters associated to the physics as described above in order to avoid numerical instabilities. We have used small integration steps, i.e., $\Delta t = 0.03$ and fixed several physical parameters, including the cell and particle masses and the spring constants, to suitable values. An exhaustive exploration of the physical parameter space will be investigated elsewhere.

Evolutionary Rules

Once our embodied, physical description of how interactions take place within our environment, we need to further extend our model by including evolutionary rules. Mechanical interactions, for example, can be understood in terms of predefined mechanisms (constrains derived from the nature of physical laws) together with sets of parameters that tune their relevance in terms of how different forces influence dynamics. In this paper we explore the impact of such parameter changes once the basic laws are already in place.

The introduction of physical constraints as part of the framework defining our cells and their interactions allows us to integrate different factors emerging at the boundaries between physics and biology. Moreover, the explicit nature of selection pressures associated with gathering nutrients from a spatially explicit, fluctuating environment allows to explicitly introduce ecological factors into play. In summary, the model takes into account physical embodiment, ecological and evolutionary constraints and allows interaction parameters among cells to change. Although no developmental programs or gene regulatory modules can emerge at this level, we will see how the degrees of freedom included in our approximation allow cell aggregates to emerge, along with innovation and the creation of niches.

At any given time, there is a set of cells $\{C_i\}$ and a set of energy particles of identical mass m belonging to different energy types, here indicated as E_k, with $k = 1,...,n_p$ (here we use $n_p = 4$ sources). Each cell C_i can in principle metabolize a subset of different sources with efficiency $\eta_{ik} \in [0,1]$. At the beginning, we start from a homogenous population of N_0 cells, each having mass $M_i = M_0$, all feeding on the same resource, namely E_1 i.e., $\eta_{i1} > 0$ and $\eta_{ij} = 0$ for all $j > 1$. At every simulation step, we introduce N_p new particles in the system (here we use $N_p = 80$, uniformly distributed over each class) starting at the top of our world. This number provides a quantitative value to the intake energy flow Φ of particles entering the system.

Eventually, cells collide with nutrient particles and consume them according to the cell-specific efficiency η_{ij} (a given nutrient particle is consumed only when this is non-zero, otherwise the particle is deflected and does not enter into the cell body). A particle of class E_k will be successfully grazed contributing to a linear mass increase:

$$M_i(t+\Delta t) = M_i(t) + \eta_{ik}m \quad (13)$$

Once consumed, the particle is removed from the system. Nutrient particles have a characteristic lifetime and degrade at rate μ_p. Cells consume a constant amount of energy $e_s > 0$ per time step to sustain themselves. A cell dies when mass M_i falls below a critical threshold $M_i < M_c$. Cell death leads to disintegration and the release of $(M_i(t)/m) \approx (M_c/m)$ residual particles to the medium. These detritus particles will be consumed by detritivores when their associated efficiency is nonzero. Detritus particles have also a characteristic lifetime and degrade with a slower degradation rate, here fixed to $\mu_p/2$.

Cells divide once their total mass is at least twice their initial mass, $M_i(t) \geq 2M_0$. After reproduction, the mother cell C_i reduces its mass by M_0, which is transferred to the daughter cell, say C_k. In other words, we have

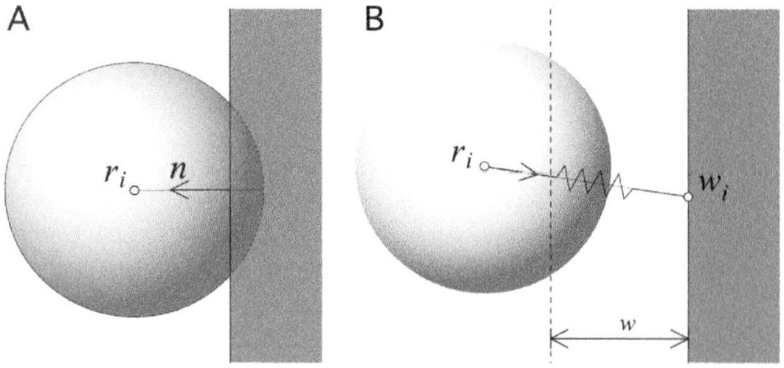

Figure 3. Possible interactions between cells and boundary walls. (A) Bouncing collision takes place when the cell interpenetrates the nearest wall any distance $\delta > 0$ along the unit normal vector n of the wall. (B) Cell-wall adhesion occurs when cell-wall distance is below some threshold w and according to adhesion probability (which depends on the cell genome, see text). In this case, a spring attaches the cell to a fixed point w_i on the wall (see text).

$$M_i \rightarrow M_i - M_0$$

$$M_k = M_0$$

However, cell division is limited by the available surrounding space. Here, we allow the mother cell to reproduce only when its number of neighboring cells (i.e., cells within a radius of size $2R_i$) is below a given density threshold N_c. The new offspring is placed at a random location close to the mother cell, and specific rules ensure that this new cell is within the boundaries of the simulated medium.

The offspring cell C_k inherits metabolic efficiencies from the mother cell C_i. Metabolic efficiencies can be slightly changed, i. e.,

$$\eta_{jk} = \eta_{ik} + v$$

where v is non-zero with probability of mutation $\mu \approx 0.02$ and zero otherwise. Here, the random v perturbation follows a Gaussian distribution with zero mean and standard deviation $\sigma_m = 0.3$. In addition, the offspring cell also inherits both cell-floor adhesion and cell-cell adhesion coefficients, respectively. Again, mutations are allowed to occur, and we have now:

$$J_k^f = J_i^f + v \tag{14}$$

$$J_k^c = J_i^c + v \tag{15}$$

The values of these parameters are always normalized between zero and one. If a given mutation is accepted and the parameters are either lower than zero or higher than one, they are fixed to zero or one, respectively. Similarly, efficiencies are normalized so that the condition $\sum_j \eta_{ij} = 1$ is always satisfied.

Results

As a result of the previous set of rules and initial condition, we have a simple ecological food web involving a set of resources and a single specialist grazer, which takes E_1 particles and grows at the expense of metabolizing them. Further degradation of particles leads to waste that is removed from the virtual tank and moreover cell death also generates an additional resource (the detritus compartment, E_d). Ecologically, this is our starting point, which will evolve as organisms change their feeding preferences and as a consequence of interactions between organisms and their physical environment.

Given the potential for evolving physical parameters as well as nutrient intake-related parameters, our cells will be able to evolve within a range of possible adaptations. For example, given the number of potential food sources to be consumed, we will observe a spread from the original parameter set defined by

$$\eta_{i1} = 1 \cap \eta_{i,j>1} = 0 \tag{16}$$

to a continuum space of efficiencies

$$\eta_i \equiv \{(\eta_{i1}, ..., \eta_{in}) \mid \eta_{ij} \in [0,1]; \ \sum_j \eta_{ij} = 1 \ \} \tag{17}$$

As we will see below, these spread leads to a predictable outcome in a first phase of the evolution process, while it leads to an unexpected innovation later on.

1.1 Transition to generalism epoch

The first trend observed in all our simulations is a tendency from the starting specialization (all cells exploiting one source with maximal efficiency) to generalism: mutations allow to exploit other resources with less efficiency but overall this is a better strategy given the finite amount of energy particles. In our model, $n_p = 4$ different types of energy particles are used. Additionally, since detritus particles resulting from cell death can also be consumed, a total of $n_p + 1$ types of particles are available.

Nutrients enter the system from the upper layer at a constant rate p and degrade into waste at a rate δ_k. Under the absence of grazers, the time evolution of the number N_k of energy particles of class k will follow a linear model $dN_k/dt = p - \delta_p N_k$ (for $k = 1, ..., n_p$) and thus each component will have an average steady value of $N_k^* = p/\delta_p$. Then, the overall number of particles at the beginning (before grazing starts) will be $N^* = \Phi p/\delta$ with $\Phi = n_p p$ is the total flow of incoming particles. Right at the beginning, our grazers exploit only one source particle (say E_1). If no mutations were involved, we would observe a trend towards an equilibrium population of both cells and class-one particles over time.

The above describes the basic dynamics that occurs at the very early stages in our model, when only one source is being exploited. However, since we are interested in the long-term evolution, we would like to study the particle-grazer population dynamics. One of the first trends easily observed in our model is the tendency towards a broader range of exploitable resources, that is, grazers become generalists. Such trend towards a more opportunistic behavior occurs when the payoff given by the exploitation of a variety of resources outweighs the loss of efficiency as more resources are grazed.

The degree of generalism of a given individual is measured by means cell entropy $H(C_i)$ defined as as the normalized diversity of efficiencies:

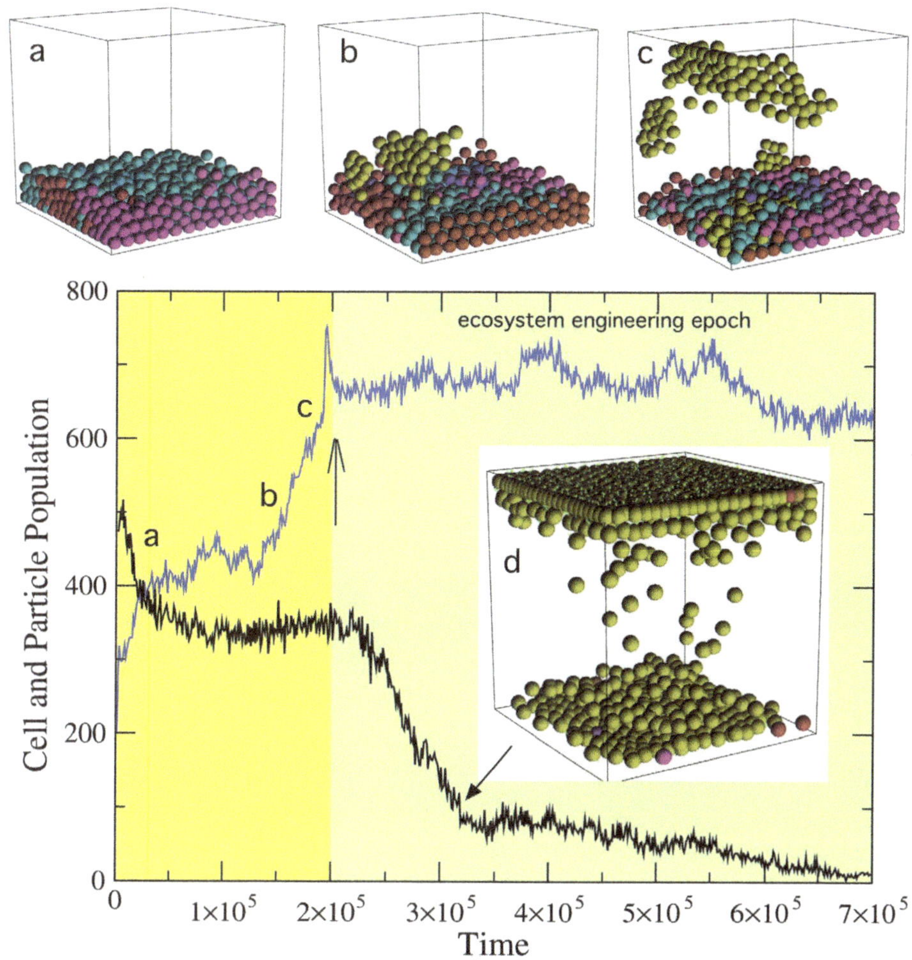

Figure 4. The time evolution of cells (blue) and particles (black) can be described with a sequence of processes. Before the transition to the inverted population state (inset) cells grow and divide as they also evolve their characteristic parameters. At the beginning, the cell population grows by adapting to the variety of energy sources and we can see a thick layer of cells (a) plotted at $t=17000$. After a while, an increase in the surface displayed by the population increases its efficiency to gather particles. Such increase is reached by evolving cell-substrate and cell-cell adhesion. The state shown in (b) is found at $t=170000$ where we can appreciate a cluster of cells which appear to be climbing the surface of the cube. In (c) the cluster close to reaching the upper floor (here $t=200000$) at this point the cell population does not experience further significant increase, but the number of particles decays to very low values. Here bright cells have high floor adhesion (where yellow indicates maximum adhesion) and darker colors correspond to cells with low adhesion or free-moving. Notice that adhesion evolves first in cells close to the wall boundaries. Here: $\Phi=25$ particles per timestep, $\Delta t=0.03$, $\delta_w=0.05$, $d_s=0.01$, $v=80$, $\xi=80$, $k=400$, $M_i=20$, $R_i=0.08$, $m_i=3.35$, $r_i=0.025$, $k_e=2$, $k_d=0.001$ (see text).

$$H(C_i;\eta_i)=-\frac{1}{\log(n_p+1)}\sum_{1\leq j\leq n_p+1}\eta_{ij}\log[\eta_{ij}] \tag{18}$$

where η_{ij} is the relative efficiency of the i-th cell when feeding on the $j-th$ nutrient resource and n_p is the number of different resources. Notice that, in the initial configuration, all cells have entropy $H(C_i)=0$. As defined, the degree of generalism will be zero when only one source is used (specialized diet) and will reach a maximum value for individuals grazing on multiple sources. Similarly, since the average efficiency for each cell is given by

$$\langle\eta_{ij}\rangle=\frac{1}{n_p(i)}\sum_j\eta_{ij} \tag{19}$$

where $n_p(i)$ indicates the number of nutrient sources exploited by C_i, given the normalization of efficiencies we have simply $\langle\eta_{ij}\rangle=1/n_p(i)$.

The simulation shows a steady increase of cell-wall attachment and cell-cell adhesion followed by a steady increase in number of cells. This is a consequence of the fact that, in order to get access to falling particles, a larger surface is needed. Cells who attach to the surface can climb up the walls. Having a moderate cell-cell adhesion also helps in avoiding them to fall down. In this way, we have formed aggregates that are moving up as cells divide. The impact of this, along with the simultaneous tendency towards a generalist grazing behavior, can be seen in figure 4. The cell population (blue curve) grows rapidly as the aggregates start to emerge and expand, eventually covering the top layer.

During the process, groups of cells, often forming layers parallel to the floor, become increasingly larger. Three snapshots involving transient steps (a-c) are also displayed in figure 4, where we have used color-coded spheres to indicate the strength of the cell-wall

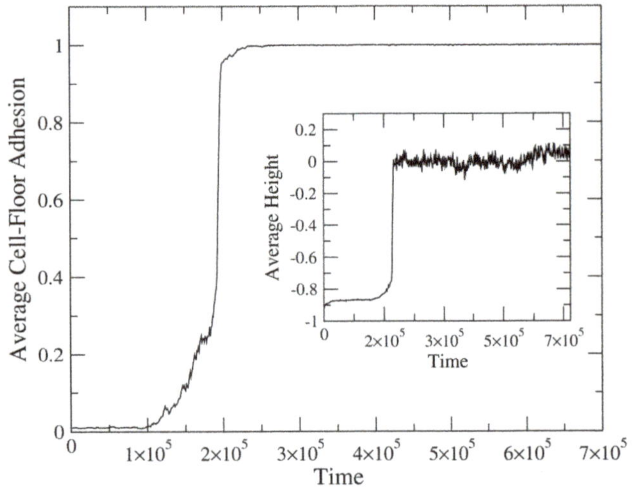

Figure 5. Sharp transition in the time evolution of average cell-floor adhesion. This transition is associated to the increasing height (inset) of the cell populations as further cells climb the walls towards the upper layer. Here the top floor is taken as the zero coordinate, whereas the bottom floor is taken as the $z=-1$ (normalized) minimum (see text).

adhesion. Lighter and spheres indicate higher and lower levels of adhesion, respectively.

1.2 Transition to ecosystem engineering epoch

As cells occupy the upper layer, the flow of nutrients further declines (filled arrow, figure 4) to reach very low levels. The kinetics of this process and how it is connected with evolving adhesion rates is summarized in figure 5, where we plot the average cell-floor adhesion (averaging over all the cell population) and the corresponding average height of cells within the cube (inset). Here we have used the top floor as the ground height ($z=0$) and the bottom one as the minimum. For convenience, we have normalized the later to $z=-1$. We can appreciate in these plots how rapidly the selection for higher adhesion occurs. As a consequence of this redistribution of cells through the upper part of the system, a whole redistribution of flows takes place, effectively triggering the emergence of a new ecological organization.

The most relevant result of our study is a key innovation made by our evolving cell populations as they "discover" that gathering particles is easier if they attach to the top floor of the system. As discussed above, cells approach the nutrient source by evolving cell-floor adhesion. This allows the population to expand its effective area for gathering nutrients, but it also favors more frequent interactions with particles, which are trapped between cells and the walls. This makes intake slightly higher than far from the boundary increases (and thus a higher fitness) because the sustained intake of nutrients. Eventually, cells with high floor adhesion ($J_i^f \approx 1$) colonize the source of nutrients in a brief burst of super-exponential growth. This is illustrated by figure 4d, where we can see a snapshot of our system soon after cells have "discovered" the roof. In terms of the population dynamics, it is also observed that the number of free nutrient particles declines as they are more efficiently found and removed from the system.

The emergence of EEN is common but requires both a high enough intake of energy (the flow of particles, Φ) and a moderate level of environmental fluctuations. The analysis of these two parameters reveals several phases in the potential types of structures that the system generates. This is illustrated in figure 6

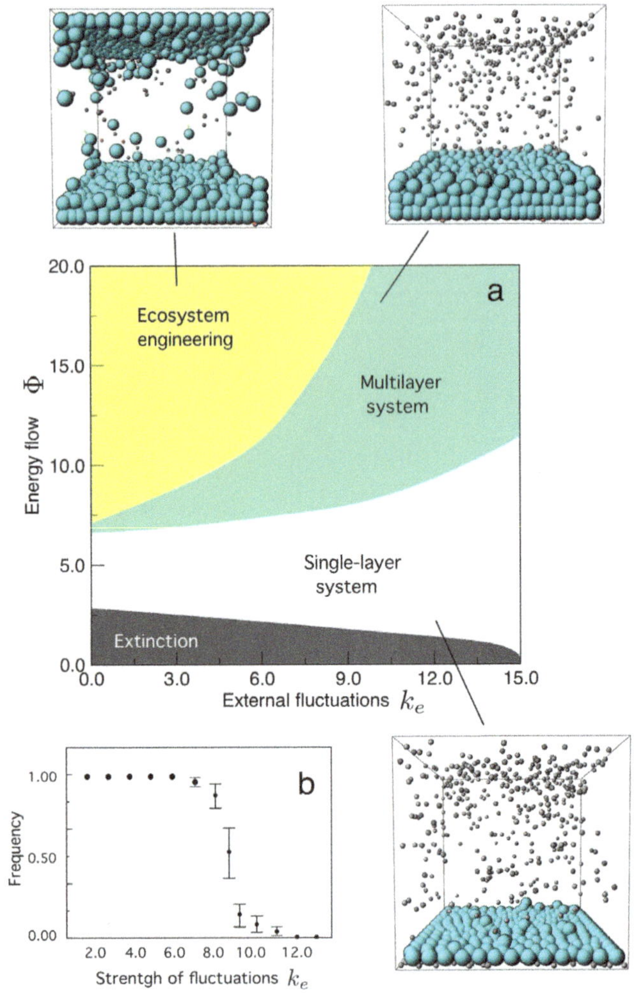

Figure 6. Interplay between energy intake and environmental fluctuations on the emergence of ecosystem engineering in CHIMERA's model. Here we have run CHIMERA using the same parameters of the previous figure but with different combinations of (k_e,Φ). Four phases are found (a) including extinction (lower part), cellular monolayers as well as multilayered systems. Examples of each scenario are indicated with 3D snapshots where the adhesion strength between cells and the boundaries is color coded. A quantitative analysis of the transition from the EEN to the multilayer phase is indicated in (b), where $\Phi=20$ has been used along with different levels of fluctuation. Ten replicas of each parameter combination were used and 2×10^5 steps used to determine the final state. The standard deviation is also shown as error bars.

where we present a summary of our exploration of the parameter space defined by (k_e,Φ). The choice of these two parameters is based on considering two relevant aspects associated to adaptation in evolving systems. The amplitude of the external fluctuations (as given by k_e, see equation 4) provides a measure of the random movements experienced by both cells and energy particles. In order to approximately determine the boundaries of these phases, the parameter space was partitioned into 25×25 pairs and evolution experiments were done for each pair.

The phase diagram reveals four phases. As it should be expected, at low flow levels the initial population cannot sustain itself. However, it is interesting to notice that the domain of extinction gets smaller for higher levels of perturbation. This is understandable when we realize what that means. Since we use a closed system, both cells and particles are influenced by external noise and, at high levels, there is a coherent movement of both types of elements, which end to aggregate in the corners of the lower level. This introduces a natural increase of encounter rates and thus allows cell survival more feasible. In a way, the high environmental noise leads to a predictable outcome due to the presence of boundary conditions.

The next phase (indicated in white color) involves a more or less stable population of cells that are confined to the floor of the system. Adhesion is high for low k_e and decreases to small (but non-zero) levels at higher levels. Here the population presents the largest area compatible with the incoming flow. On top of this phase, several layers can be stabilized. Well-defined smooth multilayers are observed on the left part of the diagram whereas they become a bit rugged on the right side, due to the presence of noise. A moderate level of cell-cell interactions allows to stabilize the system. It is within this phase that the initially multilayered system can develop a more complex organization and EEN.

An example of the type of transition that occurs between phases is illustrated in figure 6b. Here we have fixed the incoming flow to $\Phi = 20$ and determined the presence or absence of a top population of cells after 2×10^5 steps, using different k_e values. Averaging over ten replicas for each k_e, we obtain a phase transition close to the reported boundary. Around $k_e^* \approx 8$ we observe a phase change separating two clear domains associated to the presence $k_e < k_e^*$ or absence $k_e > k_e^*$ of a top layer. The error bars provide the standard deviation and are highest at the transition point, consistently with so called second-order transitions [44] where fluctuations strongly increase at criticality.

1.3 Emergence of detritivores

A major consequence is derived from this innovation. As cells in the top layer start to dominate the whole flow of nutrients, something new happens. After the colonization is completed, cells die and their waste material falls to the bottom. The resulting detritus stays for a while and is eventually removed by degradation. But some cells have already developed a mild efficiency to exploit these particles. After falling from the roof, these cells will find themselves inhabiting a niche that is rich in a given type of energy source. After a while, they develop a higher efficiency and eventually become specialists. This is illustrated in figure 7a–b, where we can see that the efficiency of detritivores grows fast after the top floor population has been established.

In the inset of figure 7a we display an example of the stationary state This transition illustrates the nonlinear impact of ecosystem engineering [31–33]. By changing the flow of nutrients, the top floor population causes a physical change in the environment due to their spatial distribution. As it occurs with freshwater phytoplankton organisms, which intercept light by placing themselves on the top of the water column or with higher plants, producing litter form dead leaves, our ecosystem provides a clever example of this scenario. By modulating the flow of nutrients, which eventually shifts from nutrient-rich to detritus-dominated flows, an effective asymmetry is generated.

The process of niche creation and the emergence of ecosystem engineering promote a new feedback towards the way detritivores behave. If cell-cell adhesion is allowed to evolve, it can be shown that the detritivore compartment evolves significant adhesion among individual cells, which can be described as a new form of

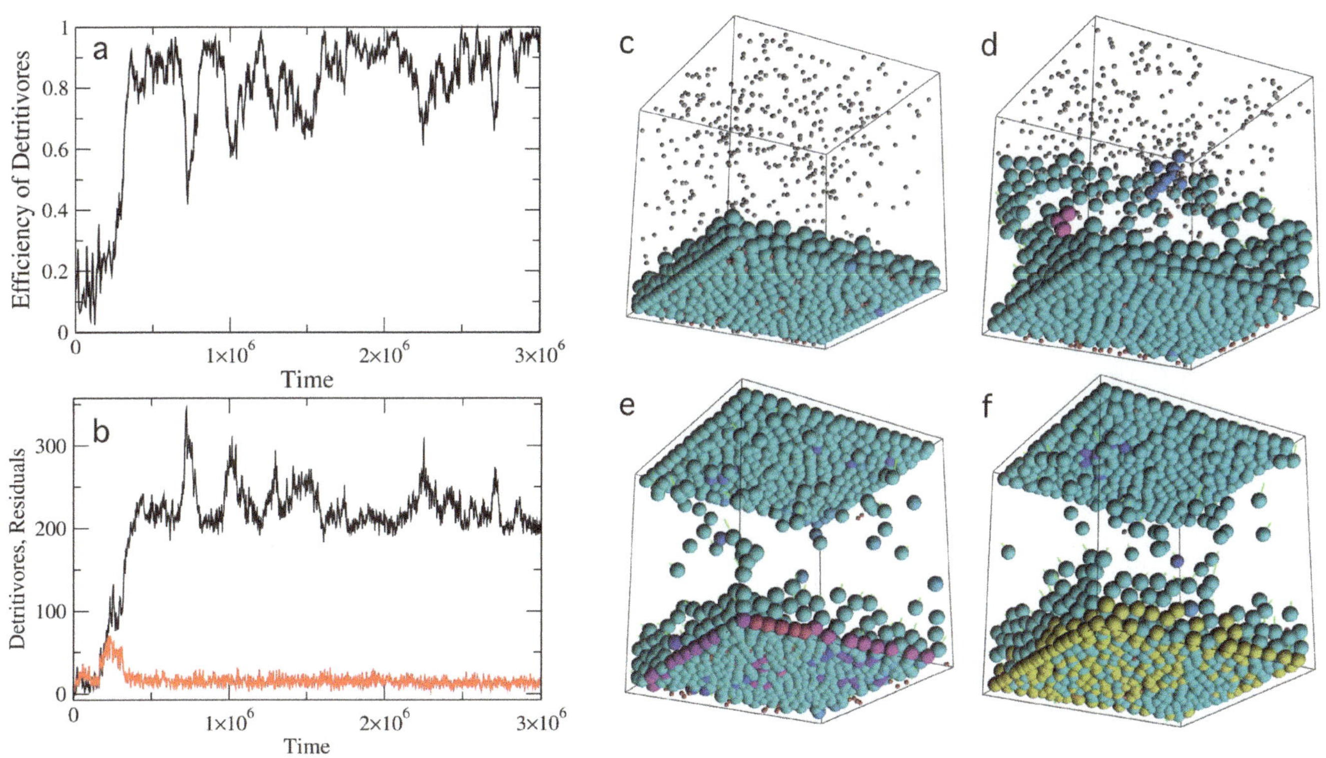

Figure 7. Dynamics of detritus grazing. In (a) the increase in efficiency rate of detritus-grazing cells is shown. In (b) we display the population dynamics of both detritivores (number of cells exhibiting some level of detritus feeding behavior) and detritus particles. In (c-f) four snapshots of the evolved system are shown. Detritivores evolve initially at the corners between wall boundaries thanks to the increased concentration of residuals. Afterwards, detritivores spread along the bottom plane and consume residuals produced by the disintegration of generalist cells coming from above. Four snapshots are shown in (c-f) at different times in the evolution of detritivores. Hot colors indicate the relative detritivore efficiency. Here: $\Phi = 25$ particles per timestep, $\Delta t = 0.03, \delta_w = 0.05, d_s = 0.01, v = 80, \xi = 80, k = 400, M_i = 20, R_i = 0.08, m_i = 3.35, r_i = 0.025, k_e = 2, k_d = 0.001$ (see text).

cell aggregate. It is interesting to notice that our aggregates do not cooperate in terms of sharing resources, but instead they emerge from selection constraints imposed by the requirement of an expanded area, which can only be obtained by attaching to the vertical surfaces. An additional advantage is obtained by living in higher locations: energy particles are falling near there and in fact, in the presence of fluctuations, it is likely that particles hitting the walls are captured by cells adhered to them. But the process has a discontinuity: by covering the roof, an accelerated transformation takes place, modifying the whole organization of the ecological assembly and allowing the emergence of an additional trophic compartment.

In figure 8 we summarize the ecological transitions experienced by our system as we cross through the different regimes. Here we have indicated the flows of matter from the external energy sources to waste. In fig. 8a, the initial state of our system is shown, with several sources of particles but only one being exploited by the single specialist. Such scenario is slowly replaced by a heterogeneous one (fig. 8b) marked by an increasing tendency to generalism: mutations affect efficiency rates and the potential for exploiting several resources. As evolution proceeds and cells develop adhesion (here indicated as small protrusions) along with a complete generalism (c). Here we indicate with $\{E_i\}$ that cells use equally all types of nutrients with the same coupling. Eventually (d) the new niche of detritus-rich particles triggers the evolution of a specialized population of detritivores . This result is predictable, provided that the abundance of incoming particles is large enough.

Discussion

In this paper we have presented a spatially-explicit model of evolution that incorporates ecology, physical embodiment and a very simple description of individuals. Our motivation was to provide an in silico platform to explore early stages of multicellular evolution under a pre-Mendelian scenario. Such pre-Mendelian world would be characterized by a higher morphological plasticity, far from the gene-regulated one of the Cambrian world [11,18]. In

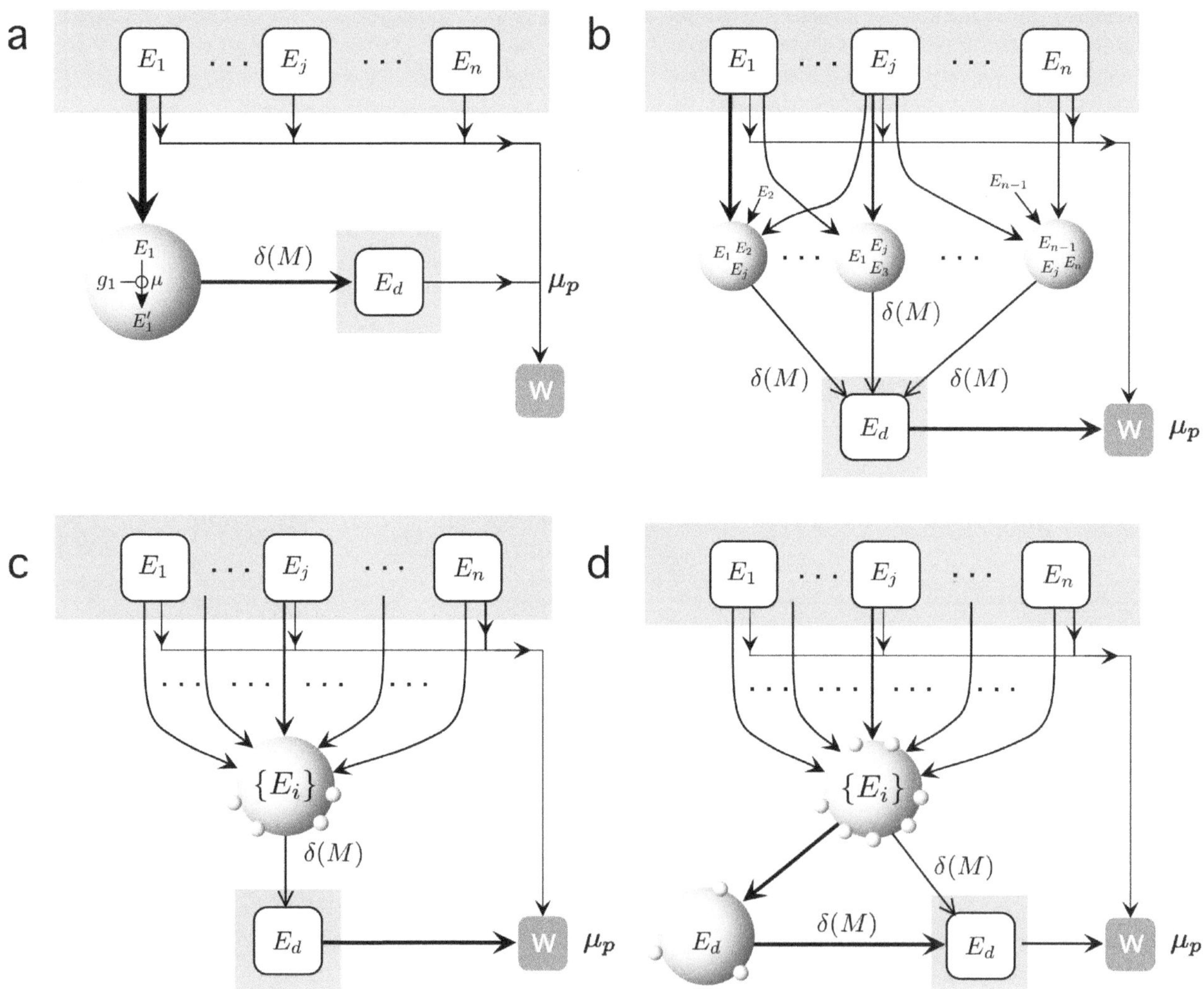

Figure 8. Chimera's food webs. (a) Flows in the initial food web, limited to a population of single-cell elements eating only one energy source (E_1) as defined by the CHIMERA model. The 1st nutrient is processed by the cells and transformed into cellular components high efficiency. At the end of the life cycle, the cell disintegrates into a number of residual particles. All molecules have similar degradation rates and get removed (right, lower box) from the system. After a while (b) an intermediate, diverse ecosystem is observed, with a large variety of cell exploiting different resources. This process proceeds until most cells become generalists, able to exploit all sources (c) whereas adhesion (indicated with smaller spheres) increases. Finally, once the population attached to the top floor has developed, another specialized population emerges at the bottom, fully composed by detritivores.

order to explore the potential repertoire of multicellular patterns that can emerge under these conditions, we have used a specific setting where a vertical flow of nutrients is introduced at the upper boundary of our system.

If the initial steps towards life took place in water, an appropriate model considering the role of physical interactions must take into account (a) events taking place within a fluid and (b) those associated to physical exchanges between individuals and individuals with the boundaries of the system (such as the sea bottom or a pond floor). By considering these basic forces, we introduce a minimal set of (possible) rules of interaction that can evolve through time. Such rules allow our artificial cells to explore their environment, interact and make decisions. But it also provides a very basic framework to explore the potential for finding patterns of pre-developmental pathways based on interactions among single-celled entities.

Despite the simplified nature of our simulation approach, which prevents (in particular) the emergence of complex regulatory networks and introduce limitations to the spatial organization of cell aggregates, several nontrivial results emerge. The evolutionary transition from the floor to the top layer leads to ecosystem engineering [27] and, as far as we know, this is the first example of such event happening in an artificial life system. By changing the actual flow of nutrients, they cause a physical change in the environment due to their physical distribution. Moreover, by doing so they also allow the emergence of a specialized, spatially segregated compartment of detritivores. Following the classification scheme from [35], our artificial creatures are autogenic engineers: they change their environment mainly via their own physical structures. Here the main structure created is a cell barrier associated to cell-floor interactions. The tempo of the transition exhibits a marked acceleration as the cell aggregates approach the top of the world. We can actually see that the cell population moves through its world by increasing in size until they detect the asymmetry associated to the top-down and start exploiting it. As it occurs with any other system out of equilibrium, our system is sustained by a gradient linked to a given flow of energy. In terms of the innovation process, the punctuated nature of the change is due to the transient time required in order to physically reach the energy-rich boundary and exploit it.

Future work should consider the explicit introduction of minimal genetic networks, the potential for cooperation in terms of nutrient sharing, an active role of our organisms in modifying their environmental conditions (thus adding an additional layer to the potential for ecosystem engineering) and variability of spatial conditions. These in silico experiments could be compared in some cases with evolutionary sequences of growth and selection observed in experiments involving transitions from single-cell to multicellular aggregates [45–49]. In this context, it can help designing and interpreting experimental approaches based on the introduction of selection favoring given traits. All these extensions of the CHIMERA model will allow us to approach relevant questions concerning the tempo and mode of the transition to multicellularity and how the different players (ecology, genetic interactions, physical embodiment and path dependence) influence the final outcome.

Finally, CHIMERA can be used to explore the effect of external events involving large extinctions and their aftermath. Recovery from mass extinction provides a unique insight into how communities are rebuild and the role played by different biotic and abiotic components in the reconstruction of a paleocommunity [50,51]. This is very well illustrated by the end-Permian extinction event [52,53] which devastated most species, to the point of near annihilation of complex life forms. The existing fossil record of the process and the patterns of change is very rich and well established [54] an illustrates the power of looking at macroevolutionary dynamics using multiple perspectives, from geological data to multispecies interactions. Some theoretical and computer models [55–57] have shown that a systems approach to these events can help to determine potential causal scenarios of recovery. But these models lack the physical embodiment that is characteristic of our model both at the individual and ecosystem levels. Since we also can trace the patterns of change in sediment particles and have a well-established segregation between trophic levels, we have a unique way of evaluating the relative contributions of ecological processes of competition or cooperation as well as the underlying evolutionary changes taking place at the individual level. These in silico experiments can help understanding the role played by altruism and cooperation and how they emerge and the nature of the resulting cooperative units [58,59].

Acknowledgments

This work started on a flight between Barcelona and New York and developed over the last years between Santa Fe, Venice, New York and the Complex Systems Lab. We thank the members of the Lab for so many useful discussions. SV thanks David Baraff for useful advices on constrained rigid-body simulation. We also thank the hospitality of the European Centre for Living Technology (ECLT) and the Santa Fe institute.

Author Contributions

Conceived and designed the experiments: RVS SV. Performed the experiments: RVS SV. Analyzed the data: RVS SV. Wrote the paper: RVS.

References

1. Carroll SB (2005) Endless forms most beautiful: The New Science of Evo Devo and the Making of the Animal Kingdom. W. W. Norton and Co.
2. Grossberg RK, Strahmann RR (2007) The Evolution of Multicellularity: A Minor Major Transition? Annual Reviews of Ecology and Evolution and Systematics 38: 621654.
3. Knoll AH (2011) The Multiple Origins of Complex Multicellularity Annual Reviews in Earth and Planetary Science 39: 217–239.
4. Lipps JH, Signor PH (1992) Origin and early evolution of the Metazoa. Plenum Press, New York.
5. Rokas A (2008) The molecular origins of multicellular transitions. Current Opinion Genetics and Development 18: 472–478.
6. King N (2004) The unicellular ancestry of animal development. Developmental Cell. 7: 313325.
7. Maynard Smith J, Szathmary E (1995) The major transitions in evolution. Oxford U. Press.
8. Niklas K (1997) The Evolutionary Biology of Plants. University of Chicago Press. Chicago.
9. Bonner JT (2001) First signals: the evolution of multicellular development. Princeton U. Press.
10. Fedonkin MA (2007) The Rise of Animals: Evolution and Diversification of the Kingdom Animalia. JHU Press.
11. Newman SA, Forgacs, G Muller GB (2006) Before programs: the physical origination of multicellular forms. International Journal of Developmental Biology 50: 289–299.
12. Eldredge N (1985) Unfinished Synthesis: Biological Hierarchies and Modern Evolutionary Thought. Oxford University Press.
13. McShea DW (2001) The hierarchical structure of organisms: a scale and documentation of a trend in the maximum. Paleobiology 27: 405–423.
14. Sebe-Pedros A, Roger AJ, Lang FB, King N, Ruiz-Trillo I (2010) Ancient origin of the integrinmediated adhesion and signaling machinery. Proceedings of the National Academy of Sciences USA. 107: 10142–10147.
15. Solé RV, Fernandez P, Kauffman SA (2003) Adaptive walks in a gene network model of morphogenesis: insights into the Cambrian explosion. International Journal of Developmental Biology 47: 685–693.

16. Fujimoto K, Ishihara S, Kaneko K (2008) Network evolution of body plans. PLoS ONE 3: e2772.
17. Tusscher KH, Hogeweg P (2011) Evolution of networks for body plan patterning; interplay of modularity, robustness and evolvability. PLoS Computational Biology 7: e1002208.
18. Newman SA, Baht R (2008) Dynamical patterning modules: physico-genetic determinants of morphological development and evolution. Physical Biology 5: 015008.
19. Graner F, Glazier JA (1992) Simulation of biological cell sorting using a two-dimensional extended Potts model. Physical Review Letters 69: 2013–2016.
20. Glazier JA, Graner F (1993) Simulation of the differential adhesion driven rearrangement of biological cells. Physical Review E 47: 21282154.
21. Hogeweg P (2000) Evolving Mechanisms of Morphogenesis: on the Interplay between Differential Adhesion and Cell Differentiation. Journal of Theoretical Biology 203: 317333.
22. Hogeweg P (2000) Shapes in the shadow: Evolutionary dynamics of morphogenesis. Artificial Life 6: 85–101.
23. Savill NJ, Hogeweg P (1997) Modeling morphogenesis: from single cells to crawling slugs. Journal of Theoretical Biology 184: 229–235.
24. Sims K (1994) Evolving 3D morphology and behavior by competition. Artificial Life 1: 353–372.
25. Pfeifer R, Lungarella M, Lida F (2007) Self-Organization, Embodiment, and Biologically Inspired Robotics. Science 318: 1088–1093.
26. Eggenberger P (1997) Evolving morphologies of simulated 3d organisms based on differential gene expression. In. Fourth Europ. Conf. Artificial Life, P. Husbands and I. Harvey, editors. 205–213. MIT Press.
27. Doursat R (2008) Organically grown architectures: creating decentralized, autonomous systems by embryomorphic engineering. In: Organic Computing, R. P. Würtz, ed: 167–200. Springer-Verlag.
28. Cummings F (2006) On the origin of pattern and form in early metazoans. International Journal of Developmental Biology 50: 193–208.
29. Kaandorp JA, Blom JG, Verhoef J, Filatov M, Postma M, et al. (2008) Modelling genetic regulation of growth and form in a branching sponge. Proceedings of the Royal Society B 275: 2569–2575.
30. Coen E, Rolland-Lagan AG, Matthews M, Bangham JA, Prusinkiewicz P (2004) The genetics of geometry.Proceedings of the National Academy of Sciences USA 101: 4728–4735.
31. Müller GB (2007) Evo-devo: extending the evolutionary synthesis. Nature Reviews Genetics 8: 943–949.
32. Forgacs G, Newman SA (2005) Biological physics of the developing embryo. Cambridge Univ. Press, Cambridge.
33. Erwin DH (2008) Macroevolution of ecosystem engineering, niche construction and diversity. Trends Ecol Evol. 23: 304–310.
34. Jones CG, Lawton JM, Shachak M (1994) Organisms as ecosystem engineers. OIKOS 69: 373–370.
35. Jones CG, Lawton JM, Shachak M (1997) Positive and negative effects of organisms as physical ecosystem engineers. Ecology 78: 1946–1957.
36. Erwin DH,Tweedt S (2012) Ecological drivers of the Ediacaran-Cambrian diversification of Metazoa. Evolutionary Ecology 26: 417–433.
37. Hastings A (2006) Ecosystem engineering in space and time. Ecology Letters 10: 153–164.
38. Palsson E (2001) A Three-dimensional model of cell movement in multicellular systems. Future Generation Computer Systems17: 835–852.
39. Palsson E, Othmer HG (2000) A model for individual and collective cell movement in Distyostlium discoideum. Proceedings of the National Academy of Sciences USA 97: 10448–10453.
40. Newman TJ (2005) Modeling Multi-cellular Systems using Sub-cellular Elements. Mathematical biosciences and engineering 2: 611–622.
41. Sandersius SA, Newman TJ (2008) Modelling cell rheology with the subcellular element model. Physical Biology 5: 015002.
42. Bell N, Yu Y, Muchs PJ (2005) Particle-Based Simulation of Granular Materials. In: SCA '05 Proceedings of the 2005 ACM SIGGRAPH Eurographics symposium on Computer animation. 77–86.
43. Ericson C (2005) Real-Time Collision Detection. Morgan Kaufmann.
44. Solé RV (2011) Phase Transitions. Princeton U. Press. Princeton.
45. Greig D, Louis EJ, Borts RH, Travisano M (2002) Hybrid speciation in experimental populations of yeast. Science 298: 1773–1775.
46. Rainey PB, Travisano M (1998) Adaptive radiation in a heterogeneous environment. Nature 394: 69–72.
47. Travisano M, Mongold JA, Bennett AF, Lenski RE (1995) Experimental tests of the roles of adaptation, chance, and history in evolution. Science 267: 87–90.
48. Lenski RE, Travisano M (1994) Dynamics of adaptation and diversification: a 10,000-generation experiment with bacterial populations. Proceedings of the National Academy of Sciences USA 91: 6808–6814.
49. Ratcliff WC, Denison RF, Borrello M, Travisano M (2012) Experimental evolution of multicellularity. Proceedings of the National Academy of Sciences USA 109: 1595–1600.
50. Erwin DH (1998) The end and the beginning: recoveries from mass extinctions. Trends in Ecology and Evolution 13: 344349.
51. Erwin DH (2001) Lessons from the past: biotic recoveries from mass extinction. Proceedings of the National Academy of Sciences USA 98: 53995403.
52. Erwin DH (1994) The PermoTriassic extinction. Nature 367: 231236.
53. Benton MJ, Twitchett RJ (2003) How to kill (almost) all life: the end-Permian extinction event. Trends in Ecology and Evolution 18: 358365.
54. Chen ZQ, Benton MJ (2012) The timing and pattern of biotic recovery following the end-Permian mass extinction. Nature Geoscience 5: 375383.
55. Solé RV, Montoya JM, Erwin DH (2002) Recovery after mass extinction: evolutionary assembly in large-scale biosphere dynamics. Philosophical Transactions of the Royal Society B 357: 697707.
56. Roopnarine PD (2006) Extinction cascades and catastrophe in ancient food webs. Paleobiology 32: 1–19.
57. Yedid G, Stredwick J, Ofria CA, Agapow PM (2012) A comparison of the e_ects of random and selective mass extinctions on erosion of evolutionary history in communities of digital organisms. PLoS ONE 7: e37233.
58. Queller DC (1997) Cooperation since life began. Quarterly Review of Biology 72: 184–188.
59. Queller DC (2000) Relatedness and the fraternal major transitions. Philosophical Transactions of the Royal Society B 355: 1647–1655.

5

Humans Strengthen Bottom-Up Effects and Weaken Trophic Cascades in a Terrestrial Food Web

Tyler B. Muhly[1,2]*, Mark Hebblewhite[3], Dale Paton[1], Justin A. Pitt[4], Mark S. Boyce[4], Marco Musiani[1]

1 Faculty of Environmental Design, University of Calgary, Calgary, Alberta, Canada, **2** Ecosystem Management, Alberta Innovates – Technology Futures, Vegreville, Alberta, Canada, **3** Department of Ecosystem and Conservation Sciences, University of Montana, Missoula, Montana, United States of America, **4** Department of Biological Sciences, University of Alberta, Edmonton, Alberta, Canada

Abstract

Ongoing debate about whether food webs are primarily regulated by predators or by primary plant productivity, cast as top-down and bottom-up effects, respectively, may becoming superfluous. Given that most of the world's ecosystems are human dominated we broadened this dichotomy by considering human effects in a terrestrial food-web. We studied a multiple human-use landscape in southwest Alberta, Canada, as opposed to protected areas where previous terrestrial food-web studies have been conducted. We used structural equation models (SEMs) to assess the strength and direction of relationships between the density and distribution of: (1) humans, measured using a density index; (2) wolves (*Canis lupus*), elk (*Cervus elpahus*) and domestic cattle (*Bos taurus*), measured using resource selection functions, and; (3) forage quality, quantity and utilization (measured at vegetation sampling plots). Relationships were evaluated by taking advantage of temporal and spatial variation in human density, including day versus night, and two landscapes with the highest and lowest human density in the study area. Here we show that forage-mediated effects of humans had primacy over predator-mediated effects in the food web. In our parsimonious SEM, occurrence of humans was most correlated with occurrence of forage ($\beta = 0.637$, $p < 0.0001$). Elk and cattle distribution were correlated with forage (elk day: $\beta = 0.400$, $p < 0.0001$; elk night: $\beta = 0.369$, $p < 0.0001$; cattle day: $\beta = 0.403$, $p < 0.0001$; cattle, night: $\beta = 0.436$, $p < 0.0001$), and the distribution of elk or cattle and wolves were positively correlated during daytime (elk: $\beta = 0.293$, $p < 0.0001$, cattle: $\beta = 0.303$, $p < 0.0001$) and nighttime (elk: $\beta = 0.460$, $p < 0.0001$, cattle: $\beta = 0.482$, $p < 0.0001$). Our results contrast with research conducted in protected areas that suggested human effects in the food web are primarily predator-mediated. Instead, human influence on vegetation may strengthen bottom-up predominance and weaken top-down trophic cascades in ecosystems. We suggest that human influences on ecosystems may usurp top-down and bottom-up effects.

Editor: Jon Moen, Umea University, Sweden

Funding: The authors' group's research on human impacts on wildlife species was funded by a substantial ($1.7 million) grant from the Natural Science and Engineering Research Council (NSERC) of Canada. The authors thank the following for their important financial contributions to this work: the Natural Science and Engineering Research Council of Canada, Shell Canada, the Institute for Sustainable Energy, Environment and Economy, the Canadian Wildlife Federation, the Glaholt Scholarship, Parks Canada, Alberta Beef Producers, Alberta Conservation Association, Alberta Tourism, Parks and Recreation, Alberta Sustainable Resource Development, Alberta Community Development, the Calgary Zoo, Alberta Ecotrust, Alberta Ingenuity, the University of Calgary, the University of Alberta, and Safari Club International. The funders had no role in study design, data collection and analysis, decision to publish, or preparation of the manuscript.

Competing Interests: Please note that the corresponding author's affiliation with Alberta Innovates – Technology Futures does not result in any competing interests and does not alter the authors' adherence to all the PLOS ONE policies on sharing data and materials.

* E-mail: tyler.muhly@albertainnovates.ca

Introduction

Food-webs may be influenced by both bottom-up effects that link plants to herbivores and higher trophic levels and by top-down effects from carnivores to plants. However, there is still significant debate about the relative importance of each in ecosystems. Recent studies on large mammal food webs in protected areas indicate that predators can have strong top-down effects on prey and indirectly on vegetation [1,2], i.e., "trophic cascades" [3]. This has led some ecologists to argue that top-down effects of top carnivores have primacy in driving food-web dynamics [4], and rising popularity of top-down forces in the popular literature [5]. In reality, top-down and bottom-up mechanisms operate simultaneously in food webs [6]. Emphasizing the importance of one mechanism over the other may falsely dichotomize how ecosystems function.

Importantly, oversimplifying ecosystem dynamics into top-down effects of predators versus bottom-up effects of plant biomass availability could be underestimating the importance of human influences in ecosystems. Human influences on food webs likely are pervasive and may operate at multiple trophic levels simultaneously. For example, humans have modified ecosystems for tens of thousands of years through bottom-up (i.e., ecosystem engineering [7]) and top-down effects (i.e., overharvest and continental-scale megafaunal extinctions [8,9]). Human impacts on ecosystems have intensified with a growing human population and demand for resources [10]. Thus, a contemporary challenge is to understand how growing human influences on ecosystems might propagate through food webs.

Here we describe and quantify human effects at multiple trophic levels of a terrestrial food web in a multiple human-use area, as opposed to protected areas where many previous terrestrial food web studies have been conducted. We show that human effects were pervasive on the studied species in the food web, and manifested through both direct and indirect pathways as they propagated across trophic levels. We show that indirect effects of humans were forage-mediated rather than predator-mediated. We conclude that pervasive effects of humans in ecosystems can perturb top-down and bottom-up effects in food-web dynamics.

Methods

The study occurred in a 9,000 km^2 area of southwest Alberta, Canada [11]. We used structural equation models (SEMs [12]) to assess the strength and direction of the spatial relationships between the occurrence of: (1) humans, measured using a density index; (2) wolves (*Canis lupus*), elk (*Cervus elaphus*) and cattle (*Bos taurus*), measured using telemetry data and resource selection functions (RSFs), and; (3) forage quality and quantity and utilization (measured at vegetation sampling plots). In selecting study species, we followed previous studies, which used human-wolf-ungulate-vegetation dynamics to illustrate the importance of top-down versus bottom-up effects in terrestrial ecosystems [1,2]. In our study, wolves were the predominant predator of the two dominant herbivores, one native, elk, and one non-native, cattle. Relationships between species occurrence were evaluated by taking advantage of pseudo-experimental temporal and spatial variation in human activity, including: (a) day versus night, and (b) in two landscapes with the highest and lowest human density in the study area. Testing these relationships across different levels of human activity allowed us to assess the sensitivity of species co-occurrence to human activity. If human effects on the food web were forage-mediated, we predicted a positive influence of humans on forage, of forage on herbivores (i.e., elk and domestic cattle), and of herbivores on wolves. Alternatively, if human effects were predator-mediated, we predicted a negative influence of humans on wolves, wolves on herbivores and herbivores on forage.

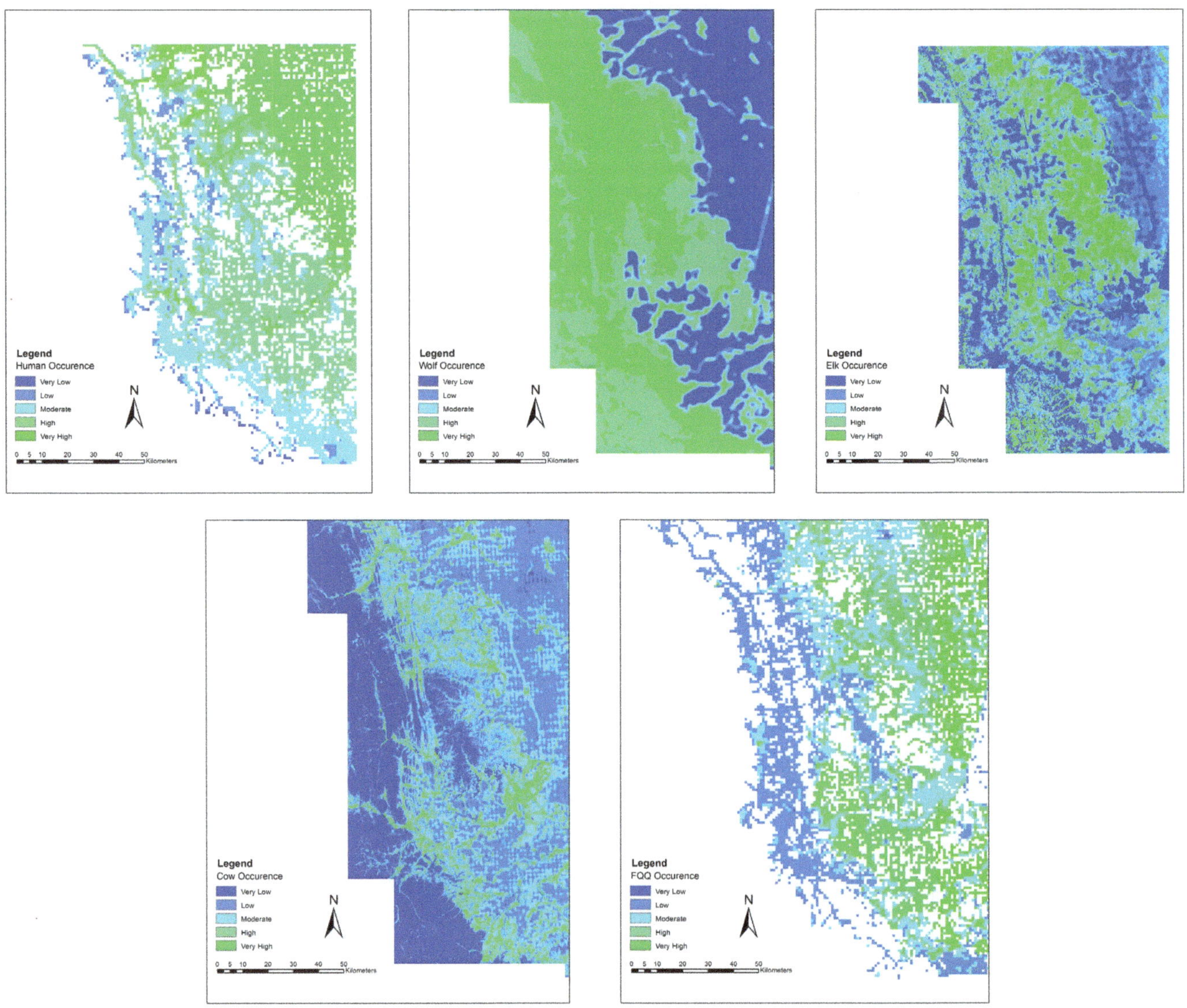

Figure 1. Spatial models of human (top left), wolf (top middle), elk (top right), cattle (bottom left) and forage quality and quantity (FQQ, bottom right) occurrence during the day in southwest Alberta, Canada. Low values (dark blue to light blue) indicate very low to moderate probability of occurrence at a location whereas high values (light blue to dark green) indicate moderate to very high probability of occurrence.

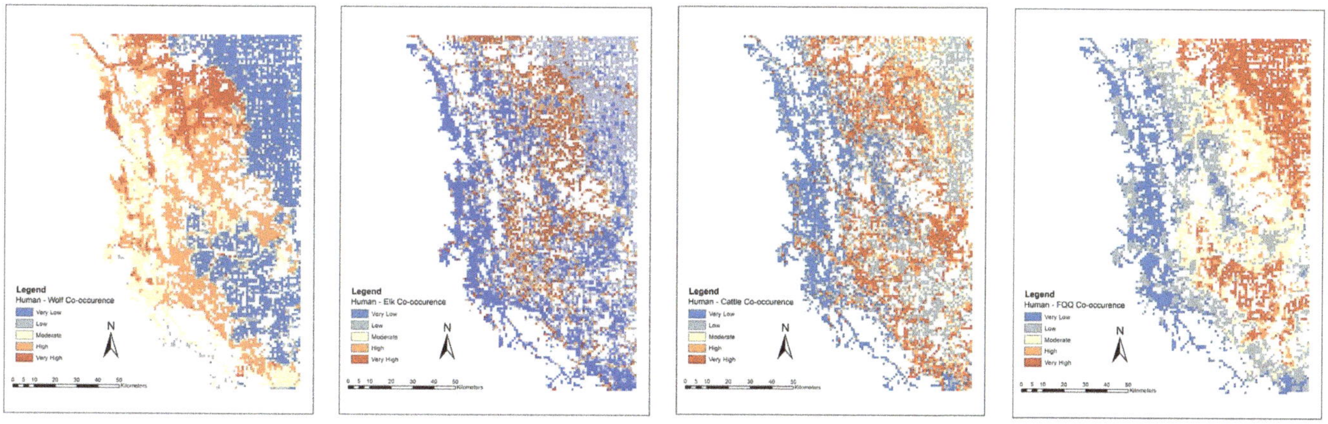

Figure 2. Co-occurrence of humans and wolves (left), humans and elk (middle left), humans and cattle (middle right) and humans and forage quality and quantity (FQQ, right), during the day along roads and trails in southwest Alberta, Canada. Low values (blue to yellow) indicate that the distribution is different (i.e., avoidance) at a location whereas high values (yellow to red) indicate that the distribution is similar (i.e., co-occurrence). For illustrative purposes roads and trails are exaggerated with a 1-km buffer.

We produced a spatial index of human density in a Geographic Information System (GIS), validated with actual counts of humans obtained using trail cameras (RECONYX Silent Image™ Model RM30, $n = 43$) and pneumatic road tube traffic counters (Diamond Traffic Products, $n = 43$) deployed on spatially randomized roads and trails across the study area in 2008. We calculated the human density index based on travel time required to access any point in the study area along existing road and trails from human population centers (>100 people), given typical travel speeds and a decay exponent of −1.45 based on typical human recreational behavior [13]. We tested whether the human density index predicted actual counts using linear regression.

We obtained satellite and Global Positioning System (GPS) telemetry location data from wolves, elk and cattle and developed population-averaged RSFs [14,15] to estimate habitat selection and the resultant predicted spatial distribution of each species. We obtained 7,462 locations from 14 wolves fitted with ARGOS satellite telemetry collars between 2004 and 2007, 267,440 locations from 62 elk fitted with GPS or ARGOS satellite telemetry collars (Lotek 4400 and 2200 series and Telonics TGW-3600) between 2007 and 2009, and 348,514 cattle location from 50 cattle GPS telemetry collars (Lotek 3300L) deployed within two separate grazing allotments between 2004 and 2007. Animal capture protocols were approved by the Universities of Alberta and Calgary and the government of Alberta (Permit Numbers: BI-2008-19, RC-06SW-001 and 23181CN).

RSFs were estimated using logistic regression in STATA 10.1 [16], where resources at telemetry locations are compared to resources at "available" locations [14,17]. Available resource locations were sampled at random in each individual animal home range for wolves and elk, estimated using a 95% kernel density estimator of location data [18], or fenced pasture for cattle. We followed the "two-stage" method to calculate population-level RSFs [15,19] where an RSF is calculated for each individual animal and these are averaged across all individuals. We produced daytime and nighttime RSFs because human counts were statistically different between night and day, and we hypothesized that wolves [20] and elk [21], might respond to these temporal changes. We validated RSF models using k-fold cross validation [22,23].

We calculated a spatial index of forage quality and quantity from remotely sensed vegetation data. We obtained a 30-m^2 spatial resolution GIS map of vegetation cover derived from Landsat data [24] and collapsed it into two ungulate food quality classes, high and low [11]. We multiplied this by the normalized difference vegetation index (NDVI) value at each pixel as an index of forage biomass [25]. This provided an index of forage quality (type) and quantity (productivity) at each pixel on the landscape.

We measured forage utilization by herbivores in 2007 and 2008 at 150 plots using the ocular estimate-by-plot method [26]. At 40 of the 150 plots, forage utilization was also assessed using the paired-subplot method [27] in which vegetation biomass on an area exposed to herbivory was compared to a nearby area where grazing was excluded with a 1.2 m×1.2 m cage. We assessed the accuracy of our visual estimates of utilization by comparing visual estimates to actual clipped biomass utilization values in a linear regression.

We overlaid our spatial models in a GIS and at each 30-m pixel along roads and trails ($n = 760{,}140$) we sampled the human index, RSF values and forage index. At each forage utilization plot we sampled RSF values for each species. SEMs were fit to these data using LISREL 8.72 [28]. Humans were modeled in SEMs as an exogenous variable, i.e., starting points of the model, whereas other species were endogenous variables, i.e., determined by pathways in the model, and that can also serve as predictors for other endogenous variables. SEMs were used in an exploratory mode where initial theoretical models of bottom-up and top-down only pathways were altered to improve the model fit by removing pathways for which the association was non-significant [29]. We assessed SEM fit using the goodness-of-fit index (GFI), adjusted goodness-of-fit index (AGFI), and root mean squared residual (RMR). A GFI>0.9, AGFI>0.9 and with a similar value to the GFI and RMR<0.05 indicate the model had a good fit to the data [30]. The best model had the lowest Akaike Information Criterion (AIC), i.e., parsimoniously fit the data, had significant interaction paths and fit the data.

Results

Species density and distribution models were predictive of actual species occurrence. We confirmed a significant linear relationship between the human density index and actual number of humans counted on roads and trails at counter locations during the day ($R^2 = 0.61$, $F = 197.78$, $p < 0.001$; Fig. 1) and night

Day

GFI=0.998, AGFI=0.985, RMR=0.015, AIC=124.280

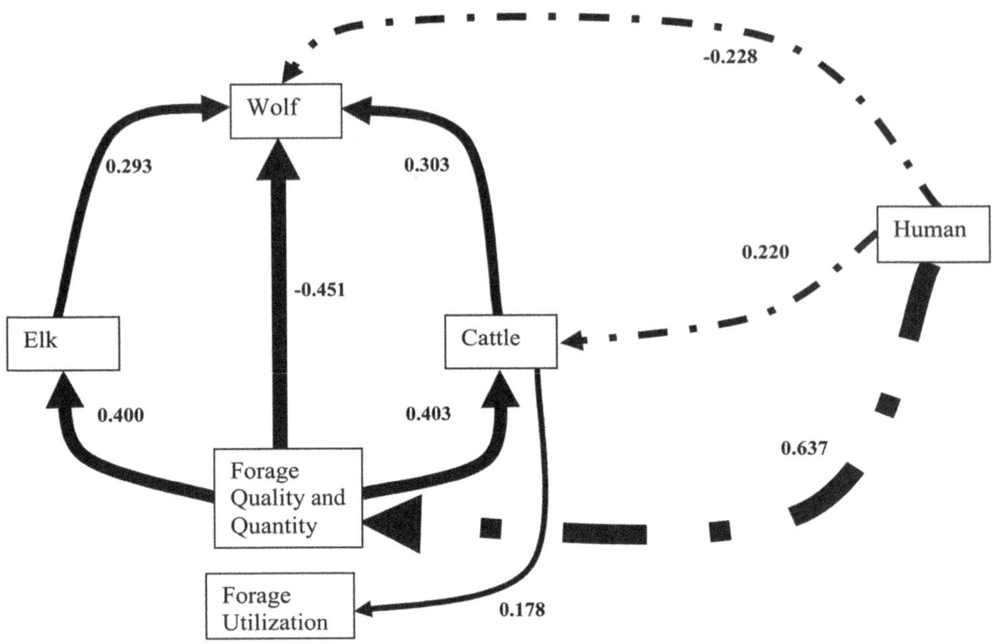

Night

GFI=0.979, AGFI=0.840, RMR=0.049, AIC= 41,543.122

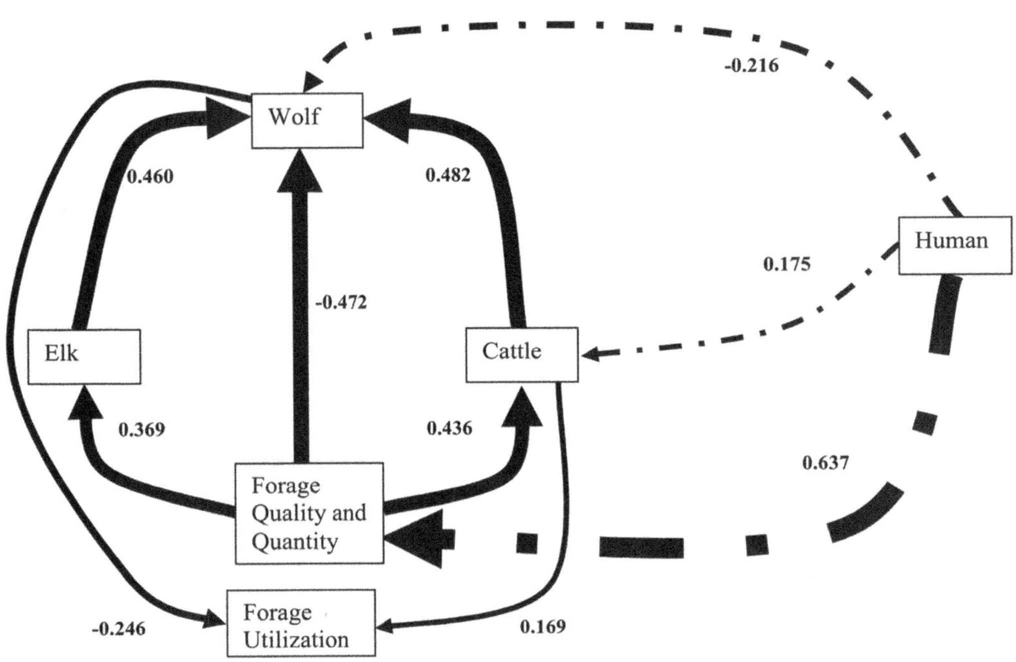

Figure 3. Structural equation model (SEM) illustrating the directions and strengths of relationships among the spatial distribution of humans, wolves, elk, cattle, and forage quality and quantity and forage utilization during the daytime and nighttime in southwest Alberta, Canada. Solid arrows indicate causal direction of the consumer-resource interaction and line thickness is proportional to relationship strength (β coefficient, also indicated). Human influences are represented by dashed-dotted lines.

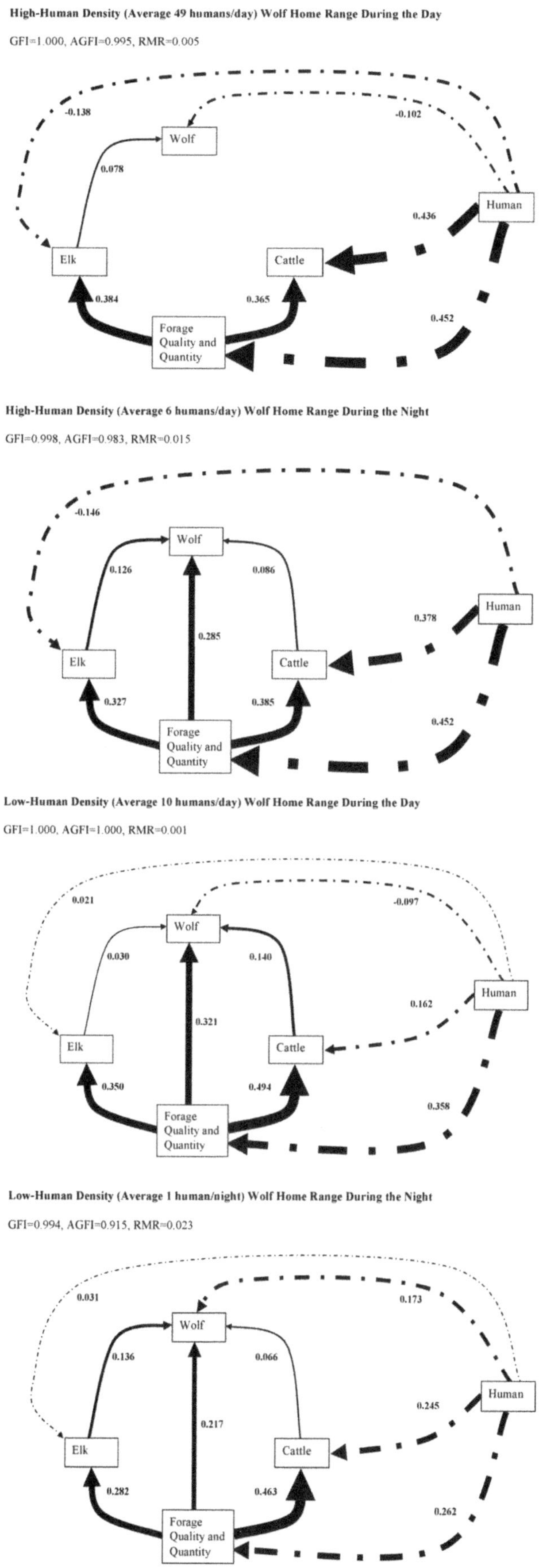

Figure 4. Structural equation model (SEM) illustrating the directions and strengths of relationships among the spatial distribution of humans, wolves, elk, cattle, and forage quality and quantity and forage utilization during the daytime and nighttime within two wolf home ranges with the highest and lowest human density in southwest Alberta, Canada.

($R^2 = 0.57$, $F = 169.85$, $p<0.001$). Population-level RSFs for wolves, elk and cattle (Table 1) were highly predictive of each species' distribution (Table 2). We found a significant linear fit ($R^2 = 0.255$, $F = 47.50$, $p<0.001$) between visual estimates of forage utilization to actual clipped biomass utilization values. We illustrated species co-occurrence with humans across the landscape by re-scaling each occurrence model index from 0 (low probability of occurrence) to 1 (high probability of occurrence) and adding species occurrence models together (Fig. 2).

Our results indicate that the predominant effect of humans was to enhance forage, augmenting herbivores (both native and non-native). In our most parsimonious SEM (Fig. 3), spatial and temporal occurrence of humans was most correlated with occurrence of forage ($\beta = 0.637$, $p<0.0001$). Similarly, elk or cattle distribution during the day and the night were correlated with forage (elk day: $\beta = 0.400$, $p<0.0001$; elk night: $\beta = 0.369$, $p<0.0001$; cattle day: $\beta = 0.403$, $p<0.0001$; cattle night: $\beta = 0.436$, $p<0.0001$; Fig. 3). We found positive relationships between the distribution of elk or cattle and wolves during daytime (elk: $\beta = 0.293$, $p<0.0001$, cattle: $\beta = 0.303$, $p<0.0001$) which strengthened during nighttime when human activity declined (elk: $\beta = 0.460$, $p<0.0001$, cattle: $\beta = 0.482$, $p<0.0001$; Fig. 3).

In addition to forage, humans had direct effects on the distribution of wolves and cattle (wolves day: $\beta = -0.228$, $p<0.0001$; wolves night: $\beta = -0.216$, $p<0.0001$; cattle day: $\beta = 0.220$, $p<0.0001$; cattle, night: $\beta = 0.175$, $p<0.0001$; Fig. 3). Finally, cattle distribution was positively related to forage utilization ($\beta = 0.178$, $p<0.05$ during the day; $\beta = 0.169$, $p<0.05$ during the night; Fig. 3). We also documented a negative, likely indirect effect of wolves on forage utilization, but only at night ($\beta = -0.246$, $p<0.05$; Fig. 3).

We found similar patterns to those above within the high and low human density ranges (Fig. 4). Furthermore, human density had a negative effect on wolf distribution during the day ($\beta = -0.102$, $p<0.0001$), but not during the night in the high-human density range. Such relationships were also negative during the day ($\beta = -0.097$, $p<0.0001$) but positive during the night ($\beta = 0.173$, $p<0.0001$) in the low-human density range. This relationship is further illustrated by the difference between day and night locations of wolves monitored in this study. During the night, wolf locations tended to occur near roads (Fig. 5), likely due to low human use over the night.

Discussion

Here we show the dominant role played by humans in influencing bottom-up vs. top-down dynamics in ecosystems. In particular, our results suggest that human effects on forage could trigger strong bottom-up effects in ecosystems. This contrasts with research conducted in protected areas or relatively intact ecosystems suggesting that human effects on food webs were primarily predator-mediated [1,2,4]. However, previous research that addressed interactions between top-down and bottom-up forces found that predator effects may be weak or absent when vegetation productivity is high, at which time bottom-up forces might prevail [31]. Thus, the mechanisms by which humans influenced the food web in our study area may have involved

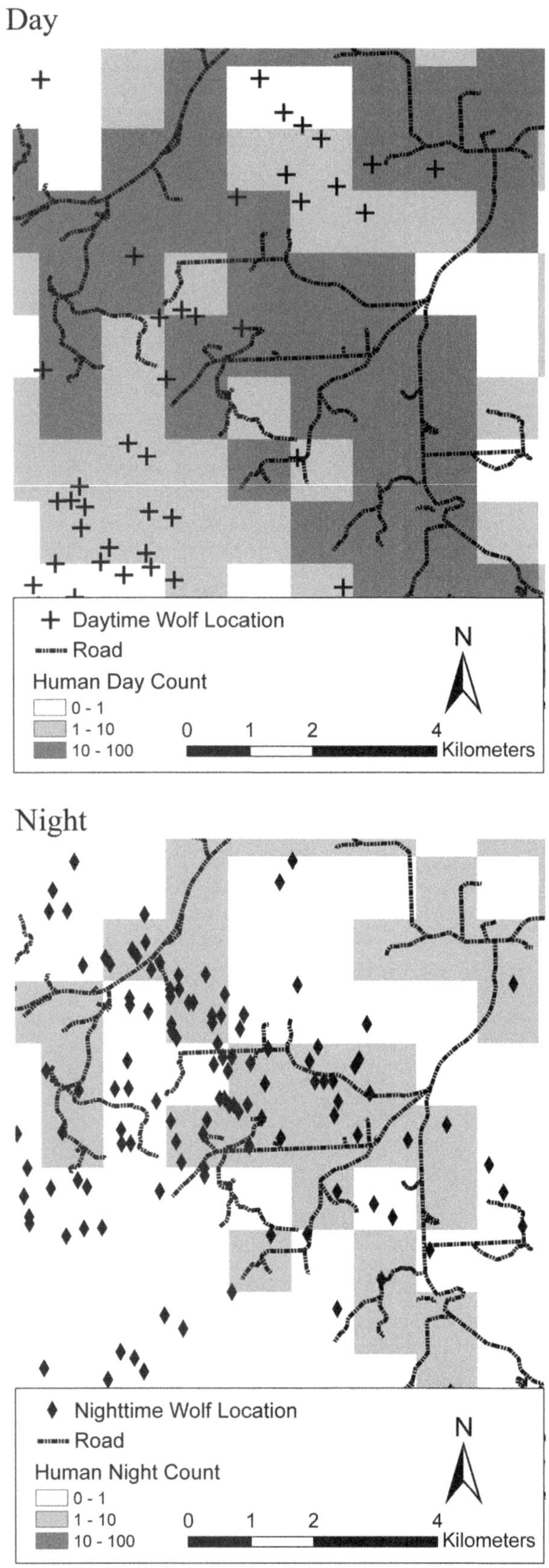

Figure 5. A sample case of wolf telemetry locations around roads with different human counts during the daytime compared to nighttime in southwest Alberta, Canada. Daytime locations are from sunrise to sunset, nighttime locations are sunset to sunrise. Human count is the average number of humans on roads and trails during the day (a), and night (b), calculated from an access index model (Apps et al. 2004). Darker pixels indicate higher human counts.

agricultural subsidies that overwhelmed top-down effects. Humans may have positively contributed to forage biomass in our study area through forage crop production and modification of forest to pasture-lands to provide forage for livestock production, an important economic activity. Indeed humans are an important reason why large portions of the world are "green" [32], as agriculture, particularly fertilization, increases nitrogen in the surrounding environment, which positively affects vegetation productivity [33]. In accordance with the mechanisms advocated in such studies, we suggest that human influences on vegetation might prompt bottom-up predominance in ecosystems.

A closer look at the pathways in our SEM also indicated that there were strong, direct influences of humans on all study species at each trophic level. For example, we found evidence of wolf avoidance of people at high levels of human use (a top-down effect). With lower human density there was no effect of humans on wolf distribution, and actually wolves selected for similar habitats as humans. The negative direct effects of humans on wolves was not surprising, because of common lethal control of wolves by humans in the study area in response to livestock predation [34].

The major implication of our results is the need to understand human effects in food webs [35].The predominant human influence on a particular food web may depend on how humans have perturbed the ecosystem [36]. In our study area, forage production was a dominant perturbation. In other areas, other types of human activity may be dominant with different implications for food webs. Understanding human effects in food webs may be particularly important in areas where predator reintroduction has been proposed as a means to restore ecosystems through top-down regulation. For example, predator reintroduction might be ineffective in restoring top-down effects in areas characterized by high densities of humans and livestock because of the predominant bottom-up effect of humans, as shown in our study. Certainly, we do not dispute the importance of top trophic levels in ecosystems, but claims that such effects have primacy worldwide [4] likely underestimate the effects of humans on ecosystems, and in particular may fail to appreciate the importance of resource-mediated effects of humans.

In the past, much emphasis has been placed on direct effects of humans on food webs, for example, through hunting or habitat change. Less emphasis has been placed on measuring indirect effects of humans. For example, conservation-oriented research has typically focused on direct influences of humans on population dynamics of declining and/or rare species. Our study approach emphasizes the importance of considering direct and indirect effects of humans on multi-species interactions and ecosystem dynamics in conservation research. We encourage the application of SEMs as a tool for comparing direct and indirect top-down vs. bottom-up effects of humans on food web dynamics across diverse landscapes. However, we caution that SEMs may allow for testing hypotheses on relationships between species occurrences but do not test the underlying mechanism of those relationships.

The human population is growing worldwide and, to meet its needs, agriculture is not expected to diminish in intensity or distribution. Subsequently, humans will have increasing impacts on ecosystems through a myriad of trophic interactions and pathways. While significant attention has been given to the top-down effects of humans on food webs, careful consideration and

Table 1. Coefficients for population-level resource selection function (RSF) models of wolves, elk and cattle during day and night in southwest Alberta, Canada.

	Wolf				Elk				Cattle			
	Day		Night		Day		Night		Day		Night	
Covariate	β	SE	β	SE	β	SE	β	SE	β	SE	β	SE
Road and Trail Density (km/km^2)	0.118	0.1946	0.384	0.1907	−0.095	0.0700	−0.107	0.0773	0.290	0.1186	0.296	0.1236
Road and Trail Density^2	−0.119	0.0478	−0.165	0.0551	−0.031	0.0144	−0.036	0.0152	−0.023	0.0232	−0.030	0.0273
Slope (°)	0.286	0.0716	0.063	0.0468	0.126	0.0099	0.122	0.0098	−0.070	0.0188	0.091	0.0327
Slope^2	−0.010	0.0022	−0.007	0.0022	−0.004	0.0003	−0.004	0.0002	−0.002	0.0009	−0.008	0.0016
Distance to Water (m)	0.066	0.0220	0.038	0.0168	−0.000	0.0000	−0.000	0.0000	−0.000	0.0001	−0.000	0.0001
Treed Wetland	Dropped		−0.001	0.0002	Dropped		Dropped		Dropped		Dropped	
Open Wetland	Dropped		Dropped		Dropped		Dropped		Dropped		Dropped	
Dense Conifer Forest	0.054	0.0215	0.036	0.0166	0.119	0.5207	0.810	0.5432	−0.836	0.2694	−0.793	0.2982
Moderately Closed Canopy Conifer Forest	0.055	0.0212	0.037	0.0164	0.567	0.5171	1.151	0.5836	−0.070	0.2330	−0.302	0.2622
Open Conifer Forest	0.059	0.0219	0.040	0.0169	1.524	0.5155	2.291	0.5878	0.572	0.2288	1.289	0.2826
Mixed Forest	0.055	0.0213	0.037	0.0165	0.739	0.5068	1.439	0.5776	0.458	0.2217	0.510	0.2462
Broadleaf Forest	0.058	0.0213	0.035	0.0164	1.320	0.5218	2.111	0.5659	1.092	0.2795	0.623	0.2630
Regen	0.071	0.0206	0.050	0.0133	0.118	0.0560	−0.384	0.0743	Dropped		Dropped	
Shrub	0.054	0.0214	0.037	0.0165	1.026	0.5268	1.696	0.5871	0.855	0.2167	1.258	0.2401
Herbaceous	0.054	0.0211	0.036	0.0164	1.332	0.5110	2.043	0.5867	0.901	0.2447	1.470	0.2426
Agricultural Field	−0.015	0.0041	−0.014	0.0082	0.554	0.0831	0.666	0.1127	Dropped		Dropped	
Barren Ground	0.050	0.0207	0.035	0.0149	0.257	0.3540	1.238	0.4835	−2.313	0.5872	−1.750	0.4711
Snow/Ice	Dropped		Dropped		−0.677	0.3451	1.621	0.4561	Dropped		Dropped	
Intercept	−188.920	72.7008	−125.986	56.0790	−2.569	0.5722	−4.185	0.6555	−0.236	0.7997	−1.913	0.7578

Table 2. Results from 5-fold and pasture cross validation of wolf, elk and cattle resource selection functions (RSFs) during day and night in southwest Alberta, Canada.

	Wolf		Elk			Cattle	
	Day	Night	Day	Night		Day	Night
Spearman r_s							
Group					**Pasture**		
1	1.000	1.000	1.000	1.000	Bob Creek – Porcupine Hills	0.726	0.808
2	0.988	0.976	1.000	1.000	Porcupine Hills – Bob Creek	0.618	0.599
3	0.964	0.988	1.000	1.000			
4	0.952	0.988	1.000	1.000			
5	0.994	0.951	1.000	1.000			
Mean	0.979	0.981	1.000	1.000		0.672	0.704
R^2							
1	0.941	0.914	0.948	0.925	Bob Creek – Porcupine Hills	0.559	0.519
2	0.863	0.794	0.945	0.911	Porcupine Hills – Bob Creek	0.122	0.025
3	0.769	0.761	0.945	0.938			
4	0.874	0.841	0.949	0.926			
5	0.857	0.839	0.946	0.924			
Mean	0.839	0.884	0.943	0.920		0.340	0.272

Spearman correlations were calculated between RSF-habitat ranks and area-adjusted frequencies on a withheld sub-sample of data (20%) 5-times. We also calculated a linear regression between observed frequency and expected RSF scores and assessed the fit. We validated the cattle RSF using a 2-pasture cross validation, where separate cattle RSFs were produced for each pasture and a Spearman rank correlation and linear regression between the two models was calculated in each pasture.

quantification of the diversity of direct and indirect effects of humans on multiple species, as we did in this study, will be necessary for effective ecosystem conservation.

Acknowledgments

We are grateful to the following for their important contributions to this work: M. Alexander, J. Berger, C. Bergman, E. Bork, T. Fung, A. Heschl, R. Kennedy, I. Laporte, T. Mack, C. Mamo, D. Marceau, G. McDermid, A. McDevitt, P. Paquet, C. Sponarski, C.C. St. Clair and B. Sundberg. We thank Dr. Hendrik Andren and one anonymous reviewer for their very helpful comments.

Author Contributions

Conceived and designed the experiments: TM MM MH MB. Performed the experiments: TM. Analyzed the data: TM. Contributed reagents/ materials/analysis tools: MB JP DP MB MM. Wrote the paper: TM MM MH MB.

References

1. Ripple WJ, Larsen EJ, Renkin RA, Smith DW (2001) Trophic cascades among wolves, elk and aspen on Yellowstone National Park's northern range. Biol Cons 102: 227–234.
2. Hebblewhite M, White CA, Nietvelt CG, McKenzie JA, Hurd TE, et al. (2005) Human activity mediates a trophic cascade caused by wolves. Ecology 86: 2135–2144.
3. Paine RT (1980) Food webs, linkage, interaction strength and community infrastructure. J Anim Ecol 49: 666–685.
4. Estes JA, Terborgh J, Brashares JS, Power ME, Berger J, et al. (2011) Trophic downgrading of planet earth. Science 333: 301–306.
5. Stolzenburg W (2008) Ecosystems unraveling. Conservation Magazine. Available: http://www.conservationmagazine.org/2008/07/ecosystems-unraveling. Accessed 28 December 2012.
6. Krebs CJ, Boutin S, Boonstra R, Sinclair ARE, Smith JNM, et al. (1995) Impact of food and predation on the snowshoe hare cycle. Science 269: 1112–1115.
7. Smith BD (2007) The ultimate ecosystem engineers. Science 315: 1797–1798.
8. Barnosky AD, Koch PL, Feranec RS, et al. (2004) Assessing the causes of late Pleistocene extinctions on the continents. Science 306: 70–75.
9. Darimont CT, Carlson SM, Kinnison MT, Paquet PC, Reimchen T, et al. (2009) Human predators outpace other agents of trait change in the wild. Proc Nati Acad Sci 106: 952–954.
10. Foley JA, DeFries R, Asner GP, Barford C, Bonan G, et al. (2005) Global consequences of land use. Science 309: 570–574.
11. Muhly TB, Alexander M, Boyce MS, Creasey R, Hebblewhite M, et al. (2010) Differential risk effects of wolves on wild versus domestic prey have consequences for conservation. Oikos 119: 1243–1254.
12. Shipley B (2002) Cause and correlation in biology, a user's guide to path analysis, structural equations and causal inference. Cambridge, UK: Cambridge University Press. 332 p.
13. Apps CD, Mclellan BN, Woods JG, Proctor MF (2004) Estimating grizzly bear distribution and abundance relative to habitat and human influence. J Wildl Manage 68: 138–152.
14. Manly BFJ, McDonald LL, Thomas DL (2002) Resource selection by animals, statistical analysis and design for field studies. Nordrecht, The Netherlands: Kluwer. 192 p.
15. Fieberg J, Matthiopoulos J, Hebblewhite M, Boyce MS, Frair JL (2010) Autocorrelation and studies of habitat selection: problem, red herring, or opportunity? Phil Trans R Soc B 365: 2233–2244.
16. StataCorp LP (2009) Stata/SE 10.1 for Windows. College Station, Texas.
17. Lele SR (2009) A new method for estimation of resource selection probability function. J Wildl Manage 73: 122–127.
18. Seaman DE, Powell RA (1996) An evaluation of the accuracy of kernel density estimators for home range analysis. Ecology 77: 2075–2085.
19. Sawyer H, Nielson RM, Lindzey F, Mcdonald LL (2006) Winter habitat selection of mule deer before and during development of a natural gas field. J Wildl Manage 70: 396–403.
20. Hebblewhite M, Merrill EH (2008) Modelling wildlife–human relationships for social species with mixed-effects resource selection models. J Appl Ecol 45: 834–844.
21. Schultz RD, Bailey JA (1978) Responses of national park elk to human activity. J Wildl Manage 42: 91–100.
22. Boyce MS, Vernier PR, Nielsen SE, Schmiegelow FKA (2002) Evaluating resource selection functions. Ecol Modell 157: 281–300.
23. Johnson CJ, Nielsen SE, Merrill EH, McDonald TL, Boyce MS (2006) Resource selection functions based on use-availability data: theoretical motivation and evaluation method. J Wildl Manage 70: 347–357.
24. McDermid GJ, Hall RJ, Sanchez-Azofeifa GA, Franklin SE, Stenhouse GB, et al. (2009) Remote sensing and forest inventory for wildlife habitat assessment. For Ecol Manage 257: 2262–2269.
25. Pettorelli N, Vik JO, Mysterud A, Gaillard JM, Tucker CJ, et al. (2005) Using the satellite derived NDVI to assess ecological responses to environmental change. TREE 20: 503–510.
26. Irving BD, Rutledge PL, Bailey AW, Naeth MA, Chanasyk DS (1995) Grass utilization and grazing distribution within intensively managed fields in Central Alberta. J Range Manage 48: 358–361.
27. Bork EW, Werner SJ (1999) Implications of spatial variability for estimating forage use. J Range Manage 52: 151–156.
28. Joreskog KG Sorbom D (2005) LISREL 8.72. Lincolnwood, IL Scientific Software International, Inc.
29. Grace JB (2006) Structural Equation Modeling and Natural Systems. Cambridge University Press, Cambridge.
30. Kelloway EK (1998) Using LISREL for Structural Equation Modeling. Sage Publications, Thousand Oaks, California.
31. Melis C, Jedrzejewska B, Apollonio M, Barton KA, Jedrzejewski W, et al. (2009) Predation has a greater impact in less productive environments, variation in roe deer, *Capreolus capreolus*, population density across Europe. Global Ecol Biogeogr 18: 724–734.
32. Polis GA (1999) Why are parts of the world green? Multiple factors control productivity and the distribution of biomass. Oikos 86: 3–15.
33. Vitousek PM, Aber JD, Howarth RW, Likens GE, Matson PA, et al. (1997) Human alteration of the global nitrogen cycle, sources and consequences. Ecol Appl 7: 737–750.
34. Musiani M, Mamo C, Boitani L, Callaghan C, Gates CC, et al. (2003) Wolf depredation trends and the use of fladry barriers to protect livestock in western North America. Cons Biol 17: 1538–1547.
35. Sinclair ARE, Ludwig D, Clark CW (2000) Conservation in the real world. Science 289: 1875.
36. Boyce MS, Anderson EM (1999) Evaluating the role of carnivores in the Greater Yellowstone ecosystem. In: Clark TW, Curlee AP, Minta SC, et al., editors. Carnivores in ecosystems: the Yellowstone experience. New Haven Connecticut: Yale University Press. 265–283.

6

Positive Feedbacks in Seagrass Ecosystems – Evidence from Large-Scale Empirical Data

Tjisse van der Heide[1]*, Egbert H. van Nes[2], Marieke M. van Katwijk[3], Han Olff[1], Alfons J. P. Smolders[4]

1 Community and Conservation Ecology Group, Centre for Ecological and Evolutionary Studies (CEES), Groningen University, Haren, The Netherlands, 2 Aquatic Ecology and Water Quality Management Group, Department of Environmental Sciences, Wageningen University, Wageningen, The Netherlands, 3 Department of Environmental Science, Institute for Wetland and Water Research, Faculty of Science, Radboud University Nijmegen, Nijmegen, The Netherlands, 4 Department of Aquatic Ecology and Environmental Biology, Institute for Water and Wetland Research, Faculty of Science, Radboud University Nijmegen, Nijmegen, The Netherlands

Abstract

Positive feedbacks cause a nonlinear response of ecosystems to environmental change and may even cause bistability. Even though the importance of feedback mechanisms has been demonstrated for many types of ecosystems, their identification and quantification is still difficult. Here, we investigated whether positive feedbacks between seagrasses and light conditions are likely in seagrass ecosystems dominated by the temperate seagrass *Zostera marina*. We applied a combination of multiple linear regression and structural equation modeling (SEM) on a dataset containing 83 sites scattered across Western Europe. Results confirmed that a positive feedback between sediment conditions, light conditions and seagrass density is likely to exist in seagrass ecosystems. This feedback indicated that seagrasses are able to trap and stabilize suspended sediments, which in turn improves water clarity and seagrass growth conditions. Furthermore, our analyses demonstrated that effects of eutrophication on light conditions, as indicated by surface water total nitrogen, were on average at least as important as sediment conditions. This suggests that in general, eutrophication might be the most important factor controlling seagrasses in sheltered estuaries, while the seagrass-sediment-light feedback is a dominant mechanism in more exposed areas. Our study demonstrates the potentials of SEM to identify and quantify positive feedbacks mechanisms for ecosystems and other complex systems.

Editor: Tamara Romanuk, Dalhousie University, Canada

Funding: The data presented in this study were collected as part of the project "Macrophytes in Estuarine Gradients (MEG)." This project was funded by the Netherlands Organization of Scientific Research - Earth and Life Sciences (NWO-ALW) as part of the framework "Land-Ocean Interactions in the Coastal Zone (LOICZ)." Website: http://www.nwo.nl/projecten.nsf/pages/1900121750. The funder had no role in study design, data collection and analysis, decision to publish, or preparation of the manuscript.

Competing Interests: The authors have declared that no competing interests exist.

* E-mail: T.van.der.Heide@rug.nl

Introduction

Ecosystems are often characterized by complex biotic and abiotic interactions. Positive feedbacks are a type of interaction that is especially relevant because such mechanisms typically cause a complex, nonlinear response of ecosystems to environmental changes. Moreover, if feedbacks are strong enough, theory suggests that these self-amplifying mechanisms may cause alternative stable states [1,2]. In such systems, small changes or disturbances may push the ecosystem beyond a critical threshold, causing a sudden shift to an alternative state [1,3]. Despite the fact that the relevance of these mechanisms has been demonstrated for a wide range of terrestrial, freshwater and marine ecosystems [1,3], identification of positive feedbacks and quantification of their importance for any particular type of ecosystem (e.g. lakes, salt-marshes, deserts) still remains difficult. Large-scale (cross-ecosystem) datasets can provide hints for bistability – for instance if there are sudden jumps in time series or if the data is bimodally distributed [2]. These types of analyses, however, do not provide any mechanistic explanation for observed trends or distributions. Experiments, on the other hand, can provide convincing evidence for alternative stable states and the underlying mechanisms [2], but cannot indicate whether the identified mechanisms are relevant for the full-scale complexity in the field situation.

Seagrass meadows are a type of ecosystem where positive feedbacks and bistability have been suggested both from small-scale experimental and modeling studies [4,5,6]. These ecosystems are of great importance to many of the World's coastal areas because they enhance biodiversity and provide important ecosystem services (e.g. storm buffering, nutrient cycling, fisheries) [7,8,9]. Over the last century, however, seagrasses have become increasingly affected by human activities, which cause severe declines that are often characterized by sudden losses [8,9]. Seagrasses are ecosystem engineers, in the sense that they significantly modify the abiotic environment of their ecosystem [10]. For instance, seagrasses relieve hydrodynamic stress by attenuating currents and waves [11,12,13] and improve light conditions by reducing suspended sediment levels [14,15,16,17,18] and nutrient concentrations [5,19]. Still, because seagrass ecosystems are typically exposed to multiple environmental stressors (e.g. light limitation, hydrodynamics, salinity) and human disturbances (e.g. siltation events, eutrophication, dredging activities) [8], assessing the importance of positive feedbacks in the field situation is very difficult.

In this study, we quantified a positive feedback between seagrasses and light conditions – probably the most important environmental variable influencing seagrass [20,21] – in systems dominated by the temperate seagrass *Zostera marina*. For this analysis, we used a combination of multiple linear regression and path analysis by structural equation modeling (SEM) [22] on a large-scale dataset containing 83 sampling locations scattered all over Western Europe. We examined the relations between nutrients, sediments, light conditions and seagrass which allowed us to identify a generic positive feedback mechanism between sediment, light conditions and seagrass in this type of ecosystem and quantify its importance relative to other factors influencing light and seagrass growth.

Methods

Data collection

Data were collected in several regions across Western Europe (Fig. 1), where seagrasses are present or were historically present. Within these regions, sites were randomly selected. Each site was sampled once in the growing season (May to September) of 2005. Water depth of the sampling locations varied between 0.5 m above to 5 m below the mean water level. At each site, we recorded the total area covered by seagrass patches and estimated the average shoot density (sh m^{-2}) within patches by counting shoots within a 0.25×0.25 m stainless steel frame (3–5 randomly selected replicates per site). From these observations, we estimated the mean shoot density by multiplying the standardized cover by shoot density. We used the light attenuation coefficient of the water column as a proxy for the light conditions at each site. Light attenuation (m^{-1}) was measured in PAR (400–700 nm) using a quantum light meter (Li-192, Li-Cor). Next, we sampled and pooled 3 replicates of surface water taken approximately 25 to 50 m apart. From these samples, we determined total nitrogen and total phosphorus, which were used as indicators for nutrient availability and growth potential of phytoplankton. Finally, we sampled and pooled 3 replicates of sediment (top 10 cm) from bare areas, i.e. areas outside seagrass meadows (at least 25 m from the edge of the meadow). We used grain size distribution of these samples as a proxy for suspendability of the sediment. We specifically chose to sample sediments outside the meadows, because these are susceptible to erosion, while sediments inside the seagrass beds are trapped and stabilized [14,17,23].

Chemical analyses

Total nitrogen and total phosphorus (μmol l^{-1}) in the surface water were measured as nitrate and ortho-phosphate after digestion with persulfate [24]. The concentration of ortho-phosphate in the digested samples was determined colorimetrically using ammonium-molybdate [25]. Nitrate was determined by sulphanilamide after reduction to nitrite in a cadmium column [26]. Grain size distribution of the sediment was measured on freeze-dried samples by laser diffraction on a Beckman Coulter particle size analyzer. We determined D10, D50 and D90 that describe the grain size (μm) at which 10%, 50%, and 90%, respectively of the volumetric fraction is smaller. All devices were calibrated according to standardized procedures provided by the manufacturers. For all analyses quality assurance measures included blanks, replicate analyses and matrix spikes. Recoveries from matrix spikes ranged from 95% to 107%. Repeated analyses did not reveal differences greater than 5%.

Data analysis

To obtain normal distribution, we applied square root transformation ($y=\sqrt{x}$) to average shoot density. All other variables that were not normally distributed were either logarithmically ($y=\log_{10}(x+1)$) or reciprocally ($y=1/x$) transformed. To

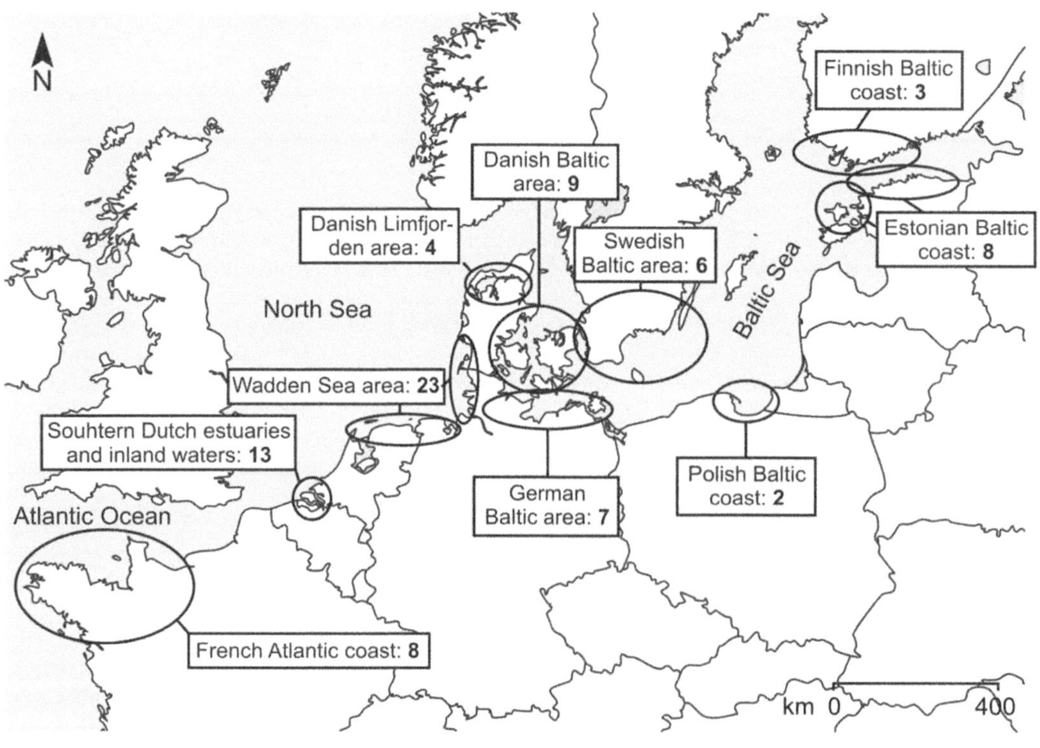

Figure 1. General overview of the geographical setting of the 83 sampled sites. All sites are located in regions where seagrass is present or has been present in the past.

test which of the surface water and sediment variables influenced light attenuation and should therefore be included in the structural equation model, we first examined the relationship between light and these variables by multiple linear regression (MLR) with stepwise backward elimination in the software package SPSS (v18, SPSS inc.). The criterion for the probability of F-to-exclude was >0.05. Subsequently, we tested for correlation between included variables using single linear regression.

Subsequently, we constructed a structural equation (SEM) model in the software package AMOS (v18, SPSS inc.) that included seagrass density, light attenuation and the variables selected by the MLR procedure: D10 and total nitrogen (see paragraph 1 of the results section). Figure 2 shows the conceptual model describing possible relationships between *D10, total nitrogen, light attenuation* and *seagrass density*. Relevance and causality of the relationships between *seagrass density – sediment grain size* and *seagrass density – total nitrogen* are not obvious a priori. Seagrasses might influence surface water total nitrogen through direct uptake of nitrate or ammonia [5,19]. However, total nitrogen might also influence seagrass density directly, for instance by means of ammonia toxicity [5,27]. Similarly, seagrasses may increase the coarseness of the sediment outside the bed by trapping small particles inside the meadow [6,17,18,23,28]. On the other hand, sediment grain size outside the bed might also correlate with seagrass density because of its general characteristics (e.g. nutrient content, organic matter) at the sampling site. Finally, possible relationships for both *seagrass density – total nitrogen* and *seagrass density – sediment grain size* could even be bidirectional, causing direct positive feedbacks. For instance, in eutrophicated systems, high seagrass density could reduce ammonium levels, which would in turn enhance seagrass growth again [5]. In contrast, trapping of small particles from the surrounding area by seagrass could cause a positive feedback in nutrient poor systems because this mechanism would increase nutrient availability in the sediment [17,29]. To elucidate the importance of these possible relations, we constructed a model with bidirectional relationships for these variables. Next, we tested all possible combinations of relationships between *seagrass density – sediment grain size* and *seagrass density – total nitrogen* using the "specification search" option in AMOS and ranked the models according to Akaike's Information Criterion (AIC) and the χ^2 test. We also tested for possible covariance between residuals of *total nitrogen* and *sediment grain size*. Unidentifiable and unstable models (i.e. models with a stability index outside the -1 to 1 range) as well as models with non-significant relations (i.e. coefficients with $p>0.05$) were excluded from the results.

Results

Our 83-site database included 34 locations with *Zostera marina* present. Average shoot densities at these sites varied from less than 1 to over 740 sh/m^2 (mean: 210 (±221 SD) sh/m^2). Of all variables (D10, D50, D90, total nitrogen, total phosphorus) only D10 – the grain size parameter describing the smallest fraction – and total nitrogen were included by the MLR procedure to explain light attenuation (Table 1). The two included variables were able to explain 61% of the variance in light attenuation. Linear regression of total nitrogen versus D10 demonstrated that there was no significant correlation between these two variables ($F=3.919$; $p=0.051$; $R^2=0.046$).

The SEM specification search procedure yielded only three stable models with significant coefficients (Fig. 3 and Table 2). All three models demonstrate a significant positive effect of sediment grain size (D10) and a negative effect of total nitrogen on the reciprocally transformed light attenuation (higher value = increased water clarity). None of the models showed significant covariance between residuals of total nitrogen and sediment grain size. Light conditions had a strong positive effect on seagrass density. Of the three models, the simplest one, that is without direct *seagrass – total nitrogen* and *seagrass – sediment* interactions, ranked third. The two highest-ranking models had a single direct relationship (but in opposite directions) between seagrass density and sediment D10, but not with total nitrogen. The highest-ranking model included a positive feedback between sediment grain size, light conditions and seagrass density. Compared to the second model, the first-ranked model demonstrated lower AIC and Chi-square values, a higher probability level and stronger significance for the regression weight between seagrass density and sediment D10 (Table 2). Despite the fact that in model 2, sediment D10 was added as direct explanatory variable for seagrass density, it was able to explain only slightly more of the variance in seagrass density (54.3%) compared to model 1 and 3 that explained 52.3% and 50.8% respectively. Additionally, results showed that the

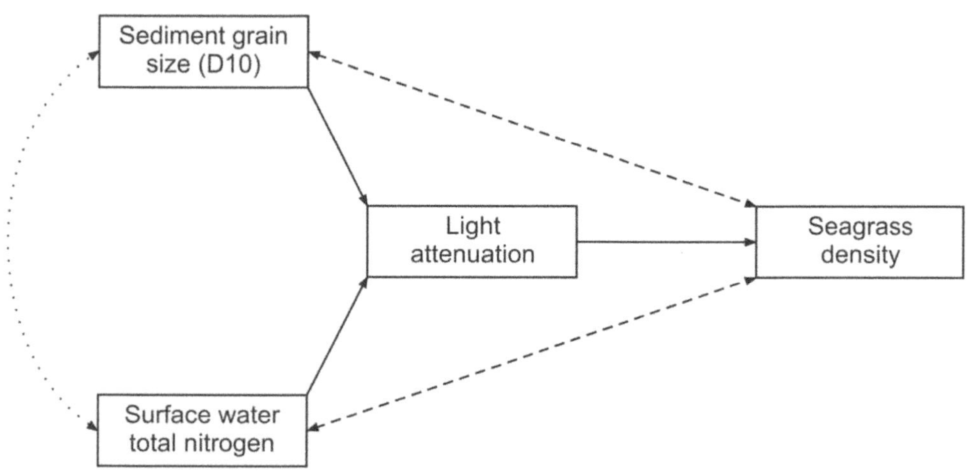

Figure 2. Diagram of a conceptual structural equation (SE) model describing possible relationships between sediment grain size (D10), total nitrogen, light attenuation and seagrass density. Bidirectional dashed arrow lines depict relationships of which the causality and relevance are not sure. We also tested for covariance between residuals of total nitrogen and sediment D10, which is indicated by the dotted arrow line.

Table 1. Results from the multiple linear regression analysis with stepwise backward selection of explanatory variables.

General model statistics			
R^2	*Adj. R^2*	*F*	*p*
0.606	0.596	61.492	<0.000
Parameter statistics			
	Coefficient	*SE*	*p*
Constant	2.606	0.366	<0.000
Sediment D10	0.009	0.001	<0.000
Total nitrogen	−1.338	0.204	<0.000

D10 (μm) was untransformed. Total nitrogen (μmol l^{-1}), total phosphorus (μmol l^{-1}), D50 (μm) and D90 (μm) were logarithmically transformed. Light attenuation (m^{-1}) was reciprocally transformed.
Total phosphorus, D50 and D90 were eliminated from the model by the selection procedure. Thus, only sediment grain size D10 and surface water total nitrogen were included to describe the dependent variable light attenuation.

standardized indirect effect of sediment D10 on seagrass density (0.315) was more important that its direct effect (0.252). Due to the positive feedback loop in model 1, it was able to explain 33.1% of all variance in sediment D10, while R^2 of light attenuation also improved slightly from 0.559 to 0.579. The standardized indirect effect of seagrass density on light in model 1 was 0.187, i.e. an increase of 1 × standard deviation (SD) in seagrass density resulted in an increase of 0.187 SD in the reciprocally transformed light attenuation. The standardized total (= indirect + direct) effect of seagrass density on itself was 0.119, while total effects of total nitrogen, sediment D10 and light attenuation on seagrass density were −0.362, 0.288 and 0.716 respectively.

Discussion

Despite the fact that the importance of positive feedback mechanisms is recognized for many types of ecosystems (e.g. lakes, deserts, coral reefs, salt marshes, seagrasses), identification or quantification of the relative importance of feedbacks in any particular kind of ecosystem is typically difficult [1,2]. By using a combination of multiple linear regression and structural equation modeling, we were able to detect and quantify a positive feedback, which indicated that seagrasses are able to trap and stabilize sufficient amounts of suspended sediments to improve water clarity and thus seagrass growth conditions.

Since our dataset covers a broad geographical area (i.e. Western Europe) with many different types of systems (i.e. open ocean conditions, sheltered estuaries, brackish inland waters) our results suggest that the identified feedback is generic for seagrass ecosystems. Nevertheless, relevance of this feedback probably varies strongly depending on the local abiotic environment. In systems like for instance the Wadden Sea, where turbidity is sediment-dominated, trapping and stabilization of silt particles by seagrass beds may be very important to sustain adequate growing conditions for seagrass itself. In such ecosystems, this positive feedback could potentially lead to alternative stable states [6]. Overall, however, eutrophication, as indicated by total nitrogen, is at least as important as sediment conditions in controlling water clarity and seagrass density. During our survey, estuaries dominated by high nutrient levels (and phytoplankton) were mainly found in the Baltic Sea, the Limfjorden and Dutch brackish inland waters. It appears that eutrophication can result in severe seagrass decline in these systems. However, because these systems are typically sheltered, hydrodynamics are generally low and sediment resuspension is therefore often limited. Hence, the risk of sudden collapse due to disturbance of the light-seagrass-sediment feedback might be less high in these systems compared to for instance the Wadden Sea area. Furthermore, it should be noted, that we found no evidence of any direct feedback between seagrass and nutrients or sediment conditions. However, this does not mean that such mechanisms are unimportant for seagrasses, but merely indicates that they are at least not sufficiently generic in the temperate systems that we studied to be identified by our analyses.

In our analyses, we used total nitrogen, total phosphorus and sediment grain size rather than for instance surface water chlorophyll A and suspended matter concentrations. Although these variables are more direct indicators of light conditions, they would be more circumstantial with regard to the ecosystem engineering mechanisms with which seagrass can modify its surroundings, i.e. reduction of nutrient levels [5,19] and trapping and stabilization of silt [6,17,23]. Also, our choice to sample sediments outside the seagrass meadows instead of inside was based on a similar reasoning. Obviously, sediments inside the meadows are generally more silty than outside, illustrating the potential of seagrasses to trap and stabilize small particles. However, in contrast to bare sediments, these trapped particles hardly erode and therefore do not influence suspended sediment concentrations anymore [14,17,18,23]. Thus, to influence light conditions, ecosystem engineering by seagrasses has to be strong enough to modify sediment grain size distribution outside the meadow by trapping silt particles inside.

Although our analysis provides a strong indication for a positive feedback in *Z. marina* ecosystems, it does not offer solid proof. Notably, the specification search yielded two other models without this feedback that were able to explain a considerable amount of variance in light conditions and seagrass density. However, from our results it seems hard to argue that the highly significant causal relation from seagrass to sediment grain size should be drawn in the opposite direction. Model 2 demonstrates that including sediment as explanatory variable for seagrass improves the fit to seagrass density by only 2.0 (model 1) to 3.5% (model 3), while the other way around seagrass is able to explain over 33% of the variance in sediment D10. Moreover, the difference in AIC between model 1 and the other two models is well over 3 (see table 2), indicating that there is considerably less support for model 2 and 3 compared to model 1 [30].

Structural equation modeling proves to be a useful tool for identifying feedback loops in complex systems like ecosystems.

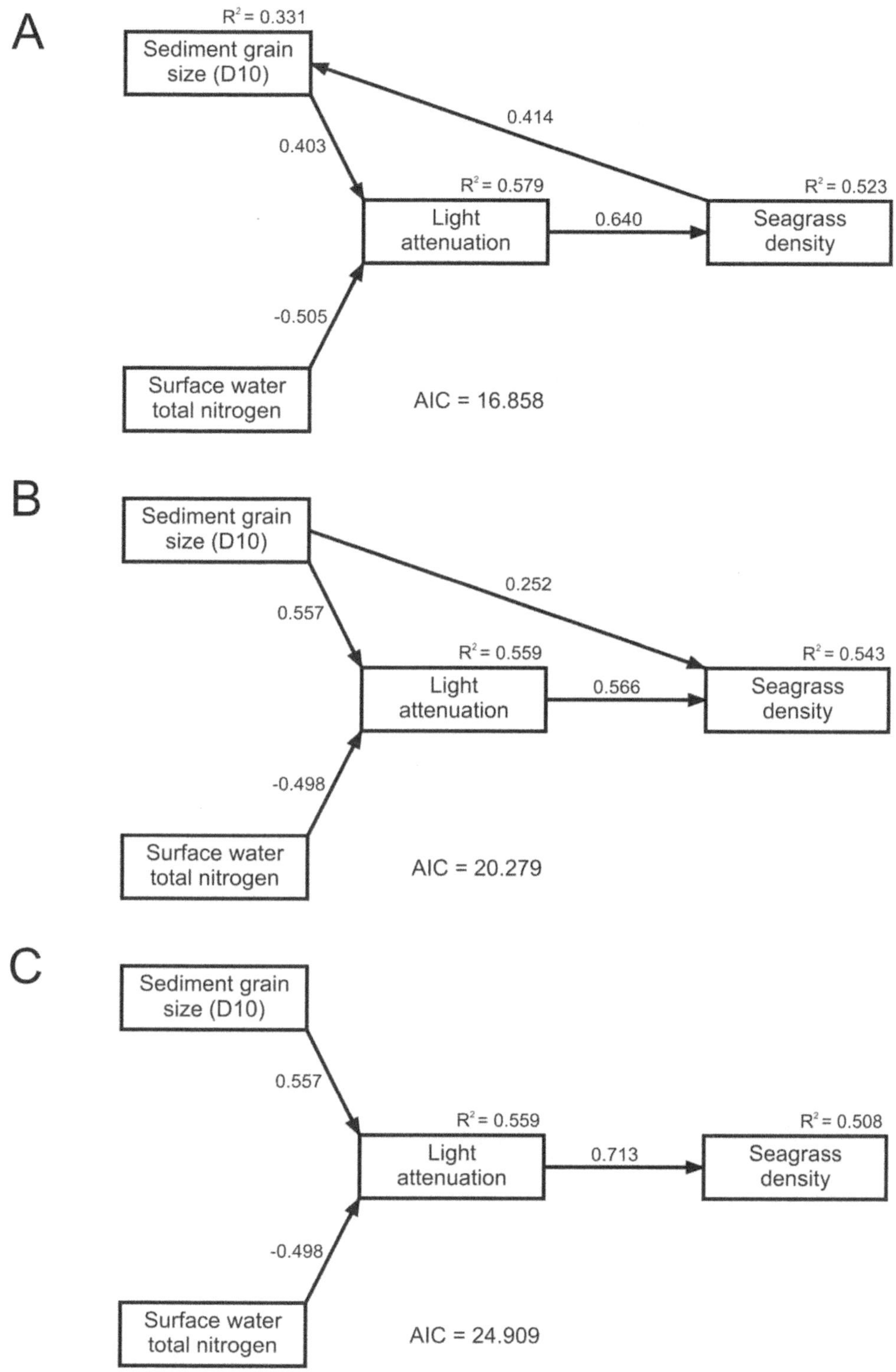

Figure 3. Diagram of the three stable and significant structural equation models. A) Model 1 provided the best fit to the data. It includes a positive feedback loop between sediment grain size (D10), light attenuation and seagrass density. **B**) The second best model (2) describes the relation between seagrass and sediment grain size in the opposite direction compared to model 1. **C**) The third, lowest-ranking model (3) was the simplest of the three model and did not include a direct relation between seagrass and sediment grain size. Note that light attenuation was reciprocally transformed and that the effect of nitrogen is therefore negative, while the effect of sediment D10 is positive (higher value for light attenuation = increased water clarity). Values above the arrow lines depict the standardized regression weights.

However, the method will not be able to determine all feedback loops, as there are problems that are not "identifiable" (i.e. cannot be solved) [31]. The model should preferably include one or more "instrumental variables" (i.e. a variable outside the feedback loop that affects only one of the variables inside the loop) or otherwise the feedback loop should at least be indirect (i.e. involving more

Table 2. Results from the structural equation modeling (SEM) specification search.

	Model 1	Model 2	Model 3
	Model fit summary		
AIC (# parameters)	16.858 (8)	20.279 (8)	24.909 (7)
χ^2	0.858	4.279	10.909
df	2	2	3
Probability level	0.651	0.118	0.012
	Squared multiple correlations (R^2)		
Light attenuation	0.579	0.559	0.559
Seagrass density	0.523	0.543	0.508
Sediment D10	0.331	–	–
	Standardized regression weights (p-value)		
Total nitrogen ≫ Light attenuation	−0.505 (<0.000)	−0.498 (<0.000)	−0.498 (<0.000)
Sediment D10 ≫ Light attenuation	0.403 (<0.000)	0.557 (<0.000)	0.557 (<0.000)
Light attenuation ≫ Seagrass density	0.640 (<0.000)	0.566 (<0.000)	0.713 (<0.000)
Seagrass density ≫ Sediment D10	0.414 (<0.000)	–	–
Sediment D10 ≫ Seagrass density	–	0.252 (0.005)	–

Seagrass density (sh m^{-2}) was square root transformed, light attenuation (m^{-1}) was reciprocally transformed, D10 (μm) was untransformed and total nitrogen (μmol l^{-1}) was logarithmically transformed.
We found no significant direct relation between total nitrogen and seagrass density. For the relation between seagrass density and sediment D10, the relation was strongest when causality was modeled from seagrass density to sediment D10. Note that light attenuation was reciprocally transformed and that effect of nitrogen is therefore negative, while the effect of sediment D10 is positive (higher value for light attenuation = increased water clarity).

than 2 variables). Obviously, the dataset should also have enough variation in conditions to determine the correlations accurately. Here, we have an indirect loop (involving 3 variables), an instrumental variable (nutrients influencing light conditions) and a dataset that includes a wide range of environmental conditions.

In conclusion, we demonstrate that by using structural equation modeling, it is possible to identify positive feedbacks in complex systems such as ecosystems. Additionally the method can provide a mechanistic explanation and quantify its importance relative to other environmental factors. However, we stress here that SEM only provides indications for positive feedback mechanisms, not for alternative stable states nor does it offer any indication for imminent shifts. Therefore, this method should preferably be used in concert with other methods that can hint for bistability like bimodality or hysteresis [2] which can provide indications for catastrophic collapse, such as critical slowing down or spatial patterns [32].

Acknowledgments

We thank J. G. C. Hopcraft and reviewers for their valuable comments on earlier versions of this manuscript. M. Versteeg and D. C. R. Hermus are acknowledged for their help with collecting samples. We are also grateful to R. Peters, M. van der Gaag, and J. Eygensteyn for their help with chemical analyses.

Author Contributions

Conceived and designed the experiments: TH AS EN MK. Performed the experiments: TH. Analyzed the data: TH AS EN. Wrote the paper: TH AS EN MK HO.

References

1. Scheffer M, Carpenter S, Foley JA, Folke C, Walker B (2001) Catastrophic shifts in ecosystems. Nature 413: 591–596.
2. Scheffer M, Carpenter SR (2003) Catastrophic regime shifts in ecosystems: linking theory to observation. Trends in Ecology & Evolution 18: 648–656.
3. Carpenter S (2001) Alternate states of ecosystems: evidence and some implications. In: Press MC, Huntly N, Levin SA, eds. Ecology: achievement and challenge. London: Blackwell. pp 357–381.
4. van der Heide T, Bouma TJ, van Nes EH, van de Koppel J, Scheffer M, et al. (2010) Spatial self-organized patterning in seagrasses along a depth gradient of an intertidal ecosystem. Ecology 91: 362–369.
5. van der Heide T, Smolders AJP, Rijkens BGA, van Nes EH, van Katwijk MM, et al. (2008) Toxicity of reduced nitrogen in eelgrass (Zostera marina) is highly dependent on shoot density and pH. Oecologia 158: 411–419.
6. van der Heide T, van Nes EH, Geerling GW, Smolders AJP, Bouma TJ, et al. (2007) Positive feedbacks in seagrass ecosystems – Implications for success in conservation and restoration. Ecosystems 10: 1311–1322.
7. Barbier EB, Koch EW, Silliman BR, Hacker SD, Wolanski E, et al. (2008) Coastal ecosystem-based management with nonlinear ecological functions and values. Science 319: 321–323.
8. Orth RJ, Carruthers TJB, Dennison WC, Duarte CM, Fourqurean JW, et al. (2006) A global crisis for seagrass ecosystems. Bioscience 56: 987–996.
9. Waycott M, Duarte CM, Carruthers TJB, Orth RJ, Dennison WC, et al. (2009) Accelerating loss of seagrasses across the globe threatens coastal ecosystems. Proceedings of the National Academy of Sciences of the United States of America 106: 12377–12381.
10. Jones CG, Lawton JH, Shachak M (1994) Organisms as Ecosystem Engineers. Oikos 69: 373–386.
11. Bouma TJ, De Vries MB, Low E, Peralta G, Tanczos C, et al. (2005) Trade-offs related to ecosystem engineering: A case study on stiffness of emerging macrophytes. Ecology 86: 2187–2199.
12. Fonseca MS, Cahalan JA (1992) A preliminary evaluation of wave attenuation by 4 species of seagrass. Estuarine Coastal and Shelf Science 35: 565–576.
13. Gambi MC, Nowell ARM, Jumars PA (1990) Flume observations on flow dynamics in Zostera marina (eelgrass) beds. Marine Ecology-Progress Series 61: 159–169.
14. Gacia E, Duarte CM (2001) Sediment retention by a mediterranean Posidonia oceanica meadow: The balance between deposition and resuspension. Estuarine Coastal and Shelf Science 52: 505–514.
15. Kemp WM, Boynton WR, Adolf JE, Boesch DF, Boicourt WC, et al. (2005) Eutrophication of Chesapeake Bay: historical trends and ecological interactions. Marine Ecology-Progress Series 303: 1–29.
16. Ward LG, Kemp WM, Boynton WR (1984) The influence of waves and seagrass communities on suspended particulates in an estuarine embayment. Marine Geology 59: 85–103.
17. Agawin NSR, Duarte GM (2002) Evidence of direct particle trapping by a tropical seagrass meadow. Estuaries 25: 1205–1209.

18. Hendriks IE, Sintes T, Bouma TJ, Duarte CM (2008) Experimental assessment and modeling evaluation of the effects of the seagrass Posidonia oceanica on flow and particle trapping. Marine Ecology-Progress Series 356: 163–173.
19. Moore KA (2004) Influence of seagrasses on water quality in shallow regions of the lower Chesapeake Bay. Journal of Coastal Research. pp 162–178.
20. Duarte CM (1991) Seagrass depth limits. Aquatic Botany 40: 363–377.
21. van der Heide T, Peeters E, Hermus DCR, van Katwijk MM, Roelofs JGM, et al. (2009) Predicting habitat suitability in temperate seagrass ecosystems. Limnology and Oceanography 54: 2018–2024.
22. Grace JB, Anderson TM, Olff H, Scheiner SM (2010) On the specification of structural equation models for ecological systems. Ecological Monographs 80: 67–87.
23. Bos AR, Bouma TJ, de Kort GLJ, Van Katwijk MM (2007) Ecosystem engineering by annual intertidal seagrass beds: sediment accretion and modification. Estuarine Coastal and Shelf Science 74: 344–348.
24. Koroleff F (1983) Simultaneous oxidation of nitrogen and phosphorus compounds by persulfate. In: Grasshoff K, Eberhardt M, Kremling K, eds. Methods of Seawater Analysis. 2 ed. Weinheim, Germany: Verlag Chemie. pp 205–206.
25. Lamers LPM, Tomassen HBM, Roelofs JGM (1998) Sulfate-induced entrophication and phytotoxicity in freshwater wetlands. Environmental Science & Technology 32: 199–205.
26. Wood ED, Armstrong FJA, Richards FA (1967) Determination of nitrate in sea water by cadmium-copper reduction to nitrite. Journal of the Marine Biological Association of the United Kingdom 47: 23–31.
27. van Katwijk MM, Vergeer LHT, Schmitz GHW, Roelofs JGM (1997) Ammonium toxicity in eelgrass Zostera marina. Marine Ecology-Progress Series 157: 159–173.
28. van Katwijk MM, Bos AR, Hermus DCR, Suykerbuyk W (2010) Sediment modification by seagrass beds: Muddification and sandification induced by plant cover and environmental conditions. Estuarine Coastal and Shelf Science 89: 175–181.
29. Barron C, Middelburg JJ, Duarte CM (2006) Phytoplankton trapped within seagrass (Posidonia oceanica) sediments are a nitrogen source: An in situ isotope labeling experiment. Limnology and Oceanography 51: 1648–1653.
30. Burnham KP, Anderson DR (2002) Model Selection and Multimodel Inference: a Practical Information - Theoretic Approach. New York, USA: Springer-Verlag.
31. Arbuckle JL (2007) Amos 16.0 User's Guide. Chicago, USA: SPSS. 637 p.
32. Scheffer M, Bascompte J, Brock WA, Brovkin V, Carpenter SR, et al. (2009) Early-warning signals for critical transitions. Nature 461: 53–59.

7

Organic Carbon Storage in China's Urban Areas

Shuqing Zhao[1]*, Chao Zhu[1], Decheng Zhou[1], Dian Huang[1], Jeremy Werner[2]

1 College of Urban and Environmental Sciences, and Key Laboratory for Earth Surface Processes of the Ministry of Education, Peking University, Beijing, China, 2 U.S. Geological Survey (USGS) Earth Resources Observation and Science (EROS) Center, Sioux Falls, South Dakota, United States of America

Abstract

China has been experiencing rapid urbanization in parallel with its economic boom over the past three decades. To date, the organic carbon storage in China's urban areas has not been quantified. Here, using data compiled from literature review and statistical yearbooks, we estimated that total carbon storage in China's urban areas was 577±60 Tg C (1 Tg $=10^{12}$ g) in 2006. Soil was the largest contributor to total carbon storage (56%), followed by buildings (36%), and vegetation (7%), while carbon storage in humans was relatively small (1%). The carbon density in China's urban areas was 17.1±1.8 kg C m^{-2}, about two times the national average of all lands. The most sensitive variable in estimating urban carbon storage was urban area. Examining urban carbon storages over a wide range of spatial extents in China and in the United States, we found a strong linear relationship between total urban carbon storage and total urban area, with a specific urban carbon storage of 16 Tg C for every 1,000 km^2 urban area. This value might be useful for estimating urban carbon storage at regional to global scales. Our results also showed that the fraction of carbon storage in urban green spaces was still much lower in China relative to western countries, suggesting a great potential to mitigate climate change through urban greening and green spaces management in China.

Editor: Ben Bond-Lamberty, DOE Pacific Northwest National Laboratory, United States of America

Funding: This study was supported by the National Basic Research Program of China on Global Change (#2010CB50600) and the National Natural Science Foundation of China (#41071050 and #31021001). The funders had no role in study design, data collection and analysis, decision to publish, or preparation of the manuscript.

Competing Interests: The authors have declared that no competing interests exist.

* E-mail: sqzhao@urban.pku.edu.cn

Introduction

Urbanization is accelerating worldwide because of rapid population growth and demographic shift from rural to urbanized areas. The proportion of the global population in urban areas rose from 29.4% to 52.1% between 1950 and 2011, and it is expected to increase to 67.2% by 2050 [1]. Although urban areas cover less than 3% of the land surface globally at present [2,3], their impacts are diverse and extended far beyond city boundaries [4,5]. The urbanization has caused serious ecological consequences such as the alteration of local to global biogeochemical cycles [6–8]. Unlike natural ecosystems which are fueled by solar energy, the maintenance of urban ecosystems relies heavily on fossil fuels. As a result, urban areas account for more than 75% of global anthropogenic carbon dioxide (CO_2) emissions [8,9].

On the other hand, urban areas have been found to be a large organic carbon pool. For example, carbon storage in urban and exurban areas of the conterminous United States accounted for 10% of its total land carbon storage in 2000 [10]. The aboveground live biomass carbon stored within the Seattle, WA region was considerably larger than the average US forest carbon stock [11]. The soils in residential turf grass of Baltimore and Denver might accumulate carbon at a rate of about two-fold higher than native ones [12]. There was a substantial amount of carbon stored within aboveground vegetation and soil in Leicester: vegetation carbon density (i.e., carbon storage per unit area) was roughly seven times the average carbon density in the county [13], and soil carbon density was roughly 1.5 times that of regional agricultural land at equivalent soil depths [14]. One important question is to what extent anthropogenic carbon dioxide emissions originated from urban areas can be offset through preserving or increasing carbon storage within urban areas themselves? Unfortunately, the urban carbon balance and its role in global carbon budget remain largely neglected.

As the world's most populous country, China has experienced rapid urbanization in parallel with its economic boom over the past three decades. From 1975 to 2006, the number of cities grew from 193 to 656 [15], and the proportion of urban population increased from 17.4% to 43.9% [16]. Increased economic growth and urbanization means greater energy consumption and therefore more CO_2 emissions. It was reported that fossil-fuel carbon emissions in China increased nearly four-fold from 1980 to 2006, making the country the world's largest contributor to CO_2 emissions for the first time [17]. By 2050, China's population is projected to be 1.3 billion with 77.3% living in cities [1]. A good understanding of China's urban carbon stock and budget is critical to mitigating climate change not just for China, but the entire world. However, to date there are no such data available.

In this study, for the first time, we estimated the organic carbon storage (including both natural and anthropogenic pools) in China's urban areas at national and regional scales for 2006 using data compiled from literature review and statistical yearbooks. We quantified the uncertainty of the estimates and performed a sensitivity analysis. A horizontal comparison with the carbon storage in the urban areas of the United States was also made to shed light on the differences in carbon storage between these two major countries and the likely controlling factors.

Materials and Methods

Study area

There are five levels of administrative units of the cities in China: (1) provincial-level cities, (2) deputy-provincial cities, (3) provincial capitals, (4) prefecture-level cities, and (5) county-level cities [18]. We made a preliminary estimation of carbon storage in 287 cities above county-level of China in a previous study [19]. In this study we considered the cities of mainland China at all levels, and the number of cities was 656 with a total area of 3.37×10^4 km^2 in 2006, which accounted for 0.35% of the country's total land area (9.60×10^6 km^2). We estimated the organic carbon storage in these 656 cities for the year 2006 at national and regional scales with more extensive literature review. The major regions of China consist of North China, Northeast China, East China, Central-south China, Southwest China and Northwest China. The coverage of each of the regions is listed in Table S1.

Data and Methods

Organic carbon in urban area is stored not only in natural pools such as vegetation and soils but also in anthropogenic ones including buildings and humans. We estimated carbon storage of these four pools in China's urban areas largely following the method of Churkina *et al.* [10], but added a procedure to estimate uncertainty and performed a sensitivity analysis.

Carbon storage in vegetation and soils

Vegetation in urban areas mainly consists of urban forest and urban grassland. In this study, we considered urban forest and grassland as one pool (i.e., urban green spaces) because the available data in China's urban areas do not make a distinction between them. We estimated vegetation carbon stock (C_{green}) in urban area with the following equation:

$$C_{green}=Area_{urban}\times\alpha\times D_{green} \quad (1)$$

Where $Area_{urban}$ represents the urban area (m^2), α represents the percentage of green space in urban area (%), and D_{green} represents the vegetation carbon density of green spaces (kg C m^{-2}).

Carbon storage in soils (C_{soil}) was calculated as a sum of carbon stored in soils beneath impervious surfaces and green spaces, assuming that all urban soils were covered by impervious surfaces and green spaces:

$$C_{soil}=Area_{urban}\times\alpha\times D_{sgreen}+Area_{urban}\times(1-\alpha)\times D_{simp} \quad (2)$$

where D_{sgreen} and D_{simp} represent the carbon density of soils beneath green space and impervious surfaces (kg C m^{-2}), respectively. Since most soil carbon surveys reach to 100 cm in depth wherever possible [20], we estimated carbon storage in urban soils to the same depth to facilitate comparison with other studies.

The information on urban area and percentage of green space in urban areas was obtained from China Urban Construction Statistical Yearbook [21]. The vegetation carbon densities in urban green spaces and carbon densities of soils beneath green spaces and impervious surfaces were compiled from literature review by provinces (municipalities or autonomous regions). For provinces where carbon density data were not available, corresponding carbon density values in the same region were used as a substitute to estimate their carbon storage (Table S1).

Carbon storage in humans

Humans and pets can also store organic carbon. We focused only on carbon storage in humans because carbon storage in pets was equivalent to less than 1% of humans [22]. Carbon storage in humans (C_{hum}) was calculated as follows:

$$C_{hum}=P_{urban}\times Weight_{ave}\times f_1\times f_2 \quad (3)$$

where P_{urban} represents the population in urban area, $Weight_{ave}$ represents the average human body weight (60 kg), f_1 and f_2 is the fraction of dry organic matter in human body (0.3) [22] and the fraction of carbon in dry organic matter (0.5) [23], respectively. The population data in urban areas were obtained from China Population and Employment Statistics Yearbook [24].

Carbon storage in buildings

In a building system, organic carbon is mainly stored in constructive materials, decorated materials, furniture, books, and foods, etc. We focused on carbon storage in constructive materials ($C_{constru}$) and furniture (C_{furn}) assuming carbon stored in other pools was relatively smaller [10]. We distinguished commercial buildings from residential buildings since wood use in the two types are different. First, we calculated the amount of wood used in constructive materials and furniture, and then we transformed the figure to carbon storage using f_2:

$$C_{constru}=A_{resid}\times f_3\times f_4\times f_2+(A_{building}-A_{resid})\times f_5\times f_4\times f_2 \quad (4)$$

$$C_{furn}=N_{set}\times N_{household}\times f_6\times(f_7\times f_8\times f_4\times f_2+f_9\times f_{10}\times f_4\times f_2) \quad (5)$$

where A_{resid} and $A_{building}$ represent the floor area of residential buildings and total buildings in urban areas (10^4 m^2), respectively; N_{set} represents the number of composite furniture owned by per household (set); $N_{household}$ represents the number of household in urban areas (household); f_3 and f_5 represent the wood use per unit of residential buildings floor area and commercial buildings floor area (0.045 and 0.055 m^3·m^{-2}) [25], respectively; f_4 represents the average bulk density of wood (0.4 t·m^{-3}) [26]; f_6 represents the number of pieces per set of composite furniture (10–30, with an average number of 20, assumed based on the communications with furniture dealers); f_7 and f_9 represent the fraction of wood furniture and steel furniture in total composite furniture (0.8 and 0.2) [27], respectively; while f_8 and f_{10} represent the wood use per unit of wood furniture and steel furniture (0.067 and 0.026 m^3·piece^{-1}) [28], respectively. The data on the floor area of residential buildings and total buildings, and the number of composite furniture owned by per household in urban areas were compiled from China Statistical Yearbook [16], China Statistical Yearbook for Regional Economy [29]. The number of household in urban areas was compiled from China Population and Employment Statistics Yearbook [24].

Estimation of Uncertainty

Most of the variables used in the calculation of carbon storage in various pools were poorly studied in urban areas. The estimates of these variables were derived from a few isolated studies or from census reports, and their uncertainty was difficult to quantify but expected to be large. In this study, we assumed the uncertainty of all variables except the soil carbon contents follows the normal distribution with mean of $\overline{v}$ and variance of $(\beta\overline{v})^2$ where β is a constant coefficient indicating the level of uncertainty was proportional to $\overline{v}$. Soil carbon content usually follows a log-normal distribution [30], which was used to represent the distribution of soil carbon content under the green cover and impervious surfaces in this study as follows:

$$v_i \sim lN(\mu_l, \sigma_l^2)$$

With

$$\mu_l = \log(\overline{v}) - k$$

$$\sigma_l = \beta \log(\overline{S})$$

or

$$\sigma_l = \beta \log(\overline{v})$$

where v_i is the estimate of a given soil carbon content variable, $\overline{v}$ is the estimated mean of the variable, $lN(\mu_l, \sigma_l^2)$ is a log-normal distribution, β is the same constant variation coefficient as above, $\overline{S}$ is the given standard deviation of v, and k is a small constant used to transform the mean of the distribution found through Monte Carlo simulation and numerical optimization. The standard deviation of soil carbon content under impervious surfaces was estimated from a national database (Table S1), which was directly used in estimating carbon storage uncertainty as a value for $\overline{S}$. For other variables, a value of 0.15 for β was used to estimate variability using equation 6, as available information was not enough to estimate their standard deviation. Using these methods, a total of 10,000 Monte Carlo simulations per province were executed for the input variables to find total carbon values and uncertainty using equations 1–5. The values for all variables in each of the 10,000 calculations were randomly and independently drawn from the normal or log-normal distributions using a Monte Carlo procedure in R [31].

Results

The total carbon storage in China's urban areas was 577±60 Tg C (95% confidence intervals, hereafter) in 2006 with a total area of 3.37×10^4 km^2 (Table 1). The average carbon density in urban areas was 17.1±1.8 kg C m^{-2}. Soil was the largest carbon pool accounting for 56.1% of the total carbon storage, followed by buildings (35.6%) and vegetation (7.3%), while humans pool was relatively small (1%) (Figure 1 and Table 2).

At the regional scale, cities in the East and Central-south regions stored the largest amount of organic carbon, with an estimate of 181±24 Tg C and 147±32 Tg C, accounting for 31.4% and 25.5% of total national urban carbon storage, respectively. The Northeast and North regions also played a significant role in organic carbon storage of China's urban areas, shared 15.4% and 12.2% of the total, respectively, while the urban areas in Southwest and Northwest regions had the lowest carbon storage (Table 1).

Carbon storage in urban areas varied significantly among different provinces (Figure 2), ranging from 0.8±0.4 Tg C in Tibet to 60.7±29 Tg C in Guangdong. In general, urban areas in the eastern provinces stored more organic carbon than western provinces. In most provinces (23 of 31), the carbon storage was below 25 Tg, whereas Guangdong accounted for 10.5% of total carbon storage in China's urban areas.

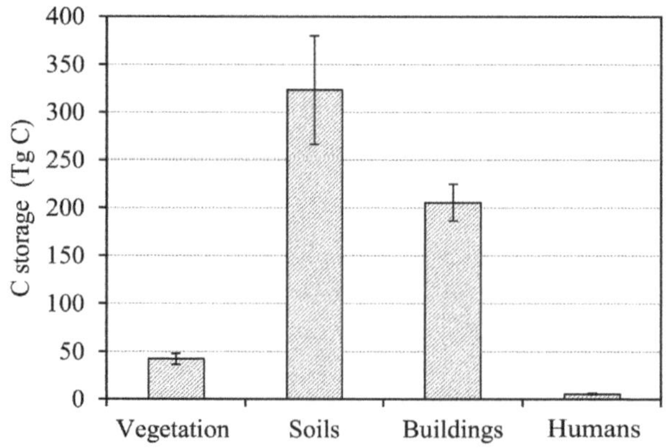

Figure 1. Carbon storage in four major pools in China's urban areas in 2006 at 95% confidence intervals (i.e., the mean±1.96× standard error).

Table 1. Carbon storage and density in urban areas of China in 2006 at national and regional scales at 95% confidence intervals.

Region	Area (km^2)	Total (Tg C)	C Density (kg C m^{-2})
North China	4775	70.2±12.8	14.7±2.7
Northeast China	4340	88.8± 38.4	20.5± 8.9
East China	10756	181.4±23.8	16.9± 2.2
Central-south China	8691	147.1±32.4	16.9±3.7
Southwest China	2929	61.7±18.2	21.1±6.2
Northwest China	2205	27.8±5.3	12.6±2.4
China	33697	577.0±60.1	17.1±1.8

Table 2. The proportion of total carbon in urban areas of China stored in four major carbon pools at regional scales and for the country as a whole for the year 2006.

Region	Vegetation (%)	Soil (%)	Humans (%)	Buildings (%)
North China	4.9	55.3	0.8	39.0
Northeast China	7.1	69.8	0.7	22.4
East China	10.3	49.9	1.0	38.8
Central-south China	6.9	55.7	1.1	36.3
Southwest China	4.8	57.5	1.0	36.7
Northwest China	3.0	53.1	1.1	42.8
China	7.3	56.1	1.0	35.6

The contribution of each pool to total carbon storage in each region was similar to the national pattern (Table 2). However, there were some differences in magnitude. The proportion of

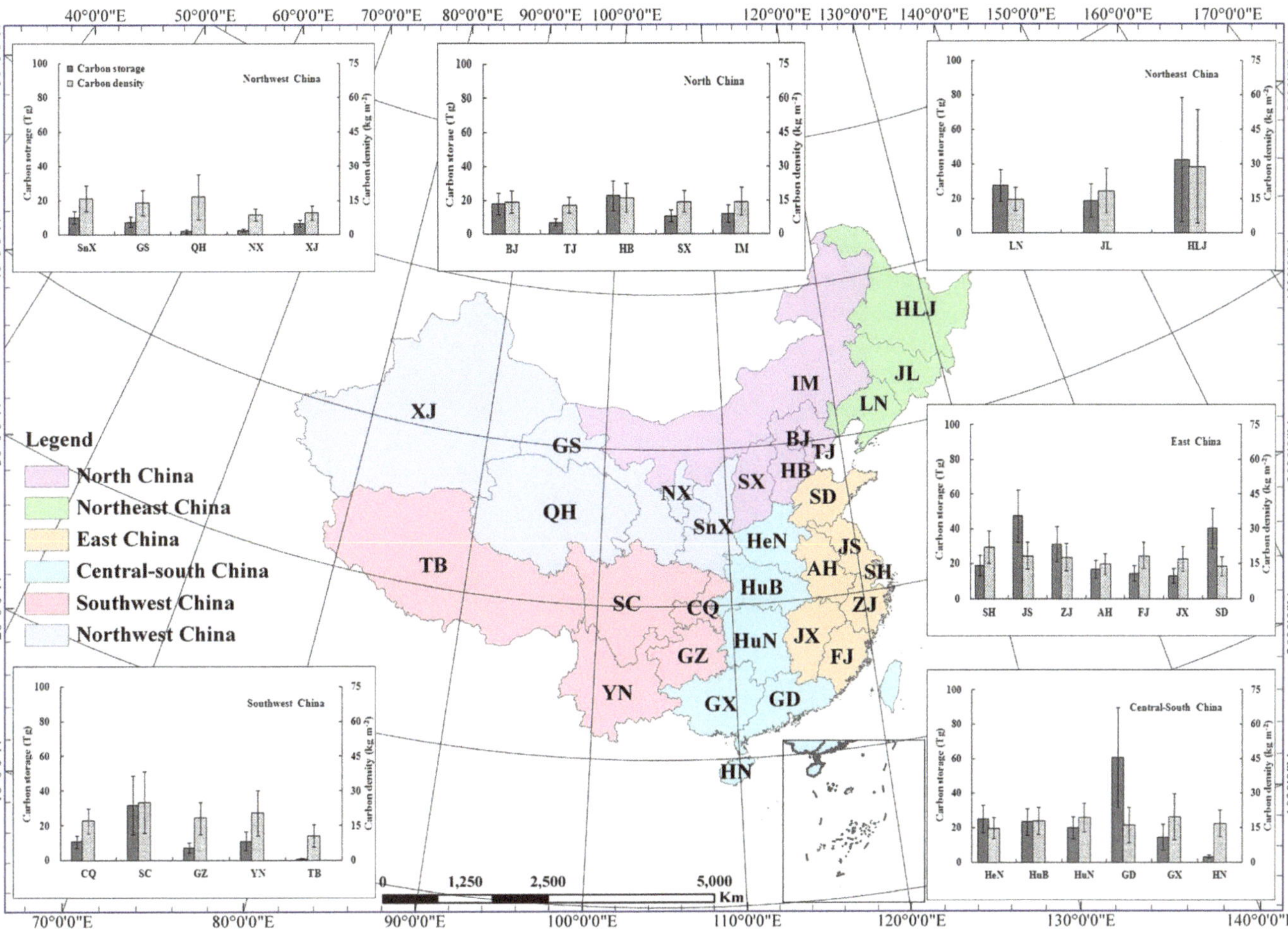

Figure 2. Carbon storage (Tg) and density (kg C m^{-2}) in the provinces (municipalities or autonomous regions) in China's urban areas in 2006 (except Taiwan, Hong Kong and Macao). Carbon storage is the sum of four pools: vegetation, soils, humans and buildings. The confidence intervals are the same as shown in Figure 1.

carbon storage in vegetation was highest in the East (10.3%) and lowest in the Northwest (3.0%). The fraction of carbon stored in soil in the Northeast (69.8%) was much higher than the national average (56.1%).

The average carbon density in urban areas increased from 12.6±2.4 kg C m^{-2} in Northwest to 21.2±6.2 kg C m^{-2} in Southwest (Table 1). The largest carbon density was found in Helongjiang at 28.9±24.6 kg C m^{-2}, and the lowest was found in Ningxia with 8.6±2.7 kg C m^{-2}, with most provinces (26 of 31) having carbon density between 12 kg C m^{-2} and 22 kg C m^{-2} (Figure 2).

Figure 3 shows the relative response or sensitivity of the total urban carbon storage on 10 percent increase of the each of the input variables. The relative sensitivities of all variables used in estimating urban carbon storage in China were smaller than 10 percent, indicating the uncertainties of these variables would be contracted in the estimated carbon storage. The most sensitive variable was urban area with a sensitivity of 6.3 percent, followed by the second group of variables (i.e., impervious area, carbon fraction in wood products, and bulk density of wood) with sensitivity ranging from 3.5 to 3.9 percent. The third group of variables with a sensitivity varying between 1.1 to 1.8 percent included wood use per unit of residential buildings floor area and commercial buildings floor area, and soil C density beneath green space. The sensitivities of the rest of the variables were all smaller than 1 percent with five of the 21 variables being negative. The sensitivity analysis clearly suggested that reducing the uncertainty in the estimated urban area, impervious area or green space, carbon fraction in wood products, and bulk density of wood might be the most cost-effective in reducing the total uncertainty in the estimated urban carbon storage in China.

Discussion

Our estimate of organic carbon stored in China's urban areas was 577±60 Tg C or ranged from 517 to 637 Tg C in 2006, while the total carbon storage in China's terrestrial ecosystems was estimated to be 77.4 Pg C (1 Pg $=10^{15}$ g) [32–34]. This suggests that urban areas accounted for 0.74% of the country's terrestrial ecosystems carbon storage, yet they represent only 0.35% of its land area. In other words, the carbon density in urban areas was 2.1 times the national average carbon density.

At both the national and regional scales, soils accounted for most of the total urban carbon storage, followed by buildings, vegetation, and humans. However, the partitioning varied regionally. Urban soils in the Northeast had the highest proportion of carbon storage across China, signifying the relatively larger impact of soil storage in this region than others. This is consistent with the finding from natural ecosystems that soils under cold climate generally have higher carbon densities [35,36]. The

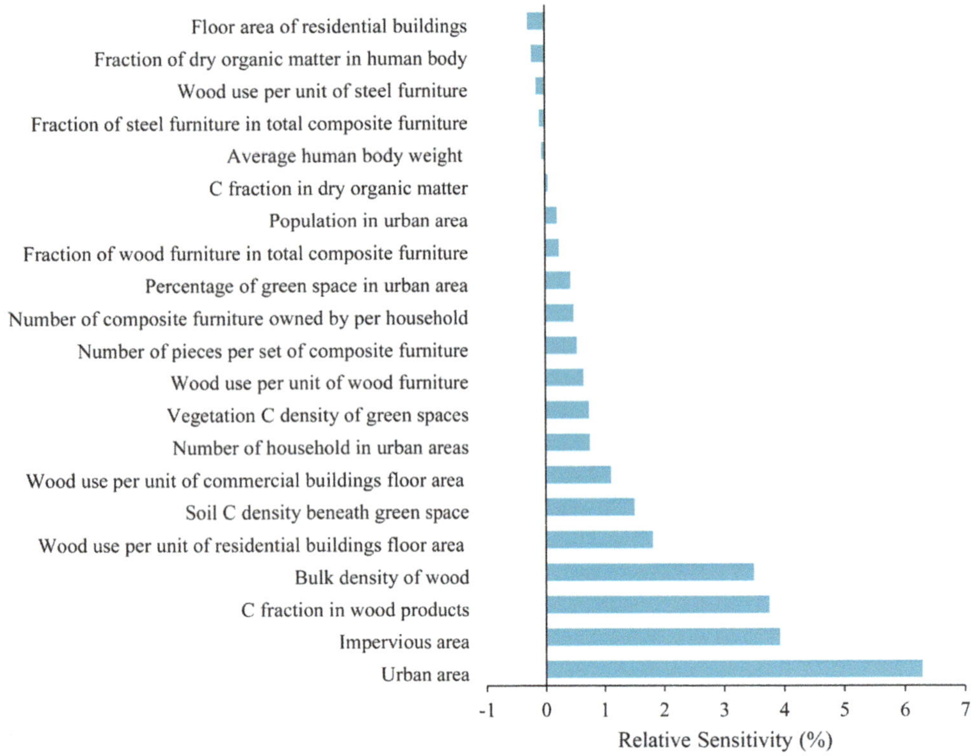

Figure 3. Sensitivity of total urban carbon storage to a 10% increase in the each of the input variables.

highest proportion of vegetation carbon storage was observed in the East where both urban green space coverage and vegetation carbon density were relatively and simultaneously high, and the lowest proportion was found in the Northwest region where both urban green space coverage and vegetation carbon density were the lowest. Regional differences in vegetation carbon storage can be attributed to climate, soil, and habitat conditions as well as urbanization process and history [33].

Urban carbon storage strongly depends on urban land area and less on carbon density (Figure 4). The regions or provinces with larger urban land areas usually stored more carbon (Table 1 and Figure 4a). East and Central-south regions accounted for 57.7% of total urban land areas and stored 56.9% of total carbon. Guangdong province was the largest in urban areas among all provinces (11.1% of national urban areas), contributing most to the national urban carbon storage (10.5%). In contrast, Northwest region only stored 4.8% of total carbon with 6.5% of total urban area. Tibet, the least urbanized province in China, had the lowest urban carbon storage. These results suggest that urban carbon storage increase in China is more likely to be caused by urban expansion. Considering the rapid urbanization rate in China, the role of urban areas in carbon cycle of terrestrial ecosystems will become more significant. Urban carbon storage is also related to its carbon density, but this relationship is much weaker than that between storage and area (Figures 4a and b).

Similarity and differences of urban carbon storage and its partitioning among various carbon pools between China and the United States might provide insights for future efforts in estimating urban carbon storage in other countries. The urban carbon storage in China estimated in this study is about 38% of that in the United States. This suggests that urban carbon densities between these two countries were comparable as the urban area in China was 36% of that in the United States (Figure 5a). The partitioning of total carbon storage in China was different from that in the United States (Figure 5b). Obviously, the contributions of vegetation pool and buildings to total urban organic carbon storage in the United States were higher than those in China. This can probably be explained by the following two factors. First, urban greening has been practiced for many years in the United States. In contrast, urban greening has just gained increasing attention in recent years in China [37]. The coverage of green spaces in the urban areas of the United States was 73% overall with 27% covered by trees and 46%, with large uncertainty, by other green covers [10,38,39]. On the other hand, the total green space (including both forests and grasses) in China's urban areas was only 31% in 2006 [21], less than half of its counterpart in the United States. Second, in addition to the difference in the number or area of buildings, differences in other factors such as wood use in constructions and furniture contributed to the disparity in carbon storage in buildings between these two countries. For example, wood use in construction materials was 18–22 kg m^{-2} in China, less than half of that (40–130 kg m^{-2}) in the United State [10,28]. Our estimate of organic carbon storage in buildings in China's urban areas was 0.34×10^6 g C per capita, which was less than thos× of industrialized countries (2.1×10^6 g C per capita) and higher than those of less industrialized countries (0.15×10^6 g C per capita) [40]. In contrast, the organic carbon stored in buildings in the United States amounted to 3.06×10^6 g C per capita [10].

In this study, we made a preliminary estimation on organic carbon storage in China's urban areas based on the data compiled from literature review and statistical yearbooks. However, the result remains uncertain mainly due to uncertainty from the input data (both in statistical data and measured data). For instance, these data did not have a good measure on the quality of green space (i.e., the type of green space, and carbon density in

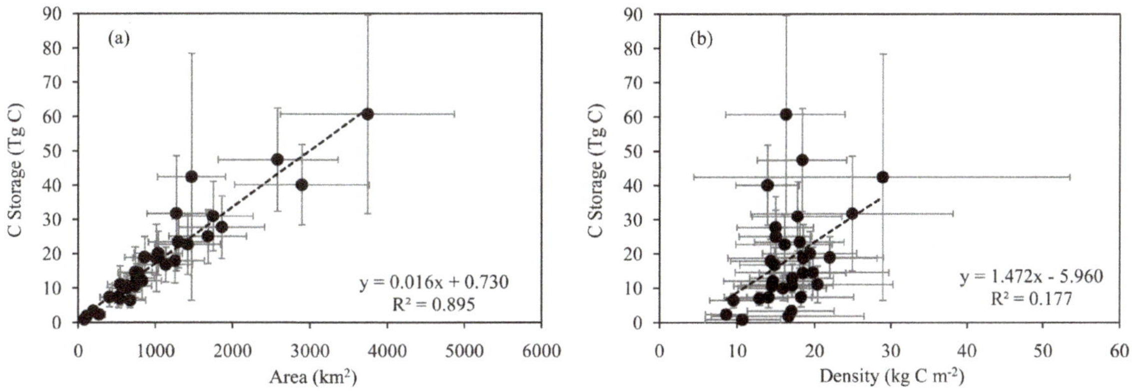

Figure 4. Relationship between urban carbon storage and urban area (a) or urban carbon density (b) at the provincial level. The confidence intervals are the same as shown in Figure 1.

vegetation and soils for each type). Most statistical data in China do not distinguish urban forest from urban grassland, although carbon density in urban forest might be very different from that in urban grassland. For example, previous studies documented that aboveground carbon density of urban forest in the United States was 8.6 kg C m^{-2}, about 170 times as large as its counterpart in urban grassland [10,41]. Wen *et al.* [42] found that biomass of urban trees and urban grassland in the city of Taizhou, East China, was 2.48 and 0.44 kg C m^{-2}, respectively. Future differentiation of urban forests from grasslands using remote sensing or inventory approaches may reduce the uncertainty.

Since population statistics in China mainly based on *hukou* (household registration system) population data which does not include floating population in cities [43], the numbers are often underestimated. However, the impact on estimating total urban carbon storage is trivial as the carbon storage in humans only accounts for 1% of the total urban carbon storage. This observation applies to cities in both China and the United States (Figure 5b), and probably applies to other cities in the world as well. Carbon storage in humans accounts for about 0.1% of the total urban carbon storage in the United States. Because of the very small contribution of humans to total urban carbon storage in general, we advocate that this term be dropped from future efforts in estimating urban carbon storage.

The slope of the strong linear relationship in Figure 4a indicates that on average the total carbon storage is 16 Tg C for every 1,000 km^2 urban area, which is very close to 15.9 Tg C per 1,000 km^2 urban area found in the United States [10]. This specific urban storage capacity might be extrapolated to other regions of the world as the relationship covered a wide range of regions from arid to moist and from developing to developed countries. Using this specific urban carbon storage capacity and the minimum and maximum estimates of global urban areas (276,000–3,524,000 km^2) [3], we calculated that the global total urban carbon storage would be within the range of 4.4 to 56.4 Pg C. The large bound was caused by the range of estimates of the global urban area. Sensitivity analysis results also suggested that urban carbon storage in China was most sensitive to total urban area (Figure 3). Therefore, reducing uncertainty in total urban area might be the most effective way to decrease uncertainty in urban carbon storage estimates.

Urban greening could be a practical approach to contribute to achieve policy targets of mitigating climate change compared with other direct greenhouse gas (GHG) emission reduction strategies [13,44]. Our results showed that the fraction of carbon storage in urban green spaces was still much lower in China than in western countries. Therefore increasing carbon storage in urban green spaces in China has great potential to mitigate carbon emissions in urban areas and also contribute to the target of reducing China's GHG emissions per unit of gross domestic product (GDP) by 40 to 45% in 2020 as compared to that of 2005 [45]. This highlights the importance of urban greening and green spaces management in China.

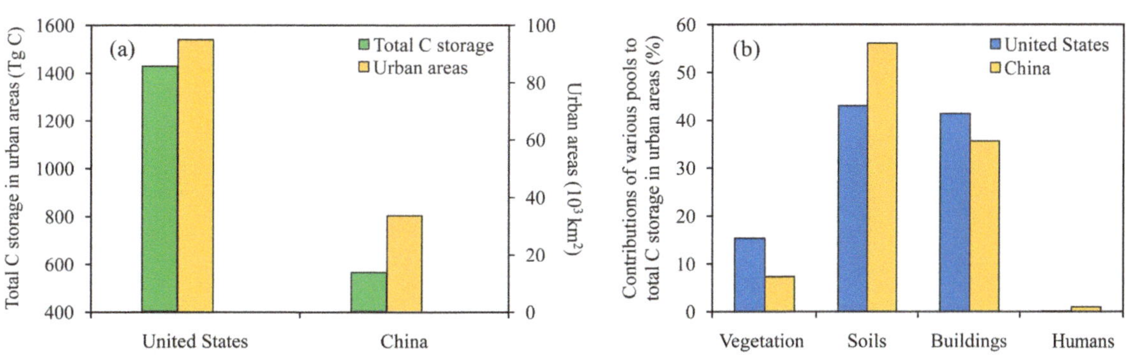

Figure 5. Comparisons of the total urban carbon storage and urban areas (a), and the fractions of four major carbon pools (b) between China and the Conterminous United States.

Conclusions

This study estimated for the first time that the organic carbon storage in the urban systems of China ranged from 517 to 637 Tg C (95% confidence interval) in soils, vegetation, buildings, and humans in the year 2006. Soil accounted for 56% of the total carbon storage, followed by buildings (36%), and vegetation (7%). The weight of humans was only about 1% of the total organic carbon storage and therefore could be ignored in the study of carbon storage in urban systems.

Several lines of evidence suggest the increasing importance of urban systems in regional and national carbon study and management. First, on a per unit area basis, the urban systems of China stored about two times the national average of all lands, which has largely been ignored in previous studies. Second, China's total urban area will likely continue to increase due to continued economic development. Third, there is still room for carbon storage enhancement in the urban green spaces of China compared with its equivalent in the United States.

Some major future research directions have been identified through uncertainty and sensitivity analysis. Uncertainty analysis indicates that urban soil carbon storage had the largest uncertainty, followed by buildings and vegetation. It is therefore important to systematically collect field samples to measure the spatial variability of soil carbon density under various cover types across geographic regions. In addition, sensitivity analysis indicates carbon storage estimate was most sensitive to the accuracy of the estimates of urban area and impervious area (or the area of green space). The quality of green space, represented by the carbon density in vegetation and soils, was also an important factor for reducing uncertainty in estimating carbon storage in China's urban systems.

Acknowledgments

We are grateful to Dr. DS Yu for providing SOC data, SG Liu, YG Chen and ZY Tang for providing constructive suggestions, and YH Yang, GL Zhang and MK Zhang for technical assistance.

Author Contributions

Conceived and designed the experiments: SQZ. Performed the experiments: SQZ CZ. Analyzed the data: SQZ CZ DCZ DH JW. Wrote the paper: SQZ CZ.

References

1. United Nations (2012) World Urbanization Prospects: The 2011 Revision. New York: United Nations Department of Economic and Social Affairs, Population Division.
2. Center for International Earth Science Information Network (CIESIN) (2004) Global Rural-Urban Mapping Project (GRUMP): Urban extents, [digital version]. Palisades, NY: CIESIN, Columbia University.
3. Schneider A, Friedl M, Potere D (2009) A new map of global urban extent from MODIS satellite data. Environmental Research Letters 4: 044003, doi:10.1088/1748-9326/4/4/044003.
4. Odum HT (1971) Environment, Power and Society. New York: Wiley-Interscience. 331 p.
5. Folke C, Jansson A, Larsson J, Costanza R (1997) Ecosystem appropriation by cities. Ambio 26: 167–172.
6. Kaye JP, McCulley RL, Burke IC (2005) Carbon fluxes, nitrogen cycling, and soil microbial communities in adjacent urban, native and agricultural ecosystems. Global Change Biology 11: 575–587.
7. Pataki DE, Alig RJ, Fung AS, Golubiewski NE, Kennedy CA, et al. (2006) Urban ecosystems and the North American carbon cycle. Global Change Biology 12: 2092–2102.
8. Grimm NB, Faeth SH, Golubiewski NE, Redman CL, Wu J, et al. (2008) Global change and the ecology of cities. Science 319: 756–760.
9. Brown LR (2001) Eco-Economy: Building an Economy for the Earth. New York: WW Norton & Company. 333 p.
10. Churkina G, Brown DG, Keoleian G (2010) Carbon stored in human settlements: the conterminous United States. Global Change Biology 16: 135–143.
11. Hutyra LR, Yoon B, Alberti M (2011) Terrestrial carbon stocks across a gradient of urbanization: a study of the Seattle, WA region. Global Change Biology 17: 783–797.
12. Pouyat RV, Yesilonis ID, Golubiewski NE (2009) A comparison of soil organic carbon stocks between residential turf grass and native soil. Urban Ecosystems 12: 45–62.
13. Davies ZG, Edmondson JL, Heinemeyer A, Leake JR, Gaston KJ (2011) Mapping an urban ecosystem service: quantifying above-ground carbon storage at a city-wide scale. Journal of Applied Ecology 48: 1125–1134.
14. Edmondson JL, Davies ZG, McHugh N, Gaston KJ, Leake JR (2012) Organic carbon hidden in urban ecosystems. Scientific reports 2: 963, DOI:10.1038/srep00963.
15. State Statistical Bureau (SSB) (2007a) China City Statistical Yearbook. Beijing: China Statistics Press.
16. State Statistical Bureau (SSB) (2007b) China Statistical Yearbook. Beijing: China Statistics Press.
17. Gregg JS, Andres RJ, Marland G (2008) China: Emissions pattern of the world leader in CO_2 emissions from fossil fuel consumption and cement production. Geophysical Research Letters 35: L08806, doi:10.1029/2007GL032887.
18. Chan KW (2010) Fundamentals of China's Urbanization and Policy. China Review-an Interdisciplinary Journal on Greater China 10: 63–93.
19. Zhu C, Zhao SQ, Zhou DC (2012) Organic carbon storage in urban built-up areas of China in 1997–2006. Chinese Journal of Applied Ecology 23:1195–1202.
20. Jobbagy EG, Jackson RB (2000) The vertical distribution of soil organic carbon and its relation to climate and vegetation. Ecological Applications 10: 423–436.
21. Ministry of housing and urban-rural development of China (MOHURD) (2007) China Urban Construction Statistical Yearbook. Beijing: China Architecture and Building Press.
22. Bramryd T (1980) Fluxes and accumulation of organic carbon in urban ecosystems on a global scale. In: Bornkamm R, Lee JA, Seaward MRD, editors. Urban Ecology. Oxford: Blackwell Scientific Publications. 3–12.
23. Birdsey RA (1992) Carbon storage and accumulation in United States forest ecosystems. Washington, D.C., US Department of Agriculture Forest Service.
24. State Statistical Bureau (SSB) (2007c) China Population and Employment Statistics Yearbook. Beijing: China Statistics Press.
25. Liu AM, Li F, Liao JG (2000) Study on the balance of supply and demand of timber and forest resources in China. Resources Science 22: 9–13.
26. Fang JY, Liu GH, Xu SL (1996a) Carbon cycling in terrestrial ecosystem of China and its influence on global climate change. In: Wang GC, Wen YP, editors. Emissions and Their Relevant Processes of Greenhouse Gases in China. Beijing: China Environmental Science Press. 129–139.
27. Xu MQ (2004) Chinese furniture industry: marketplace and resource. International Wood Industry: 3–7.
28. Hu MX, Chen TS, Lou NY (1993) Forecast for structural demand of timber in China in the year 2000. Journal of Beijing Forestry University 15: 123–131.
29. State Statistical Bureau (SSB) (2007d) China Statistical Yearbook for Regional Economy. Beijing: China Statistics Press.
30. Brejda JJ, Moorman TB, Karlen DL, Smith JL, Allan DL, et al. (2000) Distribution and variability of surface soil properties at a regional scale. Soil Science Society of America Journal 64: 974–982.
31. R Core Team (2012) R: A language and environment for statistical computing. R Foundation for Statistical Computing, Vienna, Austria. ISBN 3-900051-07-0, URL http://www.R-project.org/.
32. Fang JY, Liu GH, Xu SL (1996b) Carbon pools in terrestrial ecosystems in China. In: Wang GC, Wen YP, editors. Emissions and Their Relevant Processes of Greenhouse Gases in China. Beijing: China Environmental Science Press. 109–128.
33. Fang JY, Guo ZD, Piao SL, Chen AP (2007) Terrestrial vegetation carbon sinks in China, 1981–2000. Science in China Series D-Earth Sciences 50: 1341–1350.
34. Yang YH, Mohammat A, Feng JM, Zhou R, Fang JY (2007) Storage, patterns and environmental controls of soil organic carbon in China. Biogeochemistry 84: 131–141.
35. Post WM, Emanuel WR, Zinke PJ, Stangenberger AG (1982) Soil carbon pools and world life zones. Nature 298: 156–159.
36. Wu H, Guo Z, Peng C (2003) Distribution and storage of soil organic carbon in China. Global Biogeochemical Cycles 17: 1048, doi:10.1029/2001GB001844.
37. Jim CY, Chen WY (2009) Ecosystem services and valuation of urban forests in China. Cities 26: 187–194.

38. Nowak DJ, Rowntree RA, McPherson EG, Sisinni SM, Kerkmann ER, et al. (1996) Measuring and analyzing urban tree cover. Landscape and Urban Planning 36: 49–57.
39. Nowak DJ, Noble MH, Sisinni SM, Dwyer JF (2001) People & trees – Assessing the US urban forest resource. Journal of Forestry 99: 37–42.
40. Bramryd T (1983) Human impact on the biogeochemical cycling of carbon between terrestrial ecosystems and the atmosphere. Ecological Bulletins 35: 301–313.
41. Nowak DJ, Crane DE (2002) Carbon storage and sequestration by urban trees in the USA. Environmental Pollution 116: 381–389.
42. Wen JS, GeY, Jiao L, Deng Z P, Peng CH, et al. (2010) Does urban land use decrease carbon sequestration? – A case study in Taizhou, China. Chinese Journal of Applied Ecology 34: 651–660.
43. Chan KW (2007) Misconceptions and complexities in the study of China's cities: Definitions, statistics, and implications. Eurasian Geography and Economics 48: 383–412.
44. Escobedo F, Varela S, Zhao M, Wagner JE, Zipperer W (2010) Analyzing the efficacy of subtropical urban forests in offsetting carbon emissions from cities. Environmental Science & Policy 13: 362–372.
45. Leggett JA, Logan JS, Mackey A (2008) China's greenhouse gas emissions and mitigation policies. Washington, D.C., Congressional Research Service.

8

Effects of an Ecosystem Engineer on Belowground Movement of Microarthropods

Erin K. Cameron*, Heather C. Proctor, Erin M. Bayne

Department of Biological Sciences, University of Alberta, Edmonton, Alberta, Canada

Abstract

Ecosystem engineers affect other species by changing physical environments. Such changes may influence movement of organisms, particularly belowground where soil permeability can restrict dispersal. We investigated whether earthworms, iconic ecosystem engineers, influence microarthropod movement. Our experiment tested whether movement is affected by tunnels (i.e., burrows), earthworm excreta (mucus, castings), or earthworms themselves. Earthworm burrows form tunnel networks that may facilitate movement. This effect may be enhanced by excreta, which could provide resources for microarthropods moving along the network. Earthworms may also promote movement via phoresy. Conversely, negative effects could occur if earthworms alter predator-prey relationships or change competitive interactions between microarthropods. We used microcosms consisting of a box connecting a "source" container in which microarthropods were present and a "destination" container filled with autoclaved soil. Treatments were set up within the boxes, which also contained autoclaved soil, as follows: 1) control with no burrows; 2) artificial burrows with no excreta; 3) abandoned burrows with excreta but no earthworms; and 4) earthworms (*Lumbricus rubellus*) present in burrows. Half of the replicates were sampled once after eight days, while the other half were sampled repeatedly to examine movement over time. Rather than performing classical pairwise comparisons to test our hypotheses, we used AIC_c to assess support for three competing models (presence of tunnels, excreta, and earthworms). More individuals of Collembola, Mesostigmata, and all microarthropods together dispersed when tunnels were present. Models that included excreta and earthworms were less well supported. Total numbers of dispersing Oribatida and Prostigmata+Astigmata were not well explained by any models tested. Further research is needed to examine the impact of soil structure and ecosystem engineering on movement belowground, as the substantial increase in movement of some microarthropods when corridors were present suggests these factors can strongly affect colonization and community assembly.

Editor: Jordi Moya-Larano, Estacion Experimental de Zonas Áridas (CSIC), Spain

Funding: This research was supported by NSERC Discovery Grants to HCP and EMB and a NSERC Vanier Canada Graduate Scholarship to EKC. The funder had no role in study design, data collection and analysis, decision to publish, or preparation of the manuscript.

Competing Interests: The authors have declared that no competing interests exist.

* E-mail: ecameron@ualberta.ca

Introduction

The ability of ecosystem engineers to modify their physical environment means that they often have strong effects on other organisms [1,2]. Changes in habitat structure due to ecosystem engineering can influence resource availability, species' abundances, and community composition [1–3]. Movement of organisms is another key factor that may be affected by engineers, although it has been poorly studied. For example, post-larval dispersal of some macrofauna is greater in beds of the mussel *Mytilus edulis* L. as compared to in bare sediment patches, where turbulence and sediment flux are lower [4]. Modeling suggests that ecosystem engineers can even cause increases in their own rate of spread because of their ability to alter habitat structure [5]. They may also decrease movement, as in the case of shrubs in Mediterranean woodland that act as a physical obstacle to seed dispersal and thereby affect herbaceous plant communities [6].

The effects of ecosystem engineering on movement may be particularly important belowground, where permeability of the soil matrix limits the active movement of many groups [7,8]. Belowground movement is estimated to be approximately four times slower than aboveground locomotion for some microarthropods [9]. However, burrowing by soil macrofauna can increase porosity and create openings that might facilitate the movement of other organisms [7].

Earthworms are frequently cited examples of belowground ecosystem engineers because of their large impacts on soil structure, which can lead to cascading effects on other species, including songbirds [10], amphibians [11,12], and plants [13]. For example, earthworms and their burrowing activity can facilitate the movement and germination of giant ragweed (*Ambrosia trifida* L.) seeds [14]. Effects of earthworms vary depending on whether they dwell in litter (epigeic), in mineral soil (endogeic), or in deep burrows while feeding on leaf litter (anecic) [15]. Their impacts may be especially substantial in ecosystems where they are exotic and there are no native earthworms, as is the case in most of Canada and the northern United States [16–18].

In their native and introduced ranges, earthworms have been observed to significantly affect soil microarthropods [19–22]. Both positive and negative effects on microarthropod abundance and species richness have been reported. Impacts seem to depend on the density of earthworms, to what ecological group they belong, and the microarthropod taxa being examined. For example, microarthropods that are mainly detritivorous or microbivorous

(Collembola, Astigmata, Oribatida) have been suggested to be more negatively affected than the predatory Mesostigmata due to competition or to greater disruption of their food resources by earthworms [22,23]. Earthworms can also have variable effects on microarthropod distributions due to their effects on physicochemical properties of soil, with microarthropods being attracted to earthworms and their excreta in some cases [24–26] but not in others [25–27]. These responses appear to vary depending on the earthworm species and microarthropod species involved [26].

One of the ways in which earthworms are thought to affect microarthropods is by altering their rate of movement [22,23,28]. Some microarthropods can disperse aboveground via wind, water, phoresy, or walking [29], but their movement belowground has been less well studied. Earthworms may influence microarthropod movement via a number of mechanisms. Firstly, earthworm burrows have been suggested to act as corridors for movement of microarthropods by forming an interconnected network of pores within the soil that are easier to move through [22,23,28]. Secondly, microarthropods may be attracted to burrows by the earthworm secretions and excreta (mucus, castings, and urine) that line burrow walls [24,28] and may subsequently move through the burrows in the pursuit of microbial food resources present there [24]. Thirdly, earthworms themselves may transport microarthropods via phoresy, as appears to occur with nematodes [30,31]. Finally, earthworms could consume some microarthropod taxa or negatively affect them via disturbance of the soil or competition for microbial food resources, thereby causing a reduction in movement and abundance [20,22,32].

Here we examine the effects of an epi-endogeic earthworm (*Lumbricus rubellus* Hoffmeister, 1843) on movement of microarthropods in boreal forest soil using a microcosm experiment. *Lumbricus rubellus* lives and feeds in the upper soil layers and is a common invader in Canada and many other countries [33,34]. Our experiment tested whether movement of springtails and mites is affected by the presence of tunnels in the soil, of tunnels lined with excreta, or of earthworms themselves. We predicted microarthropod movement would be greater when there were tunnels in the soil, particularly for Mesostigmata and Collembola, which typically have a larger body size than other microarthropod taxa and hence may be more limited by soil porosity than smaller-bodied taxa [35,36]. Variation in the impacts of earthworms on movement of microarthropod taxa is one factor that could explain changes in community composition of microarthropods in response to earthworms reported in previous studies.

Materials and Methods

Ethics Statement

No specific permits were required for the described field studies, as the location where samples were collected is not privately-owned or protected. The field studies did not involve endangered or protected species. Data collected in this study will be available at https://era.library.ualberta.ca/public/home.

Experimental Design

We used a short-term microcosm experiment to investigate the effects of non-native earthworms on microarthropod movement. Each microcosm consisted of an open-ended 10 cm×4 cm×4 cm box with one 750 mL (~12.5 cm height×9 cm diameter) opaque plastic container and one 120 mL (~5 cm height×5.5 cm diameter) translucent plastic container attached to either end (Figure 1a). The 750 mL container acted as a "source" of microarthropods and the 120 mL container as a "destination". The sides of the plastic containers were cut where the box was attached to allow movement of microarthropods and earthworms across the microcosm. The bottom and sides of each box were constructed from 0.6 cm thick plywood, while the top was 0.3 cm thick acrylic plastic. We drilled two 0.5 cm diameter screw holes in the bottom and top of each box and bolted the top down after the box was filled with soil to limit movement of microarthropods on the soil surface.

Soil was collected from the organic horizon (H layer) of a trembling aspen (*Populus tremuloides* Michx.)/balsam poplar (*Populus balsamifera* L.) forest in the North Saskatchewan River valley in Edmonton, Alberta (53°32′ N 113°33′ W). Earthworms were also collected from this forested area. The soil used to fill the boxes and 120 mL destination containers was sterilized by autoclaving for 1 hour at 121°C. We mixed unsterilized soil and distilled water and sieved the slurry through a 36 μm sieve to obtain water containing microbes but no microarthropods. After allowing the autoclaved soil to cool, the sieved liquid was added to it as a food source for the earthworms and microarthropods. The 750 mL containers were filled with unsterilized soil and acted as the microarthropod source.

Two days before the start of the experiment, we filled the unattached boxes with sterilized soil and set up the treatments within them (Figure 1b). Our experiment included four treatments: 1) a "control" treatment with no earthworms and no tunnels; 2) an "artificial burrow" treatment with no earthworms but two tunnels running the length of the box made by a 0.6 cm thick wooden dowel; 3) an "abandoned burrow" treatment in which two tunnels were made using the dowel and one earthworm was introduced into each tunnel for the 30 hour period immediately prior to the experiment to lay down excreta, and then removed; and 4) an "earthworms present" treatment in which earthworms were introduced into two dowel-made tunnels 30 hours before the experiment began and remained in the microcosms during the experiment. The width of the dowel was approximately equal to the body width of the *L. rubellus* individuals used in the experiment. Although our dowel-made tunnels likely did not mimic the sinuosity of real earthworm burrows, we chose to create the tunnels in this way for all treatments in order to ensure that they were similar lengths across all replicates. For all treatments, the ends of the boxes were capped with aluminum foil that was secured with elastic bands until the microcosms were assembled. Earthworms were washed with distilled water before being added to the boxes. For the abandoned burrow and earthworms present treatments, the earthworms were removed from the boxes after 30 hours by shining a bright light at one end of the box and placing a dark cloth over the other end or by using electric shock as needed. The electric shock was administered to one end of the box using wall current. In the earthworms present treatment, the earthworms were then placed into the source container that was attached to the same box they had been removed from. This was done to test whether microarthropods might travel from the source container to the 120 mL destination container using the earthworms as phoretic hosts. Nothing was used to attract earthworms to the destination containers but, when we removed the containers for sampling, an earthworm was present in the destination container in nine of the twenty-four microcosms in the earthworm treatment.

The experiment was carried out over eight days in October 2011 in a growth chamber located at the University of Alberta with settings of 18°C, relative humidity of 40%, and day length of 14 hours. The eight day time period was chosen to reduce the possibility that reproduction of microarthropods would occur in the destination containers and affect abundances, as we were interested primarily in movement. We randomly assigned the

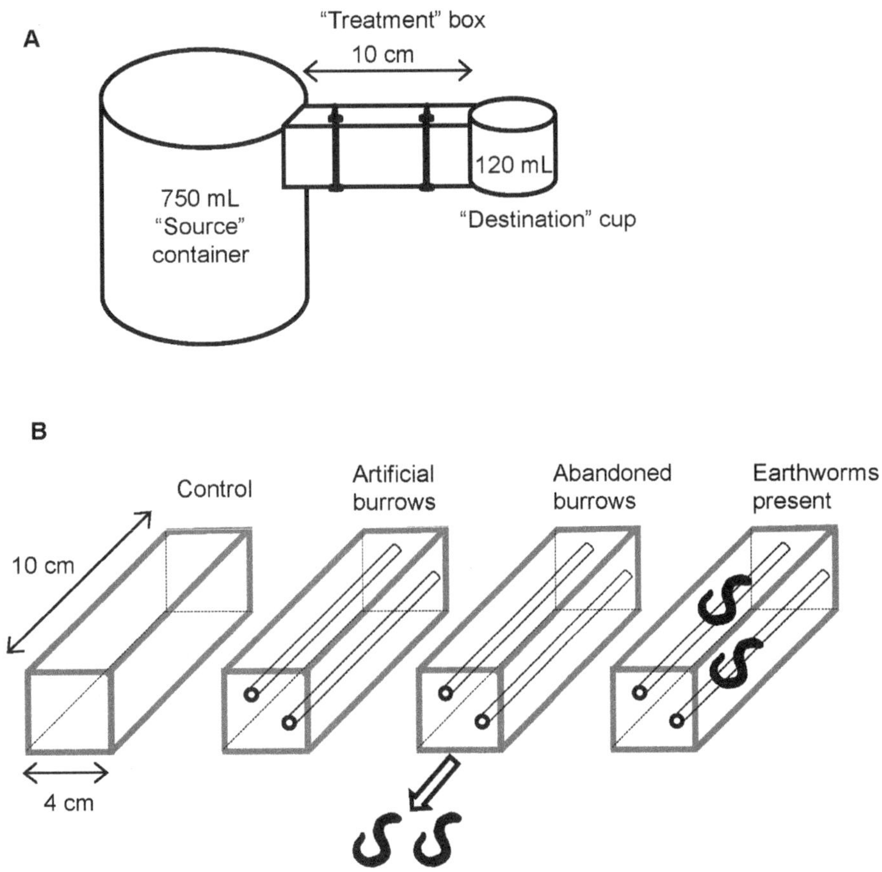

Figure 1. Experimental set-up. (a) A microcosm consisting of a 750 mL "source" container, a 10 cm long "treatment" box in which the treatments were implemented, and a 120 mL "destination" container; (b) The four treatments within the boxes, including the control treatment with no earthworms, the artificial burrows treatment with two tunnels made by a dowel, the abandoned burrows treatment in which earthworms were removed before the experiment, and the earthworms present treatment in which earthworms were present throughout the experiment.

source containers to the treatment boxes and placed the microcosms randomly within the growth chamber. Pin holes were made in the lids of the plastic containers for ventilation, as the lids were closed to prevent escape of microarthropods and earthworms. All microcosms were misted daily to maintain moist conditions.

Microarthropod Sampling

We examined both the total number of microarthropods that dispersed by the end of the experiment and the cumulative movement of microarthropods over time. Each of the four treatments had 24 replicate microcosms, with 12 replicates sampled only at the end of the experiment and 12 replicates sampled at multiple times to investigate movement over time. At each sampling time, we extracted microarthropods from the soil in the destination containers (extraction method described below). For the replicates in which the 120 mL destination container was not removed until the end of the experiment (192 hours), there were a total of 48 extractions (4 treatments$\times$12 replicates). This provided cumulative numbers of dispersers over 8 days. For the replicates sampled at multiple times, the destination container was removed from the source container at 4 h, 12 h, 24 h, 48 h, 96 h, and 192 h after the start of the experiment and immediately replaced with a new destination container filled with autoclaved soil (except following removal of the 192 h container at the end of the experiment). Thus, for example, the destination containers removed at 192 h would have had 96 h between the time of their attachment and removal to accumulate dispersing microarthropods, since the previous samples were removed at 96 h. A total of 288 extractions (4 treatments$\times$12 replicates$\times$6 time steps) were performed for this analysis and these repeated samples were used to determine whether there was an effect of treatment on rates of movement over time.

At each time step, we also extracted microarthropods from an additional 120 mL sample of autoclaved soil that was not attached to a box or source container to verify that the autoclaving had indeed sterilized the soil of microarthropods. At the end of the experiment, we extracted a 120 ml sample from each of the 96 750 mL source containers.

Microarthropods were live-extracted at the Royal Alberta Museum in Edmonton using Tullgren funnels [37]. All extractions ran for 7 days and microarthropods were preserved in 80% ethanol. We sorted microarthropods into non-astigmatan Oribatida, Astigmata, Prostigmata, Mesostigmata, and Collembola. Because there were low numbers of similarly sized (very small) Prostigmata and Astigmata, they were grouped together for statistical analyses. We did not identify the mites to finer taxonomic levels because there was inconsistent taxonomic representation among the samples. The majority of Collembola belonged to the Onychiuridae, although representatives of Entomobryidae, Isotomidae, and Hypogastruridae were present as well.

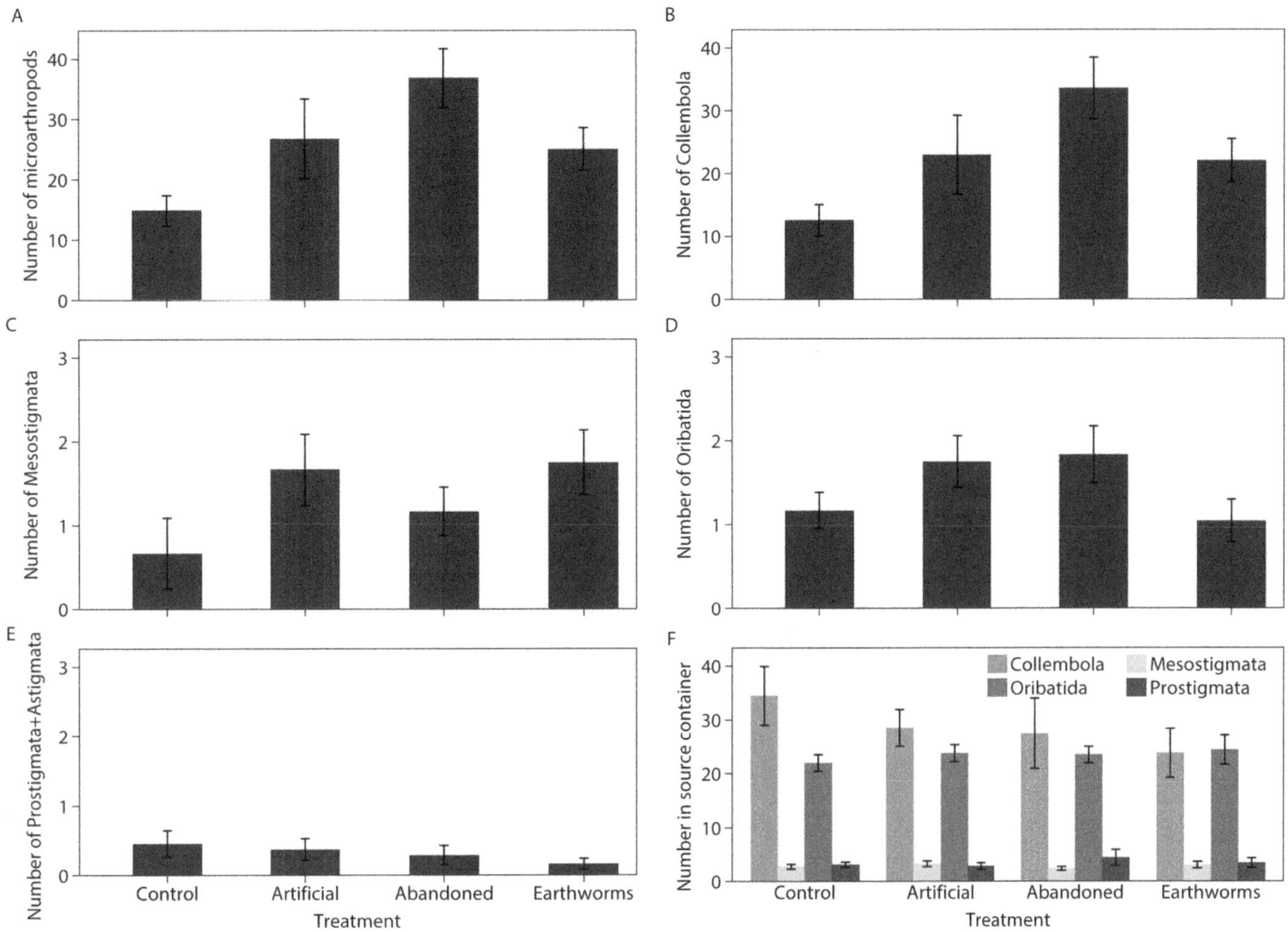

Figure 2. Total number of microarthropods in destination containers at the end of the experiment. (a) All microarthropods together (±SE) in the destination containers; (b) Collembola in the destination containers; (c) Mesostigmata in the destination containers; (d) Oribatida in the destination containers; (e) Prostigmata+Astigmata in the destination containers; and (f) Collembola, Mesostigmata, Oribatida, and Prostigmata+Astigmata in a 120 mL sample from the source containers. N = 24 replicates per treatment.

Statistical Analyses

We assessed the effects of our treatments on the total number of microarthropods dispersing into the destination container using Poisson or negative binomial regression in Stata 9.1 (Stata-Corp, College Station, Texas) depending on the distribution that best fit the data for each group. Negative binomial regression was used when data were overdispersed. Analyses were conducted for Collembola, Mesostigmata, Oribatida, Prostigmata+Astigmata, and all microarthropods together. There were 24 replicates per treatment in all of the models examined, for a total sample size of 96. In 12 of the replicates in each treatment, microarthropods were extracted at six time steps (4 h, 12 h, 24 h, 48 h, 96 h, 192 h) and the total number of microarthropods across all time steps was used in the analysis. Because microarthropod extraction occurred only at the end of the experiment in the remaining 12 replicates per treatment, all models included a variable to account for whether multiple or single extractions were performed for a given replicate microcosm.

We used dummy variable coding to re-categorize our four treatments into groups and AIC_c (Akaike's information criterion, corrected for small sample sizes) to assess support for three competing models explaining microarthropod abundance in the destination containers (see Supporting Materials S1 for an example of dummy variable coding). This approach is analogous to the use of planned comparisons within an ANOVA to test specific hypotheses [38–40]. AIC_c estimates the relative distance between a model and the mechanism that generated the observed data [41]. It relies on information theory and the principle of parsimony to provide a weight of evidence for different hypotheses. We considered models with $\Delta AIC_c < 2$ to be plausible [41]. The models examined were the presence of tunnels (earthworms present, abandoned burrow, and artificial burrow treatments vs. the control), the presence of earthworms during the experiment (earthworms vs. abandoned burrow, artificial burrow, and control), and the presence of earthworm castings/mucus (earthworms and abandoned burrow vs. artificial burrow and control) (see Supporting Materials S1 for an explanation of model parameters). We compared our three models to a global model which included a "treatment" variable (a categorical variable coding the four treatments) and a null model. All of the models, including the null model, had a variable to control for whether single vs. multiple extractions were performed. As well, negative binomial regression was used to test whether abundances of taxa differed among the source containers.

We also investigated whether our treatments influenced the cumulative total number of microarthropods dispersing over time

Table 1. Regression fit statistics for models predicting microarthropod abundance.

Taxa	Model	K	LL	ΔAIC_c	$wAIC_c$
Microarthropods	**Tunnels**	**4**	**−393.80**	**0.00**	**0.52**
	Global	**6**	**−391.68**	**0.26**	**0.45**
	Castings	4	−396.85	6.09	0.02
	Null	3	−399.67	9.55	0.00
	Worms	4	−399.59	11.59	0.00
Collembola	**Tunnels**	**4**	**−387.89**	**0.00**	**0.53**
	Global	**6**	**−385.94**	**0.59**	**0.40**
	Castings	4	−390.15	4.51	0.06
	Null	3	−392.91	7.84	0.01
	Worms	4	−392.85	9.92	0.00
Mesostigmata	**Tunnels**	**4**	**−148.57**	**0.00**	**0.53**
	Null	3	−150.71	2.11	0.18
	Worms	4	−150.09	3.05	0.12
	Global	6	−148.04	3.43	0.09
	Castings	4	−150.50	3.85	0.08
Oribatida	**Worms**	**3**	**−147.68**	**0.00**	**0.36**
	Global	**5**	**−145.58**	**0.20**	**0.32**
	Null	**2 3**	**−149.64**	**1.79**	**0.15**
	Tunnels	3	−148.72	2.09	0.13
	Castings	3	−149.64	3.91	0.05
Prostigmata +Astigmata	**Null**	**3**	**−68.34**	**0.00**	**0.31**
	Worms	**4**	**−67.33**	**0.16**	**0.28**
	Castings	**4**	**−67.51**	**0.52**	**0.24**
	Tunnels	**4**	**−68.02**	**1.56**	**0.14**
	Global	6	−67.14	4.29	0.04

Predictors included presence of tunnels (*Tunnels*), earthworm excreta (*Castings*), and earthworms (*Worms*). All models included a variable to account for whether microarthropods were extracted from a replicate at multiple times versus only once at the end of the experiment. The best model has a ΔAIC_c of zero and the highest $wAIC_c$ value. Models with $\Delta AIC_c<2$ are also considered to be plausible and are shown in bold. With *k*, number of parameters; LL, log likelihood; ΔAIC_c, difference in the Akaike's information criterion (corrected for small sample size) value between model and the most strongly supported model; $wAIC_c$, weight given by the AIC (i.e., relative strength of support for model). See Supporting Materials S1 for further explanation of parameter numbers and dummy variable coding.

(i.e., movement rates) using Poisson and negative binomial regression. This separate analysis had a sample size of 48 (12 replicates per treatment) for all models because time series data were only available for the replicates from which multiple samples were taken. A random effect was included in all models to account for the lack of independence between multiple samples taken from the same replicate over time. AIC_c was again used to rank three competing models (presence of tunnels, presence of earthworms during the experiment, and presence of earthworm castings/ mucus) and to compare them to global and null models. We examined support for these models both with time as an interacting variable and as an independent predictor.

Results

Abundance

A total of 2482 microarthropods were extracted from the destination containers, including 2186 Collembola, 126 Mesostigmata, 139 Oribatida, 27 Prostigmata, and 4 Astigmata. No microarthropods were recovered from the autoclaved soil samples that were not attached to source containers, confirming that our destination containers were indeed free of microarthropods at the beginning of the experiment. Oribatida abundance fit a Poisson distribution, while all other taxa showed evidence of over-dispersion and therefore negative binomial regression was used. According to the ΔAIC_c scores, the tunnel model was the best supported and most parsimonious model for explaining the abundances of Collembola, Mesostigmata and all microarthropods together, although the global model was also well supported for all microarthropods and Collembola (Figure 2a–e; Table 1). For Oribatida abundance, the earthworm presence ($\Delta AIC_c = 0.00$), global ($\Delta AIC_c = 0.20$), and null models ($\Delta AIC_c = 1.79$) received the most support. The null model was the best supported model for Prostigmata+Astigmata, but the worm, castings, and tunnels models also well supported with ΔAIC_c scores less than 2. In the source container samples, the abundance of each taxon did not differ significantly across treatments (Figure 2f; Collembola $\chi^2 = 1.85$, $P = 0.604$; Mesostigmata $\chi^2 = 1.87$, $P = 0.599$; Oribatida $\chi^2 = 0.970$, $P = 0.809$; Prostigmata+Astigmata $\chi^2 = 2.11$, $P = 0.551$).

Cumulative Movement Over Time

Some individuals crossed the 10 cm treatment box and reached the destination containers within the first time step (4 hours) in the artificial burrow and earthworm present treatments for Mesostigmata and in all treatments for Oribatida and Collembola (Figure 3). Therefore, the maximum movement rate for these taxa was 2.5 cm/hour. The first Prostigmata+Astigmata individuals dispersed within 12 hours of the start of the experiment in the artificial burrow treatment, resulting in a maximum movement rate of 0.83 cm/hour. Movement was slower in the control treatment for Mesostigmata and Prostigmata+Astigmata, with the first individuals arriving within 48 hours and 24 hours, respectively. Oribatida and Mesostigmata abundances were analyzed using Poisson regression and negative binomial regression was used for the remaining taxa. For all microarthropods together, the global model that included interactions between time and each of the other variables (tunnel, earthworm, and excreta) was the best supported model explaining cumulative movement over time (Table 2). The tunnel model with no interaction received the most support for Collembola. For Mesostigmata, the tunnel ($\Delta AIC_c = 0.00$), tunnel interaction ($\Delta AIC_c = 1.10$), and global ($\Delta AIC_c = 1.92$) models were best supported. The tunnel ($\Delta AIC_c = 0.00$), global ($\Delta AIC_c = 0.052$), tunnel interaction ($\Delta AIC_c = 0.34$), and earthworm models ($\Delta AIC_c = 0.42$) were the best models for Oribatida. There were too few Prostigmata+Astigmata to allow analysis of their movement over time. Assuming there was no reproduction in the source containers for any of the microarthropod groups, a disproportionately smaller number of Oribatida moved to the destination containers (about $1/10^{th}$ of the totals per treatment) than for the other groups (about 1/3-1/2 of the totals) (Figure 2f versus Figure 2b–e).

Discussion

Our results suggest that activity of *L. rubellus* earthworms can facilitate microarthropod movement. Artificial tunnels in the soil

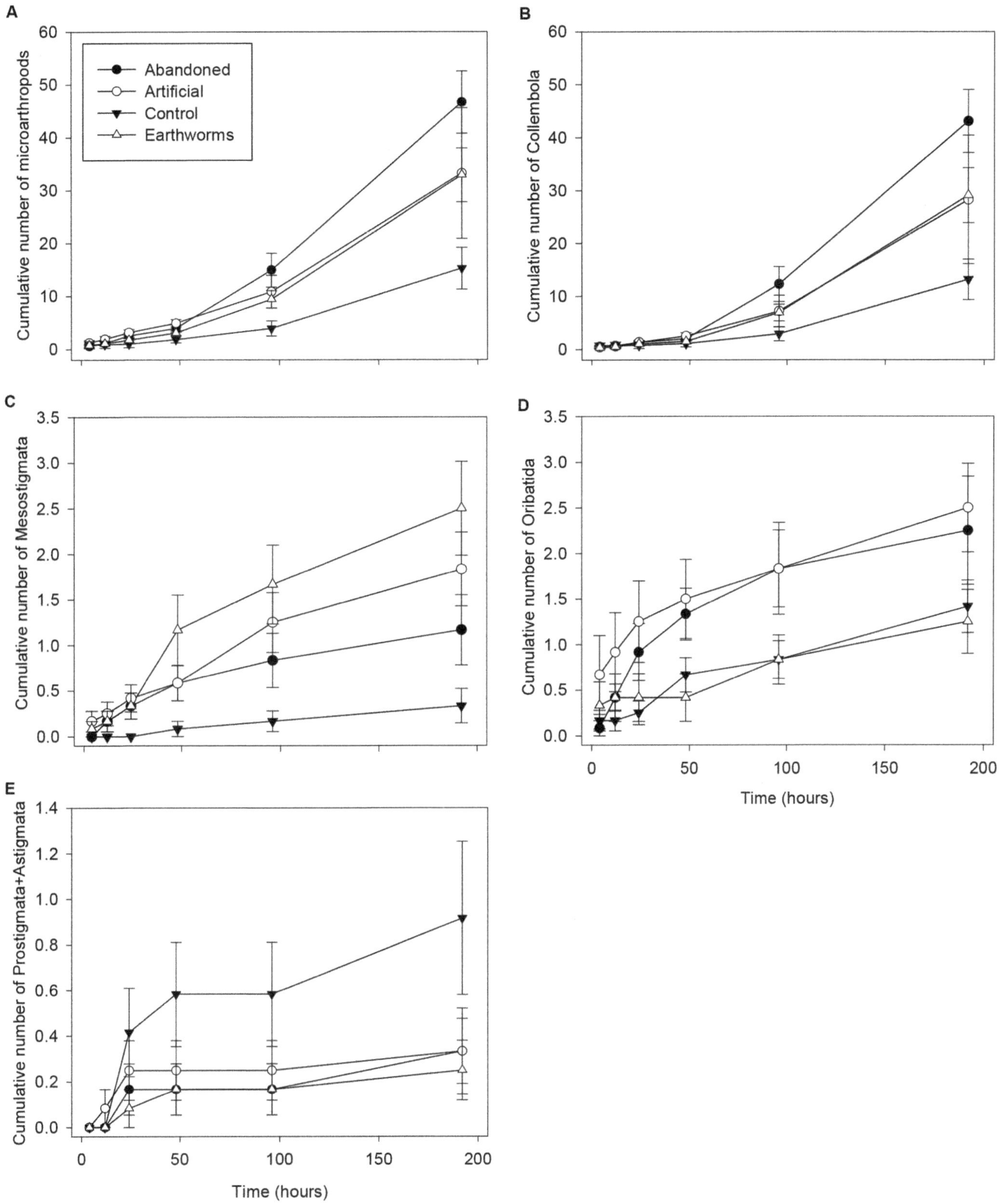

Figure 3. Cumulative number of microarthropods dispersing over time. (a) All microarthopods together (±SE); (b) Collembola; (c) Mesostigmata; (d) Oribatida; and (e) Prostigmata+Astigmata. N = 12 replicates per treatment.

increased the total number of individuals dispersing for several groups including Collembola, Mesostigmata and all microarthropods together, while the presence of earthworm excreta and earthworms themselves resulted in little additional increase.

Table 2. Regression fit statistics for models of microarthropod abundance over time.

Taxa	Model	K	LL	ΔAIC_c	$wAIC_c$
Microarthropods	**Global*Time**	**10**	**−682.19**	**0**	**0.68**
	Tunnels*Time	6	−690.02	2.91	0.16
	Tunnels	5	−690.30	3.48	0.12
	Global	7	−689.62	6.31	0.03
	Castings*Time	6	−690.48	8.03	0.01
	Castings	5	−696.92	16.73	0.00
	Worms	5	−701.08	25.05	0.00
	Worms*Time	6	−700.02	27.09	0.00
	Null	3	−869.92	358.60	0.00
Collembola	**Tunnels**	**5**	**−579.95**	**0**	**0.49**
	Tunnels*Time	6	−579.95	2.08	0.17
	Global*Time	10	−575.72	2.12	0.17
	Global	7	−578.97	2.22	0.16
	Castings	5	−587.58	15.26	0.00
	Worms	5	−592.86	25.81	0.00
	Castings*Time	6	−584.65	11.49	0.00
	Worms*Time	6	−592.13	26.45	0.00
	Null	3	−748.20	332.37	0.00
Mesostigmata	**Tunnels**	**4**	**−224.48**	**0**	**0.50**
	Tunnels*Time	**5**	**−224.00**	**1.10**	**0.29**
	Global	**6**	**−223.36**	**1.92**	**0.19**
	Global*Time	9	−222.71	6.96	0.02
	Worms	4	−229.88	10.80	0.00
	Castings	4	−230.55	12.14	0.00
	Worms*Time	5	−229.81	12.73	0.00
	Castings*Time	5	−230.55	14.20	0.00
	Null	2	−280.15	107.2406	0.00
Oribatida	**Tunnels**	**4**	**−303.43**	**0**	**0.22**
	Global	**6**	**−301.37**	**0.052**	**0.21**
	Tunnels*Time	**5**	**−302.56**	**0.34**	**0.18**
	Worms	**4**	**−303.63**	**0.42**	**0.18**
	Castings	4	−304.55	2.26	0.07
	Worms*Time	5	−303.63	2.48	0.06
	Global*Time	9	−299.79	3.23	0.04
	Castings*Time	5	−304.42	4.05	0.03
	Null	2	−339.18	67.42	0.00

Predictors included presence of openings (*Tunnels*), excreta (*Castings*), and earthworms (*Worms*). Time was included in all models, either on its own or in interaction with the other predictor variables. The best model has a ΔAIC_c of zero and the highest $wAIC_c$ value. Models with $\Delta AIC_c < 2$ are also considered to be plausible and are shown in bold. With k, number of parameters; LL, log likelihood; ΔAIC_c, difference in the Akaike's information criterion (corrected for small sample size) value between model and the most strongly supported model; $wAIC_c$, weight given by the AIC (i.e., relative strength of support for model).

Cumulative movement over time for all microarthropods together was affected by the presence of tunnels and excreta in addition to *L. rubellus*. Abundances of Oribatida and Prostigmata/Astigmata in the destination containers were not well explained by any of the variables considered, with the null models receiving the most support. However, the rate of accumulation of Oribatida over time was influenced by both tunnels and earthworms. Movement of oribatid mites was greater in the presence of tunnels, while presence of earthworms was associated with reduced movement.

Although little is known about belowground dispersal of microarthropods, corridors have been demonstrated to facilitate aboveground dispersal in moss microcosms [42,43]. Increased movement of microarthropods occurred when moss patches were connected by thin corridors of moss, particularly when the rate of emigration to the hostile surrounding matrix is low (i.e., when movement is biased to occur along the moss corridors) [44]. Our results are also consistent with several studies in which microarthropod abundance increased with soil pore volume [36,45,46]. Increased access to resources and reduced predation or competition due to greater availability of refuges have been suggested as mechanisms by which soil pore volume could lead to increased microarthropod abundances [46]. The greater soil heterogeneity created by the tunnels in our experiment might additionally contribute to increases in diversity [47].

The maximum rates of movement observed in our study, even within the control treatment, are much higher than previously recorded for microarthropods within soil in the field [48,49]. The highest potential rate of movement recorded previously was approximately 20 cm per week for some genera of Oribatida and Collembola [49], in contrast to our maximum estimated rates of 2.5 cm per hour (= 420 cm/week) for Collembola, Mesostigmata, and Oribatida and 0.83 cm per hour for Prostigmata/Astigmata (= 210 cm/week). Converting from hours to weeks undoubtedly leads to overestimation of spread rates as it is unlikely that microarthropods would continue to move in the same direction for a week. The fact that movement could only occur along the box also may have resulted in greater estimates of movement speed than would be the case in a natural system, where movement could occur in any direction. The straightness and artificial nature of the tunnels might also have resulted in greater movement rates than would be observed along burrows. Additionally, soil type can have large impacts on movement rates in nematodes [31]. If soil type has similar effects on microarthropod movement, it could account for some variation among studies. Nonetheless, the substantially greater rates of movement in our experiment suggest maximum movement rates may be much larger than previously estimated when movement is highly directional.

Microarthropod taxa can exhibit differing responses to soil pores or corridors [46]. As predicted, artificial tunnels had a stronger effect on movement of Mesostigmata and Collembola which tend to have larger body sizes than Oribatida, Prostigmata, and Astigmata [35,36]. Large soil pores may be particularly beneficial for taxa that are larger and cannot move as easily through the soil matrix [36]. These varying effects on movement of different microarthropod groups may result in altered predator-prey dynamics or competition among microarthropods and could ultimately lead to shifts in community assembly. We observed that a disproportionately smaller number of Oribatida moved to the destination containers than did the other groups (Fig. 2f versus 2b–e), resulting in a different composition of taxa in the destination containers than in the source containers; however, the treatments themselves did not seem to have a strong influence on the relative proportions of the different taxa remaining in the source containers. A more detailed analysis would be necessary to assess whether other traits besides body size (e.g., diet) can influence the responses of microarthropod taxa to tunnels.

Our artificial and abandoned burrow treatments were designed to mimic the physical changes that occur in the structure of the soil due to earthworm burrowing but not the ongoing disturbance of the soil caused by earthworms burrowing continually. Continual disturbances can be a key factor structuring microarthropod communities, and oribatid mite populations are thought to be particularly sensitive [50,51]. For example, repeated sieving and mixing of litter and soil resulted in declines in densities of most groups of oribatid mites, as well as Collembola, in a beech forest [51]. It was therefore suggested that high densities of microarthropods in some soils could be related to less mechanical disturbance by earthworms [51]. Although we found a lower cumulative number of Oribatida dispersing over time when *L. rubellus* were present in the microcosms, the presence of earthworms did not have a strong effect on the abundance of any taxa in the destination containers. However, the short duration of our experiment and the disruption of the soil during transport and construction of our microcosms may have reduced our ability to detect negative effects of earthworms per se. How earthworm effects on movement rates influence microarthropod communities over the long-term warrants further investigation.

Our study suggests that the effects of ecosystem engineers on habitat structure may strongly affect movement of other species. Movement of microarthropods was greatest when artificial tunnels were present in the soil, indicating that changes in soil structure, rather than phoresy or changes in nutrient distribution, may influence microarthropod movement belowground. The stronger influence of *L. rubellus* on movement of Collembola and Mesostigmata, as compared to Oribatida, Prostigmata, and Astigmata, suggests that alteration of movement routes is a possible mechanism driving the variable responses of different microarthropod taxa to earthworms. Little is known about movement belowground and the effects of edges and corridors on soil organisms, despite the extensive amount of research in this area in aboveground systems. Consequently, future research should further examine effects of soil structure on soil organisms, including the impacts of other earthworm species on microarthropods.

Acknowledgments

Thanks to Dr. Tyler Cobb, the Royal Alberta Museum, and the Alberta Biodiversity Monitoring Institute for use of Tullgren funnels and microscopes. Thanks as well to M. Strekies for assistance with experimental setup and microarthropod sorting and the Boutin-Bayne labs and B. Cameron for assistance with experimental setup.

Author Contributions

Provided taxonomic expertise: HCP. Conceived and designed the experiments: EKC HCP EMB. Performed the experiments: EKC. Analyzed the data: EKC EMB. Contributed reagents/materials/analysis tools: EMB HCP. Wrote the paper: EKC HCP EMB.

References

1. Jones CG, Lawton JH, Shachak M (1994) Organisms as ecosystem engineers. Oikos 69: 373–386. doi:10.2307/3545850.
2. Jones CG, Lawton JH, Shachak M (1997) Positive and negative effects of organisms as physical ecosystem engineers. Ecology 78: 1946–1957. doi:10.1890/0012-9658(1997)078[1946:PANEOO]2.0.CO;2.
3. Badano EI, Jones CG, Cavieres LA, Wright JP (2006) Assessing impacts of ecosystem engineers on community organization: a general approach illustrated by effects of a high-Andean cushion plant. Oikos 115: 369–385. doi:10.1111/j.2006.0030-1299.15132.x.
4. Commito JA, Celano EA, Celico HJ, Como S, Johnson CP (2005) Mussels matter: postlarval dispersal dynamics altered by a spatially complex ecosystem engineer. J Exp Mar Biol Ecol 316: 133–147. doi:10.1016/j.jembe.2004.10.010.
5. Cuddington K, Hastings A (2004) Invasive engineers. Ecol Model 178: 335–347. doi:10.1016/j.ecolmodel.2004.03.010.
6. Gabay O, Perevolotsky A, Shachak M (2012) How landscape modulators function: woody plant impact on seed dispersal and abiotic filtering. Plant Ecology 213: 685–693. doi:10.1007/s11258-012-0033-5.
7. Dighton J, Jones H, Robinson C, Beckett J (1997) The role of abiotic factors, cultivation practices and soil fauna in the dispersal of genetically modified microorganisms in soils. Appl Soil Ecol 5: 109–131. doi:10.1016/S0929-1393(96)00137-0.
8. Ettema CH, Wardle DA (2002) Spatial soil ecology. Trends Ecol Evol 17: 177–183. doi:10.1016/S0169-5347(02)02496-5.
9. Kaczmarek W (1978) Locomotory activity of soil fauna as a parameter of the trophic structure and of succession in forest ecosystems. Pedobiologia 18: 434–441.
10. Loss SR, Blair RB (2011) Reduced density and nest survival of ground-nesting songbirds relative to earthworm invasions in northern hardwood forests. Conserv Biol 25: 983–992. doi:10.1111/j.1523-1739.2011.01719.x.
11. Maerz JC, Nuzzo VA, Blossey B (2009) Declines in woodland salamander abundance associated with non-native earthworm and plant invasions. Conserv Biol 23: 975–981. doi:10.1111/j.1523-1739.2009.01167.x.
12. Ransom TS (2011) Earthworms, as ecosystem engineers, influence multiple aspects of a salamander's ecology. Oecologia 165: 745–754. doi:10.1007/s00442-010-1775-1.
13. Nuzzo VA, Maerz JC, Blossey B (2009) Earthworm invasion as the driving force behind plant invasion and community change in northeastern North American forests. Conserv Biol 23: 966–974. doi:10.1111/j.1523-1739.2009.01168.x.
14. Regnier E, Harrison SK, Liu J, Schmoll JT, Edwards CA, et al. (2008) Impact of an exotic earthworm on seed dispersal of an indigenous US weed. J Appl Ecol 45: 1621–1629. doi:10.1111/j.1365-2664.2008.01489.x.
15. Bouché MB (1977) Strategies lombriciennes. Ecological Bulletins 25: 122–132.
16. Gates GE (1970) Miscellanea megadrilogica VII. Megadrilogica 1: 1–14.
17. Gates GE (1982) Farewell to North American megadriles. Megadrilogica 4: 12–77.
18. Bohlen PJ, Scheu S, Hale CM, McLean MA, Migge S, et al. (2004) Non-native invasive earthworms as agents of change in northern temperate forests. Front Ecol Environ 2: 427–435.
19. Cameron EK, Knysh KM, Proctor HC, Bayne EM (2013) Influence of two exotic earthworm species with different foraging strategies on abundance and composition of boreal microarthropods. Soil Biol Biochem 57: 334–340. doi:10.1016/j.soilbio.2012.07.010.
20. McLean MA, Parkinson D (1998) Impacts of the epigeic earthworm Dendrobaena octaedra on oribatid mite community diversity and microarthropod abundances in pine forest floor: a mesocosm study. Appl Soil Ecol 7: 125–136. doi:10.1016/S0929-1393(97)00041-3.
21. Migge-Kleian S, McLean MA, Maerz JC, Heneghan L (2006) The influence of invasive earthworms on indigenous fauna in ecosystems previously uninhabited by earthworms. Biol Invasions 8: 1275–1285. doi:10.1007/s10530-006-9021-9.
22. Eisenhauer N (2010) The action of an animal ecosystem engineer: Identification of the main mechanisms of earthworm impacts on soil microarthropods. Pedobiologia 53: 343–352. doi:10.1016/j.pedobi.2010.04.003.
23. Salmon S, Geoffroy JJ, Ponge JF (2005) Earthworms and collembola relationships: effects of predatory centipedes and humus forms. Soil Biol Biochem 37: 487–495. doi:10.1016/j.soilbio.2004.08.011.
24. Salmon S, Ponge JF (1999) Distribution of *Heteromurus nitidus* (Hexapoda, Collembola) according to soil acidity: interactions with earthworms and predator pressure. Soil Biol Biochem 31: 1161–1170. doi:10.1016/S0038-0717(99)00034-6.
25. Salmon S (2001) Earthworm excreta (mucus and urine) affect the distribution of springtails in forest soils. Biol Fertil Soils 34: 304–310. doi:10.1007/s003740100407.
26. Gutierrez-Lopez M, Salmon S, Trigo D (2011) Movement response of Collembola to the excreta of two earthworm species: Importance of ammonium content and nitrogen forms. Soil Biol Biochem 43: 55–62. doi:10.1016/j.soilbio.2010.09.010.
27. Gutierrez Lopez M, Lidon JBJ, Aza DT, Fernandez Garcia R, Diaz Cosin DJ (2009) The influence of *Hormogaster elisae* (Oligochaeta, Hormogastridae) on the colonisation of defaunated soil by microarthropods in laboratory cultures. Pedobiologia 52: 163–170. doi:10.1016/j.pedobi.2008.08.001.
28. Salmon S (2004) The impact of earthworms on the abundance of Collembola: improvement of food resources or of habitat? Biol Fertil Soils 40: 323–333. doi:10.1007/s00374-004-0782-y.
29. Mitchell R (1970) An analysis of dispersal in mites. Am Nat 104: 425–431.
30. Shapiro DI, Tylka GL, Berry EC, Lewis LC (1995) Effects of earthworms on the dispersal of *Steinernema* spp. Journal of Nematology 27: 21.
31. MacMillan K, Haukeland S, Rae R, Young I, Crawford J, et al. (2009) Dispersal patterns and behaviour of the nematode *Phasmarhabditis hermaphrodita* in mineral soils and organic media. Soil Biol Biochem 41: 1483–1490. doi:10.1016/j.soilbio.2009.04.007.

32. McLean MA, Parkinson D (2000) Introduction of the epigeic earthworm *Dendrobaena octaedra* changes the oribatid community and microarthropod abundances in a pine forest. Soil Biol Biochem 32: 1671–1681.
33. Reynolds JW (1994) Earthworms of the World. Glob Biodivers 4: 11–16.
34. Hendrix PF, Callaham MA, Drake JM, Huang C-Y, James SW, et al. (2008) Pandora's box contained bait: the global problem of introduced earthworms. Annu Rev Ecol Evol Syst 39: 593–613. doi:10.1146/annurev.ecolsys.39.110707.173426.
35. Edwards CA (1967) Relationships between weights, volumes and numbers of soil animals. Progress in soil biology: Proceedings of the colloquium on dynamics of soil communities, 5–10 September, 1966. 585–593.
36. Vreeken-Buijs MJ, Hassink J, Brussaard L (1998) Relationships of soil microarthropod biomass with organic matter and pore size distribution in soils under different land use. Soil Biol Biochem 30: 97–106. doi:10.1016/S0038-0717(97)00064-3.
37. Macfadyen A (1961) Improved funnel-type extractors for soil arthropods. Journal of Animal Ecology 30: 171–184. doi:10.2307/2120.
38. Dayton CM (1998) Information criteria for the paired-comparisons problem. Am Stat 52: 144–151. doi:10.2307/2685473.
39. Burnham KP, Anderson DR, Huyvaert KP (2011) AIC model selection and multimodel inference in behavioral ecology: some background, observations, and comparisons. Behav Ecol Sociobiol 65: 23–35. doi:10.1007/s00265-010-1029-6.
40. Sugiura N (1978) Further analysis of the data by Akaike's information criterion and the finite corrections. Communications in Statistics - Theory and Methods 7: 13–26. doi:10.1080/03610927808827599.
41. Burnham KP, Anderson DR (2002) Model Selection and Multimodel Inference: A Practical Information-theoretic Approach. Springer. 512 p.
42. Gilbert F, Gonzalez A, Evans-Freke I (1998) Corridors maintain species richness in the fragmented landscapes of a microecosystem. Proc R Soc Lond Ser B-Biol Sci 265: 577–582.
43. Gonzalez A, Lawton JH, Gilbert FS, Blackburn TM, Evans-Freke I (1998) Metapopulation dynamics, abundance, and distribution in a microecosystem. Science 281: 2045–2047. doi:10.1126/science.281.5385.2045.
44. Hoyle M (2007) When corridors work: Insights from a microecosystem. Ecol Model 202: 441–453. doi:10.1016/j.ecolmodel.2006.11.008.
45. Ducarme X, Andre HM, Wauthy G, Lebrun P (2004) Are there real endogeic species in temperate forest mites? Pedobiologia 48: 139–147. doi:10.1016/j.pedobi.2003.10.002.
46. Nielsen UN, Osler GHR, Van der Wal R, Campbell CD, Burslem DFRP (2008) Soil pore volume and the abundance of soil mites in two contrasting habitats. Soil Biol Biochem 40: 1538–1541. doi:10.1016/j.soilbio.2007.12.029.
47. Wardle DA (2002) Communities and Ecosystems: Linking the Aboveground and Belowground Components. Princeton University Press. 404 p.
48. Sjogren M (1997) Dispersal rates of Collembola in metal polluted soil. Pedobiologia 41: 506–513.
49. Ojala R, Huhta V (2001) Dispersal of microarthropods in forest soil. Pedobiologia 45: 443–450. doi:10.1078/0031-4056-00098.
50. Maraun M, Scheu S (2000) The structure of oribatid mite communities (Acari, Oribatida): patterns, mechanisms and implications for future research. Ecography 23: 374–383. doi:10.1034/j.1600-0587.2000.d01-1647.x.
51. Maraun M, Salamon JA, Schneider K, Schaefer M, Scheu S (2003) Oribatid mite and collembolan diversity, density and community structure in a moder beech forest (*Fagus sylvatica*): effects of mechanical perturbations. Soil Biol Biochem 35: 1387–1394. doi:10.1016/S0038-0717(03)00218-9.

9

Crabs Mediate Interactions between Native and Invasive Salt Marsh Plants: A Mesocosm Study

Xiao-dong Zhang[1,2], Xin Jia[3], Yang-yun Chen[1], Jun-jiong Shao[1], Xin-ru Wu[1], Lei Shang[1], Bo Li[1]*

1 Coastal Ecosystems Research Station of the Yangtze River Estuary, Ministry of Education Key Laboratory for Biodiversity Science and Ecological Engineering, Institute of Biodiversity Science, Fudan University, Shanghai, China, **2** Key Laboratory of Wetland Services and Restoration, Institute of Wetland Research, Chinese Academy of Forestry, Beijing, China, **3** School of Soil and Water Conservation, Beijing Forestry University, Beijing, China

Abstract

Soil disturbance has been widely recognized as an important factor influencing the structure and dynamics of plant communities. Although soil reworkers were shown to increase habitat complexity and raise the risk of plant invasion, their role in regulating the interactions between native and invasive species remains unclear. We proposed that crab activities, via improving soil nitrogen availability, may indirectly affect the interactions between invasive *Spartina alterniflora* and native *Phragmites australis* and *Scirpus mariqueter* in salt marsh ecosystems. We conducted a two-year mesocosm experiment consisting of five species combinations, i.e., monocultures of three species and pair-wise mixtures of invasive and native species, with crabs being either present or absent for each combination. We found that crabs could mitigate soil nitrogen depletion in the mesocosm over the two years. Plant performance of all species, at both the ramet-level (height and biomass per ramet) and plot-level (density, total above- and belowground biomass), were promoted by crab activities. These plants responded to crab disturbance primarily by clonal propagation, as plot-level performance was more sensitive to crabs than ramet-level. Moreover, crab activities altered the competition between *Spartina* and native plants in favor of the former, since *Spartina* was more promoted than native plants by crab activities. Our results suggested that crab activities may increase the competition ability of *Spartina* over native *Phragmites* and *Scirpus* through alleviating soil nitrogen limitation.

Editor: John F. Valentine, Dauphin Island Sea Lab, United States of America

Funding: This research was financially supported by National Basic Research Program of China (Grant No. 2013CB430404), National Science Foundation of China (Grant No. 30930019), and the Science and Technology Commission of Shanghai Municipality (Grant No. 10JC1400700 and 12XD1400500). The funders had no role in study design, data collection and analysis, decision to publish, or preparation of the manuscript.

Competing Interests: The authors have declared that no competing interests exist.

* E-mail: bool@fudan.edu.cn

Introduction

Biological invasions are one of the most serious threats to biodiversity and ecosystem functioning [1]. The interactions between invasive and native species play a critical role in regulating the structure and dynamics of plant communities [2]. Some of the interactions are mediated by other ecosystem components (e.g., nutrient availability, herbivores) [3]. The concept of "indirect effects" in communities began to draw ecologists' attention two decades ago [4], and was soon introduced into bioinvasion studies [5]. For example, an invasive grass *Agrostis capillaris* led to a population proliferation of the herbivorous slug *Deroceras reticulatum*, imposing a negative impact on the native fern *Botrychium australe* [6]. Another example is that AM fungi strongly enhance the invader *Centaurea maculosa* to outcompete native *Festuca idahoensis* [7]. Indirect effects are implicit in several ecological theories (e.g., exploitative competition, apparent competition, indirect mutualism or facilitation, cascading effects) [4], most of which focus on the relationships between predator and prey (herbivore-primary producer) [3], [8], [9], host and parasite [10], or pollinator and plant [11], [12]. Crooks [13] reviewed the ecosystem-level consequences of abiotic resource modification (or "ecosystem engineering") by invasive species. However, the role of native soil modifiers in mediating plant interactions remains largely unknown, although a recent study reported that grazers might mediate competition between two graminoid species via altering soil nutrient status [14].

Some organisms termed "ecosystem engineer" can increase habitat complexity by altering hydrological and edaphic conditions. Ecosystem engineers influence the growth and dispersal of both native and invasive plants, sometimes facilitating successful invasions [13]. As an example, the dam building beaver (*Castor canadensis*) could increase habitat heterogeneity, and thus increases the richness of herbaceous plants in the riparian zone in central Adirondacks, NY, USA [15]. In another case, the native cushion plant *Azorella monantha* could ameliorate harsh high-elevation conditions (e.g., low temperatures) and facilitate colonization by non-native species [16]. However, there is a lack of evidence for soil modifiers to mediate interactions between native and invasive species. We propose that native soil fauna play an important role in mediating the interactions between native and invasive plants by reworking the soil and altering nitrogen availability. Crabs are common burrowing species, and their activities can change the physical conditions such as soil moisture, temperature and aeration, which may in turn influence the microbial activities such as mineralization, nitrification and denitrification [17]. Furthermore, the nutrient-rich microhabitats created by crabs may promote invasion by non-native plants, as predicted by

Davis's fluctuating resource hypothesis [18]. Invasive species are most likely to invade post-disturbance habitats, as they often benefit more from increased resource availability due to faster growth and higher resource use efficiency.

We explored how the interactions between native and invasive plants could be mediated by disturbance from soil animals in a salt marsh. A mesocosm was established to mimic salt marsh conditions in the Yangtze River estuary, where *Phragmites australis* ("*Phragmites*" hereafter) and *Scirpus mariqueter* ("*Scirpus*" hereafter) are native, and *Spartina alterniflora* ("*Spartina*" hereafter) is exotic. Crabs are the most abundant native fauna dwelling in the estuary. The crab community consists mainly of *Helice tientsinensis* ("*Helice*" hereafter), *Chiromantes dehaani*, *Parasesarma plicatum* and *Uca arcuata*, with the average density being approximately 20 ind. m^{-2} [19]. Borrowing and soil transporting by crabs accelerate the turnover of sediments from deep layers to the surface, thereby playing an important role in regulating soil conditions [20]. The study of the indirect effect of native crab activities on the interaction between invasive and native plants may provide a novel way to explore the mechanisms of plant invasion. We addressed three specific questions: (i) Do native crabs affect soil nutrient availability? (ii) Does crab disturbance to the soil influence the performance of invasive *Spartina* and native *Scirpus* and *Phragmites*? (iii) Is *Spartina* more competitive than the two natives in response to crab disturbance?

Materials and Methods

Ethics Statement

All necessary permits were obtained for the described field studies. We are authorized by the Administrative Office of Shanghai Chongming Dongtan National Nature Reserve to carry out research activities including plant and crab collection in the entire area of the reserve. In addition, the crab *Helice* was an abundant species in the reserve, and was permitted to be appropriately fished for economic income.

Study Species

All the plants and crabs used in this study were collected from the salt marsh at Chongming Dongtan in Yangtze River estuary (121° 50′ to 122° 05′ E, 31° 25′ to 31° 38′ N). *Spartina*, *Phragmites* and *Scirpus* are perennial clonal plants, expanding in salt marshes through both seed dispersal and vegetative propagation. Native *Phragmites* and *Scirpus* formed distinct zonation in Chongming Dongtan before the *Spartina* invasion. *Phragmites* occupied the high tidal zone and *Scirpus* dominated the low and intertidal zones. *Spartina* has been invading the Yangtze estuary since 1990s, and rapidly colonized tidal mudflats and salt marshes, driving a new pattern of zonation. The invasion success of *Spartina* may be attributable to its stronger tolerance to salinity than the natives, as well as greater leaf area index, higher maximal net photosynthetic rate, longer growing season, and higher use efficiency of light, water, and nitrogen [21], [22].

The crab *Helice* is one of the most abundant crab species at Chongming Dongtan, feeding on fresh plant, fauna and detritus [19]. The dietary composition of *Helice* is flexible, depending on food availability. The carapace width of adult *Helice* is ca. 20~25 mm [23]. The burrow depth of *Helice* at Chongming Dongtan depends on the belowground structures of plant tissues. The deepest burrows occur in mudflat with an average depth of 38 cm, while the shallowest occur in *Spartina* habitats with an average depth of 15 cm [23]. In this study, *Helice* was selected as the source of disturbance in the mesocosm because it dwells in all types of habitats created by the three plants at Chongming Dongtan.

Experimental Design

The experiment was conducted over two growing seasons (from May 2009 to November 2010) in a mesocosm at Fudan University in Shanghai (31°18′07′′ N, 121°29′47′′ E, China), about 10 km from the Yangtze River estuary. Forty cement plots (1×1 m^2) were randomly assigned to ten treatments, i.e., five species combinations × two crab levels (absence vs. presence), with four replicates for each treatment. The plant species combinations included the monocultures of *Spartina*, *Phragmites* and *Scirpus*, and *Spartina-Phragmites* and *Spartina-Scirpus* mixtures. At the experimental setup, we transplanted 16 *Spartina* or *Phragmites* ramets per plot in their monocultures, and alternately arranged 8 ramets of each species in mixture plots. In comparison, we transplanted 32 *Scirpus* ramets per plot in its monocultures and 16 *Scirpus* ramets in mixtures with 8 *Spartina* ramets because *Scirpus* is much smaller than the other two species. The crab-present treatment was established by placing 20 adult *Helice* individuals (with a mean carapace width of 22 mm and a sex ratio of 1 : 1) captured from Chongming Dongtan into each plot. The plots were filled with a mixture of clay (60%) and sand (40%) to a depth of 30 cm. Soil salinity in all plots was adjusted to 15, which is typical of Chongming Dongtan, by irrigating the plots with NaCl solution. In May 2009, we excavated the ramets of three species with rhizomes from Chongming Dongtan and acclimatized them in fresh water for two days prior to transplanting. The crabs were added in June 2009 after all the ramets had established. The upper edge of each plot was encompassed by slippery plastic sheets of 15 cm in width to prevent crab escaping. The few died crabs found were replaced with similar-sized healthy individuals of the same sex to maintain the disturbance intensity in crab-present plots.

Measurements of Plant Performance and Soil Nitrogen

We here focused on the interactions between *Spartina* and the two native species, so harvesting plants asynchronously may lead to a bias in estimating the species interactions. Therefore, we harvested all aboveground biomass in late October in both years, when *Spartina* had matured whereas *Scirpus* had withered and *Phragmites* was still flowering. Once the aboveground plant was reaped, the number of ramets and total fresh biomass were recorded for each plot. We randomly chose ten ramets per species from each plot to obtain the mean ramet height and fresh-mass. The ten ramets were oven-dried at 60°C to constant weight and weighed to determine mean ramet dry-mass. The total aboveground dry-mass of each plot was calculated by dividing the fresh-mass by the ratio of fresh to dry-mass of the ten collected ramets.

Soil samples were taken in May 2009 and October 2010. Five soil cores (3.4 cm in diameter) were sampled to a depth of 25 cm from each plot and mixed. Subsamples of 100 g of mixed soil from each plot were collected and sealed in ziplock bags for subsequent determination of soil properties. In October 2010, the belowground plant tissues were flushed from the remaining soil samples with running water and oven-dried at 60°C to constant weight. As it was difficult to discriminate the three species based on belowground traits, we could only estimate the plot-level belowground biomass, which was calculated by dividing the oven-dried biomass by the ratio of sampled area (i.e., the total intersection area of five soil cores) to plot area (1 m^2). The soil samples for determining chemical properties were air dried and ground to pass through a 100-mesh sieve. Fine roots were then manually removed. We used soil ammonium (NH_4^+-N) and nitrate (NO_3^-N) concentrations as indicators of nitrogen availability. Ten

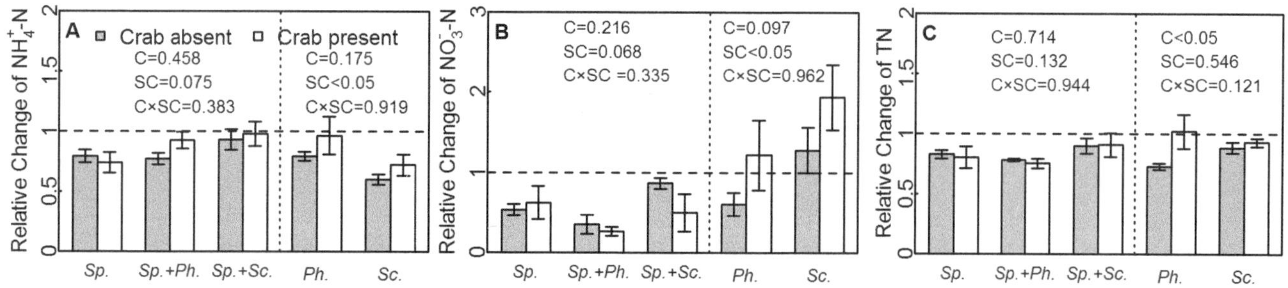

Figure 1. The relative change of soil nitrogen. The effects of crab and species combinations on the relative changes of NH_4^+-N (mean ± SE, A), NO_3^- -N (mean ± SE, B), and TN (mean ± SE, C) were tested using two-way ANOVA. The *P*-values of the effects of crabs and species combinations on relative change of nitrogen are given, where "C" stands for crab treatment and "SC" for species combination. The vertical dash line divides the species combinations into *Spartina*-present plots and *Spartina*-absent plots, and the horizontal dash line at 1 distinguishes nitrogen increase from decrease.

grams of sieved soil was extracted with 40 ml of 2 M KCl solution. The extracts were then stirred for 1 h at 30°C and filtered through medium-speed qualitative filter paper. The NH_4^+-N and NO_3^-N concentrations were determined for the filtrates with a discrete chemistry analyzer (SMARTCHEM^{TM}200, Westco Scientific Instruments, Inc., USA). Soil total nitrogen (TN) was determined for the sieved soil with a N/C analyzer (FlashEA1112 Series, Thermo Inc., Milan, Italy).

We measured the number and total volume of crab burrows in each plot to assess the disturbance intensity in 2010 after plants had been reaped. The volume of burrows was calculated from their diameters and depths under the assumption that the burrows were approximately cylindrical.

Data Analysis

We carried out all analyses using software R version 3.0.1 [24]. The relative changes of soil NH_4^+-N, NO_3^- -N and TN over the two years were calculated by dividing the concentration in October 2010 (final) by the value in May 2009 (initial). We used multivariate analysis of variance (MANOVA) to examine the variation in soil nitrogen availability among treatments at the initial state, with the overall concentrations of NH_4^+-N, NO_3^- -N and TN being dependent variables. The five species combinations were then divided into two groups to compare the effects of crabs and species combinations on soil nitrogen. The first group ("*Spartina*-present plots" hereafter) consisted of *Spartina* monocultures and *Spartina-Phragmites/Scirpus* mixtures, and the second group ("*Spartina*-absent plots" hereafter) consisted of *Phragmites* and *Scirpus* monocultures. The effects of crabs and species combinations on the overall relative changes of three soil nitrogen forms were tested using MANOVA for the two groups, respectively. Subsequently, two-way analyses of variance (ANOVA) were conducted for each of the relative changes in NH_4^+-N, NO_3^- -N and TN for the two groups, respectively. The differences in density and total volume of crab burrows among the five species combinations in crab-present plots were tested using one-way ANOVA.

Table 1. Summary for ANOVA of the linear mixed-effects model[a] for effects of species, plot types and crabs on overall plant performance[b].

Source of variance	numDf	denDf	F	P	
Species	2	99	57.24	<0.001	***[c]
Plot type	1	99	56.30	<0.001	***
Crab	1	99	50.06	<0.001	***
Species×Plot type	2	99	0.07	0.932	
Species×Crab	2	99	14.19	<0.001	***
Plot type×Crab	1	99	2.33	0.130	
Species× plot type ×crab	2	99	3.24	0.043	*

[a]More information of the linear mix model is supplied in Table S2.
[b]"plant performance" was combined of biomass per ramet, height, density and total aboveground biomass.
[c]Asterisks indicate level of significance (* <0.05, ** <0.01, *** <0.001).

A linear mixed-effects model, within the "nlme" package, was used to test the effects of crab, species and plot types on the overall plant performance. The response variables were mean height and biomass of ramets, density and total aboveground biomass. The model included "crab" (present or absent), "species" (*Spartina, Phragmites* or *Scirpus*), "plot type" (mixture or monoculture) and their interactions as fixed factors, and "year" as a random factor. For each species, mixed model ANOVAs were applied to each of the four plant traits, with "crab", "species combination" and their interaction as fixed factors and "year" as a random factor. In addition, the effects of crabs and five species combinations on the total belowground biomass in 2010 were tested using two-way ANOVA. The ratios of plant performance under crab disturbance to those without crabs (R_{crab}) were calculated to facilitate comparison of crab effects among species combinations.

We used two mixed model ANOVAs to further test whether crab activities regulate the interactions between native and the two invasive species. Only data from mixture plots were used for this purpose, the model included "crab", "species" (*Spartina* vs. *Phragmites* in the first ANOVA and *Spartina* vs. *Scirpus* in the second ANOVA) and their interaction as fixed terms and "year" as a random factor. A significant "crab × species" term indicates a change in interspecific relationship attributable to crabs. This analysis was applied to each of the four plant traits. The response variables were log-transformed to meet the assumptions of normality and homoscedasticity in all above analyses.

Results

Soil Nitrogen

Initial soil nitrogen availability did not differ among treatments (MANOVA: crab, *Pillai's trace* $_{1,30}=0.039$, $F_{3,28}=0.38$, $P=0.772$; species combination, *Pillai's* $trace_{4,30}=0.51$, $F_{12,90}=1.53$, $P=0.128$; crab × species combination: *Pillai's* $trace_{4,30}=0.19$,

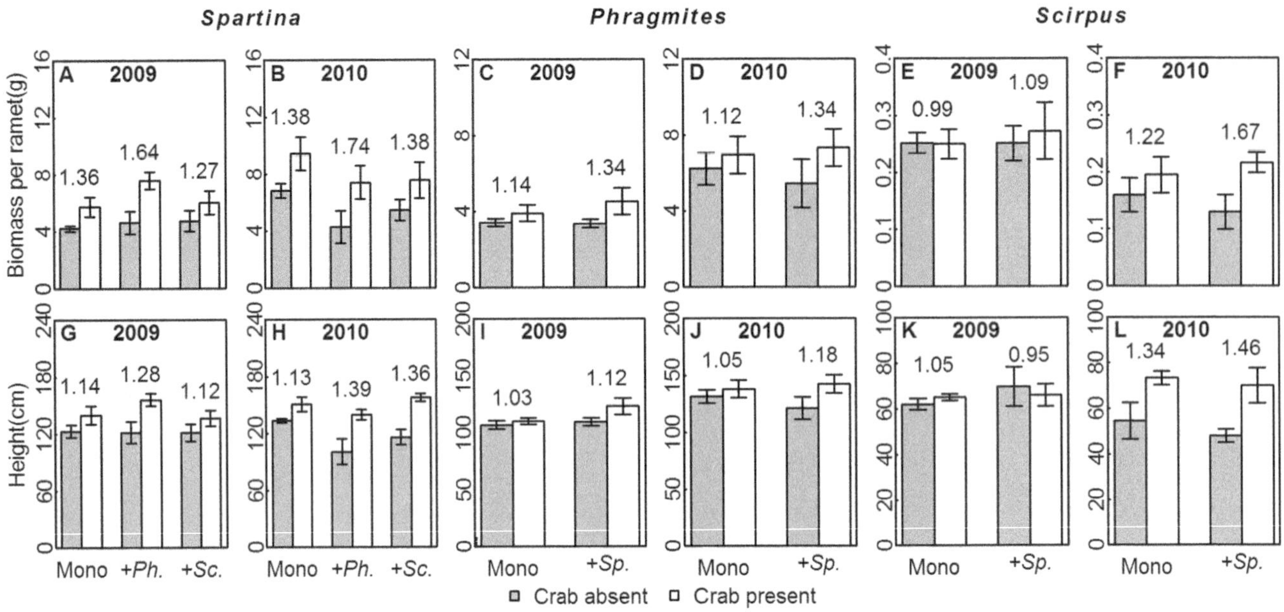

Figure 2. The plant performance at ramet-level. The biomass per ramet (mean±SE, A–F) and height (mean ± SE, G–L) of three species with/without crab disturbance in five species combinations are presented. The categories on X axis present different species combinations. "Mono", "+*Sp.*", "+*Ph.*" and "+*Sc.*" are abbreviations for "monocultures", "with *Spartina*", "with *Phragmites*" and "with *Scirpus*", respectively. The R_{crab}s of three species in specific species combinations were given above the bars.

$F_{3,12}=0.49$, $P=0.915$). Soil NH_4^+-N, NO_3^- -N and TN decreased at the end of the experiment, as indicated by <1 relative changes (Figure 1), except for the NO_3^- -N in native monocultures (Figure 1B). Final soil nitrogen was significantly affected by crabs in *Spatina*-absent plots (*Pillai's* $trace_{1,12}=0.53$, $F_{3,10}=3.73$, $P<0.05$), but not in *Spartina*-present plots (*Pillai's* $trace_{1,18}=0.15$, $F_{3,16}=0.97$, $P=0.432$). The crabs appeared to alleviate nitrogen limitation in *Spartina*-absent plots, as indicated by higher relative changes in nitrogen (see *Ph.* and *Sc.* in Figure 1A–C), although a significant crab effect was only found for TN (Figure 1C, Table S1). The species combination affected the soil nitrogen in both *Spartina*-present (*Pillai's* $trace_{1,18}=0.59$, $F_{6,34}=2.37$, $P<0.05$) and *Spartina*-absent plots (*Pillai's* $trace_{1,12}=0.57$, $F_{3,10}=4.37$, $P<0.05$), as NH_4^+-N and NO_3^- -N

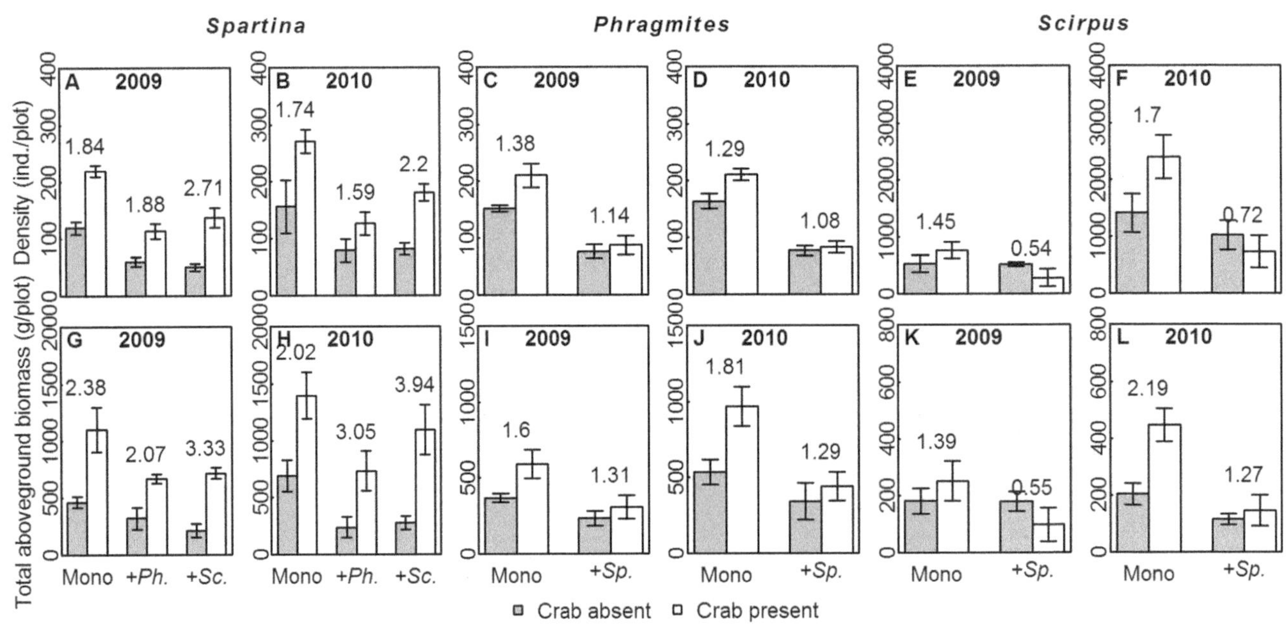

Figure 3. The plant performance at plot-level. The density (mean ± SE, A–F) and total aboveground biomass (mean±SE, G–L) per plot of three species with/without crab disturbance in five species combinations are presented. The categories on X axis present different species combinations. "Mono", "+*Sp.*", "+*Ph.*" and "+*Sc.*" are abbreviations for "monocultures", "with *Spartina*", "with *Phragmites*" and "with *Scirpus*", respectively. The R_{crab}s of three species in specific species combinations are given above the bars.

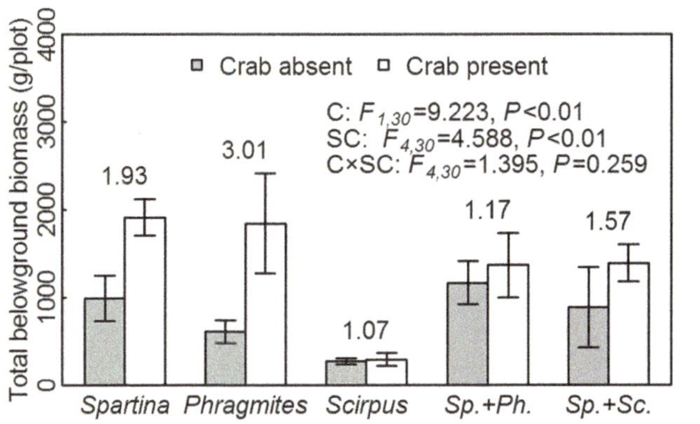

Figure 4. The total belowground biomass (mean ± SE) with/without crab disturbance in five species combinations. The categories on X axis present different species combinations. "*Sp.*", "*Ph.*" and "*Sc.*" are abbreviations for "*Spartina*", "*Phragmites*" and "*Scirpus*", respectively. The *F*-values and *P*-values of the effects of crabs and five species combinations on belowground yield are given, where "C" stands for crab treatments and "SC" stands for species combinations. The R_{crab}s of three species in specific species combinations are given above the bars.

(but not TN) varied among species combinations (Figure 1). The density and volume of crab burrows did not differ among species combinations (density: $F_{4,15} = 1.59$, $P = 0.229$; total volume: $F_{4,15} = 1.33$, $P = 0.303$).

Plant Performance

Plant performance varied with crab disturbance, species and plot types, as well as their interactions (Table 1, Table S2). All the three species in crab-present plots outperformed that in crab-absent plots (Figure 2, 3, Table S3), except for the biomass per ramet and density of *Scirpus*. In addition, the total belowground biomass of all five species combinations increased in response to crab disturbance (Figure 4). The three species, however, were promoted to different degrees, as shown by the significant "crab × species" term (Table1). The R_{crab}s of *Spartina* were generally higher than those of native species (Figure 2, 3), suggesting that *Spartina* benefited more from crab activities than native species. Furthermore, the ramet-level performance (i.e., biomass per ramet and height) of both the native and invasive plants were not affected by species combination and the interaction with crabs (Table S3), compared with the significant "crab × species combination" terms for the plot-level performance (i.e., density, Table S3). Meanwhile, the R_{crab}s of plot-level performance of three species in monocultures were generally higher than that of ramet-level performance, e.g. the R_{crab}s of total aboveground biomass of *Spartina* in monocultures (Figure 3G) were higher than that of biomass per ramet (Figure 2G). This suggests that the number, but not size, of ramets was more sensitive to crabs in different species combinations.

Interactions between Invasive and Native Plants

In mixtures, *Spartina* was facilitated more by crabs than natives (Figure 5), as demonstrated by significant "crab × species" terms for density and total aboveground biomass (Figure 5C, D, G, H, Table S4). Ramet-level performance, however, showed similar responses to crab disturbance between invasive and native plants (Figure 5A, B, E, F).

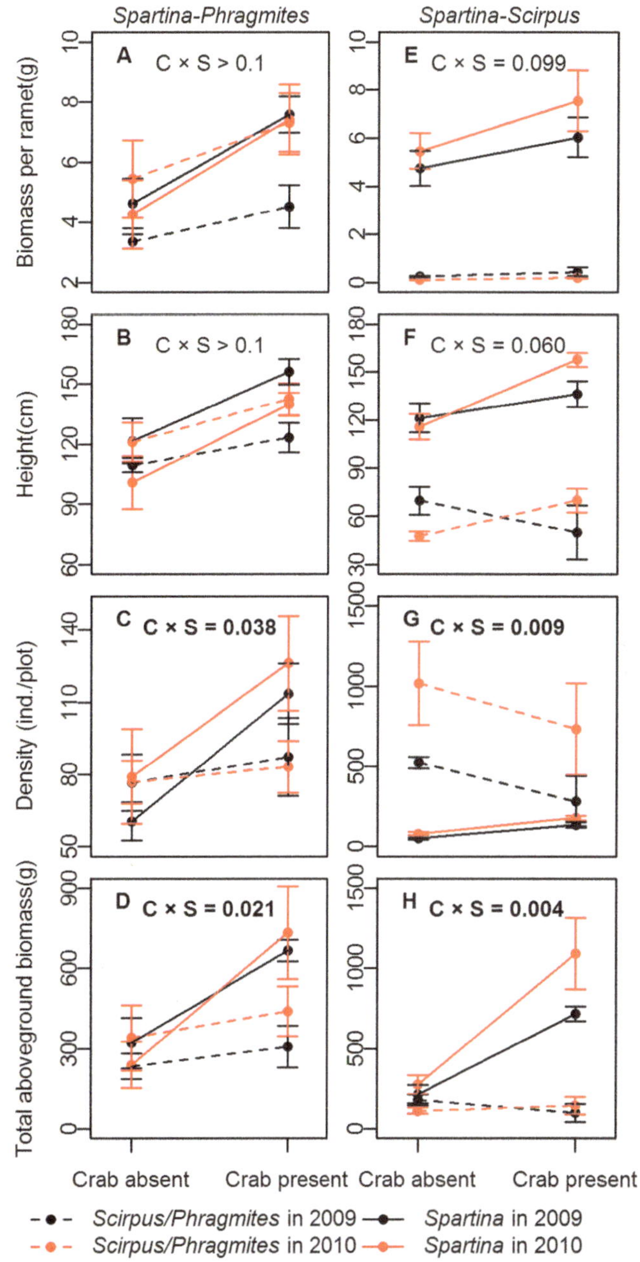

Figure 5. The performance of invasive vs. native plant in mixtures with or without crab disturbance. The biomass per ramet (mean ± SE, A and E), height (mean ± SE, B and F), density (mean ± SE, C and G) and aboveground biomass (mean ± SE, D and H) of invasive and native species in mixtures are presented. The *P*-values of interactive effects of "crab × species" are given, where "C" stands for crab treatments and "S" stands for species.

Discussion

Soil animals regulate plant communities through herbivory and soil reworking (i.e., ecosystem engineering effects). Our results suggested that crabs mediated the interactions between native and invasive plants through soil reworking. Herbivory by crabs was observed to be negligible (see monocultures in Figure 2, 3), although *Helice* might partially feeds on these plants [19]. Therefore, herbivory was unlikely to alter the relationship between invasive and native plants in this study.

As one of the important drivers of plant community structure, soil nutrient availability may determine the relative dominance of species [25]. Low nitrogen availability, which often occurs in salt marsh ecosystems [26], [27], can suppress growth and competitive ability of potentially dominant species [28], [29]. Nitrogen limitation may occur in the mesocosm, as the total inorganic nitrogen (sum of NH_4^+-N and NO_3^- -N) in the mesocosm ranged from 5 to 8 $\mu g\ g^{-1}$ and TN ranged from 0.31 to 0.67 mg g^{-1}, falling in the range observed in Chongming Dongtan salt marsh (total inorganic nitrogen from 4 to 15 $\mu g\ g^{-1}$ [30] and TN from 0.2 to 0.7 mg g^{-1} [31]). The decreases in nitrogen (Figure 1) might result from greater plant uptake than nitrogen supply by decomposition. However, mineralization could be accelerated by enhanced microbial activities, which are regulated by the physical and chemical environments. Crab activities, such as burrowing, litter fragmentation, feeding and excretion, can create better conditions for decomposition [17]. The crab effects on nitrogen were found in *Spartina*-absent plots, where the relative changes in nitrogen was higher than in crab-present plots (Figure 1B, C). The lack of crab effects on nitrogen in *Spartina*-present plots might be the result of higher nitrogen consumption by fast-growing *Spartina*, which might have offset the improved nitrogen availability by crabs. Previous studies have demonstrated that increased nitrogen storage in plant tissues may neutralize the changes in the soil nitrogen caused by fertilization, especially in nitrogen-limiting ecosystems [32–34].

The plant performance, both of aboveground and belowground parts (Figures 2–4), had been improved due to the alleviated soil conditions. Among the three species, alien *Spatina* responded most sensitively to crab disturbance, as the R_{crab}s of *Spartina* in monocultures were generally higher than the other two species (Figure 2, 3). The higher benefits that the invasive *Spartina* (C_4 photosynthetic pathway) received from crab disturbance could be attributed to the CO_2-concentrating mechanism of C_4 plants, in which each unit of nitrogen has a higher CO_2 fixation efficiency than C_3 species [35], [36]. Therefore, even slight increases in nitrogen availability may lead to marked enhancement in *Spartina* performance. Nitrogen limitation is proposed to prevent invasion, and nitrogen addition could increase the growth of invasive plants in salt marsh and desert [25], [37]. It follows that the crab activities could promote *Spartina* invasion in saltmarsh through improving nitrogen availability, as predicted by Davis [18], that the availability of limiting nitrogen will increase the vulnerability of a community to invasion.

The interactions between invasive and native species were altered by crab disturbance. In mixtures, *Spartina* benefited more from crab activities than natives, potentially leading to a competitive superiority of *Spartina* under improved nitrogen conditions (Figure 5). The enhanced competitive ability of *Spartina* in crab-present mixtures could be explained by its higher growth rate, CO_2 fixation efficiency and nitrogen consumption rate. Wang [38] also found that nitrogen addition favored invasive *Spartina* over native *Phragmites* under high-salinity conditions, while crabs played a role of improving nitrogen availability in this study, as discussed above. Therefore, we conclude that native soil reworkers (or ecosystem engineers such as crabs) are capable of altering the interactions among plant species through alleviating resource limitations.

We found that the crabs promoted the plants more effectively at the plot-level based on the significant "crab × species combinations" effects on density for all three species (Table 1), with the R_{crab}s of total aboveground biomass (Figure 3G–L) being generally higher than that of biomass per ramet (Figure 3A–F). In other words, the plants responded to crabs' modification on soils primarily through increased number, rather than size, of ramets. It was similar to Tyler's study [39], that nitrogen input could promote the spread of invasive *Spartina* sp. into previously unvegetated mudflats in estuaries of the western United States, and high nitrogen loading areas were characterized by higher total aboveground biomass and ramet density.

It should be noted that *Spartina* may use different strategies to compete with *Phragmites* and *Scirpus*. At the ramet-level, the R_{crab}s of *Spartina* in *Spartina-Phragmites* mixtures were higher than those in *Spartina-Scirpus* mixtures (Figure 2A, B, G, H), while the pattern reversed at the plot-level (Figure 3A, B, G, H). This implies that *Spartina* tended to invest more in vegetative growth to compete with *Phragmites*, whereas invested more in the clonal expansion when competing with *Scirpus*. This speculation could also be supported by the change of belowground biomass. The increase of belowground biomass in *Spartina-Phragmites* mixtures was not as distinct as that in their monocultures (Figure 4), implying that crabs might influence them more at aboveground in mixtures. Meanwhile the increase of belowground biomass in *Spartina-Scirpus* mixtures under crab disturbance might primary come from the *Spartina*, since the R_{crab}s were *Spartina* mococultures>*Spartina-Scirpus* mixtures>*Scirpus* monocultures (Figure 4).The different strategies that *Spartina* used might associated with the phylogenetic relatedness among competitors, i.e., both *Spartina* and *Phragmites* belong to Gramineae whereas *Scirpus* is a member of Cyperaceae. Species of close phylogenetic relatedness tend to have similar phenotype and niche, and thus compete more intensely [40]. *Spartina* ramets need to grow higher and larger to compete for space and light with similar sized *Phragmites* ramets. In contrast, the size superiority of *Spartina* over *Scirpus* might allow *Spartina* to invest more in belowground expansion and clonal growth.

This study suggests that native crabs may accelerate the invasion of *Spartina* through alleviating nitrogen limitation, which helps understand and predict the invasion success in natural salt marshes. However, other environmental factors should not be overlooked when extrapolating this conclusion from the mesocosm to natural ecosystems. In the Chongming Dongtan salt marsh, the role of crabs may rely on their habitat preferences. For example, crabs prefer to dwell in *Spartina* communities [41] and form a mutualistic relationship between crabs and *Spartina*, thereby promoting *Spartina* invasion. In addition, crab herbivory should not be neglected in natural ecosystems [42]. In conclusion, we suggest that native crab disturbance to soil can mediate the interspecific interactions between invasive *Spartina* and native *Phragmites*/*Scirpus* through modifying soil nitrogen availability. Although crab activities may facilitate *Spartina* invasion in salt marshes, the results obtained from the mesocosm system need to be further tested in the field.

Acknowledgments

We thank Guangxue Zhou for the mesocosm establishing and assisting in experiment management, and two anonymous reviewers provided numerous useful comments on earlier versions, which greatly improved the quality of this manuscript.

Author Contributions

Conceived and designed the experiments: XZ BL. Performed the experiments: XZ YC JS XW LS. Analyzed the data: XZ XJ. Wrote the paper: XZ XJ BL.

References

1. Pejchar L, Mooney HA (2009) Invasive species, ecosystem services and human well-being. Trends in Ecology & Evolution 24: 497–504.
2. Yurkonis KA, Meiners SJ, Wachholder BE (2005) Invasion impacts diversity through altered community dynamics. Journal of Ecology 93: 1053–1061.
3. Madrigal J, Kelt DA, Meserve PL, Gutierrez JR, Squeo FA (2011) Bottom-up control of consumers leads to top-down indirect facilitation of invasive annual herbs in semiarid Chile. Ecology 92: 282–288.
4. Sharon YS (1991) Indirect effects in community ecology: Their definition, study and importance. Trends in Ecology & Evolution 6: 206–210.
5. White EM, Wilson JC, Clarke AR (2006) Biotic indirect effects: a neglected concept in invasion biology. Diversity and Distributions 12: 443–455.
6. Sessions L, Kelly D (2002) Predator-mediated apparent competition between an introduced grass, *Agrostis capillaris*, and a native fern, *Botrychium australe* (Ophioglossaceae), in New Zealand. Oikos 96: 102–109.
7. Marler MJ, Zabinski CA, Callaway RM (1999) Mycorrhizae indirectly enhance competitive effects of an invasive forb on a native bunchgrass. Ecology 80: 1180–1186.
8. Cushman JH, Lortie CJ, Christian CE (2011) Native herbivores and plant facilitation mediate the performance and distribution of an invasive exotic grass. Journal of Ecology 99: 524–531.
9. Alberti J, Escapa M, Iribarne O, Silliman B, Bertness M (2008) Crab herbivory regulates plant facilitative and competitive processes in Argentinean marshes. Ecology 89: 155–164.
10. Poelman EH, Gols R, Snoeren TAL, Muru D, Smid HM, et al. (2011) Indirect plant-mediated interactions among parasitoid larvae. Ecology Letters 14: 670–676.
11. Vázquez DP, Simberloff D (2004) Indirect effects of an introduced ungulate on pollination and plant reproduction. Ecological Monographs 74: 281–308.
12. Mothershead K, Marquis RJ (2000) Fitness impacts of herbivory through indirect effects on plant-pollinator interactions in *Oenothera macrocarpa*. Ecology 81: 30–40.
13. Crooks JA (2002) Characterizing ecosystem-level consequences of biological invasions: the role of ecosystem engineers. Oikos 97: 153–166.
14. Medina-Roldan E, Paz-Ferreiro J, Bardgett RD (2012) Grazing-induced effects on soil properties modify plant competitive interactions in semi-natural mountain grasslands. Oecologia 170: 159–169.
15. Wright JP, Jones CG, Flecker AS (2002) An ecosystem engineer, the beaver, increases species richness at the landscape scale. Oecologia 132: 96–101.
16. Badano EI, Villarroel E, Bustamante RO, Marquet PA, Cavieres LA (2007) Ecosystem engineering facilitates invasions by exotic plants in high-Andean ecosystems. Journal of Ecology 95: 682–688.
17. Lee SY (1998) Ecological role of grapsid crabs in mangrove ecosystems: a review. Marine and Freshwater Research 49: 335–343.
18. Davis MA, Grime JP, Thompson K (2000) Fluctuating resources in plant communities: a general theory of invasibility. J Ecol 88: 528–534.
19. Qin HM, Chu TJ, Xu W, Lei GC, Chen ZB, et al. (2010) Effects of invasive cordgrass on crab distributions and diets in a Chinese salt marsh. Marine Ecology Progress Series 415: 177–187.
20. Wang JQ, Zhang XD, Jiang LF, Bertness MD, Fang CM, et al. (2010) Bioturbation of burrowing crabs promotes sediment turnover and carbon and nitrogen movements in an estuarine salt marsh. Ecosystems 13: 586–599.
21. Wang CH (2009) Effects of environmental Variation on growth, distribution of and interspecific interactions among dominant marsh plants at Chongming Dongtan. PhD dissertation, Fudan University, China.
22. Jiang LF, Luo YQ, Chen JK, Li B (2009) Ecophysiological characteristics of invasive *Spartina alterniflora* and native species in salt marshes of Yangtze River estuary, China. Estuarine, Coastal and Shelf Science 81: 74–82.
23. Wang JQ (2008) Habitat selection and ecosystem engineering effects of dominant burrowing crabs in salt marshes in the Yangtze River estuary. PhD dissertation, Fudan University, China.
24. R Core Team (2013). R: A language and environment for statistical computing. R Foundation for Statistical Computing, Vienna, Austria. Available: http://www.R-project.org/. Accessed 2013 Aug 7.
25. Brooks ML (2003) Effects of increased soil nitrogen on the dominance of alien annual plants in the Mojave Desert. Journal of Applied Ecology 40: 344–353.
26. Kiehl K, Esselink P, Bakker JP (1997) Nutrient limitation and plant species composition in temperate salt marshes. Oecologia 111: 325–330.
27. Crain C (2007) Shifting nutrient limitation and eutrophication effects in marsh vegetation across estuarine salinity gradients. Estuaries and Coasts 30: 26–34.
28. Nadelhoffer KJ, Giblin AE, Shaver GR, Laundre JA (1991) Effects of temperature and substrate quality on element mineralization in six arctic soils. Ecology 72: 242–253.
29. Robinson CH, Wookey PA, Parsons AN, Potter JA, Callaghan TV, et al. (1995) Responses of plant litter decomposition and nitrogen mineralisation to simulated environmental change in a high arctic polar semi-desert and a subarctic dwarf shrub heath. Oikos 74: 503–512.
30. Peng RH, Fang CH, Li B, Chen JK (2011) *Spartina alterniflora* invasion increases soil inorganic nitrogen pools through interactions with tidal subsidies in the Yangtze Estuary, China. Oecologia 165: 797–807.
31. Liao CZ, Luo YQ, Jiang LF, Zhou XH, Wu XW, et al. (2007) Invasion of *Spartina alterniflora* enhanced ecosystem carbon and nitrogen stocks in the Yangtze Estuary, China. Ecosystems 10: 1351–1361.
32. Lu M, Yang YH, Luo YQ, Fang CM, Zhou XH, et al. (2011) Responses of ecosystem nitrogen cycle to nitrogen addition: a meta-analysis. New phytologist 189: 1040–1050.
33. Mack MC, Schuur EA, Bret-Harte MS, Shaver GR, Chapin FS (2004) Ecosystem carbon storage in arctic tundra reduced by long-term nutrient fertilization. Nature 431: 440–443.
34. Zeng D, Li L, Fahey TJ, Yu Z, Fan Z, et al. (2009) Effects of nitrogen addition on vegetation and ecosystem carbon in a semi-arid grassland. Biogeochemistry 98: 185–193.
35. Pearcy RW, Ehleringer J (1984) Comparative ecophysiology of C3 and C4 plants. Plant, Cell & Environment 7: 1–13.
36. Dai T, Wiegert RG (1997) A field study of photosynthetic capacity and its response to nitrogen fertilization in *Spartina alterniflora*. Estuarine Coastal and Shelf Science 45: 273–283.
37. Rickey MA, Anderson RC (2004). Effects of nitrogen addition on the invasive grass *Phragmites australis* and a native competitor *Spartina pectinata*. Journal of Applied Ecology, 41: 888–896.
38. Wang CH, Lu M, Yang B, Yang Q, Zhang XD, et al. (2009) Effects of environmental gradients on the performances of four dominant plants in a Chinese saltmarsh: implications for plant zonation. Ecological Research 25: 347–358.
39. Tyler AC, Lambrinos JG, Grosholz ED (2007) Nitrogen inputs promote the spread of an invasive marsh grass. Ecological Applications 17: 1886–1898.
40. Valiente-Banuet A, Verdú M (2008) Temporal shifts from facilitation to competition occur between closely related taxa. Journal of Ecology 96: 489–494.
41. Wang JQ, Zhang XD, Nie M, Fu CZ, Chen JK, et al. (2008) Exotic *Spartina alterniflora* provides compatible habitats for native estuarine crab Sesarma dehaani in the Yangtze River estuary. Ecological Engineering 34: 57–64.
42. Coverdale TC, Altieri AH, Bertness MD (2012) Belowground herbivory increases vulnerability of New England salt marshes to die-off. Ecology 93: 2085–2094.

10

Ecosystem Engineering by Seagrasses Interacts with Grazing to Shape an Intertidal Landscape

Tjisse van der Heide[1x], Johan S. Eklöf[2,6,7], Egbert H. van Nes[3], Els M. van der Zee[4,5], Serena Donadi[2], Ellen J. Weerman[1], Han Olff[1], Britas Klemens Eriksson[2]

1 Community and Conservation Ecology Group, Centre for Ecological and Evolutionary Studies (CEES), University of Groningen, Groningen, The Netherlands, **2** Marine Benthic Ecology and Evolution, Centre for Ecological and Evolutionary Studies (CEES), University of Groningen, Groningen, The Netherlands, **3** Aquatic Ecology and Water Quality Management Group, Department of Environmental Sciences, Wageningen University, Wageningen, The Netherlands, **4** Department of Marine Ecology, NIOZ Royal Netherlands Institute for Sea Research, Den Burg, Texel, The Netherlands, **5** Animal Ecology Group, Centre for Ecological and Evolutionary Studies (CEES), University of Groningen, Groningen, The Netherlands, **6** Department of Biology and Environmental Sciences – Kristineberg, University of Gothenburg, Fiskebäckskil, Sweden, **7** Department of Systems Ecology, Stockholm University, Stockholm, Sweden

Abstract

Self-facilitation through ecosystem engineering (i.e., organism modification of the abiotic environment) and consumer-resource interactions are both major determinants of spatial patchiness in ecosystems. However, interactive effects of these two mechanisms on spatial complexity have not been extensively studied. We investigated the mechanisms underlying a spatial mosaic of low-tide exposed hummocks and waterlogged hollows on an intertidal mudflat in the Wadden Sea dominated by the seagrass *Zostera noltii*. A combination of field measurements, an experiment and a spatially explicit model indicated that the mosaic resulted from localized sediment accretion by seagrass followed by selective waterfowl grazing. Hollows were bare in winter, but were rapidly colonized by seagrass during the growth season. Colonized hollows were heavily grazed by brent geese and widgeon in autumn, converting these patches to a bare state again and disrupting sediment accretion by seagrass. In contrast, hummocks were covered by seagrass throughout the year and were rarely grazed, most likely because the waterfowl were not able to employ their preferred but water requiring feeding strategy ('dabbling') here. Our study exemplifies that interactions between ecosystem engineering by a foundation species (seagrass) and consumption (waterfowl grazing) can increase spatial complexity at the landscape level.

Editor: Richard K.F. Unsworth, Swansea University, United Kingdom

Funding: This study was financially supported by a post-doctoral grant from The Swedish Research Council for Environment, Agricultural Sciences and Spatial Planning (FORMAS) to JSE (grant no. 2008-839), and by a grant from the ZKO program from the Netherlands Organization of Scientific Research (NWO) to BKE, TVDH, EW, SD and EVDZ (grant no. 839.08.310). The funders had no role in study design, data collection and analysis, decision to publish, or preparation of the manuscript.

Competing Interests: The authors have declared that no competing interests exist.

* E-mail: T.van.der.Heide@rug.nl

Introduction

Spatial heterogeneity is important for the functioning of many different ecosystems, because it can enhance primary productivity, increase the biodiversity and carrying capacity, and stabilize the ecosystem [1–4]. Studies from a wide range of terrestrial and marine ecosystems have demonstrated that ecosystem engineers, i.e., organisms that significantly modify their abiotic environment [5,6], often determine spatial structuring in ecosystems [2]. An important factor often controlling the extent to which the system is modified is the density of the ecosystem engineer [7], while the modified environment in turn also positively or negatively affects the engineer again. In many cases, such feedback mechanisms cause an increase in the spatial patchiness of the ecosystem [8–11].

Consumer-resource interactions may also cause patchiness. For instance, plant-herbivore interactions ranging from the arctic tundra to tropical savannahs cause irregular patchy 'landscape mosaics' of intensively grazed 'lawns' of short vegetation alternating with ungrazed patches of tall vegetation in ecosystems ranging from the arctic tundra to tropical savannahs [3,12,13]. Intense grazing of the lawns combined with increased nutrient input by herbivore excrement facilitates growth of consumable and nutrient rich vegetation, while the nutrient-poor, tall vegetation excludes herbivores [12]. Similar to landscapes dominated by ecosystem engineers, these systems are driven by feedbacks. In contrast, however, these feedbacks are not characterized by biotic-abiotic interactions, but mainly driven by trophic interactions.

In this study, we investigated the mechanisms behind a spatial mosaic of low-tide exposed hummocks and waterlogged hollows on an intertidal mudflat dominated by the seagrass *Zostera noltii*, which is periodically grazed by waterfowl (Fig. 1A). Using this system as a model, we tested whether an interplay between ecosystem engineering by a foundation species and herbivore grazing activity can lead to patchiness similar to those observed for habitat modification or consumer-resource interactions alone. Intertidal seagrasses like *Z. noltii* are density-dependent ecosystem engineers in the sense that they progressively reduce hydrodynamics and accrete sediment with increasing shoot density [10,14–16]. Grazing by waterfowl is a common phenomenon in seagrass systems. In the Wadden Sea, grazing on *Z. noltii* mainly takes place in autumn by overwintering brent geese (*Branta bernicla*) and widgeon (*Anas penelope*) that migrate from the arctic tundra in

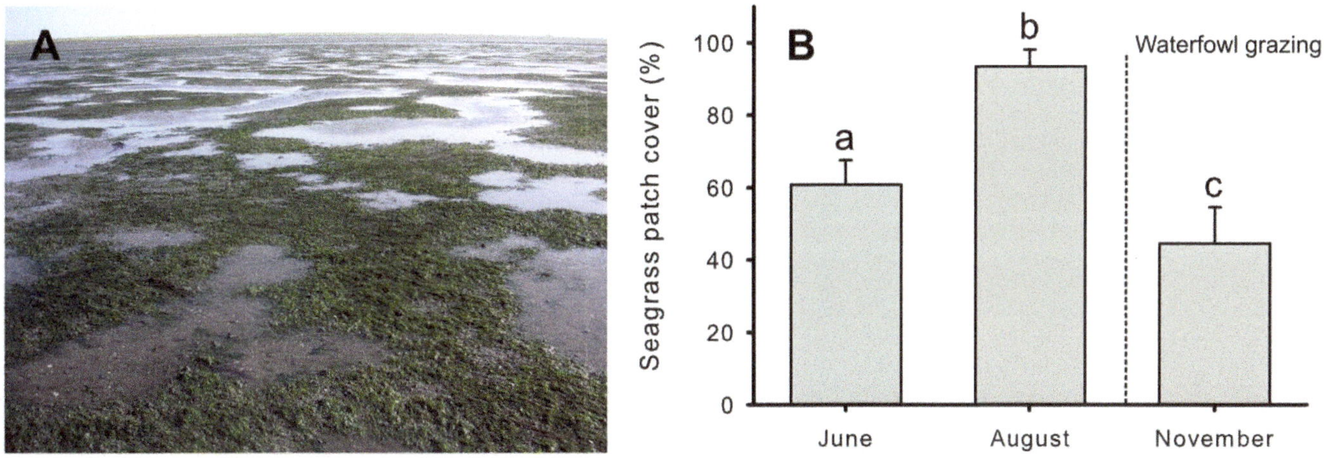

Figure 1. Low-tide exposed hummocks with seagrass alternate with waterlogged, bare hollows in June (A). Seagrass patch cover changed significantly over time (ANOVA: $F_{2,17} = 66.6$, $p < 0.001$) from about 61% in June, to over 93% in August, followed by a sudden decrease again to 44% in November due to waterfowl grazing in September and October (B). Error bars indicate SD (number of replicates = 6).

Northern Scandinavia and Siberia [17]. The birds use a number of different feeding techniques that depend on the water level. First, upending can be used when the actual water level is still relatively high; next dabbling is employed in areas with a few centimetres of water, and finally grubbing is the most common strategy on completely exposed parts [17,18]. Although waterfowl can consume significant amounts of both above- and belowground biomass, seagrass is generally not completely removed, but reduced to about 5 to 15% cover. This is because below this threshold feeding becomes energetically unprofitable for the birds, regardless of their feeding mode ('giving-up density') [17,19].

We used a combination of field surveys, a seagrass removal experiment and a spatially explicit model to identify the driving mechanisms behind the observed spatial mosaic. First, we quantified differences in sediment height between hummocks and hollows and measured the patchiness of the system and its change across the season. Second, we performed a seagrass removal experiment to test whether seagrass indeed modified its abiotic environment by sediment accumulation. Third, the effect of grazing by waterfowl on the spatial structure of the system was assessed by bird observations in September and October. Finally, to test whether the identified interactions could indeed explain the observed spatial and temporal patterns, we constructed and analyzed a spatially explicit model based on our field data.

Field Study Methods

Site Description

The study was conducted on the intertidal mudflats at Emmapolder, The Netherlands (53° 28′ 0 N, 6° 45′ 0 E) in 2009. It is one of the few areas in the Dutch Wadden Sea where *Zostera noltii* still occurs. Growth of this perennial seagrass starts in late April, peaks in summer, and ends in late autumn (October/November) with seasonal senescence [20].

Measurements and Experiments

Differences in relative bed level height between hummock and hollows were measured using a Trimble Spectra Precision LL500 Laser Level (Trimble, California, USA). We evaluated the size of seagrass patches at the site by noting the positions of seagrass edges in centimetres along six randomly selected 50 m transects in June, August and November in a 1-hectare study area. Next, seagrass cover was determined by calculating the percentage of mudflat covered by seagrass patches.

To test whether seagrass presence would indeed result in an elevated bed level, we experimentally removed all above- and belowground biomass in six 1 m^2 plots that were situated on hummocks. Next, we compared the change in bed level from the start in June (a few days after seagrass removal) to the end of the experiment in August to six untreated control plots. In order to remove only seagrass and not the sediment, plywood frames were hammered 20 cm deep along plot edges during low water. Next, the frame was filled with water and all seagrass shoots, roots and rhizomes were removed using a hand rake. After allowing suspended particles to settle, the water was slowly released and the frame gently removed. Height measurements were performed a few days after seagrass removal in June and again in August.

Finally, the effect of grazing by waterfowl was evaluated by bird observation during two low tides (~4 observation hours per tide) in September and October. For this purpose, we used a Swarovski ATM80-HD spotting scope (zoom ocular 20–60×) placed on a 9 m high dike about 350 m from the site. One observer (JSE) noted the total number of each bird species, recorded the number of feeding birds, their feeding mode and whether they were feeding on hummocks and hollows. To enable an assessment of whether the proportion of feeding birds on hummocks differed from those in hollows, we also noted the percentage cover of hollows and hummocks by visual estimation in a randomly thrown 0.25 m^2 frame (60 replicates) in the 1-hectare study area.

Statistical Analyses

Differences in seagrass patch cover between June, August and November, as well as data from the seagrass removal experiment were first tested for normality. As the data were normally distributed, change in patch size was then analyzed using one-way ANOVA. Because Levene's test showed that variances were not equal, we used a Games-Howell test for post-hoc comparisons. For the seagrass removal experiment, we used a two-tailed independent samples t-test to compare the mean rate of change in sediment height between control and removal plots, and we used a two-tailed paired samples t-test to test for differences in sediment height within treatments before and after the experiment. Bird observation data were analyzed with a Chi-square test.

Field Study Results

Height measurements in June demonstrated that hummocks were on average 5.8±1.2 cm (mean ± SD; n = 6) higher than hollows. Analysis of transects of seagrass cover in June, August and November demonstrate significant differences between all three periods (ANOVA: $F_{2,17} = 66.6$, $p<0.001$; Fig. 1B). Seagrass cover was around 61% in June, increased to over 93% in August and decreased again to just over 44% in November.

Experimental seagrass removal in 1-m^2 plots resulted in a decrease in bed level height, indicating loss of sediment stability. The bed level of the control plots showed a slight increase of 5.8±10 mm (mean ± SD), but this was not significant (paired t-test: $t_5 = 1.41$, $p = 0.218$). The bed level height in the experimental removal plots decreased significantly compared to the control plots (t-test: $t_{10} = 4.61$, $p<0.001$) and the removal plots also decreased significantly in height over time (paired t-test: $t_5 = -5.86$, $p = 0.002$) by 17.6±7.3 mm (mean ± SD).

Bird observations showed that mixed flocks of around 150 brent geese (70%) and widgeon (30%) grazed at the site in September and October, resulting in densities of around 200 individuals per hectare. Even though hollows covered only 52% of the area, over 98% of the feeding birds (80% of all) fed in hollows ($\chi^2 = 106.89$, df = 1, $p<0.001$; Fig. 2A). Visual inspection revealed that birds mostly fed by dabbling in hollows and by grubbing on hummocks. Furthermore, hollows showed distinct signs of waterfowl grazing (e.g., trampling and beak marks, uprooted plants, floating leaves) and an eventual reduction of about 90% of the original seagrass cover (Fig. 2B). In contrast, hummocks were hardly impacted.

Model Description

Results from the field study suggest that the observed two-state mosaic of hummocks and hollows resulted from an interaction between density-dependent habitat modification by seagrass and selective grazing by waterfowl. Dense seagrass stands accrete sediment, which causes elevation of the bed level compared to the bare surroundings. However, sediment accretion is disrupted in patches that are grazed by waterfowl in autumn. This results in low-tide waterlogged hollows in the meadow that are colonized by seagrass during the growth season, but are then selectively grazed again by the waterfowl – most likely because their preferred feeding strategy (dabbling) is only possible in the water-logged hollows [18].

To test whether these interactions could indeed explain the observed patchy landscape, we constructed a minimal, spatially explicit computer model based on our empirical data (see table 1 for data sources). The model describes changes in seagrass shoot density and bed level height in two differential equations:

$$\frac{dZ_{ij}}{dt} = f_S\, r\, Z_{ij}\left(1-\frac{Z_{ij}}{K}\right) - m\, Z_{ij} - G_{in} f_G \frac{Z_{ij}}{Z_{ij}+h_1} + d_Z\, \Delta Z_{ij} \quad (1)$$

$$\frac{dH_{ij}}{dt} = s\,\frac{Z_{ij}}{Z_{ij}+h_2} - e\, H_{ij} + d_H\, \Delta H_{ij} \quad (2)$$

Z_{ij} and H_{ij} are the seagrass shoot density (shoots m^{-2}) and bed level height (m) in grid cell i, j respectively. The Laplace gradient operators d_Z and d_H (day^{-1}) describe diffusion of seagrass and sediment to neighboring grid cells to mimic lateral dispersion of seagrass and to prevent unrealistically large differences in bed level height between bordering cells (i.e., very steep slopes). Function f_S (t), with t as time, describes the seasonal differences in seagrass growth:

$$f_S = e^{-0.5\left(\cos\left(\frac{t\pi/365}{0.15}\right)\right)^2} \quad (3)$$

The maximum net logistic growth rate is described by r (day^{-1}), m is the mortality rate of seagrass due to natural senescence (day^{-1}) and K is the carrying capacity (shoots m^{-2}). The maximum sedimentation rate of suspended sediment is described by s (m day^{-1}), maximum erosion rate e (day^{-1}) is calculated as the sedimentation rate s divided by the maximum bed level height H_{max} (m) and h_2 (shoots m^{-2}) is the seagrass density at which the sedimentation rate is 50% of the maximum sedimentation rate s. G_{in} is the maximum feeding rate for individual birds (sh ind^{-1} day^{-1}) and h_1 (shoots m^{-2}) is the half saturation constant for the density dependent feeding rate on seagrass. Finally, function f_G describes the local number of birds per m^2 which is dependent on

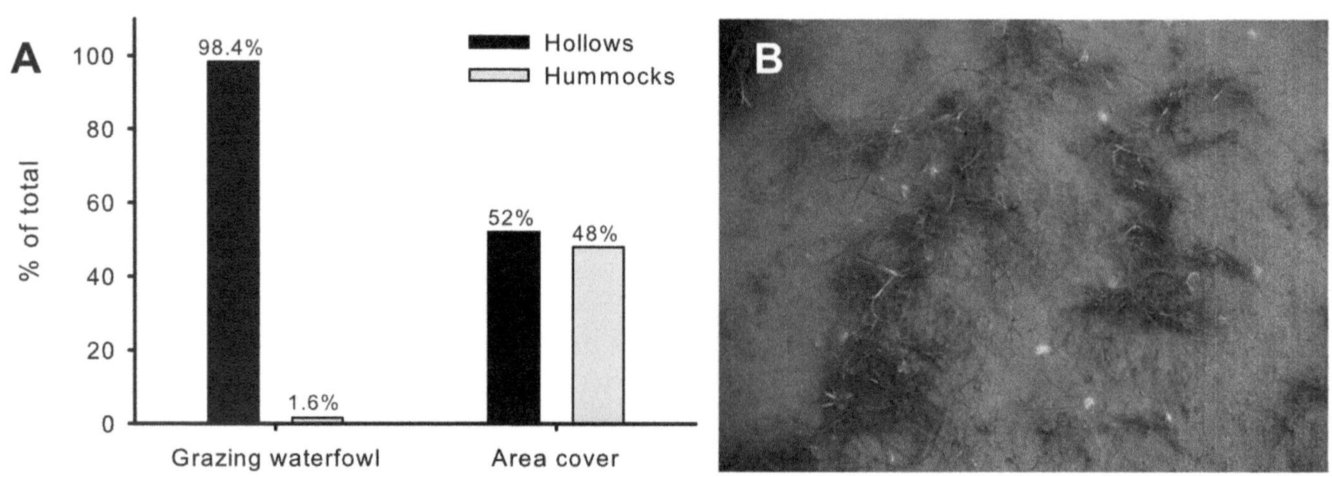

Figure 2. Over 98% of all grazing waterfowl was feeding in hollows (A), whereas hollows covered only 52% of the entire 1-hectare area ($\chi^2 = 106.89$, df = 1, p<0.001). Visual observations showed obvious grazing scars in the hollows (B): trampling and beak marks, uprooted plants, floating leaves, and a removal of seagrass up to 90% of the original biomass. Hummocks, in contrast, were minimally impacted.

Table 1. Default parameter settings of the spatially explicit model.

	Default	Unit	Description	Source
Variables				
Z		sh m^{-2}	Seagrass shoot density	
H		m	Bed height	
G_F		ind m^{-2}	Number of feeding waterfowl	
Parameters				
r	0.4	day^{-1}	Relative growth rate	*
m	0.007	day^{-1}	Mortality rate	*
K	5976	sh m^{-2}	Carrying capacity	*
s	0.00035	m day^{-1}	Sedimentation rate	*
H_{max}	0.2	m	Maximum bed height	*
e	s/H_{max}	day^{-1}	Erosion rate	
h_1	594	sh m^{-2}	Half rate constant for seagrass feeding rate	1,*
h_2	1500	sh m^{-2}	Half rate constant for sedimentation rate	*
G_{max}	0.03	ind m^{-2}	Maximum number of waterfowl	*
G_{in}	16967	sh ind^{-1} day^{-1}	Maximum intake rate for waterfowl foraging	1,*
GUD	0.1 K	sh m^{-2}	Giving-up density for waterfowl foraging	1,2
G_L	0.984		Fraction of waterfowl feeding in hollows	*,2
G_H	0.016		Fraction of waterfowl feeding on hummocks	*
H_C	0.08	m	Critical bed height for waterfowl foraging	*
d_Z	0.2 r	day^{-1}	Diffusion rate constant for seagrass	±
d_H	0.2 e	day^{-1}	Diffusion rate constant for bed height	±

(*) Obtained from field data, (1) Percival & Evans (1997), (2) Ganter (2000), (±) Gradient operators d_Z and d_H were set at 20% of the maximum growth and erosion rate respectively. Parameters measured in the field resulted either from fieldwork published in this study or from [10].

the season, seagrass density and the bed level height. Waterfowl are absent during most of the season ($f_G = 0$), but are present on a gridcell in autumn between days 243 and 334 of the year (t_y) provided that local seagrass density is above the giving-up density threshold (GUD; shoots m^{-2}):

$$f_G = G_F \quad if\ 243 < t_y < 334\ \&\ Z_{ij} > GUD \tag{4}$$

The number of feeding birds (G_F, ind m^{-2}) depends on the number of present waterfowl and the elevation of the grid cell:

$$\begin{aligned} G_F &= G_{max}\, G_H \quad if\ H_{ij} > H_C \\ G_F &= G_{max}\, G_L \quad if\ H_{ij} < H_C \end{aligned} \tag{5}$$

Where G_{max} describes the maximum number of birds (ind m^{-2}), G_H and G_L are the feeding fractions on high and low parts respectively and H_c is the critical elevation threshold (m).

To investigate whether the model could indeed generate the spatial and temporal patchiness observed in the field, we randomized seagrass density and sediment height between zero and half of their maxima (0.5 K and 0.5 E_{max} respectively) across a 200×200 grid and ran the model at default parameter settings (Table 1) for 100 years. Additionally, we performed a bifurcation analysis on the maximum number of waterfowl and the maximum sedimentation rate in a non-spatial version of the model (i.e., a 'single grid cell' system). Bistability in this non-spatial model is a necessary condition for mosaics in the spatial version of the model; i.e., alternative 'hollow' and 'hummock' states need to be possible at the same external conditions. To examine in what ranges of sedimentation rates and waterfowl densities such alternative stable states are possible, each parameter was gradually increased in small steps. After each increase, the model was allowed to stabilize and the average shoot density and sediment height across the year were recorded. Next, the same procedure was also performed in a backward manner, i.e., a gradual decrease of parameter values. We used the results from this bifurcation analysis to construct a two-dimensional plot of waterfowl numbers (G_{max}) and sedimentation rate (s). The range of parameters where alternative stable states are present in this model indicates the parameter space where mosaics are possible in the spatial model. Note that, although local bistability is a requisite for mosaics, patch size and distribution in the spatial model will also depend on the initial conditions of Z and E.

Model Results

Model simulations confirmed that the identified interactions between sediment accretion by seagrass and grazing by waterfowl could indeed explain the observed spatial and temporal patterns (Fig. 3, Movies S1 & S2). Similar to the field situation (Fig. 1B), seagrasses covered hummocks throughout the year. Hollows were colonized during the growth season, leading to full cover in summer. However, these depressions are grazed in autumn, thereby preventing these patches to accumulate sediment and returning the system to its initial patchy state.

The two-dimensional bifurcation analysis of the maximum sedimentation rate (s) and the maximum number of waterfowl

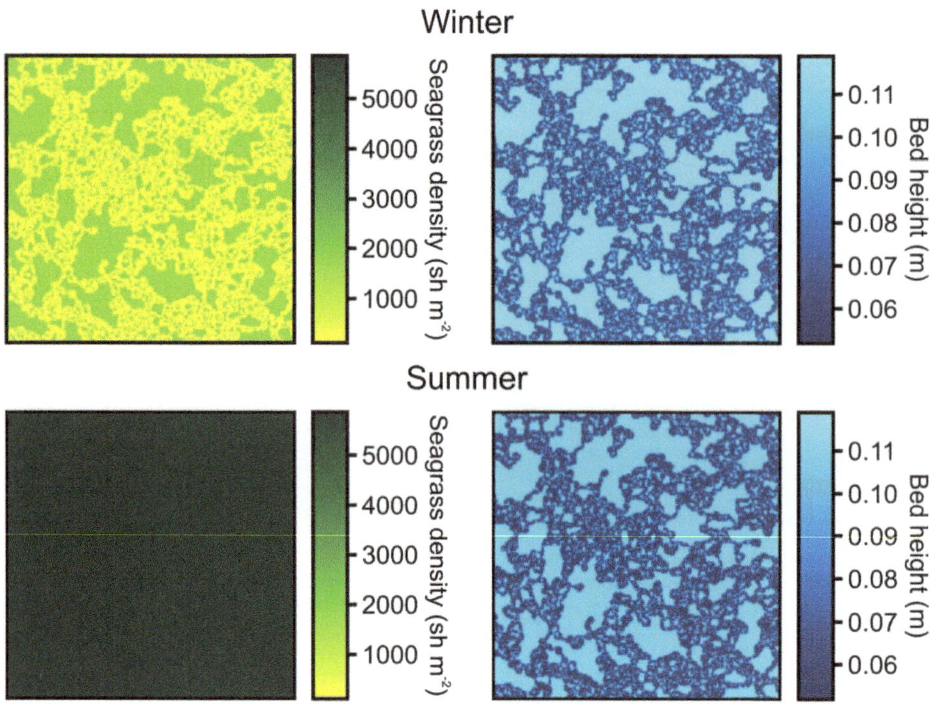

Figure 3. Model simulations at defaults settings demonstrated that interactions between sediment accretion by seagrass and selective grazing of low parts by waterfowl in autumn could indeed explain the observed temporal and spatial patterns in the field.

(G_{max}) in the non-spatial model demonstrated the possibility for spatial mosaics over a wide range of parameter settings (Fig. 4). At low sedimentation rates, bistability appears just above zero waterfowl numbers until well over the observed number of waterfowl in the field. At extremely high waterfowl numbers, the mosaic collapses to a state where the complete meadow is grazed in autumn, but is still able to recover in the next growth season. When the sedimentation rate is set extremely high, hollows can accumulate sufficient sediment in one growth season the reach a 'hummock state', thereby excluding most of the waterfowl in the meadow.

Discussion

Spatial heterogeneity in ecosystems can be caused by abiotic variability, but can also result from feedback mechanisms [8,21]. Such feedbacks emerge when organisms significantly modify their environment (i.e., ecosystem engineering) [2,21], but can also be caused by plant-herbivore, predator-prey and host-parasitoid interactions [12,13,22,23]. In this study, we report on a spatial mosaic of low-tide exposed hummocks and waterlogged hollows that results from an interaction between sediment accretion by seagrass and selective grazing by waterfowl, thereby illustrating that interactions between ecosystem engineering by a foundation species and grazing can cause spatial structuring in ecosystems. This finding is important because studies ranging from arctic to tropical environments and from terrestrial to marine ecosystems have demonstrated that spatial heterogeneity is often essential for ecosystem functioning [1–4].

Our study exemplifies how an interplay between biotic and abiotic factors can spatially structure intertidal seagrasses. The possibility of patchiness being solely driven by underlying abiotic heterogeneity or seagrass ecosystem engineering can be discarded because seagrasses colonized bare areas in summer and were observed to only retreat again due to waterfowl grazing. Similarly, our observations also reveal that waterfowl grazing alone is not sufficient to explain the observed spatial patchiness and temporal trends, as observations showed that grazing intensity was strongly driven by sediment height, which in turn was dependent on sediment accretion by seagrass. Although our study captured the most important structuring mechanisms, some processes have been disregarded or were simplified in both the model and the field experiments. Examples of factors possibly affecting the observed patchiness are bioturbation by infauna [10], local differences in current velocity and sedimentation, and stochastic events like storms or desiccation of seagrasses at low tide during days with high temperatures [24]. Another simplification is our description of grazing in the model. Here, grazing is described as a process that only removes seagrasses from the system. In reality, depending on the feeding mode (i.e., dabbling, grubbing), grazing by waterfowl also decreases the cohesiveness of the sediment, thereby increasing erosion in the impacted areas. As dabbling (the preferred feeding strategy) directly results in sediment resuspension in the hollows, this will most likely have the most pronounced effect on sediment erosion. This suggests that we may have underestimated the overall effect of waterfowl grazing on erosion of grazed patches in our model, which would in turn imply that the 'hollow-state' is in reality more resilient than in the default setting of our model. Nevertheless, our bifurcation analysis (Fig. 4) clearly demonstrates that alternative stable states exist over a much wider range of grazing intensities and sedimentation-erosion balances than those measured in the field, indicating that an under- or overestimation of grazing and/or erosion does not fundamentally alter our results.

The spatial structure of seagrass ecosystems is often attributed to abiotic factors such as wave action, currents, sediment transport and light [25–29]. Recent studies, however, have shown that seagrasses are strong ecosystem engineers that often improve their

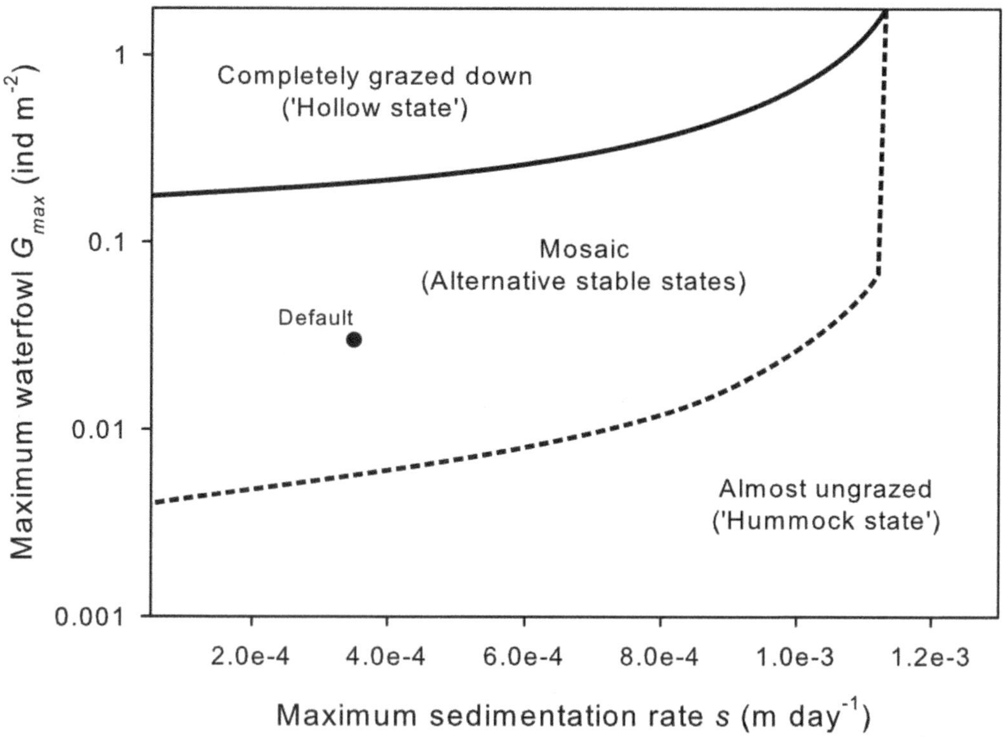

Figure 4. A two-dimensional bifurcation analyses of the maximum sedimentation rate *s* and the maximum number of waterfowl G_{max} (on a log-scale) in the non-spatial version of the model demonstrates that landscape mosaics can appear over a wide range of parameter settings. The solid line indicates the threshold between a mosaic state and the conditions where seagrass is completely grazed down in autumn. A state with an elevated bed level and low grazing is the only possible state below and right of the dashed line. Default parameter settings are indicated by the black dot. The effect of sedimentation rate was analysed in the range of 0.05 to 1.13 mm day^{-1}, bird densities were tested in the range of 0.001 to 2 ind m^{-2}. Note that formation of mosaics is not possible in the absence of sedimentation and/or birds.

own conditions, for instance by lowering nutrient levels, attenuating hydrodynamics and accumulating sediments [15,30,31]. Moreover, when such positive feedbacks interact with negative feedbacks, it may lead to spatial self-organization in seagrasses [24]. Furthermore, intensive grazing by waterfowl, turtles, dugongs, manatees and urchins has been demonstrated to have significant effects on the spatiotemporal structure and overall productivity of seagrasses as well [17,32–34]. Over the last century, seagrass meadows have been increasingly disturbed by human activities (e.g., eutrophication, siltation, dredging), resulting in dramatic and large-scale losses worldwide that were in many cases unexpected [15,35,36]. Our results support the notion that consideration of biological interactions between seagrasses and associated organisms may be crucial for conservation and restoration efforts in many seagrass ecosystems [37,38].

Spatial patchiness caused by ecosystem engineering interacts with both abiotic stress and grazing in various ecosystems [21,24,39–41]. However, grazing in these previously studied systems is not part of the structuring feedback mechanisms, and the disruption of these feedbacks by grazing therefore typically induces loss of spatial structure [9,39,40]. In contrast, the interaction with grazing is the actual cause of spatial patchiness in our system. Furthermore, in contrast to results from resource-limited systems, our model does not predict a complete collapse of the vegetation above a certain threshold for grazing [39], but rather a homogeneous state of intense periodic grazing (Fig. 4). The seagrass meadow in our model does not collapse because (1) seagrass growth is not resource-limited and (2) waterfowl grazing is periodic and does not continue below 10% of the maximum seagrass density ('giving-up density'). These results are in agreement with other studies on waterfowl grazing in intertidal seagrasses that show that seagrass survival and production are either not markedly impacted [17] or even facilitated by waterfowl grazing [33,42]. Our study therefore supports the notion that the driving mechanisms behind spatial structuring should be well understood before using patchiness as an indicator of stress in ecosystems [24,41].

Author Contributions

Conceived and designed the experiments: TVDH JSE BKE. Performed the experiments: TVDH JSE BKE EVDZ SD. Analyzed the data: TVDH EVN JSE. Wrote the paper: TVDH JSE EVN EVDZ SD EW HO BKE.

References

1. Eriksson BK, van der Heide T, van de Koppel J, Piersma T, van der Veer HW, et al. (2010) Major Changes in the Ecology of the Wadden Sea: Human Impacts, Ecosystem Engineering and Sediment Dynamics. Ecosystems 13: 752–764.
2. Hastings A, Byers JE, Crooks JA, Cuddington K, Jones CG, et al. (2007) Ecosystem engineering in space and time. Ecology Letters 10: 153–164.
3. Person BT, Herzog MP, Ruess RW, Sedinger JS, Anthony RM, et al. (2003) Feedback dynamics of grazing lawns: coupling vegetation change with animal growth. Oecologia 135: 583–592.
4. Pringle RM, Doak DF, Brody AK, Jocque R, Palmer TM (2010) Spatial Pattern Enhances Ecosystem Functioning in an African Savanna. Plos Biology 8: e1000377.
5. Jones CG, Lawton JH, Shachak M (1994) Organisms as ecosystem engineers. Oikos 69: 373–386.
6. Jones CG, Lawton JH, Shachak M (1997) Positive and negative effects of organisms as physical ecosystem engineers. Ecology 78: 1946–1957.
7. Jones CG, Gutierrez JL, Byers JE, Crooks JA, Lambrinos JG, et al. (2010) A framework for understanding physical ecosystem engineering by organisms. Oikos 119: 1862–1869.
8. Kefi S, Rietkerk M, van Baalen M, Loreau M (2007) Local facilitation, bistability and transitions in arid ecosystems. Theoretical Population Biology 71: 367–379.
9. Weerman EJ, Herman PMJ, Van de Koppel J (2011) Top-down control inhibits spatial self-organization of a patterned landscape. Ecology 92: 487–495.
10. Eklöf JS, van der Heide T, Donadi S, van der Zee EM, O'Hara R, et al. (2011) Habitat-Mediated Facilitation and Counteracting Ecosystem Engineering Interactively Influence Ecosystem Responses to Disturbance. PLoS ONE 6: e23229.
11. van Wesenbeeck BK, van de Koppel J, Herman PMJ, Bakker JP, Bouma TJ (2007) Biomechanical warfare in ecology; negative interactions between species by habitat modification. Oikos 116: 742–750.
12. Cromsigt J, Olff H (2008) Dynamics of grazing lawn formation: an experimental test of the role of scale-dependent processes. Oikos 117: 1444–1452.
13. McNaughton SJ (1984) Grazing lawns - Animals in herds, plant form, and coevolution. American Naturalist 124: 863–886.
14. Bos AR, Bouma TJ, de Kort GLJ, Van Katwijk MM (2007) Ecosystem engineering by annual intertidal seagrass beds: sediment accretion and modification. Estuarine Coastal and Shelf Science 74: 344–348.
15. van der Heide T, van Nes EH, Geerling GW, Smolders AJP, Bouma TJ, et al. (2007) Positive feedbacks in seagrass ecosystems – Implications for success in conservation and restoration. Ecosystems 10: 1311–1322.
16. van der Heide T, Van Nes EH, van Katwijk MM, Olff H, Smolders AJP (2011) Positive feedbacks in seagrass ecosystems – Evidence from large-scale empirical data. PLoS ONE 6: e16504.
17. Ganter B (2000) Seagrass (Zostera spp.) as food for Brent geese (Branta bernicla): an overview. Helgoland Marine Research 54: 63–70.
18. Fox AD (1996) Zostera exploitation by Brent geese and Wigeon on the Exe estuary, southern England. Bird Study 43: 257–268.
19. Percival SM, Evans PR (1997) Brent Geese Branta bernicla and Zostera; Factors affecting the exploitation of a seasonally declining food resource. Ibis 139: 121–128.
20. Vermaat JE, Hootsmans MJM, Nienhuis PH (1987) Seasonal dynamics and leaf growth of Zostera noltii Hornem, a perennial intertidal seagrass. Aquatic Botany 28: 287–299.
21. Rietkerk M, Dekker SC, de Ruiter PC, van de Koppel J (2004) Self-organized patchiness and catastrophic shifts in ecosystems. Science 305: 1926–1929.
22. Hassell MP, Comins HN, May RM (1991) Spatial structure and chaos in insect population-dynamics. Nature 353: 255–258.
23. Maron JL, Harrison S (1997) Spatial pattern formation in an insect host-parasitoid system. Science 278: 1619–1621.
24. van der Heide T, Bouma TJ, van Nes EH, van de Koppel J, Scheffer M, et al. (2010) Spatial self-organized patterning in seagrasses along a depth gradient of an intertidal ecosystem. Ecology 91: 362–369.
25. Bell SS, Robbins BD, Jensen SL (1999) Gap dynamics in a seagrass landscape. Ecosystems 2: 493–504.
26. Marba N, Duarte CM (1995) Coupling of Seagrass (Cymodocea-Nodosa) Patch Dynamics to Subaqueous Dune Migration. Journal of Ecology 83: 381–389.
27. Patriquin DG (1975) "Migration" of blowouts in seagrass beds at Barbados and Carriacou, West Indies, and its ecological and geological implications Aquatic Botany 1: 163–189.
28. Robbins BD, Bell SS (1994) Seagrass landscapes - A terrestrial approach to the marine subtidal environment. Trends in Ecology & Evolution 9: 301–304.
29. Fonseca M, Whitfield PE, Kelly NM, Bell SS (2002) Modeling seagrass landscape pattern and associated ecological attributes. Ecological Applications 12: 218–237.
30. Moore KA (2004) Influence of seagrasses on water quality in shallow regions of the lower Chesapeake Bay. Journal of Coastal Research: 162–178.
31. van der Heide T, Smolders AJP, Rijkens BGA, van Nes EH, van Katwijk MM, et al. (2008) Toxicity of reduced nitrogen in eelgrass (Zostera marina) is highly dependent on shoot density and pH. Oecologia 158: 411–419.
32. Heck KL, Valentine JF (2006) Plant-herbivore interactions in seagrass meadows. Journal of Experimental Marine Biology and Ecology 330: 420–436.
33. Zipperle AM, Coyer JA, Reise K, Stam WT, Olsen JL (2010) Waterfowl grazing in autumn enhances spring seedling recruitment of intertidal Zostera noltii. Aquatic Botany 93: 202–205.
34. Eklof JS, McMahon K, Lavery PS (2009) Effects of multiple disturbances in seagrass meadows: shading decreases resilience to grazing. Marine and Freshwater Research 60: 1317–1327.
35. Orth RJ, Carruthers TJB, Dennison WC, Duarte CM, Fourqurean JW, et al. (2006) A global crisis for seagrass ecosystems. Bioscience 56: 987–996.
36. Waycott M, Duarte CM, Carruthers TJB, Orth RJ, Dennison WC, et al. (2009) Accelerating loss of seagrasses across the globe threatens coastal ecosystems. Proceedings of the National Academy of Sciences of the United States of America 106: 12377–12381.
37. Eklöf JS, Alsterberg C, Havenhand JN, Sundbäck K, Wood HL, et al. (2012) Experimental climate change weakens the insurance effect of biodiversity. Ecology Letters DOI: 10.1111/j.1461-0248.2012.01810.x.
38. van der Heide T, Govers LL, de Fouw J, Olff H, van der Geest M, et al. (2012) A Three-Stage Symbiosis Forms the Foundation of Seagrass Ecosystems. Science 336: 1353–1472.
39. Kefi S, Rietkerk M, Alados CL, Pueyo Y, Papanastasis VP, et al. (2007) Spatial vegetation patterns and imminent desertification in Mediterranean arid ecosystems. Nature 449: 213–U215.
40. Lin Y, Han GD, Zhao ML, Chang SX (2010) Spatial vegetation patterns as early signs of desertification: a case study of a desert steppe in Inner Mongolia, China. Landscape Ecology 25: 1519–1527.
41. Weerman EJ, van Belzen J, Rietkerk M, Temmerman S, Kefi S, et al. (2012) Changes in diatom patch-size distribution and degradation in a spatially self-organized intertidal mudflat ecosystem. Ecology In press.
42. Nacken N, Reise K (2000) Effects of herbivorous birds on intertidal seagrass beds in the northern Wadden Sea. Helgoland Marine Research 54: 87–94.

11

Urban Tree Effects on Soil Organic Carbon

Jill L. Edmondson[1]*, Odhran S. O'Sullivan[1], Richard Inger[2], Jonathan Potter[1], Nicola McHugh[1], Kevin J. Gaston[2], Jonathan R. Leake[1]

1 Department of Animal and Plant Sciences, University of Sheffield, Sheffield, United Kingdom, **2** Environment and Sustainability Institute, University of Exeter, Penryn, Cornwall, United Kingdom

Abstract

Urban trees sequester carbon into biomass and provide many ecosystem service benefits aboveground leading to worldwide tree planting schemes. Since soils hold ~75% of ecosystem organic carbon, understanding the effect of urban trees on soil organic carbon (SOC) and soil properties that underpin belowground ecosystem services is vital. We use an observational study to investigate effects of three important tree genera and mixed-species woodlands on soil properties (to 1 m depth) compared to adjacent urban grasslands. Aboveground biomass and belowground ecosystem service provision by urban trees are found not to be directly coupled. Indeed, SOC enhancement relative to urban grasslands is genus-specific being highest under *Fraxinus excelsior* and *Acer* spp., but similar to grasslands under *Quercus robur* and mixed woodland. Tree cover type does not influence soil bulk density or C:N ratio, properties which indicate the ability of soils to provide regulating ecosystem services such as nutrient cycling and flood mitigation. The trends observed in this study suggest that genus selection is important to maximise long-term SOC storage under urban trees, but emerging threats from genus-specific pathogens must also be considered.

Editor: Ben Bond-Lamberty, DOE Pacific Northwest National Laboratory, United States of America

Funding: This work was supported by the U.K. Engineering and Physical Sciences Research Council (http://www.epsrc.ac.uk/) Sustainable Urban Environments Thematic grants EP/F007604/1 and EP/I002154/1. The funders had no role in study design, data collection and analysis, decision to publish, or preparation of the manuscript.

Competing Interests: The authors have declared that no competing interests exist.

* Email: j.edmondson@sheffield.ac.uk

Introduction

Urban ecosystems provide vital goods and services to the inhabitants of cities and towns [1]. Urban trees are especially important in providing a diverse range of ecosystem services. These include organic carbon storage [2]; flood mitigation and improved water quality [3]; filtration of atmospheric pollutants, especially removing health-damaging fine particulates such as PM10 [4], [5], and particulate-bound carcinogenic polycyclic aromatic hydrocarbons [6]; absorption of toxic gasses including O_3, SO_2 and NOX [7]; traffic noise pollution reduction [8]; and, amelioration of the urban heat island effect [9–11]. In addition, urban trees and greenspaces provide wildlife habitat and aesthetic values that further contribute benefits to human wellbeing, quality of life and health [12–14]. Trees are a ubiquitous part of cities and towns, and have been estimated to cover between 10–67% of urban and community areas in the USA [3] and 20% of Greater London, UK [5].

The recognized importance of trees for ecosystem service provision has stimulated global efforts to increase tree cover in urban areas, for example '*The Big Tree Plant*' in England run by DEFRA and the Forestry Commission [15], and the private-public run '*MillionTreesNYC initiative*' in New York, USA [16]. However, these initiatives have been conducted without a clear understanding of the effects of urban trees on provision of ecosystem services belowground. Given that approximately 75% of ecosystem carbon storage occurs in soils [17], [18], the net effect of trees on soil organic carbon (SOC) stores is particularly important since, in addition to providing a sink for atmospheric carbon dioxide fixed by photosynthesis, SOC is positively associated with regulating and supporting ecosystem services such as storm-water infiltration and nutrient holding capacity [19]. Without an assessment of the impact of tree planting on SOC and other soil properties that deliver ecosystem service benefits in an urban context it is not possible fully to understand the implications of this widespread management practice.

Trees can influence the biological, chemical and physical properties of soils directly through their deep roots and litter quality and quantity [20]. Changes in soil properties following afforestation are varied and dependent on former land-use and species, particularly whether broadleaved angiosperms or gymnosperms like pines [21], [22]. Previous studies on arable land, which has strongly depleted SOC stocks [23], have shown afforestation typically results in increased carbon sequestration both above and belowground [22]. Furthermore, the English national SOC inventory reports greater storage under woodland (primary and afforested) than pasture [24]. The UK Countryside Survey found lower soil bulk density (BD) values beneath woodland compared to improved grasslands [25], indicating a greater capacity of woodland soils to absorb sustained heavy rainfall and reduce run-off and flooding. As these countrywide survey habitats will include agricultural grasslands that have been ploughed and reseeded, they include soils that have previously been depleted in

SOC and experienced increased BD prior to tree planting, as well as ancient woodlands.

Recent studies have shown that urban soils hold much higher concentrations of SOC than typical arable fields, into the same range as semi-natural grasslands and woodlands [26], [27]. The extent to which urban trees can increase SOC stocks over those of urban grasslands is currently unknown, as is the relative importance of different major urban tree genera. This knowledge-gap is strategically important with the rising pan-European threats to keystone urban tree species from virulent pests and diseases, including Ash dieback, Oak decline and Oak Processionary Moth [28–30]. Here, we use an observational study in urban parks to examine differences in soil properties (SOC, C:N ratio, and BD) beneath three tree genera (*Acer* spp., *Fraxinus excelsior*, and *Quercus robur*) selected for their abundance and capacity to grow into large specimens, and mixed urban woodlands in comparison to adjacent urban grassland soil. We test the hypothesis that SOC stocks and C:N ratios would be increased in soil under trees compared to grassland, paralleling and positively correlating with greater above-ground carbon storage, whereas soil BD would be reduced under trees compared to grassland as seen in semi-natural ecosystems [25].

Methods

Study area

This research focussed on Leicester, a mid-sized UK city, located in the East Midlands of England (52°38′N, 1°08W). It has a human population of 310,000 [31], and covers an area of approximately 73 km^2. The region has a temperate climate, receiving 620 mm of precipitation each year and average annual daily minimum and maximum temperatures of 6.1°C and 13.9°C, respectively [32]. Soil types within the city are dominated by deep clays, deep loam and seasonally wet deep clays and loam, according to the National Soil Map for England and Wales produced by Cranfield University. The soil types sampled in the city were: Hanslope, Whimple, Salop, Beccles 3, Ragdale and Fladbury 1.

Sampling strategy

A GIS was used to select randomly urban parks for soil sampling within the city of Leicester and permission was granted for the work by the Leicester City Council. The land-use history of each park was assessed in a GIS using the series of historic Ordnance Survey maps dating back to 1887. No park had previously been built upon, one park and two country houses and grounds that went on to form parks were in existence in 1887, the remainder were agricultural land at this time. Over the following years to the present day the remainder of the parks were established as the city expanded into the surrounding agricultural landscape, with the two recent parks developed on farmland within the last 20–30 years (see Table S1 for site specific land-use history). At each park an initial assessment was made for the presence of individual trees that ranged in size from saplings to large mature specimens within our target genera, specifically isolated specimens of *Quercus robur, Fraxinus excelsior, Acer* spp. (comprised of *Acer pseudoplatanus* and *Acer platanoides*), or patches of mixed woodlands. The selected trees ranged in diameter at breast height (DBH) from 2.5 cm to 197 cm and in biomass from 1.3 kg to 61 tonnes for the largest mature *Q. robur* specimen (see Table S1). The tree genera were selected because of their importance in parks in Leicester and national abundance, *Q. robur* being the commonest tree and *F. excelsior* the second most common as an individual tree or within small patches of woodland in Great Britain [33]. *A. pseudoplatanus* is the fourth most common tree species in small patches of British woodland [33].

Where a park contained one or more of the tree genera and/or mixed woodland at least one grassland site was also identified for sampling. Each grassland site was situated in proximity to the tree sites identified, but was also over 50 m from any individual or patch of trees, to ensure that it was outside the influence of the trees. A GIS layer obtained from Leicester City Council was used to check that management at each grassland site was uniform, specifically that the regularity of mowing at all sites was approximately 25 times per year, these park grasslands were not irrigated nor did they receive any fertiliser input. The grassland sites were selected to act as a direct comparison (or paired sample) at each specific location (park) with the tree sample.

At each site, tree species, height and DBH were recorded within a 5×5 m quadrat centred on individual isolated trees within grassland or in mixed woodland. Soils were sampled in approximately 7 cm increments to 1 m depth [26], the reference depth for the national SOC inventory [21], [24], under target tree genera, mixed woodland and grasslands. Under isolated trees soil samples were taken within 1 m of the trunk to ensure that all samples were taken beneath the canopy of even the small immature trees. In total, soils were sampled beneath 12 specimens of *Quercus robur*, 11 of *Fraxinus excelsior*, 12 of *Acer* spp. In addition soils were sampled beneath mixed urban woodland at 8 sites, and urban grassland at 15 sites.

In addition to the three target tree genera specified a further six species; *Acer campestre, Corylus avellana, Crataegus monogyna, Salix caprea, Tilia* x *europaea* were identified in the mixed urban woodlands. Measured tree DBH and height were used to estimate above-ground biomass with allometric equations. For each species, where multiple equations were available they were combined to produce a generalised biomass prediction. Where species-specific equations did not exist genus level equations were used, following the methodology recently used to derive tree aboveground biomass across the city of Leicester [2]. However, allometric equations to predict biomass of urban trees specifically are scarce [34], thus those used were derived from European and North American forested ecosystems [2], [35–37].

Soil sample preparation and analysis

Soil samples were analysed for SOC, C:N ratio and BD using established procedures. In brief, soil samples were dried at 105°C for 24 hours, weighed, ball milled to homogenise, and passed through a 1 mm sieve [26]. Soil BD (g cm^{-3}) was calculated after removing the dry weight of matter greater than 1 mm [38]. Soils were analysed in duplicate for total N concentration (mg g^{-1}) in a CN analyser [26]. Inorganic carbon was removed from 2.5 g of soil sample by adding 10 ml 5.7 M HCl, samples were centrifuged at 1800 *g* for 10 minutes, supernatant discarded and dried at 105°C. Subsequent CN analysis in duplicate determined SOC concentration (mg g^{-1}) [26].

Statistical analysis

All dependant variables (SOC Concentration, SOC Density, Soil C:N & Soil Bulk density) were checked for normality and homoscedasticity, and log transformed where necessary. Each of these dependant variables was analysed using general linear mixed effects models. The maximal model included vegetation cover as a 5 level fixed factor (urban grassland, *Q. robur, F. excelsior, Acer* spp. and mixed woodland). Soil depth (at which the sample was taken) and the biomass of the individual tree in proximity to the soil sample or in mixed woodland the biomass of all trees within the 5×5 m quadrat were incorporated as covariates in the model, and urban park identity was included as a random (intercept) factor to account for differences in length of time since park establishment and geographic location within the city. Initial models included tree biomass as a variable, however this had no predictive power (as the confidence intervals spanned zero) for any measured soil property and therefore it was removed as a variable from all subsequent models. In all cases model simplification was attempted by comparing all possible simpler model subsets using Akaike's information criterion (AIC). The maximal model, containing all the predictor variables, was always found to be the top model (based on AIC) and used in the subsequent inferences. Pseudo R^2 values were calculated using the methods of Nagelkerke [39]. Mixed models were carried out using package lme4. F and p values were calculated using Satterthwaite approximations [40] to determine denominator degrees of freedom in package lmerTest. All tests were carried out in the R language and environment [41]. Raw data are available in Table S1.

Results

There were significant differences in SOC concentration and amount per soil volume (SOC density) between the three tree genera, mixed urban woodland and grassland ($F = 9.95$, $p<0.001$; and $F = 12.51$, $p<0.001$, respectively). Soil depth had a strong effect on SOC concentration ($F = 609.89$, $p<0.001$) and density ($F = 456.27$, $p<0.001$). Median SOC concentration and density were greatest beneath *F. excelsior* throughout the depth profile, followed by *Acer* spp., with no difference between *Q. robur*, mixed woodland, and grassland (Fig. 1). Model pseudo R^2, including both soil depth and tree cover, explained 59% and 57% of variation in SOC concentration and density respectively. Estimates of total SOC storage, based on summed median values for each depth category, ranged from 14–26 kg SOC m^{-2} with lowest storage beneath *Q. robur*, mixed woodland, and grassland, and the highest under *F. excelsior* (Fig. 1). The explanatory power of the mixed effects model for C:N ratio was low, with a pseudo R^2 explaining only 11% of the variation in the data (soil depth $F = 20.74$, $p<0.001$; vegetation cover $F = 7.579$, $p<0.001$). The effect of trees on BD was not significant ($F = 1.24$, $p = 0.295$). Soil depth was the most important predictor of BD ($F = 244.35$, $p<0.001$, model pseudo $R^2 = 49\%$), with increased median BD with depth from 0.99–1.59 g cm^{-3} between 0–20 cm to 80–100 cm (Fig. 2). The mixed effects models including tree biomass as a variable (excluding grassland as this cover type has no woody biomass) revealed that the effect size of tree biomass on SOC, C:N ratio and BD was small, and as confidence intervals spanned zero had no predictive power (see Table S2 for model statistics).

Discussion

Previous research in a typical UK city, Leicester, has shown that 97% of carbon present in aboveground ecosystem biomass is found in trees, with average storage increasing from 0.2 kg m^{-2} in herbaceous vegetation (most commonly urban grassland) to 28.5 kg m^{-2} in trees [2], affirming the importance of urban trees in aboveground carbon sequestration. In contrast we found that differences in SOC under trees compared to grassland were more modest and genus-specific and, surprisingly, did not occur under mixed woodland or *Q. robur*, even though several specimens of the latter had a trunk diameter of over 1.5 m. No significant differences in SOC concentration beneath urban forests and grasslands were found in Baltimore, USA [42], corresponding with

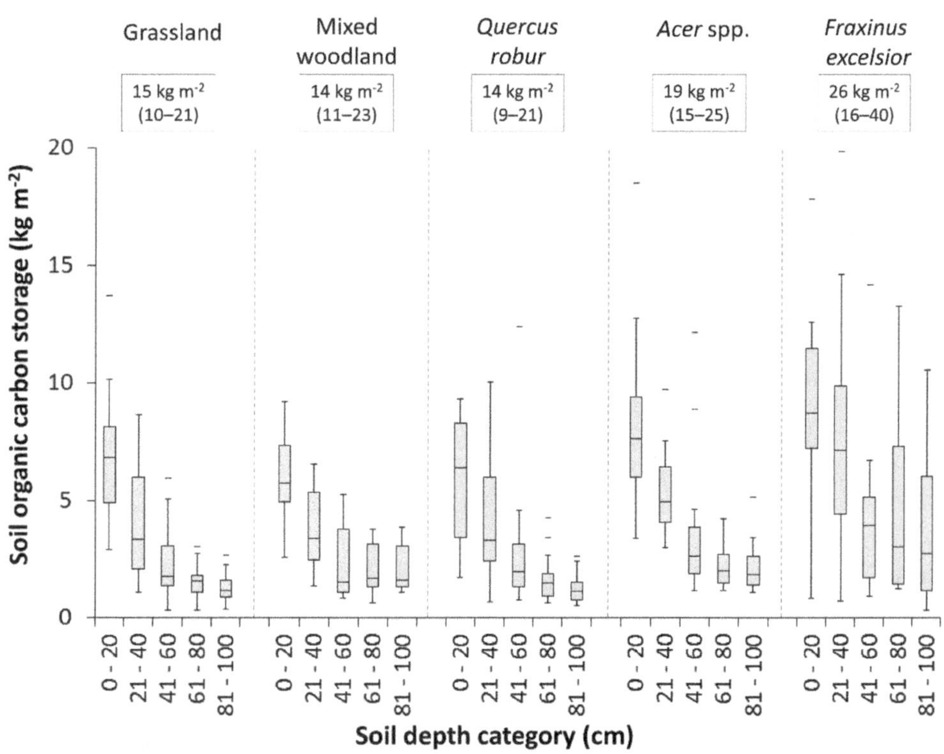

Figure 1. Soil organic carbon storage within each 20 cm depth category (summed median values are displayed in text boxes, values in parenthesis are total 25th and 75th percentiles), beneath *Quercus robur* (n = 12), *Fraxinus excelsior* (n = 11), *Acer* spp. (n = 12), mixed woodland (n = 8) and grassland (n = 15) by depth class. The horizontal line within the box indicates median, box boundaries indicate 25th and 75th percentiles, whiskers indicate highest and lowest values, horizontal lines above or below whiskers indicate outliers.

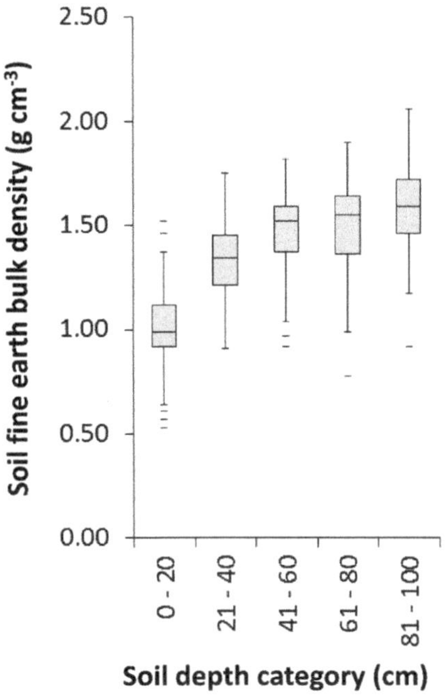

Figure 2. Soil bulk density in each 20 cm depth category. The horizontal line within the box indicates median, box boundaries indicate 25th and 75th percentiles, whiskers indicate highest and lowest values, horizontal lines above or below whiskers indicate outliers.

our findings for mixed woodlands. Nonetheless, we did find increased SOC storage under *F. excelsior* and, to a lesser extent, *Acer* spp., with gains of 11 kg m^{-2} and 5 kg m^{-2} respectively compared to the adjacent grasslands (Fig. 1).

The effects of *F. excelsior* are of particular interest in light of current concerns about the impact of ash die-back disease on this widespread and abundant species in Europe and the UK, where it is the second most abundant tree in small woodland patches and the second most common individual tree in the countryside [33]. *A. pseudoplatanus* and *F. excelsior* are mesophilic species that thrive on well-watered alluvial soils, but the former is more drought susceptible [43]. The clear SOC enrichment throughout the soil profile under *F. excelsior*, especially from 40–100 cm depth, could be attributed to several factors. This species produces a greater root mass than *A. pseudoplatanus* and *Q. robur*, and establishes its extensive deep root network more quickly than other broadleaved species [44], and is especially well adapted to clay-rich soils such as those found in Leicester. Organic carbon introduced by tree roots into clay rich subsoils will experience long residence times [45].

Compared to park grassland, none of the tree genera or mixed woodland altered soil BD, which is consistent with previous reports of the absence of effects of broadleaved trees planted in pasture [46]. However, the urban trees could still aid storm-water drainage along root channels [20]. Similarly, soil C:N ratio, one of the major controls of N availability and leaching [47], was also unaffected by trees.

These data highlight the often overlooked importance of urban park grasslands as contributors to belowground ecosystem service provision, particularly SOC storage, which we show exceeds typical values for agricultural grassland by 23% [24]. Urban park management may reduce litter inputs from trees relative to grassland as mowings are not collected but autumn leaves are removed from beneath individual trees. However, it remains unclear how important these management practices are as leaves were not removed from beneath the stands of mixed woodland we sampled, yet these showed no significant increase in SOC storage compared to adjacent grassland. Carbon inputs into the soils under grassland and park trees are likely to be strongly influenced by roots. Mean residence time of root carbon in soil is typically 2.4 times that of shoot carbon, due to higher concentrations of the more recalcitrant components such as lignin, so that SOC is often mainly derived from root inputs [48].

This observational study aimed to provide a first indication of the long-term effect of trees in urban areas on soil properties. As our approach was non-experimental we cannot be certain whether the sampled trees were planted or naturally regenerated from seed, and we have to assume that any differences in soil conditions under trees and adjacent grassland are due to the trees. However, our strategy of sampling soil under trees of very different sizes enabled us to investigate if there were any relationships between tree size (as a proxy for tree age) and soil properties across a range of urban parks. Indeed, given the range of tree sizes we studied, the trees will likely have ranged in age from a decade to several centuries- a timespan difficult to achieve in experimental manipulation studies. Perhaps one of the most surprising findings arising from this was the absence of clear effects of tree size on soil carbon storage, especially for oaks where the largest individuals were 1.6 m–2.0 m DBH.

Tree planting within urban areas is one of a range of environmental management techniques used which impact, either intentionally or unintentionally, on ecosystem service provision [1]. Although tree planting is known to increase ecosystem service provision aboveground [2–14], we now show for the first time for urban trees that this benefit does not consistently extend into the soil system as, contrary to our original hypotheses, there is no direct relationship between aboveground tree biomass and SOC concentration, soil C:N ratio and soil BD.

We demonstrate the importance of targeted tree genus selection to maximise the long-term belowground ecosystem service benefits of urban tree planting with respect to SOC. Further research should seek to elucidate the belowground effects of other common tree species on a range of soil types and climate zones across urban areas globally, better to inform urban policy and planning. Our findings that *F. excelsior* makes an important contribution to enhancing urban SOC stocks coincides with the first record of ash dieback disease spreading into our study region (Leicestershire) [49]. This raises important questions about the likely future impacts of loss of this species on SOC stocks nationally. Furthermore, it highlights the importance of long-term planning in cities and towns to couple the ecosystem service benefits of tree planting with disease and climate change resilient urban tree populations in the future.

Acknowledgments

We gratefully acknowledge Leicester City Council for access to their land.

Author Contributions

Conceived and designed the experiments: JLE NM KJG JRL. Performed the experiments: JLE OSO JP. Analyzed the data: RI. Contributed to the writing of the manuscript: JLE RI OSO JP NM KJG JRL. Designed, arranged permissions: JLE JP OSO. Carried out fieldwork: JLE JP OSO.

References

1. Gaston KJ, Ávila-Jiménez ML, Edmondson JL (2013) Managing urban ecosystems for good and services. J Appl Ecol 50: 830–840.
2. Davies ZG, Edmondson JL, Heinemeyer A, Leake JR, Gaston KJ (2011) Mapping an urban ecosystem service: quantifying above-ground carbon storage at a city-wide scale. J Appl Ecol 48: 1125–1134.
3. Nowak DJ, Greenfield EJ (2012) Tree and impervious cover in the United States. Landscape Urban Plan 107: 21–30.
4. Tiwary A, Sinnett D, Peachey C, Chalabi Z, Vardoulakis S, et al. (2009) An integrated tool to assess the role of new planting in PM_{10} capture and the human health benefits: a case study in London.
5. Tallis M, Taylor G, Sinnett D, Freer-Smith P (2011) Estimating the removal of atmospheric particulate pollution by the urban tree canopy of London, under current and future environments. Landscape Urban Plan 103: 129–138.
6. Joraeva VA, Johnson DL, Hassett JP, Nowak DJ (2002) Differences in accumulation of PAHs and metals on the leaves of *Tilia* x *euchlora* and *Pyrus calleryana*. Environ Pollut 120: 331–338.
7. Nowak DJ, Crane DE, Stevens JC (2006) Air pollution removal by urban trees and shrubs in the United States. Urban For. Urban Gree 4: 115–123.
8. Maleki K, Hosseini SM (2011) Investigation of the effects of leaves, branches and canopies of trees on noise pollution reduction. Annals of Environmental Science 5: 13–21.
9. Akbari H (2002) Shade trees reduce building energy use and CO_2 emissions from power plants. Env Pollut 116: S119–S126.
10. Hamada S, Ohta T (2010) Seasonal variations in the cooling effect of urban green areas on surrounding urban areas. Urban For Urban Gree 9: 15–24.
11. Armson D, Stringer P, Ennos AR (2012) The effect of tree shade and grass on surface and globe temperatures in an urban area. Urban For Urban Gree 11: 245–255.
12. McPherson EG, Nowak D, Heisler G, Grimmond S, Souch C, et al. (1997) Quantifying urban forest structure, function, and value: the Chicago urban forest climate project. Urban Ecosystems 1: 49–61.
13. Nowak DJ, Crane DE, Stevens JC, Hoehn RE, Walton JT, et al. (2008) A ground-based method of assessing urban forest structure and ecosystem services. Arboriculture and Urban Forestry 34: 347–358.
14. White MP, Alcock I, Wheeler BW, Depledge MH (2013) Would you be happier living in a greener urban area? A fixed-effects analysis of panel data. Psychol Sci. 24: 920–928.
15. DEFRA. The Big Tree Plant. Available: http://www.defra.gov.uk/bigtreeplant/ Accessed: 2014 Jan.
16. MillionTressNYC. Million Trees NYC. Available: http://www.milliontreesnyc.org/html/about/about.shtml Accessed: 2014 Jan.
17. Batjes NH (1996) Total carbon and nitrogen in soils of the world. Eur J Soil Sci 47: 151–163.
18. Lal R (2008) Carbon sequestration. Philos T Roy Soc B 363: 815–830.
19. Franzluebbers AJ (2002) Soil organic matter stratification ratio as an indicator of soil quality. Soil Till Res 66 95–106.
20. Day SD, Wiseman PE, Dickinson SB, Harris JR (2010) Tree root ecology in the urban environment and implications for a sustainable rhizosphere. Arboriculture and Urban Forestry 36: 193–205.
21. Jobbagy EG, Jackson RB (2000) The vertical distribution of soil organic carbon and its relation to climate and vegetation. Ecol Appl 10: 423–436.
22. Guo LB, Gifford RM (2002) Soil carbon stocks and land use change: a meta analysis. Global Change Biol 8: 345–360.
23. Lal R (2004) Soil carbon sequestration impacts on global climate change and food security. Science 304: 1623–1627.
24. Bradley RI, Milne R, Bell J, Lilly A, Jordan C, et al. (2005) A soil carbon and land use database for the United Kingdom. Soil Use Manage 21: 363–369.
25. Carey PD, Wallis S, Chamberlain PM, Cooper A, Emmett BA, et al. (2008) Countryside Survey: UK Results from 2007. NERC/Centre for Ecology & Hydrology (CEH Project Number:C03259).
26. Edmondson JL, Davies ZG, McHugh N, Gaston KJ, Leake JR (2012) Organic carbon hidden in urban ecosystems. Scientific Reports 2: 963.
27. Edmondson JL, Davies ZG, McCormack SA, Gaston KJ, Leake JR (2014) Land-cover effects on soil organic carbon stocks in a European city. Sci Total Environ 472: 444–453.
28. DEFRA (2013) Chalara management Plan. Department for Environment Food and Rural Affairs, London.
29. Forestry Commission. Bleeding Canker of Horse Chestnut. Available: http://www.forestry.gov.uk/fr/infd-6kybgv Accessed 2013 Dec.
30. Tree Health and Plant Biosecurity Taskforce (2013) Tree Health and Plant Biosecurity Task Force. Department for Environment Food and Rural Affairs, London.
31. Leicester City Council. Population of Leicester 1801–2011. Available: http://www.leicester.gov.uk/your-council-services/council-and-democracy/city-statistics/other-statistics/population-statistics/populationsince1801/ Accessed 2014 Jan.
32. Met Office. 1981–2010 averages. Available: http://www.metoffice.gov.uk/climate/uk/averages/19812010/ Accessed 2014 Jan.
33. Maskell L, Henrys P, Norton L, Smart S, Wood C (2013) Distribution of ash trees (Fraxinus excelsior) in Countryside Survey Data. Centre for Ecology and Hydrology.
34. McHale MR, Burke IC, Lefsky MA, Peper PJ, McPherson EG (2009) Urban forest biomass estimates: is it important to use allometric relationships developed specifically for urban trees? Urban Ecosystems 12: 95–113.
35. Ter-Mikaelian MT, Korukhin MD (1997) Biomass equations for sixty-five North American tree species. Forest Ecol Manag 97: 1–24.
36. Zianis D, Muukkpnen P, Mäkipää R, Mencuccini M (2005) Biomass and stem volume equations for tree species in Europe. Silva Fennice Monographs 4:63.
37. Snorrason A, Einarsson SF (2006) Single-tree biomass and stem volume functions for eleven tee species used in Icelandic forestry. Icelandic Agricultural Science 19: 15–24.
38. Edmondson JL, Davies ZG, McCormack SA, Gaston KJ, Leake JR (2011) Are soils in urban ecosystems compacted? A citywide analysis. Biol Letters 23: 771–774.
39. Nagelkerke NJD (1991) A note on a general definition of the coefficient of determination. Biometrika. 78: 691–692.
40. Satterthwaite FE (1946) An approximate distribution of estimates of variance components. Biometrics Bull 2: 110–114.
41. R Core Team (2012) R: A language and environment for statistical computing. R Foundation for Statistical Computing, Vienna, Austria. ISBN 3-900051-07-0, Available: http://www.R-project.org/.
42. Pouyet RV, Yesilonis ID, Golubiewski NE (2009) A comparison of soil organic carbon stocks between residential turf grass and native soil. Urban Ecosystems 12: 45–62.
43. Lemoine D, Peltier J-P, Marigo G (2001) Comparative studies of the water relations and the hydraulic characteristics in Fraxinus excelsior, Acer pseudoplatanus, and A.opalus trees under soil water contrasted conditions. Ann For Sci 58: 723–731.
44. Kerr G, Cahalan C (2004) A review of site factors affecting early growth of ash (Fraxinus excelsior L.). Forest Ecol Manage 188: 225–234.
45. Krull ES, Baldock JA, Skemstad JO (2003) Importance of mechanisms and processes of the stabilisation of soil organic matter for modelling carbon turnover. Funct Plant Biol 30: 207–222.
46. Berthrong ST, Jobbagy EG, Jackson RB (2009) A global meta-analysis of soil exchangeable cations, pH, carbon and nitrogen with afforestation. Ecol Appl 19: 2228–2241.
47. Dungait JAJ, Cardenas LM, Blackwell MSA, Wu L, Withers PJA, et al. (2012) Advances in the understanding of nutrient dynamics and management in UK agriculture. Sci Total Environ 434: 39–50.
48. Rasse DP, Rumpel C, Dignac M-F (2005) Is soil carbon mostly root carbon? Mechanisms for a specific stabilization. Plant Soil 269: 341–356.
49. DEFRA. Ash Die Back in Leicestershire Press Release. Available: http://www.gov.uk/government/news/ash-disease-found-in-Leicestershire Accessed 2013 Dec.

12

Using Satellite and Airborne LiDAR to Model Woodpecker Habitat Occupancy at the Landscape Scale

Lee A. Vierling[1]*, Kerri T. Vierling[2], Patrick Adam[3], Andrew T. Hudak[4]

1 Department of Forest, Rangeland, and Fire Sciences, McCall Outdoor Science School, University of Idaho, Moscow, Idaho, United States of America, **2** Department of Fish and Wildlife Sciences, University of Idaho, Moscow, Idaho, United States of America, **3** Environmental Science, University of Idaho, Moscow, Idaho, United States of America, **4** Rocky Mountain Research Station, US Forest Service, Moscow, Idaho, United States of America

Abstract

Incorporating vertical vegetation structure into models of animal distributions can improve understanding of the patterns and processes governing habitat selection. LiDAR can provide such structural information, but these data are typically collected via aircraft and thus are limited in spatial extent. Our objective was to explore the utility of satellite-based LiDAR data from the Geoscience Laser Altimeter System (GLAS) relative to airborne-based LiDAR to model the north Idaho breeding distribution of a forest-dependent ecosystem engineer, the Red-naped sapsucker (*Sphyrapicus nuchalis*). GLAS data occurred within ca. 64 m diameter ellipses spaced a minimum of 172 m apart, and all occupancy analyses were confined to this grain scale. Using a hierarchical approach, we modeled Red-naped sapsucker occupancy as a function of LiDAR metrics derived from both platforms. Occupancy models based on satellite data were weak, possibly because the data within the GLAS ellipse did not fully represent habitat characteristics important for this species. The most important structural variables influencing Red-naped Sapsucker breeding site selection based on airborne LiDAR data included foliage height diversity, the distance between major strata in the canopy vertical profile, and the vegetation density near the ground. These characteristics are consistent with the diversity of foraging activities exhibited by this species. To our knowledge, this study represents the first to examine the utility of satellite-based LiDAR to model animal distributions. The large area of each GLAS ellipse and the non-contiguous nature of GLAS data may pose significant challenges for wildlife distribution modeling; nevertheless these data can provide useful information on ecosystem vertical structure, particularly in areas of gentle terrain. Additional work is thus warranted to utilize LiDAR datasets collected from both airborne and past and future satellite platforms (e.g. GLAS, and the planned IceSAT2 mission) with the goal of improving wildlife modeling for more locations across the globe.

Editor: Adina Maya Merenlender, University of California, Berkeley, United States of America

Funding: The authors would like to thank the USGS Gap Analysis Program (agreement Grant number 08HQAG0123; K. Vierling, PI) and the NASA-sponsored Idaho Space Grant Consortium (http://www.id.spacegrant.org; graduate fellowship awarded to P. Adam) for funding. The funders had no role in study design, data collection and analysis, decision to publish, or preparation of the manuscript.

Competing Interests: The authors have declared that no competing interests exist.

* E-mail: leev@uidaho.edu

Introduction

Remote sensing data have become fundamental for mapping actual or potential species distributions [1,2] to aid conservation efforts at local to global scales [3,4,5]. Current species distribution maps draw upon a variety of information sources, including land cover, to model habitat. However, land cover classified as forest is often too broadly defined and often does not account for structural variability that can significantly alter habitat suitability for avian species [6]. While passive spectral remote sensing data products are essential for delineating horizontal habitat heterogeneity (e.g. forest fragmentation) and vegetation productivity at spatial scales relevant to many species, they are of limited use for quantifying details of vertical habitat structure crucial to animal habitat (e.g. [7]). As a result, improving maps of animal distributions is likely to require remote quantification of vertical habitat structure across wide areas [7–9].

Light detection and ranging (LiDAR) provides unique data for characterizing vertical habitat structure, and has been utilized widely to address multiple animal-habitat relationships. For instance, airborne LiDAR has been used alone or in combination with passive remote sensing data to assist habitat modeling for terrestrial birds [10–17], mammals [18,19], and invertebrates [20,21] at spatial scales relevant to numerous organisms. The basis for these advances in habitat modeling stem from the fact that the energy returned in the LiDAR signal provides information to characterize forest structural characteristics such as vegetation height, density, and volume in discrete vertical slices above ground level [22,23]. The resultant data can provide information ranging from structural characteristics of individual trees [24] to area-wide maps of forest successional status [25,26] and biomass distribution [27,28]. As a result, these vertical forest characteristics afforded from airborne LiDAR can improve our understanding and mapping of animal-habitat relationships from the stand to the landscape scale [29].

Although LiDAR has provided significant leaps forward in modeling habitat, data are typically collected via aircraft, thereby imposing limits on the extent and frequency of spatial and temporal sampling relative to orbiting sensors mounted on satellites. A satellite based LiDAR remote sensing platform may provide information about 3-D structure for animal habitat modeling critical for broad-scale application [30]. However to our knowledge no empirical study has yet tested the utility of satellite LiDAR for habitat modeling.

The Geoscience Laser Altimeter System (GLAS) LiDAR instrument aboard the ICESat satellite, in operation between 2003 and 2009, provides a novel opportunity for delineating habitat. While the primary intent of the GLAS instrument was to measure polar ice-sheet thickness [31], GLAS data has been used to characterize vertical vegetation structure [32–34] and forest biomass [35]. Similar to airborne LiDAR in the underlying physics, a photo-detector within the GLAS receiving telescope records the time-of-flight (t) between emission of the transmit laser pulse and its return to the instrument. Using the speed of light (c), distance (D) from the instrument is then computed as $D = 0.5\ tc$. Vertical vegetation structure is derived from the varying intensity at discrete heights of the returned transmit signal [34]. The major attraction of GLAS data for informing ecological questions is that (1) the data samples forest environments across the globe and (2) the data (along with computer code to manipulate it) is freely available from the U.S. National Snow and Ice Data Center (NSIDC). However, LiDAR data from GLAS differs in both spatial extent and resolution from airborne LiDAR data. While airborne LiDAR data are typically scanned at high horizontal resolution (i.e. with each laser pulse sampling a 'small footprint' $<1\ m^{-2}$ in size and typically collected at a density of >1 pulse m^{-2}) continuously across landscapes, GLAS is a 'large footprint' instrument collecting data samples with a diameter in our study area of ~64 m each and separated by a minimum of 172 m. The utility of GLAS LiDAR to describe wildlife habitat at spatial scales relevant to animals is as yet unexplored.

Advances in the global description of vertical vegetation structure are increasingly important for understanding issues associated with species distribution, diversity, and habitat connectivity (e.g. [9]). The major objective of this study was to model breeding site occupancy of a forest dependent bird species, the Red-naped Sapsucker (*Sphyrapicus nuchalis*), using both GLAS and airborne LiDAR data. We chose the Red-naped Sapsucker because of its status as an ecosystem engineer, capable of creating cavities in trees that can be used by multiple other forest avian and mammalian species (e.g. [36,37]). To our knowledge, this study represents the first to utilize spaceborne LiDAR data in assessing the habitat associations of a forest dependent species, and the first to compare GLAS and airborne-LiDAR models in this ecological context.

Methods

Study Site

Field survey sites were located in temperate mixed-conifer forests dominated by Ponderosa pine (*Pinus ponderosa*), White pine (*Pinus monticola*), Grand fir (*Abies grandis*), Western larch (*Larix occidentalis*), Douglas fir (*Pseudotsuga menziesii*), and Western red cedar (*Thuja plicata*) located at the western extent of the Clearwater Mountain Range and within the Idaho Panhandle National Forest in northern Idaho, USA (Fig. 1). Site ground elevation (hereafter, elevation) ranged from 794–1357 m. Land ownership and management across the study area included private timber concessions, private land holders, and state and federal management agencies. Permits are not explicitly required for observational bird studies on U.S. Forest Service land. We obtained verbal permission and access to state owned property from the Idaho Department of Lands. Private land holdings included Potlatch Corporation and Bennett Lumber Products, both of whom granted permission to perform research on their respective properties. The diverse ownership and management matrix has created a mosaic of forest stands that exhibit a range of successional stages that can influence the habitat selection behavior of woodpeckers [38–41].

GLAS LiDAR

The GLAS instrument is a waveform LiDAR system with 15 cm vertical resolution that uses a near infrared (1064 nm) laser. The laser records data covering a ground footprint (i.e. sample area) of ca. 64 m diameter that can vary significantly in its ellipticity over time [42]. The semimajor axis and eccentricity are recorded for each laser shot and were used to define an elliptical mask for extraction of terrain features from the airborne and GLAS data using geographic information system software. The laser footprints are spaced 172 m between centers along the satellite transect, with a between-transect spacing of about 540 m with different acquisition years in our study area (Fig. 1). We acquired GLAS data from the NSIDC, which distributes data products from 18 laser sampling campaigns. We used the GLA01 product (L1A Global Altimetry) which provides the transmitted and received waveform, and the GLA14 product (L2 Land Surface Altimetry) which provides the laser footprint geolocation and geodetic, instrument, and atmospheric corrections. The two datasets are linked by record ID and shot number. We used 'release 531' data, which specifies a level of data pre-processing, from laser campaigns 2A (Sep–Nov 2003), 2B (Feb–Mar 2004), 3A (Oct–Nov 2004), 3B (Feb–Mar 2005), 3D (Oct–Nov 2005), 3H (Mar–Apr 2007), and 3I (Oct–Nov 2007). GLAS waveforms can be influenced by interaction with clouds and signals that are saturated due to high gain settings. We therefore removed waveforms that were affected by clouds and saturation on the laser signal following Chen [43].

We modified Interactive Data Language (IDL) code provided by NASA and available on the NSIDC GLAS site [44] to extract complete laser waveform return signals (from GLA01) and waveform summary statistics (from GLA14) (Fig. 2). To compute vertical vegetation statistics from the waveform signal, we first established the ground return provided by the summary statistics in the GLA14 product. The summary statistics include up to 6 primary energy peaks generated from a Gaussian decomposition algorithm [45] to identify the major components of the signal which include both vegetation and ground returns (Fig. 2). We estimated the ground return location following Chen [46] who investigated the correlation between GLAS and airborne lidar data in forested mountainous areas and found that the largest in magnitude of the two lowest peaks corresponded most closely with ground elevations.

Before deriving canopy height statistics (Fig. 3), we filtered the waveforms by retaining signals that were greater than 4 times the standard deviation of the noise (following [47]). Maximum waveform vertical resolution at 1 ns pulse duration is 15 cm. We selected a Gaussian filter of 60 cm (4 ns) in height because we were interested in primary structural features of the canopy. We did not evaluate other filter widths, but narrower widths may allow noise to pass or may produce results that are difficult to interpret due to their vertical length scale being finer than individual branches. Canopy height was estimated from an algorithm developed by Chen [43] for a coniferous montane location in the Pacific Northwest that accounts for the effect of terrain slope on waveform vegetation indices by incorporating the maximum terrain relief distance within the laser footprint (obtained from the 10 m National Elevation Dataset [48]) and the difference between the start and end of the filtered and smoothed waveform signal. The

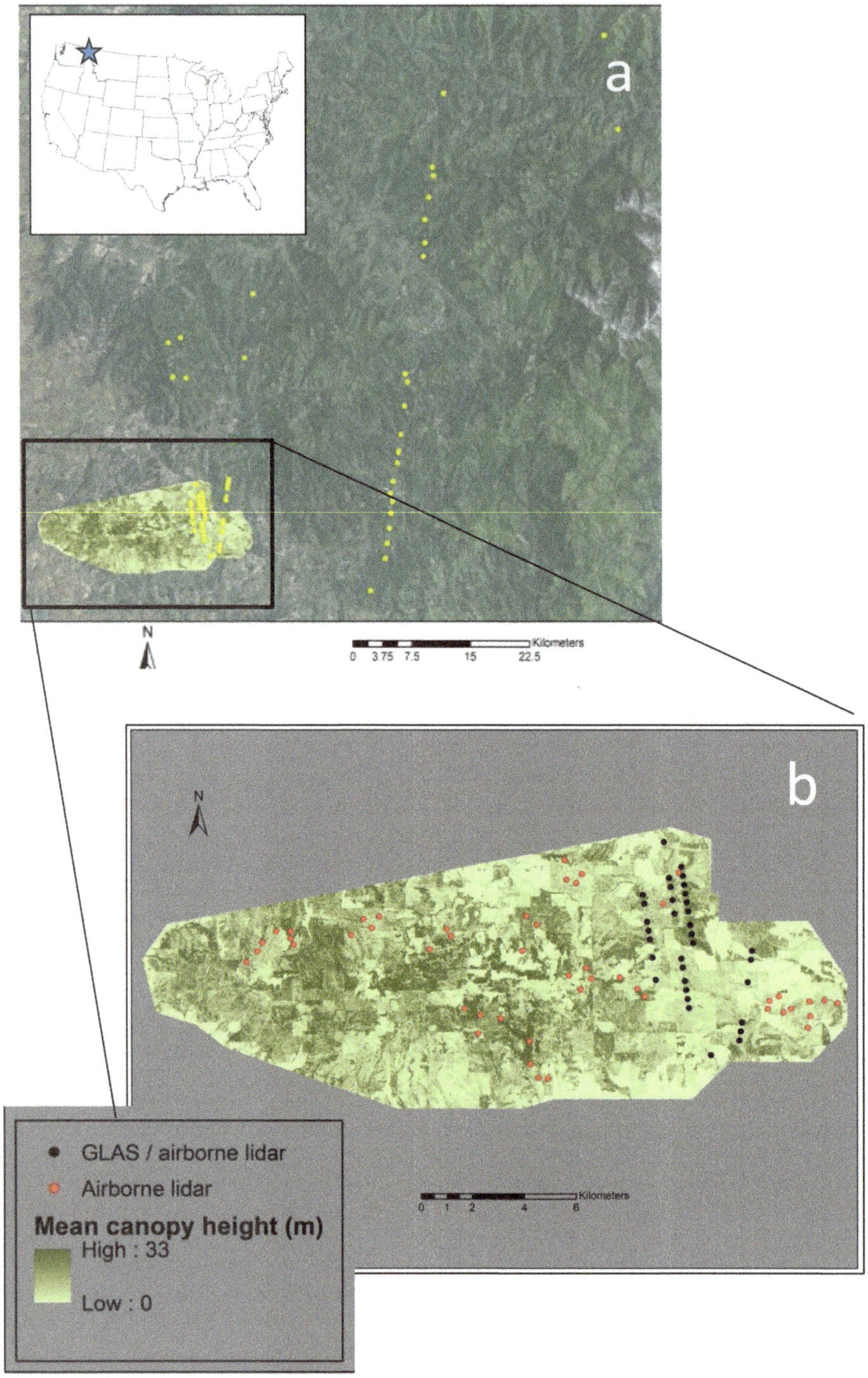

Figure 1. Map of woodpecker survey sites within GLAS LiDAR footprints (yellow dots) (a) in northern Idaho, and (b) survey sites within both GLAS and airborne LiDAR (black dots) and airborne only (red dots).

complete suite of canopy metrics derived from GLAS data is listed in Table 1.

Airborne LiDAR

LiDAR data for Moscow Mountain (Fig. 1) was acquired in July, 2009 using a discrete return airborne Leica ALS50 Phase II laser system scanning with a pulse rate of ~150 kHz at ±14° from nadir. Adjacent survey transects were flown in opposing directions with ≥50% overlap to ensure 100% terrain overlap for all transects, resulting in a mean data density of ~12 points m^{-2}. A maximum of 4 range measurements, or returns, were recorded for each laser pulse transmission. To geolocate laser point positions, aircraft attitude (pitch, roll, yaw) was recorded at 200 Hz by an onboard inertial measurement unit (IMU), while sensor position (x, y, z) was recorded at 2 Hz by an onboard differential GPS. Aircraft attitude and sensor position were indexed by GPS time to enable post-processing correction and calibration. Data were projected in UTM Zone 11 North, NAD 83 using NAVD88

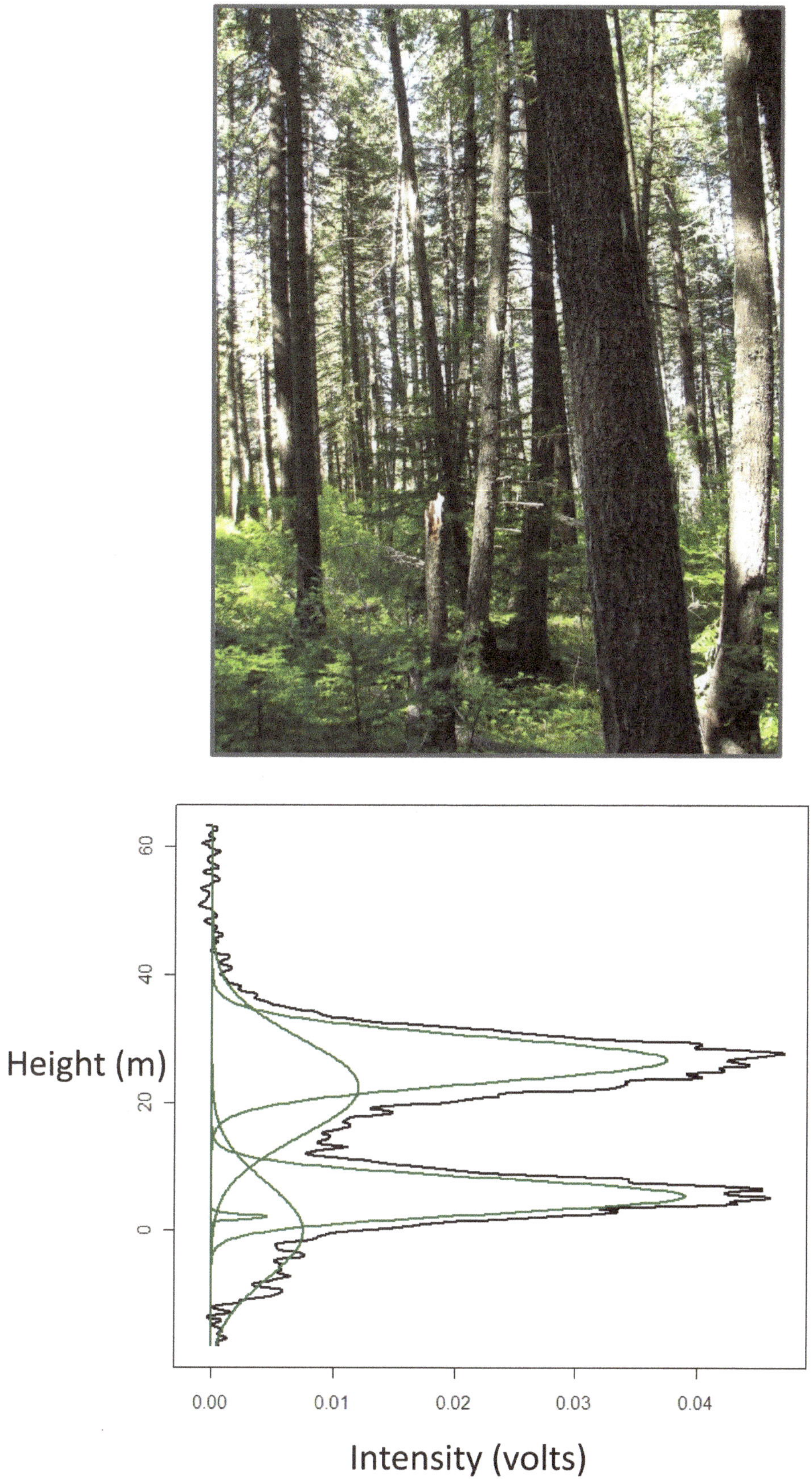

Figure 2. Example GLAS LiDAR waveform signal (bottom panel) and associated woodpecker survey location (top panel). The black line in the graphic represents the actual laser energy signal for the plot location, while the green lines represent the decomposed signal when expressed in best-fit Gaussian curves.

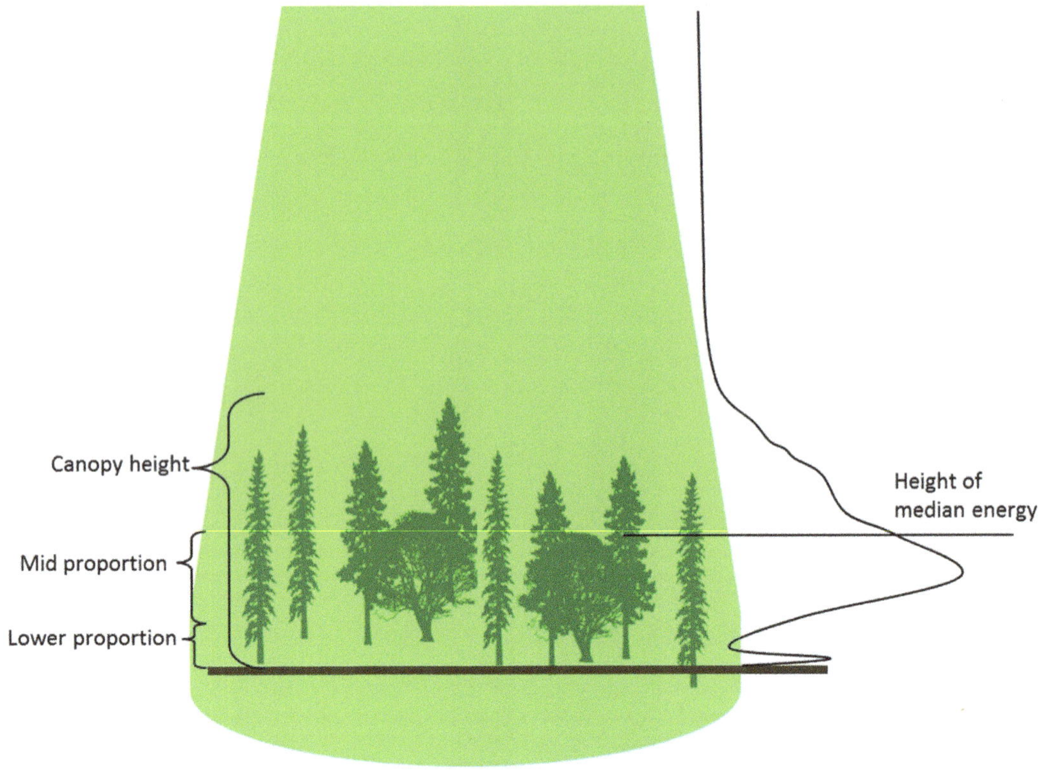

Figure 3. Cartoon depicting locations of vegetation metrics within canopy. The green cone in the figure represents the GLAS laser footprint scale, which subtended an area of roughly 64 m diameter. The black curve on the right of the figure represents a stylized GLAS energy return profile for this particular cartoon arrangement of vegetation and ground.

Table 1. GLAS LiDAR metrics considered for occupancy analysis; only the metrics denoted by double asterisks (**) were used after collinearity analysis.

Metric abbreviation	Metric
**Elev	elevation (m)*
**HLI	heat load index*
**p.lower	proportion of total return energy that represents vegetation between 0 and 3 m above ground level (agl)
**p.mid	proportion of total return energy that represents vegetation between 3 and 10 m agl
**can.height	canopy height computed from Chen (2010b) algorithm
ht.mean	mean canopy height agl
ht.var	variance of vegetation heights
ht.cv	coefficient of variation of vegetation heights
ht.med	median canopy height
ht.mad	median absolute deviation from median of canopy heights
Vdr	vertical distribution ratio (can.height – HOME)/can.height
**veg.density	canopy density – proportion of return energy representing vegetation to total return energy that includes ground returns
**HOME	height of median energy density
**FHD	foliage height diversity (1 m bin); diversity in vegetation distribution. See MacArthur & MacArthur, 1961

These metrics include 6 biotic factors (p.lower, p.mid, can.height, veg.density, HOME, and foliage height diversity (FHD)) and 2 abiotic factors (heat load index (HLI) and elevation.

*Heat load index integrates measures of site slope, aspect, and latitude, computed following [81]. Note that due to difficulties in deriving slope using GLAS data, HLI and elevation were derived from the 10-m National Elevation Database.

Table 2. Airborne LiDAR metrics considered for occupancy analysis; only metrics denoted by double asterisks (**) were used after collinearity analysis.

Metric abbreviation	Metric
**Elev	elevation (m)*
**HLI	heat load index*
p.lower	proportion of total return energy that represents vegetation between 0 and 3 m above ground level (agl)
**p.mid	proportion of total return energy that represents vegetation between 3 and 10 m agl
p.upper	proportion of total return energy that represents vegetation between 10 and 30 m + agl
VDR	Vertical distribution ratio (Goetz *et al.* 2007)
Density	canopy density – proportion of vegetation to total energy return
Hmax	maximum canopy height
Hmedian	median canopy height
Hsd	standard deviation of heights
Hmean	mean canopy height
Crr	canopy Relief Ratio (HMEAN – HMIN)/(HMAX – HMIN)
Hirq	inter-quartile range of heights
**Hmode	height of dominant mode (vegetation density) within return signal
**Hmrange	range between minimum and maximum modes
**Hskew	skewness of height profile across vegetation returns
Hkurt	kurtosis of height profile across vegetation returns
**Hvar	variance of vegetation density weighted height values
Hmad	median absolute deviation from median height
**FHD	foliage height diversity (1 m bin) see MacArthur & MacArthur, 1961

*Heat load index integrates measures of site slope, aspect, and latitude, computed following [81]. Note that while these measures are attainable using airborne lidar, to be consistent with calculations made for GLAS data HLI and elevation were derived from the 10-m National Elevation Database.

Geoid 83 for the vertical datum and NAD83 for the horizontal datum.

The Multiscale Curvature Classification algorithm [49] was used to delineate ground from non-ground coordinate corrected LiDAR data. Returns classified as ground were interpolated into a 1 m resolution DEM and subtracted from non-ground classified points to generate a topographically normalized point cloud. Metrics describing vegetation structural characteristics were derived from all points greater than 1 m in height within 20 m grid cells (see Table 2 for a list of all metrics derived by airborne LiDAR data). The 20 m grid cell resolution was used so that the airborne LiDAR metrics could be compared with and validated against measurements of forest structure collected in the field (see [28]). The resultant 20 m raster grids, each representing a vegetation height metric, were downsampled to 1 m resolution in the vicinity of the GLAS LiDAR footprints to minimize the effect of sampling incomplete pixels around the perimeter of the GLAS footprint. Grid cells within a 65 m diameter circle centered on the woodpecker survey site were aggregated to represent the mean value of each vegetation metric and to coincide with the GLAS footprint.

Woodpecker Surveys

We chose Red-naped Sapsuckers as the focus of our study for multiple reasons. As part of a cavity-excavating guild, sapsuckers are an ecologically important group owing to their status as ecosystem engineers [36]. Sapsuckers and other woodpeckers create keystone structures (i.e. cavities) that many other animals use for nesting and/or roosting [37,50–52]. In addition, cavity excavators in general rely upon a set of forest structural characteristics that may be described via LiDAR. For instance, Red-naped Sapsuckers forage in areas with a sparse understory, and in sections of the canopy that are relatively open to facilitate insect gleaning on trunks [53]. Additionally, the diameter of the nest tree, the fraction of canopy cover, and the distribution of vegetation are all important nest site attributes [54] and these are all potentially quantifiable metrics from both airborne and spaceborne LiDAR data.

Eighty two woodpecker survey sites were subsampled randomly from within the airborne LiDAR flight during 27 May–3 July 2009 from existing song bird point count locations (Fig. 1) and within seventy three GLAS spaceborne LiDAR footprints during 19 May–2 July 2010. A subset of survey sites ($n = 39$) occurred within both spaceborne and airborne LiDAR data samples (Fig. 1). To ensure that we were not double-counting the same birds, we selected sites with a minimum of 340 m spacing (equating to ~12 ha surrounding each survey location) which reflects the mean observed home range of Red-naped Sapsucker [53–55]. Surveys conducted outside of GLAS LiDAR sites were conducted to increase sample size for the airborne LiDAR derived occupancy model.

Woodpecker surveys were conducted following Drever *et al.* [56] using playback recordings to elicit responses from woodpeckers. Two identical playbacks, each comprised of a call and drum, were played for Red-naped Sapsuckers, with a two minute silence in between, and all Red-naped Sapsuckers seen or heard were recorded. Because the purpose of the surveys was to establish breeding bird-habitat relationships, we constrained the dates of the surveys to correspond to the breeding season of this species [55,56]. This timing is important because the use of callbacks is only effective while the birds are breeding and territorial. Each survey point was visited twice during the breeding season because multiple visits are necessary in order to better evaluate detection issues (i.e. when birds are really present but are recorded as absent [57]). Issues of detectability were explicitly incorporated into our analysis (see below). Surveys conducted solely within GLAS footprints were completed in the Idaho Panhandle National Forest (Fig. 1) in 2011 from 19 May–27 June. We selected survey sites by comparing the 2009 airborne LiDAR dataset to an earlier airborne LiDAR dataset flown in 2003 (see [28] for complete description of this multi-temporal analysis) to ensure that harvest operations had not taken place spanning the period from the first GLAS data collection to the date of the airborne LiDAR data acquisition.

Data Analysis

We used a t-test to compare the vegetation metrics between occupied and unoccupied sites, and considered $p \leq 0.05$ to be statistically significant. Our model development was conducted in an exploratory, rather than confirmatory, framework. We utilized this exploratory approach because Walters *et al.* [58] notes that breeding habitat relationships for Red-naped Sapsucker have been conducted where aspen (*Populus tremuloides*) and deciduous forests are important components of the forest, and our study sites contained very little deciduous forest. Preliminary data analysis revealed multi-collinearity among several GLAS and airborne LiDAR derived vegetation height and density metrics (Tables 1 and 2). We computed condition indices and variance decomposition proportions following Belsley [59] to identify the information redundant covariates. After applying conditioning diagnostics, our final set of site covariates included 6 biotic and 2 abiotic variables for the GLAS LiDAR based model (Table 1) and the airborne LiDAR based model (Table 2). We included elevation and heat load index to account for abiotic factors that may influence nest or foraging site selection. The 6 biotic variables are represented by the LiDAR-derived forest structure metrics, while elevation and heat load index were derived from the 10 m National Elevation Dataset [37].

While species presence and absence may be modeled with logistic regression, the approach assumes equal probability of detection across sites and cannot account for detection probabilities less than 1 [57,60,61]. Imperfect detection (i.e. $p<1$) produces biased estimates of latent variables that describe the state processes typically of interest [62,63]. We used a hierarchical occupancy model that explicitly accounts for detection probability that is conditional on the state process controlling species occurrence as implemented in the R package [64] 'unmarked' [65]. Due to the relatively small number of sample sizes for presence/absence modeling for each LiDAR dataset (airborne: $n=82$; GLAS: $n=73$), we pooled responses from the Moscow Mountain and the Idaho Panhandle National Forest regions for use in a single season occupancy model. Variables from Table 1 were used to model the latent occupancy state of each site. To account for detection probability, we included Julian day and a categorical measure of weather conditions (e.g. wind, cloud cover) where increasing values were associated with an increase in those weather variables.

We computed global, null, and reduced occupancy models for Red-naped Sapsuckers and evaluated relative model performance using Akaike's Information Criterion corrected for small sample size (AICc) [66]. We allowed for all linear combinations of predictor variables, and excluded models where the only predictor variables were the detectability covariates, requiring at least one process variable (vegetation, heat load index, or elevation) to be present. This method generated a list of models that struck a balance between including all possible habitat structure combinations, while excluding ecologically irrelevant models [67].

Global (full) model fit and checks for overdispersion were evaluated using a likelihood ratio test procedure to compute the deviance between the saturated model and the full model. The deviance was then divided by the residual degrees of freedom ($n-p=72$) to assess fit and potential overdispersion. We performed all first-order subsets regression to determine the top candidate models; while many studies utilize $\Delta AICc<2$, we opted to utilize a more conservative $\Delta AICc<6$ for final model selection evaluation [68,69]. Where model selection uncertainty was high (e.g. weights were <0.3) [66], we utilized a model-averaging approach. Parameter estimates in model averaging represent the sum of the product of the parameter estimates in each model with their associated model weight. As model weight tends towards zero, so too does the value of the parameter estimate. The model-averaging approach generates a new set of parameters that are used to replace the original parameter estimates in the global model. The procedure is recommended when all-subsets modeling is performed [70] and when strong support is lacking for a single model [66,71].

Relative variable importance was computed from model averaged parameter estimate weights to ascertain the contribution of each variable to the averaged model [66]. We believe our approach to model development addresses the concerns raised by Murtaugh [72] and Dahlgren [73], specifically with regards to shrinkage of estimates derived from model averaging that contribute the least amount of information to the model.

Model Cross Validation and Predictive Power

We assessed model performance using the Area Under the Curve (AUC) from the Receiver Operator Characteristic (ROC) curve where presence/absence probability cut-off values were determined using *k*-means cluster analysis by partitioning the predicted probabilities into 2 groups. We generated AUC scores first with unvalidated data, and then generated AUC scores using leave-one-out cross validation (LOOCV) to assess the degree of over-fitting and model sensitivity. AUC values close to 0.5 represent a model whose predictions of presence/absence are no different than random.

Results

We detected Red-naped Sapsucker responses at 48% (39 out of 82) of sites surveyed within area covered by airborne LiDAR and at 28% (28 out of 73) of sites surveyed within area covered by spaceborne LiDAR. Of the 155 total survey sites, 39 were common to both airborne and spaceborne LiDAR surveys. Fifteen of 39 Red-naped Sapsucker responses were reported at the coincident airborne and spaceborne LiDAR sites (44%). There were no statistical differences between the occupied and unoccupied sites (Table 3).

Model Fit and Parameter Estimation

GLAS LiDAR. Global model fit was good ($D=82.8$) when compared with the residual degrees of freedom ($rdof=63$) as indicated by the over-dispersion parameter, $\hat{c}$ ($\hat{c}=1.3$). Because $\hat{c}\approx 1$, there is no reason to doubt the fit of the global model, nor does over-dispersion appear to be large enough to warrant using a quasi AICc framework [65]. Due to high model selection uncertainty as indicated by AICc weights<0.3 (Table 4), we proceeded with a full model averaging approach for models with $\Delta AIC \leq 6$. Neither the original global model nor the averaged

Table 3. Means and standard errors of GLAS and airborne LiDAR variables used in sapsucker models.

Metrics	Occupied sites (n = 28)		Unoccupied sites (n = 45)	
GLAS (spaceborne LiDAR; n = 73)	**Mean**	**SE**	**Mean**	**SE**
Elevation (m)	950	21.1	974	17.8
HLI (heat load index)	0.842	0.014	0.871	0.009
p. lower	0.226	0.025	0.205	0.018
p.mid	0.347	0.026	0.389	0.021
Can.height (m)	34.1	2.06	32.2	1.41
Veg.density	0.897	0.014	0.905	0.011
HOME	10.87	1.02	10.96	0.669
FHD	2.092	0.015	2.11	0.012
	Occupied sites (n = 39)		**Unoccupied sites (n = 43)**	
Airborne LiDAR (n = 82)	**Mean**	**SE**	**Mean**	**SE**
Elevation (m)	956	13.9	941	12.8
HLI (heat load index)	0.875	0.008	0.845	0.014
p.mid	0.205	0.019	0.239	0.022
Hmode	2.59	0.736	3.80	0.965
Hmrange	18.4	1.00	16.5	1.00
Hskew	1.51	0.244	1.35	0.340
Hvar	56.9	6.33	47.6	5.09
FHD	1.33	0.014	1.34	0.038

model produced parameter estimates with acceptable confidence as the confidence intervals all included zero (Table 5).

Airborne LiDAR. Global model fit for Red-naped Sapsuckers indicated that the selection and number of parameters were adequate to model the data compared to that of a saturated model. The deviance was 76.6 and the overdispersion parameter (*c-hat*) = 1.05 with 72 residual degrees of freedom ($dof = n - p$). Model selection uncertainty in the airborne LiDAR derived models was high according to AICc model weights so we performed full model averaging including models with AICc≤6 (Tables 6 and 7).

Foliage height diversity (FHD) was highly influential to the averaged model (RVI = 0.95) (Table 8) and positively associated with Red-naped Sapsucker occupancy. Other variables that were important for this species included heat load index, the distance between the lowest and highest vegetation biomass (hmrange), and the distribution of distribution throughout the canopy (hskew). The positive association with hmrange indicated that occupancy increased with a greater vertical distance between the minimum and the maximum nodes of vegetation biomass. Visually, this would be represented by a gap in the vegetation structure at an intermediate height between the biomass mode nearest to the ground (the minimum mode likely representing the shrub layer) and the biomass mode in the canopy (the maximum mode likely representing the canopy). The positive relationship with hskew indicated that there is more vegetation closer to the ground than in the upper canopy. The proportion of vegetation in the mid-story (2.5–10 m) was positively associated with Red-naped Sapsucker presence but contributed the least to the averaged model (RVI = 0.32). Wind was a detectability covariate, and occupancy was negatively associated with high wind.

Model Cross Validation and Predictive Power

Due to the high uncertainty associated with our spaceborne models, we did not proceed with an assessment of model performance based on spaceborne LiDAR. The AUC score for the airborne data indicated that the global model performed better than random, with AUC scores ranging between 0.6 and 0.72 (Fig. 4).

Discussion

This study was novel for two reasons. First, this study addressed whether GLAS spaceborne LiDAR data could be used for wildlife habitat application in the case of one bird species known to be an ecosystem engineer important to creating habitat for other forest species in the Inland Northwest, USA. Secondly, we were able to compare GLAS and airborne LiDAR models developed at the same spatial scale; no studies to date have compared these LiDAR data sets in the context of wildlife habitat assessment. LiDAR data used in this study were averaged over an area of approximately 0.3 ha centered on the survey point. While it is unlikely that the survey points were coincident with the centers of the home ranges of individual birds, the 0.3 ha sampling unit coincided with the size of the GLAS footprint. Thus, we limited our analyses to this spatial scale in order to facilitate comparisons between the two LiDAR data sets as applied to Red-naped Sapsucker habitat.

The model results based upon GLAS spaceborne LiDAR were weak, which may be due to several reasons. First, we had relatively small sample sizes; although we gathered data across three years, fewer birds were detected during 2011. This was a wet summer, and few birds were noted in our expanded GLAS survey sites. Additionally, a scale mismatch may have occurred between the spatial distribution and extent of GLAS footprints and variables

Table 4. Occupancy models for Red-naped Sapsuckers with a ΔAICc≤6 based on GLAS LiDAR.

Model #	Elev (m)	FHD	HOME	Canopy height	HLI	p.lower	p.mid	Vegetation density	Julian Day	sky	df	logLik	AICc	delta	weight
160	−67.05	37.45	−98.89	203.13	−48.58			31.84			8	−66.60	151.45	0.00	0.177
224	−93.36	61.76	−157.13	285.28	−63.16		−9.75	57.39			9	−65.85	152.56	1.11	0.101
382	−68.41		−153.43	156.54	−77.74	−100.44	−2.72		−0.33		9	−65.97	152.79	1.34	0.090
928	−78.24	43.22	−114.44	236.73	−56.85			36.39	−0.36	−0.19	10	−64.69	152.92	1.47	0.085
672	−77.98	43.53	−115.08	236.15	−56.31			37.07		−0.19	9	−66.16	153.17	1.72	0.075
221			−144.38	174.87	−66.23		−78.37	76.93			7	−68.88	153.47	2.02	0.064
176	−147.04	107.85	−110.06	294.86		69.83		62.13			8	−68.05	154.35	2.90	0.041
894	−106.34		−221.14	247.94	−119.19	−147.19	10.61		−0.32	−0.20	10	−65.49	154.54	3.08	0.038
206	−68.28		−135.07	132.80			−49.41	79.31			7	−69.64	155.01	3.55	0.030
477			−91.61	127.66	−41.94		−45.09	41.18	−0.29		8	−68.44	155.14	3.68	0.028
733			−177.87	216.25	−82.75		−96.36	94.85		−0.18	8	−68.46	155.16	3.71	0.028
192	−62.66	−15.09	−137.55	140.90	−82.43	−106.11		−6.70			9	−67.31	155.47	4.02	0.024
448	−68.42	12.38	−135.16	178.96	−69.03	−64.18		10.00	−0.33		10	−65.97	155.50	4.04	0.023
736	−87.09	70.01	−209.57	296.30	−60.45		−46.08	78.18		−0.21	10	−66.02	155.59	4.14	0.022
223		−33.28	−140.63	197.60	−104.06		−78.12	63.89			8	−68.78	155.81	4.36	0.020
462	−51.77		−103.44	101.11			−38.33	60.88	−0.28		8	−68.79	155.83	4.38	0.020
992	−189.14	121.41	−364.00	606.65	−138.12		−55.46	137.20	−0.32	−0.22	11	−64.77	155.87	4.42	0.019
989			−129.90	157.61	−59.85		−70.45	69.10	−0.29	−0.19	9	−67.55	155.96	4.51	0.019
512	−89.08	41.11	−166.08	248.49	−72.57	−44.97	−1.72	37.27	−0.37		11	−64.95	156.23	4.78	0.016
479		−16.27	−132.75	175.36	−74.68		−77.53	63.72	−0.29		9	−67.75	156.36	4.90	0.015
16	−58.55	40.27	−52.28	88.17							6	−71.58	156.44	4.98	0.015
640	−76.42	2.10	−166.48	180.24	−84.73	−102.70	−3.51			−0.21	10	−66.75	157.04	5.59	0.011
704	−56.05	0.02	−107.61	137.96	−63.41	−64.87		3.77		−0.21	10	−66.79	157.13	5.68	0.010
718	−53.67		−107.68	105.14			−39.86	63.27		−0.15	8	−69.47	157.19	5.74	0.010
768	−62.97	30.52	−102.80	185.63	−49.88	−13.95	−1.48	28.47		−0.19	11	−65.43	157.19	5.74	0.010
238	−50.93		−102.06	107.28		7.77	−41.17	63.69			8	−69.53	157.30	5.85	0.009

Table 5. Model averaged parameter estimates, standard errors, and 95% confidence intervals using GLAS LiDAR data.

Parameter		Estimate	Std. Error	Lower CI	Upper CI
Occupancy	elev	69.6	116	−157	296
	FHD	110	291	−461	681
	HLI	56.5	196	−327	440
	hmode	−59.4	87.4	−231	112
	hmrange	82.8	179	−268	434
	hskew	112	316	−507	731
	hvar	−50.7	103	−253	152
	p.mid	19.7	74.5	−126	166
Detection	wind	−0.26	0.20	−0.64	0.13

Table 7. Original global model parameter estimates, standard errors, and 95% confidence intervals using airborne LiDAR data.

Parameter		Estimate	Std. Error	Lower CI	Upper CI
Occupancy	Elev*	78.3	248	61.7	112
	HLI*	30.3	91.3	16.9	47.9
	p.mid	17.5	98.0	−17.6	41.2
	FHD*	114	301	92.9	146
	Hmode*	−63.2	166	−80.2	−53.0
	Hmrange*	92.5	343	48.0	138
	Hskew*	92.5	268	61.2	141
	Hvar*	−53.7	212	−79.1	−27.9
Detection	wind	−0.25	0.20	−0.65	0.12

*Confidence intervals do not include zero.

that influence Red-naped Sapsucker habitat selection, which is likely to have affected both the GLAS results as well as the performance of the airborne LiDAR models. Indeed, red-naped Sapsucker may respond to habitat features at multiple spatial scales [58], and recently Sadoti and Vierling [74] noted that features at the home range scale (~9.1 ha) were most important in a multi-scale analysis of selection. Multiple studies have been conducted on Red-naped Sapsucker breeding habitat in aspen forests [74,75], but features that might influence Red-naped Sapsuckers in conifer-dominated forests and the associated important spatial scales are relatively unknown [76]. Other studies have noted that woodpeckers select habitat features at multiple spatial scales, and these spatial scales range from nest tree characteristics to landscape features [74,75,77,78]. We were constrained by the spatial distribution and extent of GLAS footprints, and were thus unable to explore the effects of multiple scales. Furthermore, factors within the GLAS footprints that were not detectable might have heavily influenced Red-naped Sapsucker occupancy (e.g. the presence of fungal conks; [75]).

The most important structural variables that influenced Red-naped Sapsucker selection based on airborne LiDAR data included foliage height diversity, the distance between major components in the canopy profile, and the amount of vegetation near the ground. When considered jointly, these three metrics describe a vegetation profile that is uniform throughout its height except for a larger contribution of vegetation near the ground. Red-naped Sapsuckers have been observed to inhabit montane conifer forests, and Walters [53] notes that increased foraging preference was observed in areas of the canopy with fewer branches. Conifer stands with an even distribution of foliage along the height of the forest may represent branch structure that is more open and accessible for foraging. Additionally, a larger contribution of vegetation near the ground, likely representing shrub growth and could support insects. The presence of shrub growth likely supports an increased insect population, and Red-naped Sapsuckers will utilize a combination of flycatching and gleaning to forage during the breeding season [58,74].

The presence of Red-naped Sapsuckers was also positively associated with increasing elevation and heat load index, although their effect sizes were the smallest. Higher heat indices indicate a preference for sites with warmer combinations of ground slope and aspect. Studies have noted tree cavity

Table 6. All subsets models for Red-naped Sapsuckers with ΔAIC≤6 based on airborne LiDAR data.

Model #	Elev	FHD	HLI	hmode	hmrange	hskew	hvar	p.mid	wind	df	logLik	AICc	delta	weight
384	70.50	115.97	27.05	−66.44	82.84	83.16	−54.48		−0.25	10	−79.57	182.24	0.00	0.44
512	78.33	113.98	30.32	−63.20	92.53	92.54	−53.74	17.49	−0.25	11	−79.47	184.70	2.47	0.13
448	73.00	90.95	24.19	−49.88	55.74	111.96		23.54	−0.30	10	−81.13	185.35	3.12	0.09
120	21.96	107.30	65.65		73.64	88.37	−21.12			8	−84.19	186.36	4.12	0.06
188	108.37	193.01		−73.42	140.18	327.12		64.45		8	−84.34	186.66	4.42	0.05
7		−28.76	153.00							4	−89.12	186.76	4.52	0.05
63		339.22	196.97	37.61	145.38	371.68				7	−85.96	187.43	5.20	0.03
263		−34.82	186.06						−0.19	5	−88.53	187.84	5.61	0.03
56	34.05	170.41	111.76		106.84	144.86				7	−86.22	187.96	5.72	0.03
135		−29.03	133.33					5.20		5	−88.60	187.99	5.76	0.02
37			142.39			33.49				4	−89.76	188.04	5.81	0.02
149			95.41		−54.03			−58.84		5	−88.64	188.08	5.84	0.02
124	63.46	76.63		−51.78	63.56	51.64	−34.86			8	−85.08	188.13	5.89	0.02

Table 8. Relative variable importance (RVI) from model averaging.

Parameter	RVI
FHD	0.95
HLI	0.93
hmrange	0.88
hskew	0.88
elev	0.82
hmode	0.77
wind	0.69
hvar	0.65
p.mid	0.32
julDay	–
sky	–

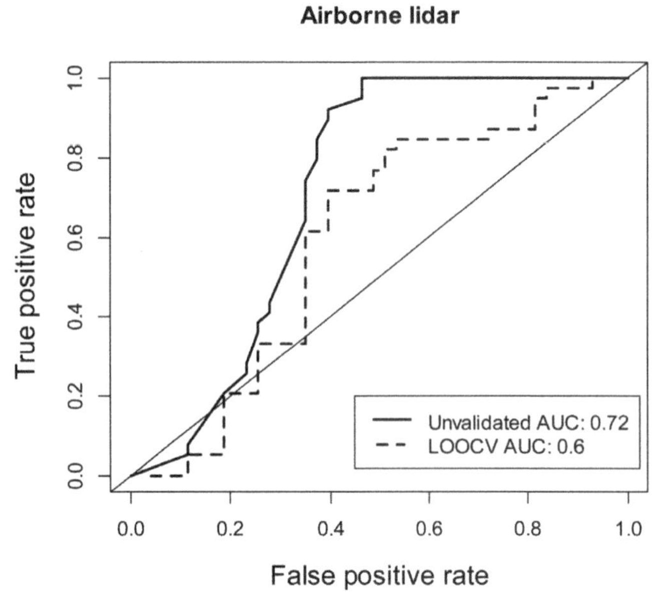

Figure 4. Receiver operator curves with unvalidated and leave-one-out (LOOCV) cross validation for airborne LiDAR.

entrances located on southern aspects in British Columbia [53], and Colorado and Wyoming [79], to no apparent directional preference in Montana [54]. However, while Martinuzzi *et al.* [18] found that slope and aspect could affect the presence of forest snags near the study area, we are not aware of any studies that indicate slope or aspect preference for the nest tree location within the landscape.

Our results demonstrated that small-footprint airborne LiDAR data were able to provide ecologically meaningful information relative to occupancy for Red-naped Sapsuckers when aggregated and analyzed at a scale similar to that sampled by the large-footprint GLAS satellite LiDAR. This finding indicates that the forest structural metrics calculable from airborne LiDAR were likely more suitable for modeling habitat of this bird than GLAS-based structural metrics at this spatial scale. However, because not all samples contained coincident airborne and satellite lidar, this finding requires further testing at additional field sites and wildlife species, and underscores the need for denser spatial sampling in future satellite-based LiDAR missions for application to mapping and modeling wildlife distributions (see [30]). While the large footprint and non-contiguous nature of GLAS data poses significant challenges for wildlife distribution modeling, these data do provide useful information on ecosystem vertical structure, particularly in areas of gentle terrain (e.g. [80]). We therefore encourage the remote sensing and wildlife communities to join efforts so that additional progress can be made to incorporate LiDAR datasets collected from both airborne and past and future satellite platforms (e.g. GLAS, and the planned IceSAT2 mission) with the prospect of improving wildlife biodiversity and single-species modeling for more locations across the globe.

Acknowledgments

This study benefited from discussions with Drs. Fred Hummerich, Dale Goble, Amy Neuenschwander, Laura Duncanson, Sebastian Martinuzzi, Jody Vogeler, and members of the University of Idaho Geospatial Laboratory for Environmental Dynamics. We thank two anonymous reviewers for their constructive comments on eariler versions of the manuscript. We also acknowledge Potlatch Corporation, Bennett Lumber Products, and the Idaho Department of Lands for their gracious permission to conduct research on their property.

Author Contributions

Conceived and designed the experiments: LV KV PA. Performed the experiments: PA. Analyzed the data: LV KV PA AH. Contributed reagents/materials/analysis tools: LV KV PA AH. Wrote the paper: LV KV PA AH.

References

1. Innes JL, Koch B (1998) Forest biodiversity and its assessment by remote sensing. Global Ecology and Biogeography 7: 397–419.
2. Turner W, Spector S, Gardiner N, Fladeland M, Sterling E, et al. (2003) Remote sensing for biodiversity science and conservation. Trends in Ecology & Evolution 18: 306–314.
3. Scott JM, Davis F, Csuti B, Noss R, Butterfield B, et al. (1993) GAP analysis - a geographic approach to protection of biological diversity. Wildlife Monographs 123: 1–41.
4. Rodrigues ASL, Andelman SJ, Bakarr MI, Boitani L, Brooks TM, et al. (2004) Effectiveness of the global protected area network in representing species diversity. Nature 428: 640–643.
5. Nagendra H (2001) Using remote sensing to assess biodiversity. International Journal of Remote Sensing 22: 2377–2400.
6. Scott JM, Peterson CR, Karl JW, Strand E, Svancara LK, et al. (2002) A Gap Analysis of Idaho: Final Report. Moscow: Idaho Cooperative Fish and Wildlife Research Unit.
7. MacArthur R, MacArthur JW (1961) On bird species-diversity. Ecology 42: 594–598.
8. Vierling KT, Vierling LA, Gould WA, Martinuzzi S, Clawges RM (2008) LiDAR: shedding new light on habitat characterization and modeling. Frontiers in Ecology and the Environment 6: 90–98.
9. Elith J, Leathwick JR (2009) Species Distribution Models: Ecological Explanation and Prediction Across Space and Time. Annual Review of Ecology, Evolution and Systematics. 40: 677–697.
10. Hinsley SA, Hill RA, Gaveau DLA, Bellamy PE (2002) Quantifying woodland structure and habitat quality for birds using airborne laser scanning. Functional Ecology 16: 851–857.
11. Hill RA, Hinsley SA, Gaveau DLA, Bellamy PE (2004) Predicting habitat quality for Great Tits (Parus major) with airborne laser scanning data. International Journal of Remote Sensing 25: 4851–4855.
12. Goetz S, Steinberg D, Dubayah R, Blair B (2007) Laser remote sensing of canopy habitat heterogeneity as a predictor of bird species richness in an eastern temperate forest, USA. Remote Sensing of Environment 108: 254–263.

13. Boelman NT, Asner GP, Hart PJ, Martin RE (2007) Multi-trophic invasion resistance in Hawaii: Bioacoustics, field surveys, and airborne remote sensing. Ecological Applications 17: 2137–2144.
14. Clawges R, Vierling K, Vierling L, Rowell E (2008) The use of airborne LiDAR to assess avian species diversity, density, and occurrence in a pine/aspen forest. Remote Sensing of Environment 112: 2064–2073.
15. Goetz SJ, Steinberg D, Betts MG, Holmes RT, Doran PJ, et al. (2010) LiDAR remote sensing variables predict breeding habitat of a Neotropical migrant bird. Ecology 91: 1569–1576.
16. Lesak AA, Radeloff VC, Hawbaker TJ, Pidgeon AM, Gobakken T, et al. (2011) Modeling forest songbird species richness using LiDAR-derived measures of forest structure. Remote Sensing of Environment 115: 2823–2835.
17. Vogeler JC, Hudak AT, Vierling LA, Vierling KT (2013) Lidar-derived canopy architecture predicts Brown Creeper occupancy of two western coniferous forests. The Condor 115: 1–9.
18. Martinuzzi S, Vierling LA, Gould WA, Falkowski MJ, Evans JS, et al. (2009) Mapping snags and understory shrubs for a LiDAR-based assessment of wildlife habitat suitability. Remote Sensing of Environment 113: 2533–2546.
19. Nelson R, Keller C, Ratnaswamy M (2005) Locating and estimating the extent of Delmarva fox squirrel habitat using an airborne LiDAR profiler. Remote Sensing of Environment 96: 292–301.
20. Muller J, Brandl R (2009) Assessing biodiversity by remote sensing in mountainous terrain: the potential of LiDAR to predict forest beetle assemblages. Journal of Applied Ecology 46: 897–905.
21. Vierling KT, Bässler C, Brandl R, Vierling LA, Weiß I, et al. (2011) Spinning a laser web: predicting spider community characteristics using LiDAR. Ecological Applications 21: 577–588.
22. Lefsky MA, Cohen WB, Parker GG, Harding DJ (2002) Lidar remote sensing for ecosystem studies. BioScience 52: 19–30.
23. Dubayah RO, Drake JB (2000) Lidar remote sensing for forestry. Journal of Forestry 98: 44–46.
24. Falkowski M, Smith AM, Hudak AT, Gessler PE, Vierling LA, et al. (2006) Automated estimation of individual conifer tree height and crown diameter via two-dimensional wavelet analysis of lidar data. Canadian Journal of Remote Sensing 32: 153–161.
25. Falkowski MJ, Evans JS, Martinuzzi S, Gessler PE, Hudak AT (2009) Characterizing forest succession with lidar data: An evaluation for the Inland Northwest, USA. Remote Sensing of Environment 113: 946–956.
26. Martinuzzi S, Gould WA, Vierling LA, Hudak AT, Nelson RF, et al. (2013) Quantifying Tropical Dry Forest Type and Succession: Substantial Improvement with LiDAR. Biotropica 45: 135–146.
27. Gonzalez P, Asner GP, Battles JJ, Lefsky MA, Waring KM, et al. (2010) Forest carbon densities and uncertainties from Lidar, QuickBird, and field measurements in California. Remote Sensing of Environment 114: 1561–1575.
28. Hudak AT, Strand EK, Vierling LA, Byrne JC, Eitel JUH, et al. (2012) Quantifying aboveground forest carbon pools and fluxes from repeat LiDAR surveys. Remote Sensing of Environment 123: 25–40.
29. Seavy NE, Viers JH, Wood JK (2009) Riparian bird response to vegetation structure: a multiscale analysis using LiDAR measurements of canopy height. Ecological Applications 19: 1848–1857.
30. Bergen KM, Goetz SJ, Dubayah RO, Henebry GM, Hunsaker CT, et al. (2009) Remote sensing of vegetation 3-D structure for biodiversity and habitat: Review and implications for LiDAR and radar spaceborne missions. Journal of Geophysical Research-Biogeosciences 114: G00e06, DOI: 10.1029/2008JG000883.
31. Schutz BE, Zwally HJ, Shuman CA, Hancock D, DiMarzio JP (2005) Overview of the ICESat Mission. Geophysical Research Letters 32: L21S01, DOI: 10.1029/2005GL024009.
32. Lefsky MA (2010) A global forest canopy height map from the Moderate Resolution Imaging Spectroradiometer and the Geoscience Laser Altimeter System. Geophysical Research Letters 37: L14501, DOI: 10.1029/2010GL043622.
33. Harding DJ, Carabajal CC (2005) ICESat waveform measurements of within-footprint topographic relief and vegetation vertical structure. Geophysical Research Letters 32: L21S10, DOI: 10.1029/2005GL023471.
34. Lefsky MA, Harding DJ, Keller M, Cohen WB, Carabajal CC, et al. (2005) Estimates of forest canopy height and aboveground biomass using ICESat. Geophysical Research Letters 32: L22S02, DOI: 10.1029/2005GL023971.
35. Nelson R, Boudreau J, Gregoire TG, Margolis H, Naesset E, et al. (2009) Estimating Quebec provincial forest resources using ICESat/GLAS. Canadian Journal of Forest Research 39: 862–881.
36. Jones CG, Lawton JH, Shachak M (1994) Organisms as Ecosystem Engineers. Oikos 69: 373–386.
37. Daily GC, Ehrlich PR, Haddad NM (1993) Double keystone bird in a keystone species complex. Proceedings of the National Academy of Sciences of the United States of America 90: 592–594.
38. Aitken KEH, Martin K (2008) Resource selection plasticity and community responses to experimental reduction of a critical resource. Ecology 89: 971–980.
39. Bull EL, Holthausen RS (1993) Habitat Use and Management of Pileated Woodpeckers in Northeastern Oregon. The Journal of Wildlife Management 57: 335–345.
40. Drever MC, Martin K (2010) Response of woodpeckers to changes in forest health and harvest: Implications for conservation of avian biodiversity. Forest Ecology and Management 259: 958–966.
41. Ripper D, Bednarz JC, Varland DE (2007) Landscape Use by Hairy Woodpeckers in Managed Forests of Northwestern Washington. Journal of Wildlife Management 71: 2612–2623.
42. Abshire JB, Sun XL, Riris H, Sirota JM, McGarry JF, et al. (2005) Geoscience Laser Altimeter System (GLAS) on the ICESat mission: On-orbit measurement performance. Geophysical Research Letters 32: L21S02, DOI:10.1029/2005GL024028.
43. Chen Q (2010) Retrieving vegetation height of forests and woodlands over mountainous areas in the Pacific Coast region using satellite laser altimetry. Remote Sensing of Environment 114: 1610–1627.
44. NASA (2012) National Snow and Ice Data Center - ICESat/GLAS Data. Available: http://nsidc.org/. Accessed 5 October 2010.
45. Brenner A, Zwally H, Bentley C, Csatho B, Harding D, et al. (2003) Derivation of range and range distributions from laser pulse waveform analysis for surface elevations, roughness, slope, and vegetation heights. Algorithm Theoertical Basis Document Version 4.1, NASA Goddard Space Flight Center.
46. Chen Q (2010) Assessment of terrain elevation derived from satellite laser altimetry over mountainous forest areas using airborne LiDAR data. ISPRS Journal of Photogrammetry and Remote Sensing 65: 111–122.
47. Lefsky MA, Keller M, Pang Y, de Camargo PB, Hunter MO (2007) Revised method for forest canopy height estimation from Geoscience Laser Altimeter System waveforms. Journal of Applied Remote Sensing 1: 1–18.
48. USDA (2010) Geospatial Data Gateway. Available: http://gcmd.nasa.gov/records/USDA_Geo_Gateway.html. Accessed 15 June 2010.
49. Evans JS, Hudak AT (2007) A multiscale curvature algorithm for classifying discrete return LiDAR in forested environments. IEEE Transactions on Geoscience and Remote Sensing 45: 1029–1038.
50. Cooke HA, Hannon SJ (2011) Do aggregated harvests with structural retention conserve the cavity web of old upland forest in the boreal plains? Forest Ecology and Management 261: 662–674.
51. Gentry DJ, Vierling KT (2008) Reuse of Woodpecker Cavities in the Breeding and Non-Breeding Seasons in Old Burn Habitats in the Black Hills, South Dakota. American Midland Naturalist 160: 413–429.
52. Martin K, Aitken KEH, Wiebe KL (2004) Nest Sites and Nest Webs for Cavity-Nesting Communities in Interior British Columbia, Canada: Nest Characteristics and Niche Partitioning. The Condor 106: 5–19.
53. Walters EL (1996) Habitat and space use of the Red-naped Sapsucker, ~Sphyrapicus nuchalis~, in the Hat Creek valley, south-central British Columbia [Master's Thesis]. Victoria, BC: Univ. of Victoria.
54. McClelland BR, McClelland PT (2000) Red-naped sapsucker nest trees in northern rocky mountain old-growth forest. The Wilson Bulletin 112: 44–50.
55. McClelland BR (1977) Relationships between hole-nesting birds, forest snags, and decay in western larch-Douglas fir forests of the northern Rocky Mountains [PhD dissertation]. Missoula: Univ. of Montana.
56. Drever MC, Aitken KEH, Norris AR, Martin K (2008) Woodpeckers as reliable indicators of bird richness, forest health and harvest. Biological Conservation 141: 624–634.
57. MacKenzie DI, Nichols JD, Lachman GB, Droege S, Royle JA, et al. (2002) Estimating site occupancy rates when detection probabilities are less than one. Ecology 83: 2248–2255.
58. Walters EL, Miller EH, Lowther PE (2002) Red-naped Sapsucker (Sphyrapicus nuchalis) In: Poole A, editor. The Birds of North America Online. Ithaca: Cornell Lab of Ornithology. Available: http://bna.birds.cornell.edu/bna/species/663b; doi:10.2173/bna.663. Accessed 10 January 2012.
59. Belsley D (1991) Conditioning diagnostics : collinearity and weak data in regression. New York: John Wiley. 396 p.
60. Royle JA, Nichols JD (2003) Estimating abundance from repeated presence–absence data or point counts. Ecology 84: 777–790.
61. McCallum DA (2005) A conceptual guide to detection probability for point counts and other count-based survey methods. USDA Forest Service General Technical Report PSW-GTR-191: 754–761.
62. Gu WD, Swihart RK (2004) Absent or undetected? Effects of non-detection of species occurrence on wildlife-habitat models. Biological Conservation 116: 195–203.
63. Mazerolle MJ, Desrochers A, Rochefort L (2005) Landscape characteristics influence pond occupancy by frogs after accounting for detectability. Ecological Applications 15: 824–834.
64. R Development Core Team (2012) R: A language and environment for statistical computing. Vienna, Austria: R Foundation for Statistical Computing.
65. Fiske IJ, Chandler RB (2011) Unmarked: An R Package for Fitting Hierarchical Models of Wildlife Occurrence and Abundance. Journal of Statistical Software 43: 1–23.
66. Burnham K, Anderson D (2002) Model selection and multimodel inference: a practical information-theoretic approach: Springer Verlag. 488 p.
67. Dochtermann NA, Jenkins SH (2011) Developing multiple hypotheses in behavioral ecology. Behavioral Ecology and Sociobiology 65: 37–45.
68. Richards SA (2005) Testing ecological theory using the information-theoretic approach: Examples and cautionary results. Ecology 86: 2805–2814.
69. Richards SA (2008) Dealing with overdispersed count data in applied ecology. Journal of Applied Ecology 45: 218–227.
70. Lukacs P, Burnham K, Anderson D (2010) Model selection bias and Freedman's paradox. Annals of the Institute of Statistical Mathematics 62: 117–125.

71. Burnham KP, Anderson DR, Huyvaert KP (2011) AIC model selection and multimodel inference in behavioral ecology: some background, observations, and comparisons. Behavioral Ecology and Sociobiology 65: 23–35.
72. Murtaugh PA (2009) Performance of several variable-selection methods applied to real ecological data. Ecology Letters 12: 1061–1068.
73. Dahlgren JP (2010) Alternative regression methods are not considered in Murtaugh (2009) or by ecologists in general. Ecology Letters 13: E7–E9.
74. Sadoti G, Vierling KT (2010) Nonideal habitat selection by a North American cavity excavator: pecking up the wrong tree? Canadian Journal of Zoology 88: 527–535.
75. Lawler JJ, Edwards TC (2006) A variance-decomposition approach to investigating multiscale habitat associations. Condor 108: 47–58.
76. Wagner MA (2011) Habitat selection by Red-breasted Sapsucker (Sphyrapicus ruber) in southeast Alaska old-growth forest [Master's thesis]. Humboldt: Humboldt State University. 58 p.
77. Lawler J, Edwards T (2002) Landscape patterns as habitat predictors: building and testing models for cavity-nesting birds in the Uinta Mountains of Utah, USA. Landscape Ecology 17: 233–245.
78. Vierling KT, Lentile LB, Nielsen-Pincus N (2008) Preburn characteristics and woodpecker use of burned coniferous forests. Journal of Wildlife Management 72: 422–427.
79. Crockett AB, Hadow HH (1975) Nest site selection by Williamson's and Red-naped sapsuckers. Condor 77: 365–368.
80. Lefsky MA (2010) A global forest canopy height map from the Moderate Resolution Imaging Spectroradiometer and the Geoscience Laser Altimeter System. Geophysical Research Letters 37: Article Number: L15401 DOI: 10.1029/2010GL043622.
81. McCune B, Keon D (2002) Equations for potential annual direct incident radiation and heat load. Journal of Vegetation Science 13(4): 603–606.

13

Testing the Effects of an Introduced Palm on a Riparian Invertebrate Community in Southern California

Theresa Sinicrope Talley[1]*, Kim-Chi Nguyen[1], Anthony Nguyen[2]

1 Scripps Institution of Oceanography, University of California San Diego, La Jolla, California, United States of America, 2 Department of Biology, University of San Diego, San Diego, California, United States of America

Abstract

Despite the iconic association of palms with semi-arid regions, most are introduced and can invade natural areas. Along the San Diego River (San Diego, California, USA), the introduced Canary Island date palm (*Phoenix canariensis*) forms dense patches among native riparian shrubs like arroyo willow (*Salix lasiolepis*). The structural differences between the palm and native shrubs are visually obvious, but little is known about palm's effects on the ecosystem. We tested for the effects of the palm on a riparian invertebrate community in June 2011 by comparing the faunal and environmental variables associated with palm and willow canopies, trunks and ground beneath each species. The palm invertebrate community had lower abundance and diversity, fewer taxa feeding on the host (e.g., specialized hemipterans), and more taxa likely using only the plant's physical structure (e.g., web-builders, oak moths, willow hemipterans). There were no observed effects on the ground-dwelling fauna. Faunal differences were due to the physical and trophic changes associated with palm presence, namely increased canopy density, unpalatable leaves, trunk rugosity, and litter accumulations. Palm presence and resulting community shifts may have further ecosystem-level effects through alteration of physical properties, food, and structural resources. These results were consistent with a recent study of invasive palm effects on desert spring arthropods, illustrating that effects may be relatively generalizable. Since spread of the palm is largely localized, but effects are dramatic where it does occur, we recommend combining our results with several further investigations in order to prioritize management decisions.

Editor: Simon Thrush, National Institute of Water & Atmospheric Research, New Zealand

Funding: No external funding was received for this study. Some resources (supplies and research assistance) were donated by Paul Dayton through NOAA Grant nos. NA08OAR4170669 and NA10OAR4170060, California Sea Grant Project R/ENV-209, via NOAA's National Sea Grant College Program, U.S. Dept. of Commerce. The statements, findings, conclusions and recommendations are those of the researchers and do not necessarily reflect the views of the aforementioned organizations. The funders had no role in study design, data collection and analysis, decision to publish, or preparation of the manuscript.

Competing Interests: The authors have declared that no competing interests exist.

* E-mail: tstalley@ucsd.edu

Introduction

Images of many semi-arid and/or semi-tropical regions would not be complete without picturesque palms, but in areas such as California these palms are mostly non-native. Tropical icons, such as the fan palms (*Washingtonia* spp.) and the date palms (*Phoenix* spp.), are commonly used in landscaping [1] and are making their way throughout relatively moist natural areas [2], such as desert springs in California [3] and riparian corridors of California and Florida [4–5]. In California desert springs, two introduced palms (*W. filifera* and *P. dactylifera* L.) outcompete native vegetation for sunlight, space, and water; and thick layers of wide, fibrous palm frond litter prevent new plants from growing [3], [6–7]. These desert invaders were also associated with lower abundances and diversity of arthropods, and fewer herbivores than on native plant species [3]. The effects of palm invasions in riparian corridors, however, are uncertain despite the higher prevalence of these ecosystems and their vulnerability to invasion [8].

The Canary Island date palm (*P. canariensis*) is a particularly popular landscaping species in warm climates throughout the world [1], especially California [1], [9–10], making it a common invader of moist ecosystems within these regions, such as along the San Diego River. Unassisted large-scale spread is relatively slow, especially compared to more aggressive invaders (e.g., the giant reed grass, *Arundo donax*), but localized spread and dense patch formation are common along Southern California rivers due to the high propagule pressure of horticultural sources, the high fecundity of established adults [2], [11–12] and moderate salt tolerance [13–14] allowing it access to estuarine river reaches [15].

On the San Diego River floodplain, the Canary Island date palm (hereafter "the palm") is found at mid-riparian elevations and often in association with arroyo willow (*Salix lasiolepis*) [15]. In areas where palm establishment has occurred, the structural changes due to differences in the canopy, trunk and litter are visible at a glance (Fig. 1) and likely influence associated fauna by changes to primary production food sources and/or living space (e.g., [3], [16–18]). The long, tough, and dense fronds of the palm canopy contrast with the airy, complex networks of willow branches and small soft leaves (Fig. 1A), a similar contrast to that observed between palms and native desert spring shrubs [3]. The palm trunk is extremely rugous with frond bases that capture detritus, while that of the willow is relatively smooth and branching (Fig. 1B). The ground beneath the palm is carpeted with the tough, fallen fronds while there is a thinner layer of small-leaf litter in various states of decomposition lies beneath the willow (Fig. 1C). Palm presence, therefore, potentially alters the type, amount, and quality of available primary production, including increasing relative amounts of tough, little-digestible leaves and leaf litter, decreasing herbaceous understory cover, and increasing detritus trapped on the rugous trunk (e.g., [3], [19–20]). Living space (physical structure) is also likely altered with palm presence due to the seemingly denser, larger, simpler (less branching) canopy, more trunk surface rugosity, and greater accumulations of litter and debris (e.g., [21–22]).

Goals and Objectives

The main goal of this study was to test the effects of the invasion of Canary Island palm on a riparian invertebrate community. Invertebrates often closely track environmental changes and can reveal the ways in which changes to physical properties and basal resources, such as those that occur with plant invasions, may affect the greater ecological community [23]. Our goal was met by testing for differences (or similarities) in invertebrate community variables (abundance, diversity, composition) associated with the canopy, trunk, and ground beneath the canary island palm and the native arroyo willow.

Materials and Methods

This study was conducted during June 2011 in a 9.7 ha area of riparian woodland within the Mission Valley Preserve, a public park located on the lower San Diego River (32°45′44.56″N, 117°11′37.72″W). Permission to perform studies in this area was granted by the City of San Diego, owners and managers of the land, and the San Diego River Park Foundation, managers of the park. No other special permits were required, and no protected species were involved. Fifteen pairs of Canary Island palm and nearby (10–20 m apart) arroyo willow shrubs were selected. Pairs were at least 50 m apart. The canopy, stems and ground beneath the dripline of each woody plant were sampled for invertebrate community and environmental variables.

Invertebrate Community

Canopy and trunk dwelling fauna. Each plant canopy and trunk was sampled mid-day (10 am – 2 pm) over two consecutive sunny, warm (24–26 deg C) days using a leaf blower that had been modified to vacuum [24]. Each canopy and trunk was sampled from the ground by continuously passing the vacuum over green foliage (canopy) or live main and secondary stems (trunk) for 30 seconds. When possible, invertebrates were identified, enumerated and released in the field. Remaining samples were transferred to zipped plastic bags and kept cool until frozen in the lab. Invertebrates were sorted from plant material and debris, enumerated and identified to the lowest taxonomic group possible using a stereomicroscope.

Ground-dwelling fauna. Ground-dwelling fauna were sampled using pitfall traps, each consisting of a 15 cm-diameter ×15 cm-deep bucket with a lid suspended 2 cm above the opening to reduce escape of flying invertebrates and predation by

Figure 1. Structural differences between an introduced palm and native willow. Photographs highlighting the structural differences in the canopies (A), trunks (B), and litter (C) of the introduced Canary Island date palm (left) and the native arroyo willow (right) along the San Diego River, California, USA. The front of the palm trunk had been trimmed revealing the branching structure – see side of plant for denseness of canopy. Note ground under willow is patchy with plants and thin patches of leaf litter while thick litter accumulations occur under palm with no understory growth. Photos are from July 2011.

larger animals, and green leaves placed in the bottom as a refuge to reduce risks of desiccation and within-trap predation. One pitfall trap per plant was set in the ground beneath the dripline and left in the field for 6 days. When possible, invertebrates were identified, enumerated and released in the field. The remaining samples were transferred to zipped plastic bags and kept cool until frozen in the lab. Invertebrates were sorted from plant material and debris, enumerated and identified to the lowest taxonomic group possible and/or separated by the morpho-species concept using a stereomicroscope.

Environmental Variables

The differences in structure and resulting physical and biological properties between each woody species was quantified by measuring the physical properties of each plant, the biomass of plant litter, and the percent ground cover and light attenuation under each canopy.

Plant physical properties taken per plant consisted of height, measured using a clinometer, canopy area (longest diameter × perpendicular diameter) measured with measuring tape, and trunk diameter at ground level, measured with a diameter tape. Five subsamples of leaf toughness, measured with a penetrometer, and trunk rugosity were taken per plant and averaged. Trunk rugosity was estimated using a 100 cm long string, run vertically and in contact with the surface of each trunk at about chest height. The distance between each end of the string was then measured and subtracted from the 100 cm total length to give an estimate of amount of rugosity (e.g., length of string run over bumps). Rugosity index is presented as a percent (length of string run over bumps/total length of string).

The biomass of plant litter for each plant was estimated by clipping (if needed) and collecting all litter within three 0.25 m^2 quadrat subsamples, weighing the biomass in the field with a handheld scale, and averaging the three subsamples. Percent ground cover under the dripline of each plant was estimated by averaging visual estimates of cover within three 0.25 m^2 quadrats (% open space, % annual grasses, % litter). Photosynthetic-light attenuation of each plant canopy was measured using an Apogee Quantum handheld light meter during mid-day (10 am –2 pm) over two consecutive sunny days. Three measures were taken at each plant, standardized to full sunlight (under-canopy light [μmol photons m^{-2} s^{-1}]/full sunlight [μmol photons m^{-2} s^{-1}]), and averaged.

Statistics

All data were log (x+1) transformed or arcsin square root transformed (proportion data) before analyses [25]. Differences in environmental variables, and invertebrate total abundance, diversity (no. species per sample and H'), and proportions of major taxonomic groups (usually Order) between palm and arroyo willow were performed using paired t-tests in JMP® 9 Statistical Software.

Multivariate analyses were carried out on invertebrate assemblages of the palm and willow plants using Primer Software [26] and species lists with counts (no abundance cut-off, standardization or weighting was used). A total of 83 canopy species, 63 trunk species and 24 ground-dwelling species were found and used for all of the multivariate analyses. Bray-Curtis similarity indices of log (x+1) transformed, unstandardized, unrelativized data were calculated to compare communities and relate them to environmental variables. There were 5.7% zeros and a coefficient of variation (CV) of 52% for the canopy faunal matrix, 12% zeros and CV = 76% for the trunk faunal matrix and 38% zeros and CV = 99% for the ground faunal matrix. Community similarities and differences were explored by non-metric multidimensional scaling (MDS; see [27]) on the Bray-Curtis similarity indices. Six different random staring points with up to 1,000 steps were used. The stress values from the six runs were examined for stability to determine whether a global solution had been found. Only analyses with stress values of <0.2 were used. Stress is a measure of how well the solution (in this case the two-dimensional MDS plots) represents the distances between the data (roughly, the percentage of variance not explained by all dimensions in the MDS.) [27] suggests values <0.1 are good and <0.2 are useful.

Significance testing for differences in faunal composition between palm and willow plants and between pair number, or site (i.e., location), were done using an analysis of similarity (ANOSIM) procedure. This is a randomized permutation test based on rank similarities of samples [27]. The significance level of resultant pairwise comparisons were adjusted by Bonferroni correction for the number of comparisons made. Analyses of invertebrate dissimilarities between palm and willow (faunal composition among pair numbers was not significant), and the particular taxa contributing to the dissimilarity, were carried out using SIMPER on the Bray-Curtis matrices [27]. The SIMPER results specify which taxa are responsible for the ANOSIM results by comparing the average abundances of taxa between palm and willow assemblages. The average dissimilarity between samples from the two groups (palm and willow) is computed and then broken down into contributions from each species. Those species with high average terms relative to the standard deviation are important in the differentiation of willow and palm assemblages.

Relationships between the environmental variables and species diversity and total abundance were tested with forward, stepwise multiple regressions using JMP® 9 Statistical Software. Variables included in the model met the criteria of $p \leq 0.05$ and $r^2 \geq 0.04$. Tests of the environmental variables that best explain community differences were conducted using the Bray-Curtis similarity indices with the BEST Analysis in Primer Software using the BVSTEP method (criteria: rho >0.95, delta rho <0.001) with fixed starting variables and a Euclidean distance resemblance measure. BVSTEP sequentially adds environmental variables, keeping those that best explain faunal community patterns and eliminating those that explain least. Several iterations of the test are run with a random selection of variables to ensure that the best match is found [26], [27]. The environmental variables tested are listed in Table 1.

Results

Physical and Biological Environments

The palm had on average a 24% larger canopy area and 3 times wider trunk than the willow (Table 1). The greater aboveground stature and dense fronds of the palm compared with willow resulted in 7% less light penetration through the canopy, 20% more litter accumulation, and 25% less grass cover (Table 1). Palm trunks were also nearly 10 times as rugous (amount of non-branch structure on stems) as those of the willow (Table 1).

Invertebrate Abundance and Diversity

The palm had only about 60% of the invertebrate abundance and 60–70% of the species per sample than the willow (Table 2). Diversity including evenness (H') of the palm invertebrate assemblages was 70–80% of those in the willow, although this value (H') was low for the ground-dwelling faunal community of both plants (Table 2).

Table 1. Differences as revealed by paired t-test analyses in physical and biological environmental variables associated with the introduced Canary Island date palm and native arroyo willow found along the lower San Diego River, California in June 2011.

	Willow	Palm	Paired t-test statistics			
variable tested	mean ± 1SE	mean ± 1SE	P	t	df	n
Canopy properties						
height	4.2±0.3 m	4.7±0.3 m	0.130	1.6	14	15
canopy area	50.0±6 m^2	61.6±3 m^2	0.046	2.2	14	15
light attenuation	93±0.4%	97±0.3%	<0.001	10.2	14	15
leaf toughness	0.12±0.01 kg cm^{-2}	2.0±0.2 kg cm^{-2}	<0.001	9.9	14	15
Trunk properties						
trunk diameter	41±4 cm	134±14 cm	<0.001	6.2	14	15
trunk rugosity	4±0.7%	38±1%	<0.001	17.6	14	15
Understory properties						
litter biomass	178±30 g 0.25 m^{-2}	340±39 g 0.25 m^{-2}	<0.001	4.4	14	15
% grass cover	41±9%	16±7%	0.018	2.7	14	15
% litter cover	41±9%	61±8%	0.084	1.9	14	15
% pen space	18±4%	23±6%	0.450	0.8	14	15

Table 2. Differences as revealed by paired t-test analyses in invertebrate abundance and diversity associated with A.) canopies, B.) trunks, and C.) ground beneath the introduced Canary Island date palm and native arroyo willow along the lower San Diego River, California, USA in June 2011.

A. Canopy

	Willow	palm	Paired t-test statistics			
	mean ± 1SE	mean ± 1SE	P	t	df	n
total abundance	44.4±6.2	28.9±5.7	0.045	2.2	14	15
no. species	13.0±1.1	8.8±0.7	0.010	3.0	14	15
H'	2.1±0.08	1.7±0.10	0.046	2.2	14	15

B. Trunk

	Willow	palm	Paired t-test statistics			
	mean ± 1SE	mean ± 1SE	P	t	df	n
total abundance	19.9±2.8	11.5±1.7	0.038	2.3	14	15
no. species	8.7±0.7	5.4±0.7	0.015	2.8	14	15
H'	1.8±0.1	1.3±0.1	0.040	2.2	14	15

C. Ground

	Willow	palm	Paired t-test statistics			
	mean ± 1SE	mean ± 1SE	P	t	df	n
total abundance	50.3±17.4	33.3±14.1	0.36	0.94	14	15
no. species	3.8±0.3	2.8±0.2	0.02	2.51	13	14
H'	0.8±0.1	0.6±0.1	0.49	0.71	13	14

total abundance: average number of individuals per sample.
no. species, H': average number of species/H' per sample.

Faunal Community Composition

Major taxonomic groupings. The palm canopies and/or trunks had higher proportions of arachnids, lepidopterans, neuropterans and psocopterans than willow, while willow canopies and trunks had higher proportions of coleopterans and hymenopterans (Table 3). None of the ground-dwelling taxa differed between palm and willow, with arachnids, hymenoptera (mostly the Argentine ant) and coleopteran dominating the ground beneath both plants (Table 3).

Species compositions. Many of the differences between the faunal assemblages associated with palm and willow occurred at lower taxonomic levels (morpho-species groups), although there was still no difference in the ground-dwelling invertebrate communities beneath the palm and willow (ANOSIM Global P = 0.58). The palm and willow faunal communities differed for inhabitants of the canopies (ANOSIM Global P = 0.001; SIMPER 81% dissimilar) and inhabitants of trunks (ANOSIM Global P = 0.002; SIMPER 82% dissimilar) (Fig. 2). About 60% of the dissimilarity in canopy communities was due to willow canopies housing higher abundances of parasitoid wasps (Hymenoptera: pompillid, brachonid and ichneumonid), the introduced Argentine ant (*Linepithema humile*, Hymenoptera), leaf hoppers (Hemiptera, Cicidellidae: Idiocerus sp. and a type of deltocephanline hopper), a jumping plant louse (Hemiptera: Psyllidae), a minute brown scavenger beetle (Coleoptera, Latrididae), an inch worm caterpillar (Leptidoptera: Geometridae), and an array of midges (Diptera); while palm canopies hosted a bark louse (adult and nymphal Psocoptera), a hemerobiid lace wing (Neuroptera), and a crab spider (Aranae: Thomsidae) (Fig. 2).

About 60% of the dissimilarity in trunk communities was due to willow hosting higher abundances of the introduced Argentine ant, several dipterans (a dolichopodid and two midges), one leaf hopper (Hemipteran: Detocephaline hopper), a jumping plant louse (Hemipteran: Psyllidae), a caddisfly (Trichoptera), and two types of pompilid wasps; and palm trunks hosting higher abundances of the oak leaf-roller moth (*Epinotia emarginana*), a bark louse (Psocoptera), a leaf hopper (Idiocerus sp., Cicinellidae) that feeds on cottonwoods and willows, and a cobweb spider (Aranae: Theridiidae) (Fig. 2).

Table 3. Average (± 1SE) proportions (%) of major invertebrate taxonomic groups (insect order or chelicerata class) found in the canopy, trunk and ground beneath the Canary Island date palm and the native arroyo willow along the San Diego River, California, USA in June 2011.

	Canopy		**Trunk**		**Ground**	
taxonomic group	**Willow**	**palm**	**willow**	**palm**	**willow**	**palm**
Arachnida	3±1	13±4**	5±2	15±4**	33±10	54±12
Coleoptera	5±1	4±1	4±2*	0	9±5	3±2
Collembola	0	2±1	0	0	0	0
Dermaptera	0	0	0	0	<1	0
Diplopoda	0	0	0	<1	<1	<1
Diptera	6±2	9±2	26±6	14±3	<1	0
Hemiptera	29±5	28±7	19±5	22±7	1±1	0
Hymenoptera	28±5***	9±2	30±6**	13±4	51±12	42±12
Lepidoptera	2±1	2±1	6±2	24±7**	1±1	<1
Isopoda	0	0	0	0	3±3	<1
Neuroptera	<1	4±1**	0	0	0	0
Odonata	0	<1	0	0	0	0
Orthoptera	0	0	<1	0	<1	1±1
Psocoptera	6±4	27±7***	4±2	8±4	0	0
Thysanoptera	0	1±1	1±1	<1	0	0
Trichoptera	<1	<1	4±2	1±1	0	0

Differences in the proportion of each taxon between palm and willow as revealed by paired t-test analyses are indicated. Paired t-test statistics:
*$0.05 \le p \le 0.10$, $t = 1.8$–2, $df = 14$, $n = 15$
**$0.01 \le p \le 0.05$, $t = 2.1$–3, $df = 14$, $n = 15$
***$p < 0.01$, $t = 3.1$–7, $df = 14$, $n = 15$

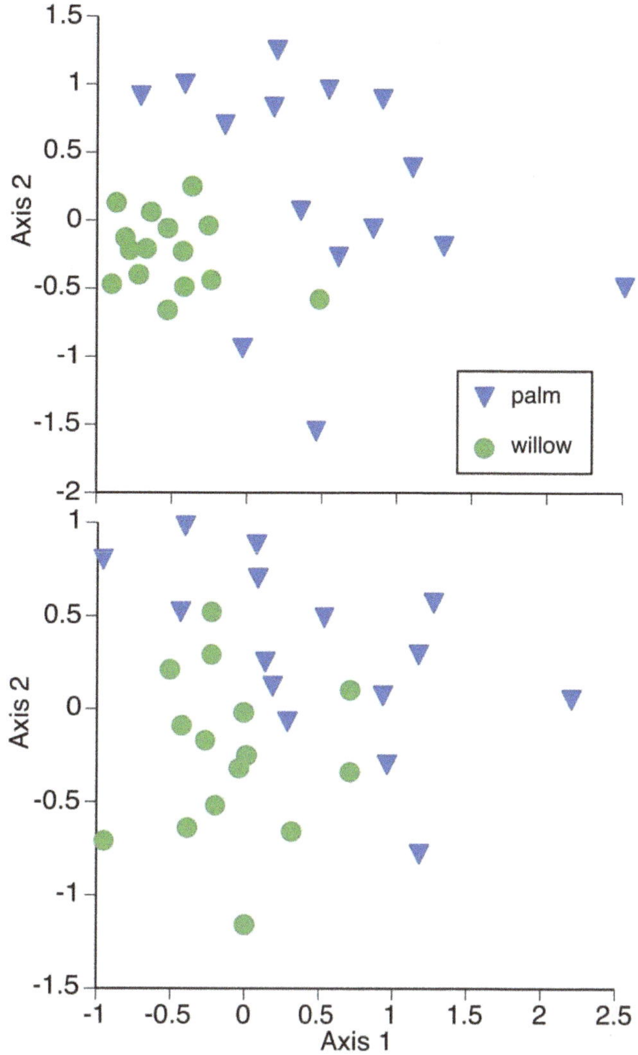

Figure 2. Differences in the invertebrate communities associated with an introduced palm and native willow. Multidimensional scaling plots of the invertebrate communities associated with the canopies (A.) and trunks (B.) of the introduced Canary Island date palm and native arroyo willow along the San Diego River, California, USA. Data are species lists with counts from June 2011. Stress values, or the overall measure of quality of fit of the MDS to the data, are stress = 0.17 for the canopy and stress = 0.19 for the trunk graph. Canopy community differences were correlated with the tougher leaves and rougher trunk of palm than willow, and trunk community differences were correlated with the rougher trunks of palm.

Environmental Influence on Fauna

The larger area and/or denser canopies associated with palm were associated with lower total abundances of canopy fauna (Table 4A) and diversity (H') of canopy and trunk fauna (Table 4A,B). The larger, more rugous trunk structure of the palm was associated with lower total abundances of trunk fauna and lower diversity (no. species) of canopy and trunk fauna (Table 4A,B). Canopy-community differences were mostly attributed to the greater toughness of leaves on palm than willow (BVSTEP $R = 0.22$), while trunk community differences were attributed to both the more rugose trunks and tougher leaves on palm than willow (BVSTEP $R = 0.49$).

Discussion

Influence of the Palm on Invertebrate Communities

The presence of the Canary Island date palm was associated with lower riparian invertebrate abundance and diversity, especially for canopy and trunk dwelling communities, and shifted the composition of these communities from a mix of beetles, midges, leaf hoppers, parasitoid wasps, and spiders to communities dominated by bark lice, the adult oak-roller moth and spiders. The palm, however, seemed to discourage inhabitation by the introduced Argentine ant despite the ant's affinity for nesting in plant litter [28]. The greener undergrowth associated with willow may have provided the moisture required by the ant during the dry season [28]. The palm's dense and wide canopy, tough leaves, rugous trunk, and accumulations of frond litter were all variously linked to the declines in diversity and abundance and the community shift, a finding consistent with the desert spring invasion [3]. The structure provided by palm trunks allows for detrital accumulations and, likely, epiphytes, such as fungus [20], [29], which may have attracted the bark lice [30]. More trunk rugosity may also have provided structure for hunting and web-building predators like neuropterans and spiders [22], and refuge or roosts for the adult leaf-roller moth and leafhoppers that specialize on native riparian shrubs (e.g., [31], J. Powell and R. Gill pers. comm.). Willow, on the other hand, may have provided a mix of trophic and structural resources conducive to a more diverse community than that on palm. The softer, more palatable leaves of willow (i.e, more available food), and the complex

Table 4. Results of forward, stepwise multiple regressions showing relationships between environmental and faunal variables associated with the canopies (A) and trunks (B) of the introduced Canary Island date palm and native arroyo willow found along the San Diego River, June 2011.

A. Canopy							
Response variable	**R^2**	**P**	**F**	**n**	**df**	**Independent variable(s)**	**+/−**
Abundance (no. ind. sample^{-1})	0.14	0.039	4.7	30	1,28	% open ground	-
Diversity (H′)	0.25	0.005	9.3	30	1,28	canopy area	-
Diversity (no. species sample^{-1})	0.27	0.003	10.2	30	1,28	trunk rugosity	-
B. Trunk							
Response variable	**R^2**	**P**	**F**	**n**	**df**	**Independent variable(s)**	**+/−**
Abundance (no. ind. sample^{-1})	0.19	0.015	6.8	30	1,28	trunk rugosity	-
Diversity (no. species sample^{-1})	0.29	0.002	11.3	30	1,28	trunk rugosity	-
Diversity (H′)	0.21	0.012	7.3	30	1,28	light attenuation	-

Note that although all environmental variables shown in Table 1 were used in the analyses, only one variable per test ended up meeting the model criteria of $p \geq 0.05$ and $r^2 \geq 0.03$.

network of smooth trunks and branches [32–34] may have fueled this balance. Further, the sparser canopy cover of willow allowed enough sunlight through to fuel undergrowth, and greater plant species in an area may increase food types and complexity [35]. Common riparian vegetation in Southern California includes other winter deciduous woody species, such as mulefat (*Baccharis salicifolia*), sandbar willow (*Salix hindsiana*), and black willow (*Salix gooddingii*) [36], all of which have soft, palatable leaves and similar structure to arroyo willow, making it likely that they differ from the palm in similar ways as arroyo willow.

Neither the piles of accumulated, fibrous leaf litter nor any of the other characteristics of the palm seemed to affect the ground dwelling invertebrate community during this study. It may be that the ground beetles, spiders and Argentine ant, which all prefer cover and were found throughout the study area, obtained suitable cover and/or food from both the accumulated palm litter and the living grass and fine willow leaf layers beneath the willow during this time period [37–38]. The use of pitfalls also biases the catch toward larger-bodied, more-mobile arthropods [39–41], so while differences between palm and willow were not observed for these groups, smaller-bodied, less-mobile taxa (e.g., wood lice, smaller beetles) may not have been well assessed with this method. Furthermore, it is likely that some of the larger animals inhabiting this ecosystem, such as lizards and rodents, may be more affected by the differences in conditions on the ground (e.g., litter biomass and complexity, light and temperature levels, understory growth) [42–43] than the invertebrates.

Ecosystem-level Effects of Palm

The presence of the palm alters the ecosystem directly and indirectly through both trophic and structural means [3], [44]. For example, the low food quality and novel structure of the palm resulted in less herbivory, and more fungivory (e.g., by bark lice), predation (e.g., spiders and lace wings), and other uses of structural resources (e.g., roosting by adult oak-roller moths and willow-eating leaf hopper) than on the willow. In total, the presence of the palm itself and the subsequent shift in the invertebrate taxa and their functions can alter physical conditions (e.g., light levels, temperature), primary production and nutrient turnover, the base of the aboveground food web (i.e., from leaves to fungus and detritus), and the pathway connecting basal and upper trophic levels (i.e., the species and trophic groups present) [3], [17], [22], [45].

These effects may extend to vertebrates within the ecosystem through direct interactions (e.g., [45]) or indirect interactions via the invertebrate fauna. Indirectly, shifts in invertebrate prey abundance and types may affect insectivorous vertebrates, such as lizards and birds. For example, the Federally endangered Least Bell's Vireo (*Vireo bellii pusillus*), an obligate riparian species in this region, exclusively feeds on insects such as hemipterans, beetles and, in particular, caterpillars [46–47]. Shifts in insect composition or, more likely, decreased abundances as observed on palm may impact birds such as this one. Direct effects of palm presence on vertebrates would particularly influence those that depend upon native shrubs for nesting habitat and food capture (e.g., [48]). For example, the Least Bell's Vireo nests about one meter above the ground in native riparian shrubs, frequently willow [49–50], requires dense and diverse understory to conceal the nests and foraging activities, and captures prey by picking insects off the bark or leaves and/or hovering [50–51]. These are trophic and non-trophic activities that could all be impeded by the impenetrable, detritus-covered, tough (downright sharp!) structure of palm canopies, trunks and litter (e.g., [50]).

How Generalizable are Palm Effects?

Despite the different nature of our California study systems– desert springs and a coastal riparian system– the findings in this study were similar to those of [3]. Both studies found that introduced palm reduced arthropod diversity and abundance, and shifted compositions to fewer herbivores, and more predators and fungivores because of similar palm-induced changes to physical properties (e.g., more shading), and to structural and trophic resources (e.g., less canopy complexity, lower food quality). Further, [3] found that the effects of palm on arthropods did not vary much with site or season, despite variations in native plant compositions across sites and that their study site was Death Valley, one of the most annually extreme environments in the world. Given that palm effects are stronger than the temporal and regional spatial variations examined across these studies, it is likely that similar effects of palm on invertebrates will be observed in other ecosystems as well. Predictions of susceptible ecosystems and regions are therefore useful.

As seen in California, moist ecosystems within arid and semi-arid climates may be susceptible to palm invasion and similar effects on the physical environment and invertebrate community [3] (this study). Many ecosystems within tropical and subtropical climates are suitable for introduced palms because of the combination of warm temperatures and moisture (e.g., [12], [52]). Within these regions, palms may have the most dramatic effects where native vegetation is most different. For example, areas dominated by palatable mid-story and/or ground cover plants, such as grassland, scrub [3], (this study), or broadleaf tropical forest [52]. Further, effects of invasion may be most dramatic in "island" ecosystems, where there are relatively small patches of ecosystem surrounded by a highly contrasting matrix. These "islands" are vulnerable to invasion and degradation (due to edge effects), and to native species declines (due to stochasticity), and lack of recolonization because of their isolated nature (e.g., [53–54]). Island ecosystems susceptible to palm invasion may be true islands [45], desert springs surrounded by an arid landscape [55–56], or almost any fragmented or encroached ecosystem that receives moisture from stream, groundwater or tidal flows (e.g., inland and coastal wetlands, canyons bottoms, and fragmented riparian areas) (e.g., [2], (this study)). In particular, urban ecosystems are at risk of invasion because of the combination of high propagule pressure and artificial moistening of an otherwise dry landscape through irrigation and runoff. Further, controlled river flows and urban runoff in populated areas have shifted many arid- and semi-arid rivers from historic ephemeral flows to perennial low flows [57]. This shift reduces the frequency and intensity of flood and scour events that would clear out invasives, and creates the moist environments favored by many introduced plants [57], such as palms.

Future Directions and Recommendations

Further testing of the effects of palm presence on the invertebrate community during other times of the year, and the effects of both palm presence and its associated invertebrates on common and threatened riparian vertebrates, and on other ecosystem processes (e.g., decomposition rates, nutrient cycling), are needed to assess the threat level of these relatively under-studied invaders. Understanding the effects of the palm at this point, while it mostly exists in localized patches, is important for making management decisions. A species does not have to be abundant to have a relatively large impact on its ecosystem (e.g., ecosystem engineers, [58]) and, if this is the case with the palm, now would be the time to initiate removal and control strategies [59–60]. The California Coastal Commission currently prohibits the planting of [2] listed plants in new coastal developments, but existing developments are plentiful and may need to be addressed. Control recommendations include source removal efforts, or propagule removal efforts if removal is too expensive for some landowners and land managers. However, total removal would be ideal to avoid the current effects of living palms and "ghost palm" effects, where dead palm material persists for many years reducing habitat quality [3], [11]. Another control approach would be to discourage the further planting of this species in developments adjacent to riparian corridors or other moist ecosystems where short-distance dispersal of the abundant propagules is sure to occur. This sort of effort would require a coordinated outreach effort among the nurseries offering this species, the public, ecologists and environmental managers [61].

Study of the rates and patterns of spread of the palm are also needed. Anecdotal data from Southern California reveals mostly localized spread of the palm around parent plants [2], but with somewhat recent expansion of the distribution, in particular into the estuarine reach of the river [15]. Information on spread rate and patterns, coupled with information from this study on the local effects, could help to predict larger-scale effects of spread. Although the palm is not currently one of the most aggressive invaders in the waterway, it has a prolific seed set and is displaying salinity tolerance, two traits that could contribute to successful spread in this coastal river system (e.g., tamarisk; [17]). Many species start in a lag phase and unforeseen conditions trigger rapid spread [60]. More thoroughly understanding the effects of palm invasion will be important for prioritizing management decisions.

Acknowledgments

We would like to thank Mona Wang, Eduardo Hernandez and Erick Ruiz for their enthusiastic help in the field and lab, and both Drew Talley (University of San Diego) and Paul Dayton (Scripps Institution of Oceanography) for their generosity in the use of lab space and equipment. Special thanks to Lynn Kimsey of the Bohart Entomological Museum at the University of California, Davis, Jerry Powell of the Essig Museum of Entomology at the University of California, Berkeley and Ray Gill of the California Department of Food and Agriculture for information about the taxonomy and natural history of many of our insects, in particular the tortricid moths (J.P.) and leafhoppers (R.G.). We are grateful to Shannon Quigley-Raymond (San Diego River Park Foundation) and Tracey Walker (City of San Diego City Park and Recreation Department) for helpful background information and permission to access the study site.

Author Contributions

Conceived and designed the experiments: TST AN. Performed the experiments: TST KCN AN. Analyzed the data: TST KCN. Contributed reagents/materials/analysis tools: TST. Wrote the paper: TST KCN. Sorted and identified species: TST KCN AN.

References

1. Zona S (2008) The horticultural history of the Canary Island date palm (*Phoenix canariensis*). Garden History 36: 301–309.
2. Cal-IPC (2006) California Invasive Plant Inventory. Cal-IPC Publication 2006–02. Berkeley, CA: California Invasive Plant Council. Available: http://www.cal-ipc.org. Accessed 2012 Feb 6.
3. Holmquist JG, Schmidt-Gengenbach J, Slaton MR (2011) Influence of invasive palms on terrestrial arthropod assemblages in desert spring habitat. Biological Conservation 144: 518–525.
4. Burkhart B, Kelly M (2005) Which weeds dominate southern California urban riparian systems? California Invasive Plant Council News 13: 4–5, 12.
5. Gilman EF, Watson DG (2009) *Phoenix canariensis*: Canary Island Date Palm. University of Florida IFAS Extension. Available: http://edis.ifas.ufl.edu/st439. Accessed 8 February 2012.
6. Cornett JW (1987) Naturalized populations of the desert fan palm, *Washingtonia filifera*, in Death Valley National Monument. In: Hall CA Jr, Doyle-Jones V (Eds). White Mountain Research Station Symposium: Plant Biology of Eastern California, Vol 2. Los Angeles: White Mountain Research Station. 167–174.
7. Cornett JW (2010) Desert Palm Oasis. Nature Trails Press, Palm Springs, CA.
8. Hood WG, Naiman RJ (2000) Vulnerability of riparian zones to invasion by exotic vascular plants. Plant Ecology 148: 105–114.
9. Trent H, Seymour J (2010) Examining California's first palm tree: The Serra palm. The Journal of San Diego History 56: 105–120.
10. Tuck R (2012) History of the Canary Island date palm in California. Sunset Palms. Available: http://www.sunsetpalms.net/information.php?go=4. Accessed 12 February 2012.
11. Kelly M (2007) Why palms made it into the Inventory. California Invasive Plant Council News 15: 4, 13, 15.
12. Florida Exotic Pest Plant Council (2009) List of Invasive Plant Species. Wildland Weeds 12: 13–116.

13. Wu L, Dodge L (2005) Landscape Plant Salt Tolerance Selection Guide for Recycled Water Irrigation. A Special Report for the Elvenia J Slosson Endowment Fund, University of California, Davis article: 40.
14. Tanji KK, Rollins L, Suyama P, Farris C (2007) Choose salt-tolerant plants. Salinity Management Guide. WateReuse Foundation. Available: http://www.salinitymanagement.org/Salinity%20Management%20Guide/cp/cp_7_table-3.html. Accessed 2012 Feb 12.
15. San Diego River Park Master Plan (2010) City of San Diego. Available: http://www.sandiego.gov/planning/programs/parkplanning/pdf/sdriverparkpdf/sdrpmasterplanbodyfull.pdf. Accessed 2012 Feb 6.
16. Talley TS, Levin LA (2001) Modification of sediments and macrofauna by an invasive marsh plant. Biological Invasions 3: 51–68.
17. Whitcraft CR, Levin LA, Talley D, Crooks JA (2008) Utilization of invasive tamarisk by salt marsh consumers. Oecologia 158: 259–272.
18. Zuefle ME, Brown WP, Tallamy DW (2008) Effects of non-native plants on the native insect community of Delaware. Biological Invasions 10: 1159–1169.
19. Hickman JC (1993) The Jepson Manual. Higher plants of California. Berkeley: University of California Press.
20. Lopez-Llorca LV, Carbonell T, Salinas J (1999) Colonization of plant waste substrates by entomopathogenic and mycoparasitic fungi - a SEM study. Micron 30: 325–333.
21. Strong DR, Lawton JH, Southwood TRE (1984) Insects on Plants: Community Patterns and Mechanisms. Cambridge, MA: Harvard University Press.
22. Pearson DE (2009) Invasive plant architecture alters trophic interactions by changing predator abundance and behavior. Oecologia 159: 549–558.
23. Schowalter TD (2011) Insect Ecology: An Ecosystem Approach, 3rd ed. San Diego, CA: Elsevier. 650 p.
24. Buffington ML, Radak RA (1998) A comparison of vacuum sampling versus sweep-netting for arthropod biodiversity measurements in California coastal sage scrub. Journal of Insect Conservation 2: 99–106.
25. Zar JH (2009) Biostatistical Analysis, 5th ed. Englewood, NJ: Prentice Hall. 960 p.
26. Clarke KR, Warwick RM (2001) Change in marine communities: an approach to statistical analysis and interpretation, 2d ed. Plymouth: PRIMER-E.
27. Clarke KR (1993) Non-parametric multivariate analyses of changes in community structure. Australian Journal of Ecology 18: 117–143.
28. Ward PS (1987) Distribution of the introduced Argentine ant (*Iridomyrmex humilis*) in natural habitats of the lower Sacramento Valley and its effects on the indigenous ant fauna. Hilgardia 55: 1–16.
29. Gomez-Vidal S, Lopez-Llorca LV, Jansson H-B, Salinas J (2006) Endophytic colonization of date palm (*Phoenix dactylifera* L.) leaves by entomopathogenic fungi. Micron 37: 624–632.
30. Triplehorn CA, Johnson NF (2005) Borror and DeLong's Introduction to the Study of Insects, 7th ed. Belmont, CA: Brooks/Cole Publishing. 864 p.
31. Powell JA, Hogue CL (1979) California Insects. Los Angeles: University of California Press. 388 p.
32. Nadkarni NM (1994) Diversity of species and interactions in the upper tree canopy of forest ecosystems. American Zoologist 34: 70–78.
33. Humphrey JW, Hawes C, Peace AJ, Ferris-Kaan R, Jukes MR (1998) Relationships between insect diversity and habitat characteristics in plantation forests. Forest Ecology and Management 113: 11–21.
34. Siemann E, Tilman D, Haarstad J, Ritchie M (1998) Experimental tests of the dependence of arthropod diversity on plant diversity. American Naturalist 152: 738–750.
35. Keer G, Zedler JB (2002) Salt marsh canopy architecture differs with the number and composition of species. Ecological Applications 12: 456–473.
36. Sawyer JO, Keeler-Wolf T (1995) A Manual of California Vegetation. Sacramento, CA: California Native Plant Society.
37. Uetz GW (1979) The influence of variation in litter habitats on spider communities. Oecologia 40: 29–42.
38. Bultman TL, Uetz GW (1984) Effect of Structure and Nutritional Quality of Litter on Abundances of Litter-dwelling Arthropods. American Midland Naturalist 111: 165–172.
39. Spence JR, Niemelä JK (1994) Sampling carabid assemblages with pitfall traps: The madness and the method. The Canadian Entomologist 126: 881–894.
40. Work TT, Buddle CM, Korinus LM, Spence JR (2002) Pitfall trap size and capture of three taxa of litter-dwelling arthropods: Implications for biodiversity studies. Environmental Entomology 31: 438–448.
41. Hancock MH, Legg CJ (2011) Pitfall trapping bias and arthropod body mass. Insect Conservation and Diversity. doi: 10.1111/j.1752-4598.2011.00162.x.
42. Doyle AT (1990) Use of riparian and upland habitats by small mammals. Journal of Mammalogy 71: 14–23.
43. Vitt LJ, Caldwell JP (1994) Resource utilization and guild structure of small vertebrates in the Amazon forest leaf litter. Journal of Zoology 234: 463–476.
44. Dennis P, Young MR, Gordon IJ (1998) Distribution and abundance of small insects and arachnids in relation to structural heterogenetiy of grazed, indigenous grasslands. Ecological Entomology 23: 253–264.
45. Senior K (2010) Seabirds boycott Palmyra's palms. Frontiers in Ecology and the Environment 8: 62.
46. Chapin EA (1925) Food habits of the vireos. U S Department of Agriculture Bulletin 1355. Washington, DC: USDA.
47. Bent AC (1950) Life histories of North American wagtails, shrikes, vireos, and their allies. Bulletin of the United States National Museum 197. New York: Dover Publications.
48. Gaines DA (1977) The valley riparian forests of California: Their importance to bird populations. In: Anne Sands, ed Riparian forests in California: Their ecology and conservation. University of California, Davis, Inst of Ecol Publ No 15. 57–87.
49. Olson TE, Gray MV (1989) Characteristics of least Bell's vireo nest sites along the Santa Ynez River. In: Abell DL, editor. Proceedings of the California Riparian Systems Conference: protection, management, and restoration for the 1990s; September 22–24; Davis, CA. Gen Tech Rep PSW-110. Berkeley, CA: Pacific Southwest Forest and Range Experiment Station, Forest Service, USDA. 278–284.
50. Kus B (2002) Least Bell's Vireo (*Vireo bellii pusillus*). In: The Riparian Bird Conservation Plan: a strategy for reversing the decline of riparian-associated birds in California. Stinson Beack, CA: California Partners in Flight. Available: http://www.prbo.org/calpif/htmldocs/species/riparian/least_bell_vireo.htm. Accessed 2012 Feb 18.
51. Miner KL (1989) Foraging ecology of the Least Bell's Vireo, *Vireo bellii pusillus*. Master's thesis, San Diego State University.
52. Svenning JC (2002) Non-native ornamental palms invade a secondary tropical forest in Panama. Palms: Invasive Palms in Panama 46(2): 1–3.
53. MacArthur RH, Wilson EO (1967) The theory of island biogeography. Princeton, NJ: Princeton University Press.
54. Fleishman E, Ray C, Sjogren-Gulve P, Boggs CL, Murphy DD (2002) Assessing the roles of patch quality, area, and isolation in predicting metapopulation dynamics. Conservation Biology 16: 706–716.
55. Fleishman E, Murphy DD, Sada DW (2006) Effects of environmental heterogeneity and disturbance on the native and non-native flora of desert springs. Biological Invasions 8: 1091–1101.
56. Perla BS, Stevens LE (2008) Biodiversity and productivity at an undisturbed spring in comparison with adjacent grazed riparian and upland habitats. In: Stevens LE, Meretsky VJ, editors. Aridland Springs in North America: Ecology and Conservation. Tucson, AZ: University of Arizona Press. 230–243.
57. Grimm NB, Faeth SH, Golubiewski NE, Redman CR, Wu J, et al. (2008) Global change and the ecology of cities. Science 319: 756–760.
58. Hastings A, Byers JE, Crooks JA, Cuddington K, Jones CG, et al. (2007) Ecosystem engineering in space and time. Ecology Letters 10: 153–164.
59. Hobbs RJ, Humphries SE (1995) An integrated approach to the ecology and management of plant invasions. Conservation Biology 9: 761–770.
60. Crooks JA (2005) Lag times and exotic species: the ecology and management of biological invasions in slow motion. Ecoscience 12: 316–329.
61. Reichard SH, White P (2001) Horticulture as a pathway of invasive plant introductions in the United States. Bioscience 51: 103–113.

14

Insects as Stem Engineers: Interactions Mediated by the Twig-Girdler *Oncideres albomarginata chamela* Enhance Arthropod Diversity

Nancy Calderón-Cortés[1]*, Mauricio Quesada[1], Luis H. Escalera-Vázquez[2]

1 Centro de Investigaciones en Ecosistemas, Universidad Nacional Autónoma de México, Morelia, Michoacán, México, **2** Departamento de Zoología, Instituto de Biología, Universidad Nacional Autónoma de México, México D.F., México

Abstract

Background: Ecosystem engineering may influence community structure and biodiversity by controlling the availability of resources and/or habitats used by other organisms. Insect herbivores may act as ecosystem engineers but there is still poor understanding of the role of these insects structuring arthropod communities.

Methodology/Principal Findings: We evaluated the effect of ecosystem engineering by the stem-borer *Oncideres albomarginata chamela* on the arthropod community of a tropical dry forest for three consecutive years. The results showed that ecosystem engineering by *O. albomarginata chamela* had strong positive effects on the colonization, abundance, species richness and composition of the associated arthropod community, and it occurred mainly through the creation of a habitat with high availability of oviposition sites for secondary colonizers. These effects cascade upward to higher trophic levels. Overall, ecosystem engineering by *O. albomarginata chamela* was responsible for nearly 95% of the abundance of secondary colonizers and 82% of the species richness.

Conclusions/Significance: Our results suggest that ecosystem engineering by *O. albomarginata chamela* is a keystone process structuring an arthropod community composed by xylovores, predators and parasitoids. This study is the first to empirically demonstrate the effect of the ecosystem engineering by stem-boring insects on important attributes of arthropod communities. The results of this study have important implications for conservation.

Editor: Guy Smagghe, Ghent University, Belgium

Funding: This study was supported by Inter-American Institute for Global Change Research (IAI) CRN II # 021 by the US National Science Foundation (grant GEO-0452325), and by grants from the National Council of Science and Technology (CONACyT), México (CONACyT 31826-N, U50863Q; SEMARNAT-CONACyT 2002-C01-0597 and 2002-C01-0544, and CONACyT sabbatical fellowship to MQ) and the Dirección General de Asuntos del Personal Académico at the Universidad Nacional Autónoma de México (UNAM) (grants # IN221305 and IN224108). CONACyT (scholarship # 164921), the Graduate Program in Biological Sciences, UNAM, and the Dirección General de Estudios en Posgrado (DGEP-UNAM) provided financial support for graduate studies to NCC. The funders had no role in study design, data collection and analysis, decision to publish, or preparation of the manuscript.

Competing Interests: The authors have declared that no competing interests exist.

* E-mail: ncalderon@oikos.unam.mx

Introduction

One of the central issues in ecology is to understand the mechanisms that structure ecological communities. Even though direct pairwise interactions (e.g. competition and predation) play a major role in explaining the structure of many biological communities (e.g. [1,2]), other interactions can also be important [3,4]. One interaction with important consequences on biological communities and biodiversity is the relationship between organisms that modify or create new habitats with those organisms that use these new habitats, a process called "ecosystem engineering" [5], or habitat modification [6]. Ecosystem engineers are species that control the availability of resources for other species by causing physical state changes in biotic or abiotic materials [5]. Because some ecosystem engineers create habitats where entire communities establish, they are also called "foundation species" [6] or "keystone engineers", when the impact of the ecosystem engineer is higher than its abundance [7]. Currently, an increasing number of studies have experimentally demonstrated that some species act as ecosystem engineers (e.g. beavers, salmons, pocket-gophers) affecting communities and ecosystems (reviewed in [8,9]).

Several insect herbivores manipulate their host-plants to build a variety of structures, which are secondarily occupied by organisms other than the original constructor. Hence, these herbivores can act as ecosystem engineers [10]. However, the role of insects as ecosystem engineers has only been experimentally evaluated for leaf-rollers [11,12], gall makers [13] and leaf miners [14,15]. These studies indicate that ecosystem engineering by insect herbivores influence overall abundance, species richness, and composition of arthropod communities by providing new habitats for other herbivores that are used for shelter (from natural enemies and adverse microclimates) and for more nutritious food [4,10–15]. Ecosystem engineering effects can propagate to higher trophic levels, triggering cascades of interactions including trophic, antagonistic and mutualistic interactions [4,11,13].

One insect guild comparatively less studied in this regard is represented by stem-borers, which are insects that develop (for at least part of their life cycle) in wood, bark or woody stems of plants

[16]. Many of them begin their life cycle as eggs laid under bark by free-living adult females; the larvae feed on the wood inside stems and eventually emerge as adults to repeat the cycle [16]. The stem-boring larvae produce complex systems of cavities that can be secondarily occupied by other arthropods [10,17–20]. This suggests that the guild of stem-borers includes several species that can act as ecosystem engineers. However, empirical studies evaluating the effects of stem engineering on arthropod communities are currently lacking.

Stem-boring insects play important functional roles in forest ecosystems, as they contribute to nutrient cycling [21–23], alteration of tree architecture [23,24], resource regulation [25], and alteration of the composition and hydrology of forests [25,26]. Therefore, the study of factors structuring their communities has important implications for forest conservation.

Here we present the results of a field experiment designed to evaluate the effect of ecosystem engineering by the stem-boring beetle, *Oncideres albomarginata chamela* (Cerambycidae: Lamiinae), on the arthropod community associated with detached branches of *Spondias purpurea* (Anacardiaceae). The study was carried out for three consecutive years. *O. albomarginata chamela* actively manipulates its host plant through a process consisting of two steps. First, adult females of *O. albomarginata chamela* preferentially girdle and detach reproductive branches of *S. purpurea,* before the reproductive season of the tree [27], when reproductive branches have accumulated the maximum concentration of non-structural carbohydrates [28] and nitrogen [27]. Second, adult females make incisions and gnaw egg niches along the detached branches for oviposition [27]. Therefore, *O. albomarginata chamela* females provide a high quality environment for offspring development [29]. Incidentally, these females also provide a suitable environment for secondary colonization [30], particularly for insects that oviposit opportunistically in cracks and crevices in the bark or cortex of plants [16,30].

Based on this evidence, we hypothesized that the modification of tree branches by *O. albomarginata chamela* plays a key role in the establishment of a new arthropod community and promotes interactions with positive effects on arthropod abundance and diversity. To test this hypothesis, we simulated *O. albomarginata chamela* physical modification of *S. purpurea* branches, and compared the community composition, frequency of colonization, abundance and species richness of secondary arthropod colonizers between non-engineered and engineered branches (both artificially and naturally modified branches).

Methods

Ethics statement

All animals were handled according to relevant national and international guidelines. Insects were reared at natural conditions at the study site, and released *in situ* after the experiment. The animal work was approved by the authorities of Chamela Biological Station, Universidad Nacional Autónoma de México (National Autonomous University of Mexico), and by national authorities of Secretaría de Medio Ambiente y Recursos Naturales (Secretary of Environment and Natural Resources; SEMARNAT, permission SGPA/DGVS/05876/10).

Study system

Oncideres albomarginata chamela Chemsak and Gisbert is a longhorn beetle (Cerambycidae) that detaches branches 2–3 cm in diameter from the tropical tree *Spondias purpurea* L. (Anacardiaceae), and oviposits in them [27]. Alternative but less used host plants of *O. albomarginata chamela* include: *Comocladia engleriana* Loes (Anacardiaceae), *Mangifera indica* L. (Anacardiaceae), *Amphipterygium adstringens* Schide ex Schlecht (Rubiaceae), *Bursera* Jacq. ex L. spp. (Burseraceae), *Ceiba pentandra* (L.) Gaertn (Bombacaceae), *Urera* (L.) Gaud. sp. (Urticacaceae) and *Delonix regia* (Bojer ex Hook) Raf. (Fabaceae) [31]. *O. albomarginata chamela* is distributed in Mexico in the states of Jalisco, Nayarit, Guerrero, Oaxaca, Chiapas and Veracruz [32], but *O. albomarginata* Thomson is distributed in México, Central America (Nicaragua) and South America (British and French Guiana, Venezuela) [33]. The body length of *O. albomarginata chamela* is 17–31 mm and 6.5–12 mm wide [32]. The reproductive period of this species is from October to February; eggs hatch and larvae develop inside detached branches until the adults emerge in low densities 6–8 months later. Adult females of *O. albomarginata chamela* are the only herbivores at the study site that detach branches of *S. purpurea* and immediately oviposit in them [27], but after a certain period of time other species of stem-boring beetles (mainly non-girdling species) take advantage of the detached branches and oviposit in them as well.

S. purpurea is a common dioecious tree of the tropical dry forest of Mexico [28]. The ratio of male and female trees of *S. purpurea* in the population at the study site is 1:1 [27]. This species can reach 15 m in height and almost 80 cm in diameter at the base; leaves are compound with 5 to 12 elliptic-acute leaflets from 2 to 4 cm in length [34]. Flowers are red, sessile, unisexual and dimorphic between males and females [34]. Trees are deciduous with flowering and fruiting occurring from December to May, and leaves are maintained from June to November [34].

Study site

The study was conducted in the Chamela-Cuixmala Biosphere Reserve at Chamela Biological Station, UNAM (19°30′N, 105°03′W) located on the Pacific coast of Jalisco, Mexico, from December 2006 to January 2010. The vegetation is tropical dry forest with a mean annual rainfall of 707 mm and a dry season that extends from November to June [35].

Experimental design

In order to evaluate the effect of ecosystem engineering by *O. albomarginata chamela* on the arthropod community, during December 2006 to January 2008, we conducted a field experiment consisting of three treatments ($N \approx 50$ branches/treatment): *O. albomarginata chamela* engineered and colonized-branches (OE), artificially simulated engineered branches (SE), and non-engineered branches (NE). For treatment OE, we collected branches of *S. purpurea* naturally detached and colonized by *O. albomarginata chamela* on December 2006. This treatment was used as control to provide baseline data on the arthropod community associated with *S. purpurea* branches, and to analyze the effects of the ecosystem engineer presence. For treatment SE, branches exhibiting similar characteristics (reproductive branches from 2–3 cm in diameter) to those detached and colonized by *O. albomarginata chamela* were artificially cut off from *S. purpurea* trees. We artificially simulated the structural modification of branches made by adult females of *O. albomarginata chamela*, by making numerous incisions with scissors (every 5 mm) on the bark of these branches. Treatment NE consisted of simply artificially detached reproductive branches of 2–3 cm in diameter of *S. purpurea* with no manipulation. We called this treatment "non-engineered branches" because mechanical factors, such as wind, water stress, mechanical branch damage, among others, detach a great proportion of branches and twigs from trees in the study site. Specifically, broken branches (2–20 cm in circumference) constitute the most important component (43%) of the forest total above-ground dead phytomass in Chamela tropical dry forest [36]. Thus, broken branches can represent non-engineered but available habitats. All branches were marked, and they were left hanging on the source-tree for 45 days (December 2006 to February 2007) to allow the colonization of secondary opportunistic species. Our preliminary analysis indicated that 30–45 days (during that period of the year) is when most insect borers colonize *S. purpurea* detached branches. The gender of each source-tree was registered. To control for the size of the branches used for each treatment, we measured the diameter at the point of branch cutting with an electronic caliper (Mytutoyo Inc). To control for adult female host selection, we cut off two branches for treatments NE and SE from the same tree where *O. albomarginata chamela* had previously detached and colonized branches. Additionally, the treatments were conducted in the same host plants to control for any related chemical attractive signals emitted by the same tree, as well as to control for any other factors associated with the nutritional value of host trees. Therefore, the branches of the three treatments had the same probability to be located by secondary colonizers. After 45 days, all branches were enclosed in mesh bags (<0.5 mm mesh size) to prevent further colonization and escape of the established fauna. Branches collected in mesh bags were placed in an open room at the study site, and maintained at local environmental conditions. Emerging arthropods from each branch were recorded monthly from March 2007 to January 2008, and released. We measured the total length of 20–40 adults of each insect species to estimate the size of the secondary colonizers. The exact same experiment was repeated for two more years: December 2007 to January 2009, and December 2008 to January 2010. Taxonomic identification of species that emerged was carried out by the beetle specialist Dr. Felipe A. Noguera and using available taxonomic literature [37–39].

Data analysis

First, we compared the diameter (at the point of branch cutting) of detached branches to determine differences in sizes between treatments, through one way Analysis of Variance (ANOVA) using PROC ANOVA [40]. Our results indicated that branch diameter did not differ significantly between treatments (2007: $F_{2,\ 154} = 1.564$, $P = 0.213$; 2008: $F_{2,\ 155} = 1.068$, $P = 0.346$; 2009: $F_{2,\ 178} = 0.263$, $P = 0.769$). In a previous study, Uribe-Mú and Quesada [27] found that branch gender had no effect on *O. albomarginata chamela* larval performance. Therefore we expected that branch gender had no effect on the number of secondary colonizers emerging from *S. purpurea* branches. This was confirmed when we analyzed the variation associated with branch gender through a Generalized Linear Model using a GENMOD procedure [40], in which the number of secondary colonizers that emerged from *S. purpurea* branches was used as the response variable, and tree gender as the independent variable. We used a Poisson distribution with a logarithmic link function for the analysis and corrected for overdispersion of data. Tree gender had no significant effect (2007: $\chi^2 = 0.37$, $P = 0.5428$; 2008: $\chi^2 = 1.42$, $P = 0.2327$; 2009: $\chi^2 = 2.09$, $P = 0.1479$) and was not included in further analyses.

Non-metric multidimensional scaling (NMDS) was used as an ordination procedure to determine differences in community composition among OE, SE and NE branches. The NMDS analysis was based on ranked Bray-Curtis dissimilarity distances [41]. Differences in community composition between treatments were tested using an analysis of similarity (ANOSIM), which uses 1000 random reassignments of species to groups and determines whether the group assignments were significantly different from those generated by chance. NMDS and ANOSIM analyses were performed with the software PRIMER 5.2.9 for windows (PRIMER-Ltd, Plymouth, U.K.). Multiple comparisons in ANOSIM were made using a sequential Bonferroni correction [42].

To evaluate the effect of *O. albomarginata chamela* on the frequency of colonization of *S. purpurea* branches by secondary colonizers, each species was quantified as being present or absent. Data were analyzed using a Generalized Linear Mixed Model that implements a generalization of the standard linear model allowing the incorporation of random effects [43]. We used the GLIMMIX procedure in SAS statistical software with a binomial distribution, and a logit link function specified for the dependent variable [40]. Branch condition (colonized *vs.* non-colonized) was the response variable. Treatment, year and their interaction were included as fixed variables, while tree identity and its interaction with treatment as random effects. We used a Least Square Means (LSMeans) test for *a posteriori* comparisons [40].

Secondary xylovores showed two general traits in size and developmental time. These are key life-history traits in insects related to fitness, habitat selection, oviposition strategies, and response to natural enemies [44]. Therefore, we used them to define two putative life forms: a) species with small body size and short developmental time (life form I); and b) species with large body size and longer developmental time (life form II). Natural enemies were analyzed separately. A Generalized Linear Mixed Model was used to evaluate the effect of *O. albomarginata chamela* on the abundance of the secondary colonizers (SAS, GLIMMIX procedure) [40]. This model used: (i) the number of adult secondary colonizers that emerged from branches as the response variable, (ii) treatment, year and their interaction as fixed variables, and (iii) plant identity and its interaction with treatment as random effects. We used a Poisson distribution with a logarithmic link function in the analysis. The degrees of freedom of F-tests for the fixed effects were adjusted using the Satterthwaite method. To control for overdispersion, we applied a Poisson error distribution to the model. We used LSMeans tests for *a posteriori* comparisons [40].

To determine the impact of ecosystem engineers on species richness of engineered-habitats, we used a Generalized Linear Mixed Model (SAS, GLIMMIX procedure) [40]. We used the same model applied for the abundance analysis, but in this case the number of species that emerged from *S. purpurea* branches was the response variable. An increased number of species is expected as a random consequence of larger pools of individuals [45]. Therefore, to examine whether treatment differences in the species richness of secondary colonizers were driven by differences in the abundance of secondary colonizers, we constructed rarefaction curves for each treatment. We used cumulative species per branch including all branches sampled during the three study years (EcoSim 7.0, 10,000 iterations) [46].

Results

Effect of habitat engineering on community composition

In total, 28,301 secondary colonizers emerged from 478 detached branches of *S. purpurea* in three consecutive years of study. These included at least 25 species from eight families (Table 1), of which Bostrichidae (Coleoptera) was the most abundant, comprising 76% (±10 SD) of the overall natural arthropod community (Table 1), and Cerambycidae was the most diverse (9 spp.; Table 1). The natural arthropod community consisted of xylovore and predatory beetles, and parasitic wasps (Table 1). In addition to secondary colonizers that use *S. purpurea* branches for oviposition and offspring development, other "inquiline" species (which eventually arrived to *S. purpurea* branches, but did not oviposit in them) were recorded. These species included: termites, ants, pseudoscorpions, spiders, crickets and silverfish. However, given that inquilines emerged in very low

Table 1. Secondary colonizers that emerged from *Spondias purpurea* branches detached and colonized by *Oncideres albomarginata chamela*.

Family	Abundance (%)	Species	Size (mm)
XYLOVORE BEETLES			
Bostrichidae	75.76 (±9.8)	*Amphicerus* (LeConte) sp. ‡	11.13 (±1.43)
		Bostrychopsis (Lesne) sp. †	3.58 (±0.23)
		Dendrobiella (Casey) sp. *†	5.49 (±0.25)
		Melalgus (Dejean) sp. ‡	11.56 (±1.42)
		Micrapate (Casey) sp. †	3.56 (±0.18)
		Prostephanus truncatus (Horn) †	3.40 (±0.22)
		Xylobiops (Casey) sp. †	3.71 (±0.24)
Curculionidae	4.10 (±3.1)	*Hypothenemus* (Weswoot) spp. †	1.59 (±0.27)
(Scolytinae)		*Pityophthorus* (Eichhoff) sp. †	2.00 (±0.25)
Lyctidae	7.07 (±3.7)	*Lyctus* (Fabricius) sp. †	3.11 (±0.38)
Buprestidae	1.45 (±1.2)	*Acmaeodera* (Eschscholtz) sp. †	6.27 (±0.54)
		Agrilus (Curtis) sp. †	4.43 (±0.35)
Cerambycidae	3.25 (±2.6)	*Ataxia alpha* (Chemsak and Noguera)* §	14.30 (±1.44)
		Estoloides chamelae (Chemsak and Noguera)* ‡§	12.47 (±1.17)
		Eutrichillus comus (Bartes) ‡	8.08 (±0.47)
		Lagocheirus obsoletus (Thomson) ‡	13.52 (±1.53)
		Lissonotus flavocinctus (Dupont)* ‡§	13.53 (±2.63)
		Poliaenus hesperus (Chemsak and Noguera) ‡	8.69 (±0.60)
		Sphaenothecus maccartyi (Chemsak and Noguera) ‡§	14.61 (±1.46)
		Sphaenothecus trilineatus (Dupont) ‡	21.42 (±1.52)
		Trachyderes mandibularis (Serville)* ‡§	21.97 (±0.79)
NATURAL ENEMIES			
Predator beteles			
Histeridae	7.58 (±3.2)	*Teretriosoma nigrescens* (Lewis) †	2.26 (±0.16)
Cleridae	0.30 (±0.2)	*Enoclerus quadrisignatus* (Say.) ‡	10.40 (±0.64)
Parasitic waps			
Hymenoptera	0.48 (±0.06)	ND §	ND

*Not recorded in 2007;
†Life form I;
‡Life form II;
§Not recorded in non-engineered branches (NE); ND = not determined. Abundance values are means across the three years (±SD).

numbers and were not present every year, we did not consider them in further analyses.

Secondary xylovores of life form I (Table 1) began to emerge one month after branches were enclosed in mesh bags, with a maximum emergence peak recorded in May. Secondary xylovores of life form II and natural enemies (Table 1) emerged throughout the year, but their maximum emergence peaks were observed in September and July, respectively. *O. albomarginata chamela*, the species with the greatest size (23.58 mm ±2.24), was the last species to emerge (September to December). These emergence patterns were consistent across years.

There were significant differences in the composition of the community of secondary colonizers between treatments ($R = 0.425$, $n = 478$, $P < 0.01$; Figure 1). However, the strongest differences in community composition were between non-engineered (NE) and engineered (OE and SE) branches (NE vs. OE: $R = 0.691$, $P < 0.01$; NE *vs.* SE: $R = 0.574$, $P < 0.01$; SE vs. OE: $R = 0.098$, $P < 0.01$).

Effect of habitat engineering on branch colonization frequency by secondary colonizers

Data analyses were performed separately by families, with the exception of Bostrichidae, which were analyzed in two species groups because they exhibit two different life forms (I and II; Table 1). The results indicated a highly significant effect of treatment for all families, with the exception of Cleridae; a significant effect of year for all families except for Buprestidae and Hymenoptera; and a significant interaction between treatment and year for Bostrichidae life form II, Lyctidae, Buprestidae and Histeridae (Table 2). All secondary colonizers significantly colonized engineered branches (treatments OE and SE) more frequently than non-engineered branches (NE; Figure 2), with the exception of Buprestidae for which significant differences were found only between *O. albomarginata chamela*-colonized branches (OE) and non-engineered branches (NE) in 2007 and 2008 (Figure 2). The comparison between OE and SE treatments showed variation across years and groups of secondary colonizers

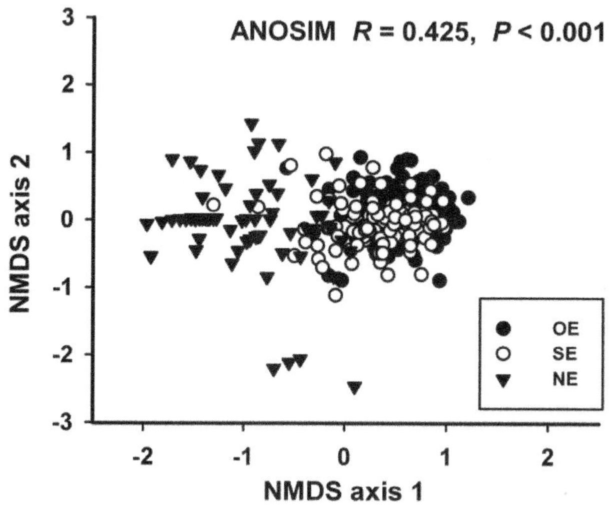

Figure 1. Arthropod community composition in detached *Spondias purpurea* branches. OE = *O. albomarginata chamela*-engineered branches; SE = simulated-engineered branches; and NE = non-engineered branches. Each point is a two-dimensional (axis 1 and axis 2) representation of the arthropod community composition on an individual branch based on global non-metric multidimensional scaling (NMDS) analysis (stress = 0.19).

(Figure 2), but in general there were no significant effects related to the presence of *O. albomarginata chamela.*

Effect of habitat engineering on the abundance of secondary colonizers

There was a highly significant effect of treatment, year and the interaction between treatment and year on the abundance of all groups of secondary colonizers: life form I xylovores ($F_{2,\ 319}$ = 366.04, P<0.0001; $F_{2,\ 319}$ = 74.44, P<0.0001; $F_{4,\ 475}$ = 6.17, P<0.0001); life form II xylovores ($F_{2,\ 475}$ = 31.34, P<0.0001; $F_{2,\ 421}$ = 28.54, P<0.0001; $F_{4,\ 475}$ = 3.21, P = 0.0129); and natural enemies ($F_{2,\ 465}$ = 127.51, P<0.0001; $F_{2,\ 406}$ = 30.23, P<0.0001; $F_{4,\ 475}$ = 5.03, P = 0.0006). The three groups of secondary colonizers showed the following pattern of abundance: OE>SE>NE, where engineered *vs.* non-engineered branches (OE and SE *vs.* NE) showed significant differences for the three groups of secondary colonizers (Figure 3). However, the OE *vs.* SE comparison only showed a significant difference for life form I xylovores in 2007 and 2008, for life form II xylovores in 2008 and 2009, and for natural enemies for 2007 (Figure 3). The abundance of all secondary colonizers in non-engineered branches (NE) was 95% (2007), 93% (2008) and 96% (2009) lower than the abundance of secondary colonizers in *O. albomarginata chamela*-colonized branches (OE) (Figure 3).

Effect of habitat engineering on the species richness of secondary colonizers

There was a strong effect of treatment ($F_{2,\ 478}$ = 367.7, P<0.0001), year ($F_{2,\ 478}$ = 45.77, P<0.0001), habitat engineering (NE *vs.* SE and NE *vs.* OE; Figure 4) and the presence of the ecosystem engineer (OE *vs.* SE; Figure 4) on the species richness of secondary colonizers that emerged from *S. purpurea* branches, with no significant interaction between treatment and year ($F_{4,\ 478}$ = 0.64, P = 0.6310). The results showed the OE>SE>NE pattern of species richness, consistent across years (Figure 4). NE branches showed 85% (2007), 80% (2008), and 82% (2009) fewer species than OE branches (Figure 4). Rarefaction curves showed that the observed differences in cumulative species richness persisted even when samples were rarefied to similar abundances of individuals (Figure 5).

Discussion

Several insect herbivores can create new habitats and alter habitat resource availability for other organisms, by modifying the structural and/or nutritional properties of plant tissues [4,10,14,15]. *O. albomarginata chamela* actively manipulates its host plant by: (i) girdling and detaching branches, and (ii) gnawing eggs niches and incisions into the bark or stems [27]. These modifications were key factors for the establishment and development of an arthropod community composed by xylovores (Bostrichidae, Scolytinae, Buprestidae, Lyctidae and Cerambycidae) and natural enemies (Histeridae, Cleridae and Hymenoptera).

Benefits of stem-boring engineering to secondary colonizers

The reported benefits of insect ecosystem engineering to secondary colonizers include: shelter from harsh abiotic factors, avoidance of natural enemies, and modification of resource quality

Table 2. Effect of habitat engineering on branch colonization frequency by secondary colonizers.

Family	Treatment			Year			Treatment x Year		
	df	F	P	df	F	P	df	F	P
Bostrichidae (Life form I)	2, 307	21.81	<0.0001	2, 307	5.72	0.0036	4, 307	0.47	0.7609
Bostrichidae (Life form II)	2, 259	22.21	<0.0001	2, 259	10.37	<0.0001	4,259	0.1584	<0.0001
Curculionidae (Scolytinae)	2, 307	32.52	<0.0001	2, 307	16.14	<0.0001	4, 307	1.94	0.1040
Lyctidae	2, 307	101.31	<0.0001	2, 307	17.82	<0.0001	4, 307	4.05	0.0032
Buprestidae	2, 307	23.72	<0.0001	2, 307	0.15	0.8619	4, 307	3.30	0.0114
Cerambycidae	2, 259	53.70	<0.0001	2, 259	46.94	<0.0001	4, 259	1.21	0.3078
Histeridae	2, 307	60.73	<0.0001	2, 307	19.55	<0.0001	4, 307	6.48	<0.0001
Cleridae	1, 153	2.73	0.1003	2, 153	9.11	0.0002	2, 153	1.44	0.2408
Hymenoptera	1, 153	16.73	<0.0001	2, 153	1.16	0.3150	2, 153	0.55	0.5762

Data analyses were performed through a generalized linear model with a binomial distribution and a logit link function using a GLIMMIX procedure in SAS.

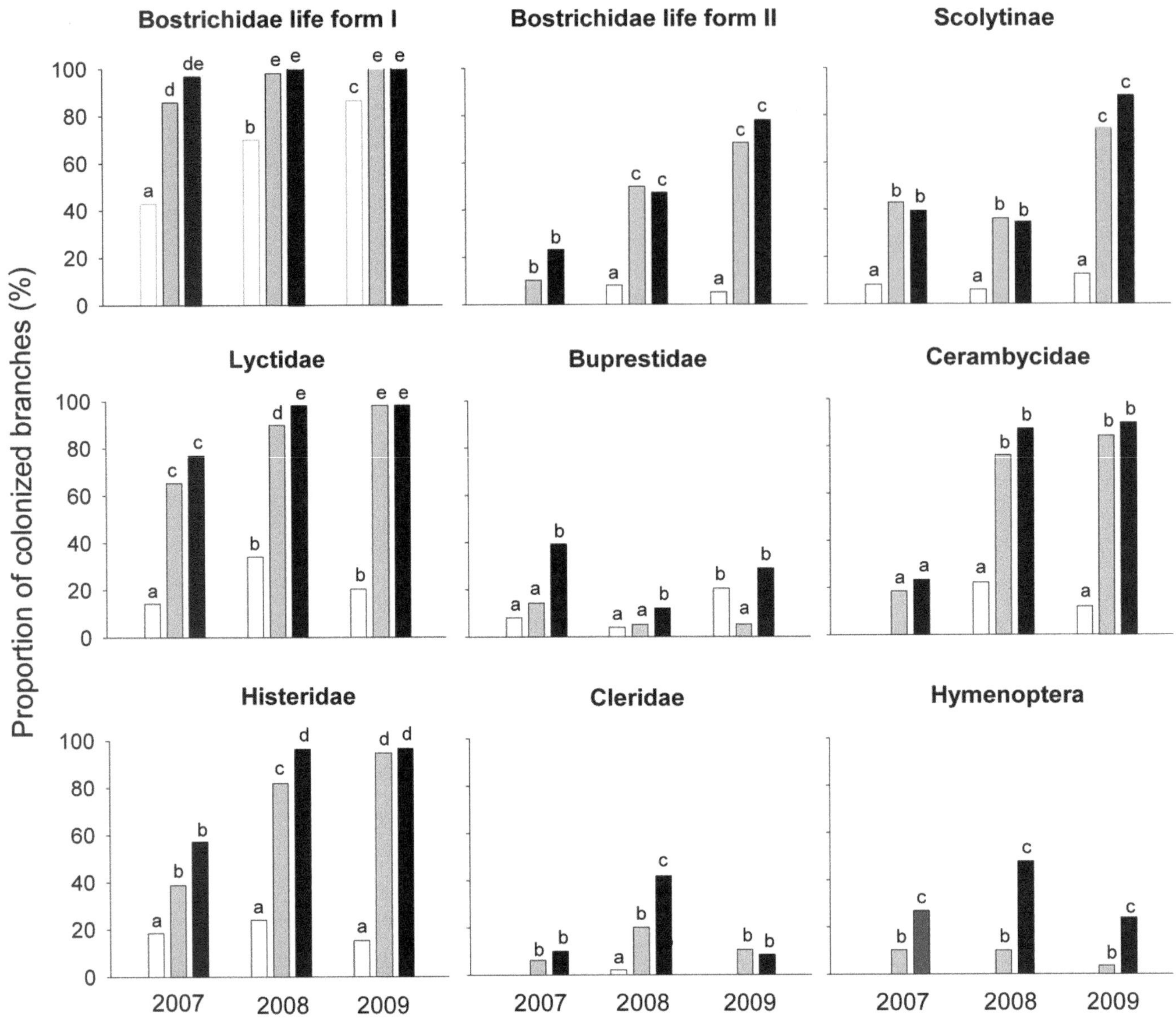

Figure 2. Branch colonization frequency by secondary colonizers. Non-engineered branches are represented by white bars (NE); simulated-engineered branches by gray bars (SE); and *O. albomarginata chamela*- engineered branches by black bars (OE). Values are the percentage of branches colonized. Different letters indicate significant differences ($P<0.05$) between the frequencies of branch colonization of the treatments.

[4,10,47]. However, the importance of each benefit differs among insect guilds. For example: leaf rolls and leaf mines are colonized for shelter rather than for the food they contain [11,15], whereas galls provide shelter, protection from natural enemies and high quality food resources [10].

The main benefits of the stem-boring engineering by *O. albomarginata chamela* to secondary colonizers were related to the creation of a habitat with high availability of oviposition sites, because branches without incisions (non-engineered) were poorly colonized. Moreover, artificially-engineered branches were colonized by a similar arthropod community that colonized branches naturally detached by *O. albomarginata chamela*. This confirms that incisions made by *O. albomarginata chamela* adult females along the detached *S. purpurea* branches are used by other arthropod species as oviposition sites. Availability of oviposition sites offers three benefits to secondary colonizers because they can: (i) save costs of searching for suitable oviposition sites; (ii) diminish the "excavation costs" of the initial stem penetration [10]; and (iii) reduce exophytic predation during the oviposition period (*sensu* [48]).

Additionally, the presence of *O. albomarginata chamela* had a significant impact on the abundance of xylovore species in some years (31–50%), as well as on species richness of the arthropod community in the removed *S. purpurea* branches (15%). The increase of nutrient availability by deposition of faecal pellets is one of the potential benefits of the ecosystem engineer to secondary colonizers [5,10,49]. *O. albomarginata chamela* larvae digest cellulose [50], transforming complex structural carbohydrates into simple sugars, which can be eliminated with faecal pellets. Thus, it is possible that this insect supplies partially digested food to secondary colonizers. However, future studies are needed to confirm this hypothesis.

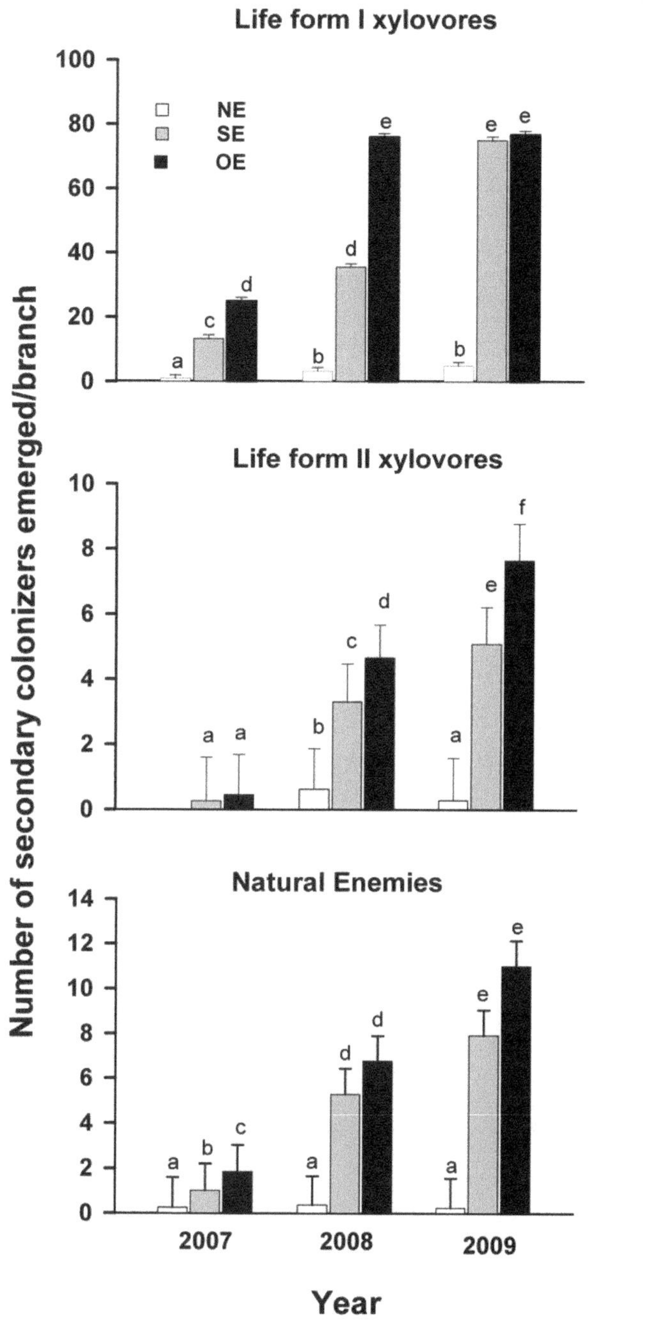

Figure 3. Abundance of secondary colonizers that emerged from *Spondias purpurea* detached branches. Bars indicate LSMeans (±SE) of the number of secondary colonizers that emerged per *S. purpurea* branch in the three studied years. White bars indicate non-engineered branches (NE); gray bars indicate simulated-engineered branches (SE); and black bars indicate *O. albomarginata chamela*-colonized branches (OE). Different letters indicate significant differences between treatments (P<0.05).

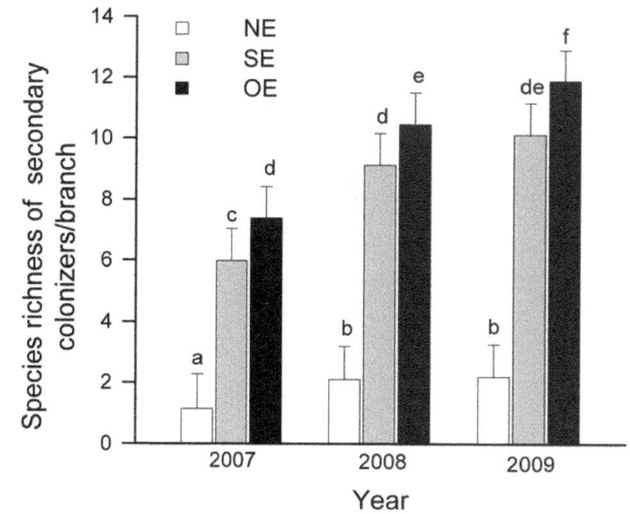

Figure 4. Species richness of secondary colonizers that emerged from *Spondias purpurea* branches. LSMeans (±SE) of the number of species per *S. purpurea* branch in the three studied years; white bars indicate non-engineered branches (NE); gray bars indicate simulated-engineered branches (SE); and black bars indicate *O. albomarginata chamela*-engineered branches (OE). Different letters indicate significant differences between treatments (P<0.05).

Effects of stem-boring engineering on arthropod community

Our study demonstrated that ecosystem engineering by *O. albomarginata chamela* had strong positive effects on its associated arthropod community. The abundance and species richness of xylovore insects were higher in engineered branches than in non-engineered branches, possibly due to greater quantity and quality of habitat and food resources provided by engineered branches [51]. This is consistent with previous studies reporting that higher colonization and performance, following the improvement of resource quality, increase the abundance and species richness of insect herbivores [52–54]. Another mechanism that promotes high species richness, is the increase of the abundance of rare resources or combinations of resources that are required by specialist species [51,54,55]. Five cerambycid species were restricted to engineered branches in the three studied years (Table 1). These species were among the larger xylovore colonizers (Life form II; Table 1). Larger insect species produce larger eggs [44], indicating that they

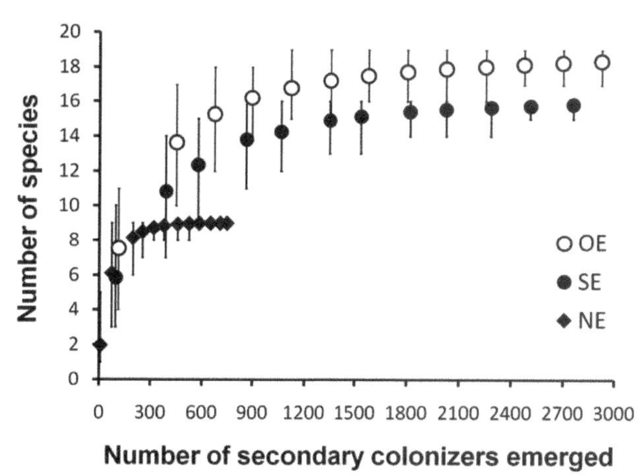

Figure 5. Rarefaction curves plotting the number of species of secondary colonizers *vs.* the number of individuals sampled in detached *Spondias purpurea* branches. NE = non-engineered branches; SE = simulated-engineered branches; and OE = *O. albomarginata chamela-* engineered branches. Bars represent 95% confidence intervals obtained from 10 000 re-sampling iterations. Bars that overlap the mean for alternate treatments indicate that treatments were not significantly different (P>0.05).

require larger egg-niches. Species of the genus *Oncideres* gnaw large egg niches (4–5 mm in width) in which oviposit [17], whereas cerambycid species that do not have the ability to gnaw egg niches, wander over the hosts probing the bark with the ovipositor for cracks and crevices in which they oviposit [30]. Therefore, it is possible that the size of egg niches gnawed by *O. albomarginata chamela* females allow these species to oviposit in them. These findings suggest that the increased species richness in engineered branches can be a consequence of the greater abundance of specific egg niches required by specialist species.

The increase in abundance and number of secondary xylovores, which represent potential prey and hosts for natural enemies, in turn may influence the abundance and species richness of natural enemies and result in bottom-up effects [3,4]. In this study, the overall abundance and species richness of natural enemies was higher in engineered branches than in non-engineered branches. Specifically, only one (Histeridae: *Teretriosoma nigrescens*) of the three natural enemies was consistently recorded in non-engineered branches (Figure 2). *T. nigrescens* preys upon some bostrichid beetles [56], which were the main species that colonized non-engineered branches. However, there were five cerambycid species that did not colonize non-engineered branches. Parasitic wasps are reported as one of the main natural enemies of cerambycid beetles [57]. Thus, the absence of colonization of parasitic wasps in non-engineered branches could be related to the reduced colonization by cerambycid beetles.

Our results confirm the notion that changes in the composition of the xylovore community cascade upward to higher trophic levels through bottom-up effects.

Implications of stem-boring ecosystem engineering for biodiversity

On average, ecosystem engineering by *O. albomarginata chamela* was responsible for nearly 95% of the abundance of secondary colonizers and 82% of the species richness. These results are consistent with the positive effects on arthropod diversity reported for other insect ecosystem engineers [11–13,15]. However, ecosystem engineering by *O. albomarginata chamela* had greater effects on species richness than leaf-roller caterpillars (14–84%) [11,12], gall-makers (32%) [13], and leaf-miners [15], possibly because ecosystem engineering by this species allowed the establishment of an entire arthropod community, and regulated the structure of this community. Therefore, based on Paines "keystone" concept [58], ecosystem engineering by *O. albomarginata chamela* can be considered a keystone process (*sensu* [7]).

There are two explanations for this keystone process: the existence of a highly structured community, and the degree of specialization (i.e. interaction strength) between the secondary colonizers and the engineered habitat [58]. The arthropod community associated with branches engineered by *O. albomarginata chamela* is a highly structured community, because it consists of organisms with different life history traits and trophic positions. In addition, our study suggests that the xylovore community associated with *S. purpurea* branches might be specialists in branches girdled and detached by *O. albomarginata chamela.* Furthermore, the known host plants for the Cerambycidae and Scolytinae species emerging from *S. purpurea* branches completely correspond to the alternate host plants of *O. albomarginata chamela,* and to the host plants of other girdling-beetles in the study site, such as *Oncideres rubra* and *Taricanus zaragozai* [32,59]. Some of these cerambycid species, as well as most species of Bostrichidae and Buprestidae in the *S. purpurea* branches, also use branches girdled by other beetle species in different tropical and subtropical regions [17,18,20,60].

Conclusions

The importance of interactions mediated by insects in shaping herbivore communities is becoming widely recognized [4]. However most studies have focused on herbivore-induced changes to plant chemical composition (reviewed in [4]) and only few to plant-structural modifications made by insect engineers (reviewed in [10]). This study provides evidence that interactions mediated by ecosystem engineering may be a common factor enhancing species richness and structuring communities of borer insects. Therefore, our findings have important implications for conservation, because through the understanding of the mechanisms underlying ecosystem engineering it is possible to develop effective strategies of ecosystem management [61].

Acknowledgments

We are grateful to: Chamela Biological Station (UNAM) for collecting facilities; G. Sánchez-Montoya for help in the field; Dr. Felipe A. Noguera for insect identification; K. Oyama, A. González-Rodríguez, S. Martén-Rodríguez, P. Hanson and two anonymous reviewers for comments and suggestions. This study was performed in partial fulfillment of the requirements of the Ph.D. degree of NCC at the Graduate Program in Biological Sciences, Universidad Nacional Autónoma de México (UNAM).

Author Contributions

Conceived and designed the experiments: NCC MQ. Performed the experiments: NCC LHEV MQ. Analyzed the data: NCC MQ LHEV. Contributed reagents/materials/analysis tools: MQ. Wrote the paper: NCC MQ. Provision of software used for data analyses: MQ.

References

1. Tilman D (1982) Resource competition and community structure. New Jersey: Princeton University Press. 296 p.
2. Hairston NG (1989) Ecological experiments, purpose, design, and execution. Cambridge: Cambridge University Press. 388 p.
3. Morin PJ (1999) Community ecology. Malden: Blackwell Science. 424 p.
4. Ohgushi T (2005) Indirect interaction webs: herbivore-induced effects through trait change in plants. Annu Rev Ecol Evol Syst 36: 81–105.
5. Jones CG, Lawton JH, Shachak M (1994) Organisms as ecosystems engineers. Oikos 69: 373–386.
6. Stachowicz JJ (2001) Mutualism, facilitation, and the structure of ecological communities. BioScience 51: 235–246.
7. Power ME, Tilman D, Estes JA, Menge BA, Bond WJ, et al. (1996) Challenges in the quest for keystones. BioScience 46: 609–620.
8. Wright JP, Jones CG (2004) Predicting effects of ecosystem engineering on patch-scale species richness from primary productivity. Ecology 85: 2071–2081.
9. Hastings A, Byers IE, Crooks JA, Cuddington K, Jones CG, et al. (2007) Ecosystem engineering in space and time. Ecol Lett 10: 153–164.
10. Marquis RJ, Lill JT (2007) Effects of arthropods as physical ecosystem engineers on plant-based trophic interaction webs. In: Ohgushi T, Craig TP, Price PW, eds. Ecological communities: plant mediation in indirect interactions webs. New York: Cambridge University Press. pp 246–274.
11. Martinsen GD, Floate KD, Waltz AM, Wimp GM, Whitham TG (2000) Positive interactions between leafrollers and other arthropods enhance biodiversity on hybrid cottonwoods. Oecologia 123: 82–89.
12. Lill JT, Marquis RJ (2003) Ecosystem engineering by caterpillars increases insect herbivore diversity on white oak. Ecology 84: 682–690.
13. Waltz AM, Whitham TG (1997) Plant development directly and indirectly affects arthropod community structure: opposing impacts of species removal. Ecology 78: 2133–2144.
14. Johnson SN Mayhew PJ, Douglas AE, Hartley SE (2002) Insects as leaf engineers: can leaf-miners alter leaf structure for birch aphids? Funct Ecol 16: 575–584.
15. Kagata H, Ohgushi T (2004) Leaf miner as a physical ecosystem engineer: secondary use of vacant leaf mines by other arthropods. Ann Entomol Soc Am 97: 923–927.
16. Lieutier F, Day KR, Battisti A, Grégoire JC, Evans HF (2004) Bark and wood boring insects in living trees in Europe, a synthesis. Dordrecht: Springer. 569 p.

17. Polk KL, Ueckert DN (1973) Biology and ecology of a mesquite twig girdler, *Oncideres rhodosticta*, in west Texas. Ann Entomol Soc Am 66: 411–417.
18. Hovore FT, Penrose RL (1982) Notes of cerambycidae co-habiting girdles of *Oncideres pustulata* LeConte (Coleptera:Cerambycidae). Southwest Nat 27: 23–27.
19. Di Iorio OR (1996) Cerambycidae y otros Coleoptera de Leguminosae cortadas por *Oncideres germari* (Lamiinae: Onciderini) en Argentina. Rev Biol Trop 44: 551–561.
20. Feller IC, Mathis WN (1997) Primary herbivory by wood boring insects along an architectural gradient of *Rhizophora* mangle. Biotropica 29: 440–451.
21. Amman GD (1977) The role of the mountain pine beetle in lodgepole pine ecosystems: impact on succession. In: Mattson WJ, ed. The Role of Arthropods in Forest Ecosystems. New York: Springer-Verlag. pp 3–18.
22. Schowalter TD (1981) Insect herbivore relationship to the state of the host plant: biotic regulation of the ecosystem nutrient cycling through ecological succession. Oikos 37: 126–130.
23. Feller IC (2002) The role of herbivory by wood-boring insects in mangrove ecosystems in Belize. Oikos 97: 167–176.
24. Martínez AJ, López-Portillo A, Eben A, Golubov J (2009) Cerambycid girdling and water stress modify mesquite architecture and reproduction. Popul Ecol 51: 533–541.
25. Duval BD, Whitford WG (2008) Resource regulation by a twig-girdling beetle has implications for desertification. Ecol Entomol 33: 161–166.
26. Feller IC, McKee KL (1999) Small gap creation in Belizean mangrove forests by a wood-boring insect. Biotropica 31: 607–617.
27. Uribe-Mú C, Quesada M (2006) Preferences, patterns and consequences of attack on the dioecious tropical tree *Spondias purpurea* (Anacardiaceae) by the insect borer *Oncideres albomarginata chamela* (Cerambycidae). Oikos 112: 691–197.
28. Bullock SH (1992) Seasonal differences in nonstructural carbohydrates in two dioecious monsoon-climate trees. Biotropica 24: 140–145.
29. Forcella F (1982) Why twig-girdling beetles girdle twigs. Naturwissenschaften 69: 398–400.
30. Hanks LM (1999) Influence of the larval host plant on reproductive strategies of cerambycid beetles. Annu Rev Entomol 44: 483–505.
31. Chemsak JA, Noguera FA (1993) Annotated checklist of the Cerambycidae of the Estacion de Biologia Chamela, Jalisco, Mexico (Coleoptera), with descriptions of new genera and species. Folia Entomol Mex 89: 55–102.
32. Noguera FA (1993) Revisión taxonómica del género *Oncideres* Serville en México (Coleoptera: Cerambycidae). Folia Entomol Mex 88: 9–60.
33. Duffy EAJ (1960) Monograph of the Immature Stages of Neotropical timber Beetles (Cerambycidae). London: British Museum (Natural History). 327 p.
34. Bullock SH, Solís-Magallanes JA (1990) Phenology of canopy trees of a tropical deciduous forest in Mexico. Biotropica 22: 22–35.
35. Bullock SH (1988) Rasgos del ambiente físico y biológico de Chamela, Jalisco, México. Folia Entomol Mex 77: 5–17.
36. Maass MJ, Martinez-Irizar A, Patiño C, Sarukhan J (2002) Distribution and annual net accumulation of above-ground dead phytomass and its influence on throughfall quality in a Mexican tropical deciduous forest ecosystem. J Trop Ecol 18: 821–834.
37. Gerberg EJ (1957) A revision of the New World species of powderpost beetles belonging to the family Lyctidae. U S Dept Agr Tech Bull 1157: 1–55.
38. Arnett RH (1963) The beetles of the United States: a manual for identification. Washington D.C, The Catholic University of America Press. 1112 p.
39. Binda F, Joly LJ (1991) Los Bostrichidae (Coleoptera) de Venezuela. Bol Entomol Venez 6: 83–133.
40. SAS (2010) SAS Software, version 9.2. Cary, NC: SAS Institute.
41. Faith DP, Minchin PR, Belbin L (1987) Compositional dissimilarity as a robust measure of ecological distance. Vegetatio 69: 57–68.
42. Rice WR (1989) Analyzing tables of statistical tests. Evolution 43: 223–225.
43. Littell RC, Milliken GA, Stroup WW, Wolfinger RD, Schabenberger O (2006) SAS for Mixed Models. Cary, NC: SAS Institute. 380 p.
44. Nylin S, Gotthard K (1998) Plasticity in life-history traits. Annu Rev Entomol. pp 63–83.
45. Gotelli NJ, Colwell RK (2001) Quantifying biodiversity: procedures and pitfalls in the measurement and comparison of species richness. Ecol Lett 4: 379–391.
46. Gotelli NJ, Entsminger GL (2004) EcoSim: Null models software for ecology. Version 7. Acquired Intelligence Inc. & Kesey-Bear. Jericho, VT 05465. Available: http://garyentsminger.com/ecosim/index.htm.
47. Damman H (1993) Patterns of interaction among herbivore species. In: Stamp NE, Casey TM, eds. Caterpillars: ecological and evolutionary constraints on foraging. New York: Chapman and Hall. pp 132–169.
48. Aukema BH, Raffa KF (2002) Relative effects of exophytic predation, endophytic predation, and intraspecific competition on a subcortical herbivore: consequences to the reproduction of *Ips pini* and *Thanasimus dubius*. Popul Ecol. pp 483–491.
49. Daleo P, Fanjul E, Mendez-Casariego A, Silliman BR, Bertness MD, et al. (2007) Ecosystem engineers activate mycorrhizal mutualism in salt marshes. Ecol Lett 10: 902–908.
50. Calderón-Cortés N, Watanabe H, Cano-Camacho H, Zavala-Páramo G, Quesada M (2010) cDNA cloning, homology modelling and evolutionary insights into novel endogenous cellulases of the borer beetle *Oncideres albomarginata chamela* (Cerambycidae). Insect Mol Biol 19: 323–336.
51. Abrams PA (1995) Monotonic or unimodal diversity-productivity gradients: what does competition theory predict? Ecology 76: 2019–2027.
52. Martinsen GD, Driebe EM, Whitham TG (1998) Indirect interactions mediated by changing plant chemistry: beaver browsing benefits beetles. Ecology 79: 192–200.
53. Awmack CS, Leather SR (2002) Host-plant quality and fecundity in herbivorous insects. Annu Rev Entomol 47: 817–844.
54. Utsumi S, Ohgushi T (2009) Community-wide impacts of herbivore-induced plant regrowth on arthropods in a multi-willow species system. Oikos 118: 1805–1815.
55. Srivastava DS, Lawton LH (1998) Why more productive sites have more species: an experimental test of theory using tree-hole communities. Am Nat 152: 510–529.
56. Helbig J, Schulz FA (1996) The potential of the predator *Teretriosoma nigrescens* Lewis (Coleoptera: Histeridae) for the control of *Prostephanus truncatus* (Horn) (Coleoptera: Bostrichidae) on dried cassava chips and cassava wood. J Stored Prod Res 32: 91–96.
57. Kenis M, Hilszczanski J (2004) Natural enemies of Cerambycidae and Buprestidae. In: Lieutier F, Day KR, Battisti A, Grégoire JC, Evans HF, eds. Bark and Wood-Boring Insects in Living Trees in Europe, a Synthesis. Dordrecht: Springer. pp 484–498.
58. Paine RT (1966) Food web complexity and species diversity. Am Nat 100: 65–75.
59. Equihua MA, Atkinson TH (1986) Annotated checklist of bark and ambrosia beetles (Coleoptera: Scolytidae and Platypodidae) associated with a tropical deciduous forest at Chamela, Jalisco, Mexico. Fla Entomol 69: 619–635.
60. Ramírez-Martínez M, de Alba-Avila A, Ramírez-Zurbia R (1994) Discovery of the larger grain borer in a tropical deciduous forest in México. J Appl Entomol 118: 354–360.
61. Byers JE, Cuddington K, Jones CG, Talley TS, Hastings A, et al. (2006) Using ecosystem engineers to restore ecological systems. Trends Ecol Evol 21: 493–500.

15

Conditional Responses of Benthic Communities to Interference from an Intertidal Bivalve

Carl Van Colen[1]*, Simon F. Thrush[2,3], Magda Vincx[1], Tom Ysebaert[4]

1 Ghent University, Department of Biology, Marine Biology Research Group, Ghent, Belgium, **2** National Institute of Water and Atmospheric Research, Hamilton, New Zealand, **3** University of Auckland, School of Environment, Auckland, New Zealand, **4** Netherlands Institute for Sea Research (NIOZ-Yerseke), Yerseke, The Netherlands

Abstract

Habitat-modifying organisms that impact other organisms and local functioning are important in determining ecosystem resilience. However, it is often unclear how the outcome of interactions performed by key species varies depending on the spatial and temporal disturbance context which makes the prediction of disturbance-driven regime shifts difficult. We investigated the strength and generality of effects of the filter feeding cockle *Cerastoderma edule* on its ambient intertidal benthic physical and biological environment. By comparing the magnitude of the effect of experimental cockle removal between a non-cohesive and a sheltered cohesive sediment in two different periods of the year, we show that the outcome of cockle interference effects relates to differences in physical disturbance, and to temporal changes in suspended sediment load and ontogenetic changes in organism traits. Interference effects were only present in the cohesive sediments, though the effects varied seasonally. Cockle presence decreased only the density of surface-dwelling species suggesting that interference effects were particularly mediated by bioturbation of the surface sediments. Furthermore, density reductions in the presence of cockles were most pronounced during the season when larvae and juveniles were present, suggesting that these life history stages are most vulnerable to interference competition. We further illustrate that cockles may enhance benthic microalgal biomass, most likely through the reduction of surface-dwelling grazing species, especially in periods with high sediment load and supposedly also high bioturbation rates. Our results emphasize that the physical disturbance of the sediment may obliterate biotic interactions, and that temporal changes in environmental stressors, such as suspended sediments, may affect the outcome of key species interference effects at the local scale. Consequently, natural processes and anthropogenic activities that change bed shear stress and sediment dynamics in coastal soft-sediment systems will affect cockle-mediated influences on ecosystem properties and therefore the resilience of these systems to environmental change.

Editor: Philippe Archambault, Université du Québec à Rimouski, Canada

Funding: CVC is a postdoctoral fellow of the Flemish Fund for Scientific Research FWO (FWO-1.2.380.11.N.00). Additional financial support for this study came from the special research fund of Ghent University through project BOF-GOA 01GA1911W and a Marie Curie International Fellowship to SFT (FP7-PEOPLE-2007-4-2-IIF-ENV221065). The funders had no role in study design, data collection and analysis, decision to publish, or preparation of the manuscript.

Competing Interests: The authors have declared that no competing interests exist.

* E-mail: carl.vancolen@ugent.be

Introduction

Coastal soft-sediment habitats such as estuarine tidal flats represent some of our most valued ecosystems and are characterized by a high biomass of invertebrate benthic organisms that sustain coastal foodwebs [1,2]. However, these productive systems increasingly experience disturbances with the expanding human population and exploitation of coastal areas [3,4]. For example, changes in sedimentation regimes and water-borne suspended sediment concentrations (SSC) are associated with changes in land-use and engineering activities in the coastal zone, and increased rates of sediment loading adversely affect coastal soft-sediment ecosystem biodiversity and ecological value [5]. Suspended sediments may deposit or remain suspended depending on the physical forces associated with waves and currents, which may vary spatially and temporally and hence affect the balance between deposition and (re)suspension. Nevertheless, both processes influence sediment community composition and functioning. Numerous studies illustrate that terrigenous sediment deposition affects coastal soft-sediment community composition, dynamics, species behavior and ecosystem functioning [6,7,8]. Furthermore, suspended sediments decrease the fitness of filter feeding organisms directly by decreasing filtration rate and absorption efficiency that reduces net energy intake [9], and indirectly by increasing turbidity and hence decreasing water column primary production, i.e. food resource availability. Moreover, changes in sediment dynamics that impact the condition and behavior of key species that are involved in maintaining ecosystem resilience, such as habitat-modifying or ecosystem engineering organisms, *sensu* Jones et al. [10], e.g. [11], have the potential to shift the interactions within an ecosystem [12], thereby altering ecosystem stability. This is likely to have far-reaching consequences since the resilience of ecosystems represents an insurance against potentially adverse changes in the delivery of ecosystem goods and services.

In general, experimental field studies indicate strong context-dependency in the strength of ecosystem engineering effects mediated by environmental setting. For example, Norkko et al. [13] illustrate that facilitation strength of a filter feeding bivalve on benthic communities decreased with increased suspended sediment concentration in the water column, while Thrush et al. [14] indicate that the negative effects of a deposit feeding bivalve on juvenile bivalves increased with hydrodynamic stress. In addition, the magnitude of effects may also depend on the local community composition and the strength of specific species interactions. For example, species living in the upper few centimeters of the sediment are particularly affected by sediment erosion [15] and organisms that affect sediment erodibility may therefore particularly affect survival and distribution of juveniles which are unable to burrow deep, thereby compromising natural demographic population processes. The examples above highlight the importance of integrating spatio-temporal dependency in experimental designs to investigate how stressor-mediated changes in behavior of habitat-modifying organisms affect ecosystem properties.

The filter feeding cockle *Cerastoderma edule* (Mollusca: Pelecypoda) lives buried in the top few centimeters of coastal sediments where it can occur in densities >1000 ind.m^{-2} [16,17]. Field observations indicate a shift towards a less diverse assemblage dominated by deep-living species where *C. edule* is abundant [18]. This species modifies its habitat by altering nutrient processing, depleting the water column of food and larvae, and by changing composition and resuspension of the sediment via vertical and horizontal movements, namely shaking and valve clapping [19–21]. Furthermore, mesoscosm experiments reveal that enhanced suspended sediments may affect the outcome of cockle interactions within the ecosystem because suspended sediments stimulate shell shaking and valve clapping [22] with implications for biodiversity and ecosystem functioning (e.g. interference competition and sediment resuspension). Consequently, *C. edule* can be viewed within a network of direct and indirect interactions that support ecosystem functioning and biodiversity which are mediated by bio-physical processes that define intrinsic ecological dynamics.

There is growing evidence that interactions between the impact of stressors and key species-mediated ecological dynamics can initiate threshold responses and increase the risk of a regime shift ([23] and references therein). Consequently, gaining insights in stressor-mediated outcomes of ecosystem engineering interactions will contribute to a better assessment of the stability of functioning in soft-sedimentary habitats that are subject to environmental change, i.e. resilience. This study investigated the strength and generality of the effects of *C. edule* on its intertidal physical and biological environment and contrasted these effects between two different environmental settings. We compared the magnitude of the effect of experimental cockle removal between cohesive and non-cohesive sediments during the recruitment and post-recruitment period of the same year. Comparing the outcome of cockle interference between a strongly physically stressed non-cohesive sediment versus a more stable cohesive sediment should enable interpretation of the importance of environmental physical forcing and differences in community composition in determining cockle effects. Inclusion of temporal variability in the study design allows the assessment of effects on temporally available juveniles, and the importance of temporal changes in suspended sediment concentrations in governing cockle interference effects. We expected that cockle effects on sediment physical and biological properties would be most pronounced in periods when juvenile organisms are present and suspended sediment concentrations are highest, though physical disturbance of the sediment bed resulting from hydrodynamic forces may overwhelm local interference effects.

Materials and Methods

Study sites: physical conditions and ambient cockle populations

Cockle densities were manipulated in 2011 at two study sites, which are characterized by contrasting physical and biological properties (see results and below). One study site was located in the mid-intertidal zone of the Paulinaschor, in the Westerschelde estuary, whereas the other study site was located in the mid-intertidal zone of the tidal flat near Viane in the Oosterschelde estuary (Figure S1 in File S1). Emersion time at both study sites varied between 50–60%, but the study site at Viane was more exposed to the predominant southwestern winds and experienced higher tidal current velocities (30–40 cm.s^{-1}) than the study site at Paulinaschor (20–25 cm.s^{-1}). Suspended sediment concentrations (SSC) were higher at Paulinaschor (8.56–165 mg.L^{-1}), then at Viane (2.71–54.1 mg.L^{-1}) over the course of the experiment. In general, SSC was about twice as high in autumn and winter than in spring and summer at both study sites (Figure 1). Sediment at Paulinaschor was cohesive and consisted predominantly of silt (median grain size = 49.4 μm), whereas sediment at Viane was non-cohesive consisting predominantly of fine and medium sand (median grain size = 184.9 μm). In addition, the non-cohesive sediment contained less pore water and lower organic matter derived from primary production, i.e. chloroplastic photopigment equivalents (CPE) (Figure 2). Study sites are further referred to as cohesive site (Paulinaschor) and non-cohesive site (Viane).

The adult cockle population in 2011 consisted of two cohorts (recruited in 2009 and 2010) at both sites. Average adult density was 16.9±1.6 SE ind.0.25 m^{-2} at the non-cohesive site and 18.0±7.5 ind.0.25 m^{-2} at the cohesive site.

Experimental design and sample processing

Four paired 0.5 m×0.5 m (i.e. 0.25 m^2) experimental plots were selected along a 25 m transect which was randomly placed parallel to the tide line in the mid intertidal zone of both study sites.

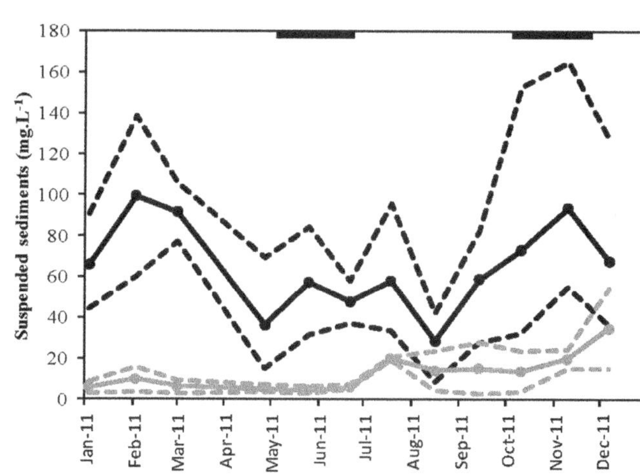

Figure 1. Temporal variation in suspended sediment concentration (SSC) at both study sites in 2011. Black bars indicate the two experimental periods. SSC were retrieved from the public url: www.waterbase.nl. SSC at Paulina (black lines) was calculated as the average value from recordings at Terneuzen, Vlissingen and Hansweert, whereas SSC at Viane (grey lines) was calculated as the average value from recordings at Zijpe and Wissenkerke; see Figure S1 in File S1 for indication of sampling locations. Solid lines indicate average concentrations, dashed lines the minimum and maximum concentration recorded at a sampling location.

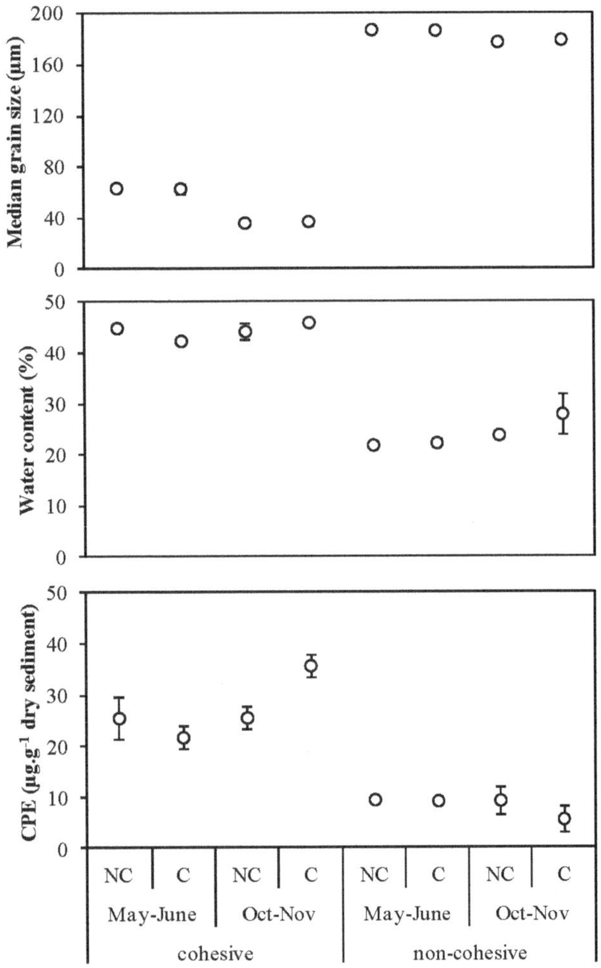

Figure 2. Spatio-temporal effects of cockle treatment on physical sediment properties. Sediment median grain size, water content, and chloroplastic photopigment equivalent (CPE) concentration, shown as mean ± SE for plots without (NC) and with (C) cockles in the two contrasting sediments during both experimental trials, i.e. during recruitment (May–June) and post-recruitment (October–November).

Distances between pairs of plots was 5 m. Cockle densities were manipulated by incubating each plot for 10 days under a nylon 1 mm mesh-sized net, that was placed flush with the sediment surface, using metal frames. This technique chased cockles to the sediment surface so that they could easily and effectively be removed. During the same low tide, 17 adult alive cockles (i.e. average ambient adult density) that were collected at the study site were added into one randomly assigned plot per pair, yielding one plot with cockles (C) and one plot without cockles (NC) per pair. Cockles were added using a 0.5×0.5 m frame with 25 10×10 cm quadrats, ensuring a homogenous distribution of the study organisms at the start of the experiment. We performed the experiment in two different periods of the year; i.e. during recruitment (May–June, frame deployment on April 18th and 19th) and post-recruitment (October–November, frame deployment on September 22nd and 23th) at both study sites [24,25]. The average size of added cockles was 28.9±0.1 SE mm at the non-cohesive site for the trial during the recruitment period, and 28.0±0.1 SE mm for the post-recruitment trial. At the cohesive site, the size of added cockles was 28.6±0.1 SE mm for the trial during recruitment, and 29.4±0.1 SE mm for the post-recruitment trial.

Two days after removal of the frames, we collected samples in the outer rim of each NC plot and from the adjacent ambient unmanipulated sediment to assess potential artifacts on sediment physical and biological properties associated with deployment of the frames and nets. Deployment of frames did not significantly influenced sediment particle size, water content, redox potential discontinuity layer depth, and macrofauna community composition (except adult cockles) during both study periods at both sites, though CPE was slightly lower in the plots as compared to the ambient sediment at the start of the post-recruitment trial, i.e. respectively −6.9 $\mu g.g^{-1}$ and −1.9 $\mu g.g^{-1}$ at the cohesive and the non-cohesive site (Table S1 in File S1).

Cockle effects on sediment physical and biological properties, and the number of adult cockles present in the plots, were determined after six weeks. The number of adult cockles was assessed by sieving the inner 30 cm diameter of each plot. Adult cockle densities in the inner 30 cm diameter of the C plots after six weeks was, on average, respectively 4.8 and 5.0 during the recruitment trial and the post-recruitment trial in the non-cohesive sediments, and respectively 3.8 and 4.0 in the cohesive sediments during the recruitment and post-recruitment trial. Cockle densities thus equaled 13–17 individuals per C plot across all treatments and trials, while only one adult cockle was found in all NC plots. Cockle effects on benthic communities were determined from three 10 cm (i.d.) perspex corers that were randomly collected to a depth of 10 cm in the inner 30 cm diameter of each C and NC plot. These samples were pooled and fixed on a neutralized 8% formaldehyde-tapwater solution. Subsequently, samples were sieved over a 1 mm mesh in the laboratory, and the organism retained on the sieve were counted and identified to the lowest possible taxonomic level. Similarly, we collected three 3.2 cm (i.d.) perspex corers to a depth of 5 cm from the inner 30 cm of each plot in order to quantify sediment properties. From each corer we stored the upper 0.5 cm of sediment on dry ice and at −20°C (for sediment composition and water content, two samples) and −80°C (for CPE concentrations, one sample) upon arrival in the laboratory awaiting further processing. Samples were weighed wet, lyophilized in the dark and weighed again to yield absolute water content (%). Sediment composition was determined from the dry sediment using Malvern laser diffraction and CPE, i.e. sum of chlorophyll a and its degradation products (phaeopigments), was determined following Jeffrey et al. [26] using HPLC of the supernatant that was extracted from the freeze-dried sediment by adding 10 ml 90% acetone.

Data analysis

Two-way Analysis of Similarities [27] was used to compare community composition between study sites and experimental trials, using the samples from the ambient sediment that were collected two days after removal of the frames.

We compared the magnitude of the effect of cockle presence (C) versus experimental cockle removal (NC) in two contrasting sediment types. Given the general higher variability in non-cohesive sediments and the difference in dominant species presence between both sediment types, we analyzed the effects of cockle interference using a two-way factorial generalized linear model for each sediment type separately. This model included the fixed factors Cockle (present, removed) and Timing (recruitment, post-recruitment period), as timing of the experiment was deliberately chosen to represent seasonal changes in SSC and presence of juveniles which may affect the outcome of cockle interference on sediment properties. Gamma error structures, using a log-link function were used for models of total density and species density, while normal error structures were used for general community attributes (species richness, Pielou's evenness, and Simpson's index of diversity) and physical sediment properties. Densities are expressed as individuals per 0.024 m^{-2}, i.e. the total surface of the three pooled corers. We removed typical meio-, hyper- and epifaunal taxa from the biotic dataset since these organisms were not sampled qualitatively. Log_e transformation was applied to meet the assumption of within group variance homogeneity (Levene test), if needed. Alternatively, if Log_e transformation was not successful in producing homogeneity of variance, we conducted two-way factorial permutational analysis of variance after data similarities were calculated using Euclidean distances, and checked for multivariate dispersion [28].We assessed cockle effects on the density of the five most dominant species from each site, which comprised 91 and 71% of the total density at the cohesive and non-cohesive site, respectively. Since we expected that surface-dwelling organisms that feed on the surface sediment and that have rather limited dispersal capacities during their adult life stage would be particularly influenced by cockles, we additionally analyzed cockle effects on the diversity and density of this pool of organisms separately. These surface deposit feeders that have limited dispersal capacities as compared to the highly mobile amphipods are further referred to as LDSDF.

Results

Species assemblage of non-cohesive and cohesive sediments

Benthic community composition differed significantly among study sites and over time (ANOSIM, site $R = 0.995$, $p<0.001$; time $R = 0.289$, $p = 0.003$). Overall, Polychaeta (4 species), and Crustacea (3 species) comprised 88% of the individuals present at the non-cohesive study site, while Polychaeta (8 species) and Oligochaeta spp. were the most dominant taxa at the cohesive study site comprising, respectively, 88 and 7% of the total number of individuals. At the cohesive site, LDSDF comprised 86% of the total community, while only 59% of the organisms were LDSDF at the non-cohesive site.

During both trials, a higher total density and species richness were present in the cohesive sediments, but total community diversity and evenness was higher in the non-cohesive sediments (Figure 3, a–d). Similarly, LDSDF species richness and total density were higher in the non-cohesive sediments, while LDSDF evenness was higher during both trials, and LDSDF diversity was higher during in May–June, in the non-cohesive sediments (Figure 3, e–f). The five most dominant species at the cohesive site were the LDSDF *Aphelochaeta marioni*, *Pygospio elegans* and juvenile recruits of *Macoma balthica*, and the subsurface deposit feeding species *Heteromastus filiformis* and Oligochaeta spp. In contrast, the LDSDF species *Scoloplos armiger* and *Pygospio elegans*, the mobile species *Hydrobia ulvae* and *Urothoe poseidonis*, and the filter feeding recruits of *C. edule* were the most dominant species at the non-cohesive site.

Cockle effects in cohesive sediments

Median grain size decreased during the post-recruitment trial, independent of cockle treatment (Figure 2). Cockles significantly affected physical and biological sediment properties, though the effects often varied temporally. Cockle presence significantly enhanced CPE during the post-recruitment trial in October–November (Figure 2) (Table S2 in File S1). Furthermore, total macrobenthos and LDSDF density was reduced when cockles were present, though this decrease was more pronounced during the recruitment trial in May–June (Figure 3). Cockle presence significantly decreased the density of *A. marioni*, *P. elegans* and juveniles of *M. balthica* during at least one of the trials (Table S2 in File S1). Density of *A. marioni* was reduced by 35% during the recruitment trial, while density of *M. balthica* recruits was reduced by 65% during the post-recruitment trial (Figure 4). The density of *P. elegans* was, on average, reduced by 45% in May–June and by 51% in October–November. No detectable effects were apparent for Oligochaeta spp. and *H. filiformis*. Diversity and evenness of the LDSDF community was lowest in the C plots at the end of the post-recruitment trial (Figure 3).

Cockle effects in non-cohesive sediments

None of the measured physical and biological sediment properties was significantly influenced by the cockle treatment in the non-cohesive sediments (Table S3 in File S1). However, some sediment properties revealed significant temporal variation, independent of cockle presence. Similar to the cohesive sediments, median grain size decreased during the post-recruitment trial (Figure 2). In general, the density of the most dominant species was lower during the post-recruitment trial in October–November, except for *Hydrobia ulvae* (Figure 4). The latter species was absent in May–June, but was the most dominant member of the community in October–November. However, these changes in species density did not significantly affected overall and LDSDF community density, diversity and evenness.

Discussion

We have examined the generality of the interaction between a filter feeding infaunal bivalve, *C. edule*, and its ambient biotic and physical environment. By manipulating cockle densities in two contrasting sediment types (i.e. a sheltered cohesive tidal flat *vs.* a physically stressed non-cohesive tidal flat) in two different periods (i.e. during recruitment and post-recruitment phase) we were able to relate the significance of cockle effects to physical disturbance, and to temporal changes in the presence of species life stages and likely changes in cockle behavior.

Firstly, we demonstrate the temporal dependence of the strength and outcome of cockle effects on physical and biological sediment properties. We suggest two aspects as being governing drivers of the observed temporal effect: (1) species-specific life-stage dependent vulnerability to cockle interference, and (2) changing cockle behavior associated with increasing SSC. We found reduced densities of the dominant surface-dwelling species when cockles were present in the cohesive sediments. Lower densities may have resulted from both direct interference competition and active post-settlement dispersal [29,30]. The absence of an effect of cockle removal on the subsurface-dwelling Oligochaeta spp. and *H. filiformis* suggests that cockle effects on benthic communities in cohesive sediments are mainly mediated by interference effects of cockle bioturbation in the surface sediment layer. Furthermore, densities of the LDSDF *A. marioni* and *P. elegans* were numerically less reduced in the C plots during the post-recruitment trial, than during the recruitment trial which suggests that larvae and recently recruited juveniles of these species are more vulnerable to cockle interference as compared to 4–5 months old juveniles. This is particularly clear for *A. marioni* which was reduced by 35% during the recruitment trial, while the population density of this species was not affected by cockle presence or removal during the post-recruitment trial. This species has a direct larval benthic development (www.marlin.ac.uk) and is thus potentially highly vulnerable to disturbance of the sediment during its larval stage. Average suspended sediment concentrations increased at the

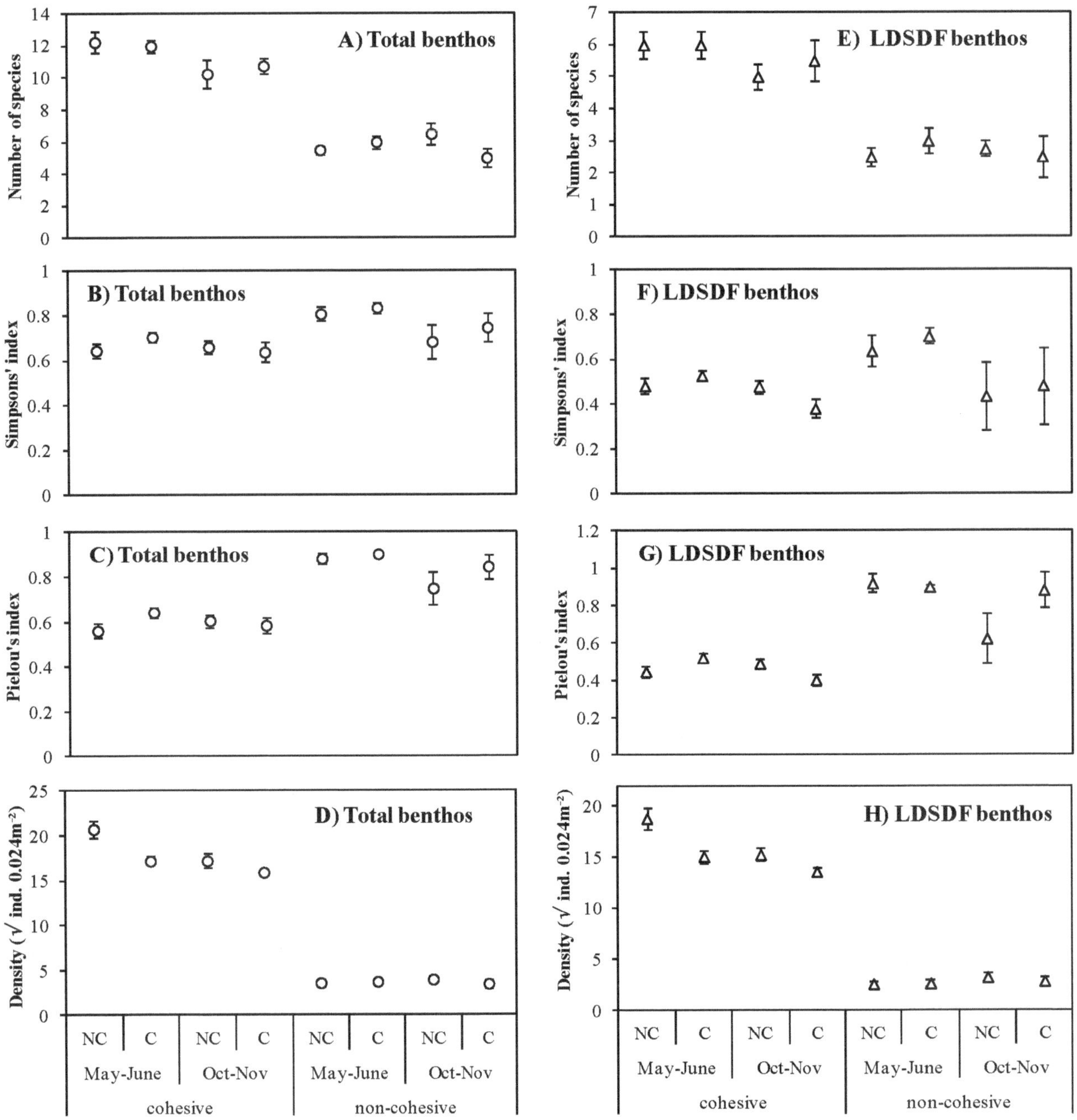

Figure 3. Spatio-temporal effects of cockle treatment on biological sediment properties. Species richness, diversity, evenness and total density for the total community (A–D) and the less-mobile surface deposit feeders (LDSDF, E–H); shown as mean ± SE for plots without (NC) and with (C) cockles in the two contrasting sediments during both experimental trials, i.e. during recruitment (May–June) and post-recruitment (October–November). Note that presented densities are square root transformed.

cohesive study site from 47.8–57.3 $mg.L^{-1}$ in May–June to 73.3–93.9 $mg.L^{-1}$ in October–November (i.e. post-recruitment), with a recorded maximum concentration of 165.0 $mg.L^{-1}$ in November. This is a yearly phenomenon [31,32] and is most likely associated with enhanced rainfall and associated soil erosion in the Sheldt catchment area in this period of the year [32]. Ciutat et al. [22] empirically demonstrated that *C. edule* increases its frequency of valve clapping with increasing SSC in order to eject excess sediments from its mantle cavity. This behavior generally increases linearly with the increase in SSC until ~300 $mg.L^{-1}$ [22]. Populations of *P. elegans* and juveniles of *M. balthica* were significantly reduced when cockles were present in the period with enhanced SSC and thus supposedly also higher cockle bioturbation rates. Both species rely substantially on microalgal carbon for their diet at Paulinaschor [18,33]. Hence, the observed higher CPE concentration in the C plots during the post-recruitment trial may therefore result from enhanced cockle bioturbation-mediated decreases in density, and thus grazing pressure, of *P. elegans* and *M. balthica*. The two numerically dominant species in the community, *A. marioni* and *P. elegans*, thus exhibited a different response to cockle presence during the post-recruitment trial in autumn. Because no species were outcompeted

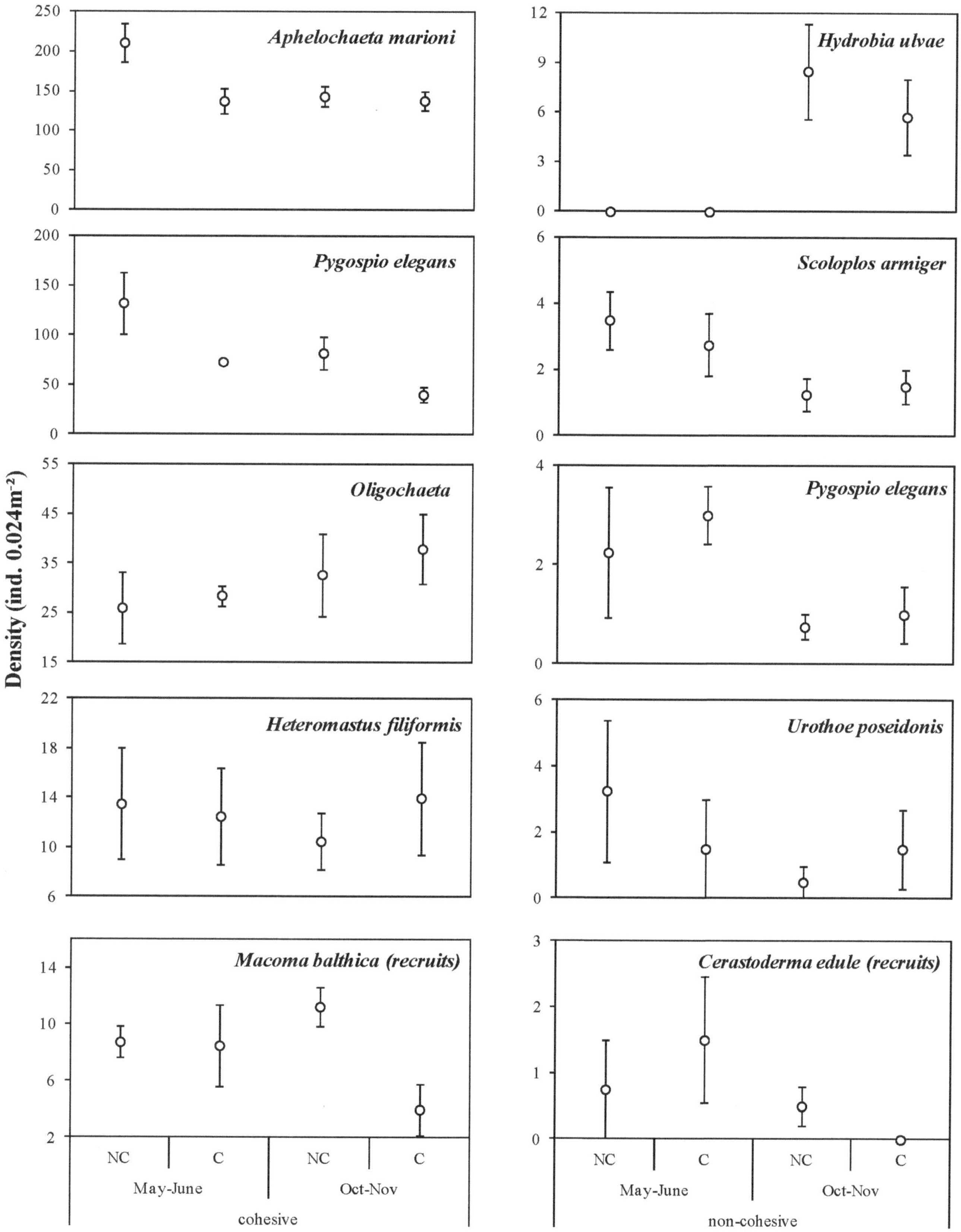

Figure 4. Temporal effects of cockle treatment on the density of the five most dominant species per study site. Densities are shown as mean ± SE for plots without (NC) and with (C) cockles during both experimental trials, i.e. during recruitment (May–June) and post-recruitment (October–November) in the cohesive sediments (left panel) and in the non-cohesive sediments (right panel).

by cockles (i.e. total species number remained constant in C and NC plots) but, of the two species, only *P. elegans* was inhibited by cockle presence during the post-recruitment trial, evenness and diversity decreased during this trial.

Secondly, our study demonstrates the potential for physical disturbance resulting from hydrodynamic forces to overwhelm the effects of cockles on physical and biological sediment properties. Above a threshold, physical factors are considered relatively more important than biotic interactions in mediating community organization (e.g. [34,35]). Though *C. edule* clearly has the potential to affect its physical and biological environment, we conclude that the physical disturbance of the sediment bed associated with hydrodynamic forces at the non-cohesive study site, which has consequences for organism dispersal and species population demographics [15,36], likely obliterated the ecosystem engineering of *C. edule* on the surface sediment layers. In general, differences in benthic biological community attributes and physical sediment properties were greater between the two contrasting sediment types during both experimental trials, than between when cockles were present or removed (Figures 2, 3). This illustrates that physical disturbance resulting from wave action and tidal currents is a governing process of estuarine benthic diversity and sediment properties (e.g. [18,37]). The stability of bed sediment depends on the balance between hydrodynamic forces that cause erosion and the forces within the sediment that resist it, with consolidated sediment that contain fine particles (e.g. mud, i.e. particles <63 μm) typically having a lower propensity to be eroded then unconsolidated, non-cohesive sediments (Grabowski et al. [38], and references therein).

In conclusion, our results emphasize that physical disturbance of the sediment bed due to hydrodynamic forcing, can constrain the outcome of biotic interactions at the local scale, and that temporal changes in environmental stressors, e.g. suspended sediment concentrations, may affect the outcome of key species interference effects, e.g. through changes in key species behavior and life-stage specific vulnerability to interference from bioturbation. As a consequence, the obtained results suggest that inconsistencies among previous studies that examined the effect of cockle loss or manipulated cockle densities on benthic communities (see e.g. [39–41]) may result from differences in the strength and outcome of interference effects depending on the spatial and temporal context. Experiments conducted at multiple sites and times should therefore be considered preferable in order to generalize results of how key species affect ecosystem properties [42], therewith enhancing our understanding of ecosystem resilience in the context of spatio-temporal variable stressors. Consequently, the obtained insights into the context-dependent effects of cockles on biotic and abiotic sediment properties in this study are relevant to assess shallow coastal ecosystem stability, in particular for those systems that are subjected to changing sediment dynamics. This study illustrates that processes that alter bed shear stress and sediment load in coastal cohesive soft-sediment systems (e.g. changes in hydro-morphology, storm frequency and intensity, terrigeneous sediment run-off, etc.) will affect cockle-mediated influences on ecosystem properties, impairing the resilience of ecosystem functioning.

Supporting Information

File S1 Figure S1, Location of the study sites (transect lines in white) in the mid intertidal zone at Paulinaschor and Viane, and locations for monthly samplings of suspended sediment concentrations (red circles) in the subtidal channel. Table S1, Three-way factorial permutational anova results of the effect of plot incubation (Factor: Location, i.e. in or out the plot) on physical (water content, chloroplastic photopigment equivalent concentrations, median grain size) and biological (community structure) sediment properties at the two sites (Factor: Site) for the two experimental trials (Factor: Trial). Boldfaced p-values indicate significant effect at p<0.05. Table S2, Two-way factorial analysis of variance results (p-values) of the effect of cockle removal or presence (Factor: Treatment) on physical and biological sediment properties during the two experimental trials (Factor: Trial) at the cohesive study site. Boldfaced p-values indicate significant effect at p<0.05. Table S3, Two-way factorial analysis of variance results (p-values) of the effect of cockle removal or presence (Factor: Treatment) on physical and biological sediment properties during the two experimental trials (Factor: Trial) at the non-cohesive study site. [P]results from permutational analysis of variance. Boldfaced p-values indicate significant effect at p<0.05.

Acknowledgments

We thank Jeroen Van Daelen, Jos Van Soelen, Douwe van den Ende, Mershad Taheri, Elisabeth Debusschere, Niels Viane, Bart Beuselinck, Anne Willems, Lander Blommaert and Eline Remue for assistance in the fieldwork. Jurgen Verstraeten is acknowledged for construction of the incubation frames and Dirk Van Gansbeke and Bart Beuselinck for execution of the physico-chemical analyses.

Author Contributions

Conceived and designed the experiments: CVC TY SFT. Performed the experiments: CVC. Analyzed the data: CVC. Contributed reagents/materials/analysis tools: CVC MV TY. Wrote the paper: CVC SFT MV TY.

References

1. Costanza R, dArge R, deGroot R, Farber S, Grasso M, et al. (1997) The value of the world's ecosystem services and natural capital. Nature 387: 253–260.
2. Snelgrove PVR (1999) Getting to the bottom of marine biodiversity: Sedimentary habitats - Ocean bottoms are the most widespread habitat on Earth and support high biodiversity and key ecosystem services. Bioscience 49: 129–138.
3. Airoldi L, Beck MW (2007) Loss, status and trends for coastal marine habitats of Europe. Oceanography and Marine Biology, Vol 45 45: 345–405.4.
4. Lotze HK, Lenihan HS, Bourque BJ, Bradbury RH, Cooke RG, et al. (2006) Depletion, degradation, and recovery potential of estuaries and coastal seas. Science 312: 1806–1809.
5. Thrush SF, Hewitt JE, Cummings V, Ellis JI, Hatton C, et al. (2004) Muddy waters: elevating sediment input to coastal and estuarine habitats. Frontiers in Ecology and the Environment 2: 299–306.
6. Rodil IF, Lohrer AM, Chiaroni LD, Hewitt JE, Thrush SF (2011) Disturbance of sandflats by thin terrigenous sediment deposits: consequences for primary production and nutrient cycling. Ecological Applications 21: 416–426.
7. Lohrer AM, Thrush SF, Lundquist CJ, Vopel K, Hewitt JE, et al. (2006) Deposition of terrigenous sediment on subtidal marine macrobenthos: response of two contrasting community types. Marine Ecology-Progress Series 307: 115–125.
8. Woodin SA, Wethey DS, Hewitt JE, Thrush SF (2012) Small scale terrestrial clay deposits on intertidal sandflats: Behavioral changes and productivity reduction. Journal of Experimental Marine Biology and Ecology 413: 184–191.
9. Navarro JM, Widdows J (1997) Feeding physiology of Cerastoderma edule in response to a wide range of seston concentrations. Marine Ecology-Progress Series 152: 175–186.
10. Jones CG, Lawton JH, Shachak M (1994) Organisms as Ecosystem Engineers. Oikos 69: 373–386.
11. Braeckman U, Van Colen C, Soetaert K, Vincx M, Vanaverbeke J (2011) Contrasting macrobenthic activities differentially affect nematode density and diversity in a shallow subtidal marine sediment. Marine Ecology-Progress Series 422: 179–191.

12. Thrush SF, Hewitt JE, Lohrer AM (2012) Interaction networks in coastal soft-sediments highlight the potential for change in ecological resilience. Ecological Applications 22: 1213–1223.
13. Norkko A, Hewitt JE, Thrush SF, Funnell GA (2006) Conditional outcomes of facilitation by a habitat-modifying subtidal bivalve. Ecology 87: 226–234.
14. Thrush SF, Hewitt JE, Cummings VJ, Green MO, Funnell GA, et al. (2000) The generality of field experiments: Interactions between local and broad-scale processes. Ecology 81: 399–415.
15. Hunt HL (2004) Transport of juvenile clams: effects of species and sediment grain size. Journal of Experimental Marine Biology and Ecology 312: 271–284.
16. Van Colen C, Montserrat F, Vincx M, Herman PMJ, Ysebaert T, et al. (2010) Long-term divergent tidal flat benthic community recovery following hypoxia-induced mortality. Marine Pollution Bulletin 60: 178–186.
17. Gam M, de Montaudouin X, Bazairi H (2010) Population dynamics and secondary production of the cockle Cerastoderma edule: A comparison between Merja Zerga (Moroccan Atlantic Coast) and Arcachon Bay (French Atlantic Coast). Journal of Sea Research 63: 191–201.
18. Van Colen C, Meulepas G, De Backer A, van der Wal D, Vincx M, et al. (2010) Diversity, trait displacements and shifts in assemblage structure of tidal flat deposit feeders along a gradient of hydrodynamic stress. Marine Ecology Progress Series 406: 79–89.
19. Troost K, Kamermans P, Wolff WJ (2008) Larviphagy in native bivalves and an introduced oyster. Journal of Sea Research 60: 157–163.
20. Rossi F, Gribsholt B, Middelburg JJ, Heip C (2008) Context-dependent effects of suspension feeding on intertidal ecosystem functioning. Marine Ecology-Progress Series 354: 47–57.
21. Montserrat F, Van Colen C, Provoost P, Milla M, Ponti M, et al. (2009) Sediment segregation by biodiffusing bivalves. Estuarine Coastal and Shelf Science 83: 379–391.
22. Ciutat A, Widdows J, Pope ND (2007) Effect of Cerastoderma edule density on near-bed hydrodynamics and stability of cohesive muddy sediments. Journal of Experimental Marine Biology and Ecology 346: 114–126.
23. Thrush SF, Hewitt JE, Dayton PK, Coco G, Lohrer AM, et al. (2009) Forecasting the limits of resilience: integrating empirical research with theory. Proceedings of the Royal Society B 276: 3209–3217.
24. Coosen J, Seys J, Meire PM, Craymeersch JAM (1994) Effects of sedimentological and hydroydnamical changes in the intertidal areas of the Oosterschelde estuary (SW Netherlands) on distribution, density and biomass of five common macrobenthic species: Spio martinensis (Mesnil), Hydrobia ulvae (Pennant), Arenicola marina (L.), Scoloplos armiger (Muller) and Bathyporeia sp. Hydrobiologia 282:283: 235–249.
25. Van Colen C, Montserrat F, Vincx M, Herman PMJ, Ysebaert T, et al. (2008) Macrobenthic recovery from hypoxia in an estuarine tidal mudflat. *Marine ecology progress series* 372: 31–42.
26. Jeffrey SW, Mantoura RFC, Wright SW (1997) Phytoplankton pigments in oceanography. United Nations International, Paris
27. Clarke KR, Gorley RN (2006) Primer v6: User manual/tutorial. Primer-E, Plymouth Marine Laboratory, Plymouth, U.K.
28. Anderson MJ, Gorley RN, Clarke KR (2008) Permanova+ for primer: guide to software and statistical methods. Primer-e: Plymouth, UK.
29. Hiddink JG, Kock RP, Wolff WJ (2002) Active pelagic migrations of the bivalve Macoma balthica are dangerous. Marine Biology 140: 1149–1156.
30. Van Colen C, Lenoir J, De Backer A, Vanelslander B, Vincx M, et al. (2009) Settlement of *Macoma balthica* larvae in response to benthic diatom films. Marine Biology 156: 2161–2171.
31. Van Damme S, Struyf E, Maris T, Ysebaert T, Dehairs F, et al. (1995) Spatial and temporal patterns of water quality along the estuarine salinity gradient of the Scheldt estuary (Belgium and the Netherlands): results of an integrated monitoring approach. Hydrobiologia 540: 26–45.
32. Chen MS, Wartel S, Van Eck B, Van Maldegem D (2005) Suspended matter in the Sheldt esturary. Hydrobiologia 540: 79–104.
33. Van Colen C, Montserrat F, Vincx M, Herman PMJ, Ysebaert T, et al. (2010) Macrobenthos recruitment success in a tidal flat: feeding trait dependent effects of disturbance history. Jorunal of Experimental Marine Biology and Ecology 385: 79–84.
34. Menge BA, Sutherland JP (1987) Community Regulation - Variation in Disturbance, Competition, and Predation in Relation to Environmental-Stress and Recruitment. American Naturalist 130: 730–757.
35. Bertness MD, Callaway R (1994) Positive Interactions in Communities. Trends in Ecology & Evolution 9: 191–193.
36. Commito JA, Thrush SF, Pridmore RD, Hewitt JE, Cummings VJ (1995) Dispersal dynamics in a wind-driven benthic system. Limnology and Oceanography 40: 1513–1518.
37. Le Hir P, Roberts W, Cazaillet O, Christie M, Basoullet P, et al. (2000) Characterization of intertidal flat hydrodynamics. Continental Shelf Research 20: 1433–1459.
38. Grabowski RC, Droppo IG, Wharton G (2011) Erodibility of cohesive sediment: The importance of sediment properties. Earth-Science Reviews 105: 101–120.
39. Flach EC (1996) The influence of the cockle, Cerastoderma edule, on the macrozoobenthic community of tidal flats in the Wadden Sea. Marine Ecology-Pubblicazioni Della Stazione Zoologica Di Napoli I 17: 87–98.
40. Cesar CP, Frid CLJ (2009) Effects of experimental small-scale cockle (Cerastoderma edule L.) fishing on ecosystem function. Marine Ecology-an Evolutionary Perspective 30: 123–137.
41. Wijnhoven S, Escavarage V, Herman PMJ, Smaal A, Hummel H (2011) Short and mid-long term effects of cockle-dredging on non-target macrobenthic species: a before-after-control impact experiment on a tidal mudflat in the Oosterschelde (The Netherlands). Marine Ecology 32: 1–13.
42. Hewitt JE, Thrush SF, Dayton PK, Bonsdorf E (2007) The effect of spatial and temporal heterogeneity on the design and analysis of empirical studies of scale-dependent systems. The American Naturalist 169: 398–408.

Habitat-Mediated Facilitation and Counteracting Ecosystem Engineering Interactively Influence Ecosystem Responses to Disturbance

Johan S. Eklöf[1*¤a¤b], Tjisse van der Heide[2], Serena Donadi[1], Els M. van der Zee[3,4], Robert O'Hara[5], Britas Klemens Eriksson[1]

1 Department of Marine Benthic Ecology and Evolution, Center for Ecology & Evolutionary Studies, Groningen University, Groningen, The Netherlands, **2** Community and Conservation Ecology Group, Center for Ecology & Evolutionary Studies, Groningen University, Groningen, The Netherlands, **3** Animal Ecology Group, Center for Ecology & Evolutionary Studies, Groningen University, Groningen, The Netherlands, **4** Department of Marine Ecology, NIOZ Royal Netherlands Institute for Sea Research, Den Burg, Texel, The Netherlands, **5** Biodiversity and Climate Research Center, Frankfurt, Germany

Abstract

Recovery of an ecosystem following disturbance can be severely hampered or even shift altogether when a point disturbance exceeds a certain spatial threshold. Such scale-dependent dynamics may be caused by preemptive competition, but may also result from diminished self-facilitation due to weakened ecosystem engineering. Moreover, disturbance can facilitate colonization by engineering species that alter abiotic conditions in ways that exacerbate stress on the original species. Consequently, establishment of such counteracting engineers might *reduce* the spatial threshold for the disturbance, by effectively slowing recovery and increasing the risk for ecosystem shifts to alternative states. We tested these predictions in an intertidal mudflat characterized by a two-state mosaic of *hummocks* (humps exposed during low tide) dominated by the sediment-stabilizing seagrass *Zostera noltii*) and *hollows* (low-tide waterlogged depressions dominated by the bioturbating lugworm *Arenicola marina*). In contrast to expectations, seagrass recolonized both natural and experimental clearings via lateral expansion and seemed unaffected by both clearing size and lugworm addition. Near the end of the growth season, however, an additional disturbance (most likely waterfowl grazing and/or strong hydrodynamics) selectively impacted recolonizing seagrass in the largest (1 m^2) clearings (regardless of lugworm addition), and in those medium (0.25 m^2) clearings where lugworms had been added nearly five months earlier. Further analyses showed that the risk for the disturbance increased with hollow size, with a threshold of 0.24 m^2. Hollows of that size were caused by seagrass removal alone in the largest clearings, and by a weaker seagrass removal effect exacerbated by lugworm bioturbation in the medium clearings. Consequently, a sufficiently large disturbance increased the vulnerability of recolonizing seagrass to additional disturbance by weakening seagrass engineering effects (sediment stabilization). Meanwhile, the counteracting ecosystem engineering (lugworm bioturbation) reduced that threshold size. Therefore, scale-dependent interactions between habitat-mediated facilitation, competition and disturbance seem to maintain the spatial two-state mosaic in this ecosystem.

Editor: Tamara Natasha Romanuk, Dalhousie University, Canada

Funding: The study was financed by FORMAS (grant no. 2008-839), and by the ZKO project "Wadden engine: A human-driven regime shift through the loss of ecosystem engineers" from The Netherlands Organization for Research (NWO). The funders had no role in study design, data collection and analysis, decision to publish, or preparation of the manuscript.

Competing Interests: The authors have declared that no competing interests exist.

* E-mail: johane@ecology.su.se

¤a Current address: Department of Marine Ecology, Göteborg University, Göteborg, Sweden
¤b Current address: Department of Systems Ecology, Stockholm University, Stockholm, Sweden

Introduction

One of the most studied but also debated issues in ecology is the relative importance of factors affecting how organisms and ecosystems respond to disturbance [1,2,3,4]. One factor which may have a fundamental impact is the size of point disturbances; following a disturbance that exceeds a threshold size, local processes often change, recovery slows down, and communities may even develop into alternative stable states [5]. Such scale-dependent responses have typically been explained by weakened competitive exclusion from surrounding individuals, which increases the chance that previously inferior competitors can recruit into and dominate the center of disturbed areas [5,6,7]. For instance, ice-scour in hard bottom rockweed communities can trigger shifts to domination by mussels or fucoid macroalgae, if the disturbance is so large that the "whiplash" from surrounding rockweed cannot exclude competitors [8]. Importantly, such shifts are mediated by large disturbances, but ultimately depend on competition and space preemption.

Theory and observation suggests scale-dependent ecosystem shifts can also be caused by increased abiotic stress on recolonizing individuals, when the removal of an "ecosystem engineering" species simultaneously removes the self-facilitation required for recovery. Such disturbance responses should be most common in abiotically stressed environments, where communities depend on facilitation from ecosystem engineers. These are species that via presence and/or function change abiotic conditions, concentrate resources and/or alleviate local stress, which induces positive organism-environment feedback [9,10,11]. For instance, attenuation of waves and currents by submerged vegetation creates sheltered and calm microenvironments necessary for their own recruitment [12]. Importantly, such effects typically *exceed* engineers both spatially and temporally, i.e. extends beyond engineer patch edges and outlive engineers [13]. As a consequence, engineers may facilitate recolonization of conspecifics into areas impacted by small disturbances; a phenomenon known as short-range facilitation [14,15,16]. If, however, the disturbed area spatially exceeds the range of the facilitation, increased abiotic stress will prevent recovery; so called long-range inhibition [17]. One example is increased hydrodynamic stress within large gaps in canopy-forming submerged vegetation [18,19]. Due to lack of recovery, the impacted area can in theory be colonized by other species that tolerate or even benefit from the altered abiotic conditions. However, the ecosystem shift (from a state with the engineer to a state without the engineer) will occur even if competitors do not colonize, because increased stress is the factor preventing recovery.

Neither of these mechanism takes into account that those inferior competitors that may recruit into a disturbed area are often ecosystem engineers that modify abiotic conditions in other ways than the original community. Establishment of such engineers may actually counteract the spatial facilitation from un-impacted individuals in surrounding areas, and *increase* stress on recolonizing members of the original community. There are many examples of how contrasting engineering effects speeds up competitive exclusion of "counteracting" engineers via "habitat-mediated competition", which contributes to the formation of spatial mosaics of engineered patches [11,20]. For example, the "trophic group amensalism" hypothesis suggests that bioturbators locally suppress suspension-feeders by smothering them [21], and the "biomechanical warfare" hypothesis that antagonistic, counteracting engineers – e.g. sediment-destabilizing bioturbators and sediment-stabilizing plants – locally exclude each other partly by influencing abiotic conditions in ways that benefits conspecifics but impacts antagonists [22]. Because engineering effects (including habitat-mediated competition) exceed engineers spatially and temporally, we suggest establishment of counteracting engineers may in fact *reduce* the threshold disturbance size associated with increased risk for changes in recovery and shifts to alternative states increases. This is because counteracting engineering should *exacerbates* stress on recolonizing individuals, and thereby slow down recovery.

We tested these predictions using a field survey and a removal-addition experiment in the eastern Dutch Wadden Sea (Netherlands). The study area was a near-shore intertidal mud flat characterized by a striking mosaic of two alternating habitat types (Fig. 1A): low tide exposed humps (from here on *hummocks*) dominated by the sediment-stabilizing seagrass *Zostera noltii* Hornem (from here on *seagrass*), alternating with low tide water-logged pools or depressions (*hollows*) dominated by the sediment-destabilizing lugworm *Arenicola marina* L. (from here on *lugworms*). Seagrasses can facilitate themselves by stabilizing sediments and reducing erosion and turbidity through attenuation of water flow [12]. In contrast, lugworms are bioturbators that de-stabilize sediments through burrowing and feeding, which in combination with high hydrodynamic activity increases erosion of fine particles [23]. This typically leads to self-facilitation since lugworms prefer sandy substrates [11], but also to competitive exclusion of marine plants which prefer more muddy sediments [22,24].

After experimentally clearing seagrass and/or adding lugworms from areas ranging in size from 0 to 1 m^2, we predicted that (i) the strength of seagrass ecosystem engineering and rate of recovery *decreases* with increasing size of disturbance, (ii) lugworm establishment success *increases* with size of disturbance on seagrass due to weakened spatial influence from surrounding seagrass, and (iii) lugworm engineering in disturbed areas *reduces* the threshold disturbance size needed to slow down seagrass recovery by *exacerbating* erosion caused by the removal of seagrass.

Methods

Study setting

The study was conducted on the intertidal mudflats at Emmapolder (53°28′ 0 N, 6°45′ 0 E); one of a few areas in the Dutch Wadden Sea where *Zostera noltii* still occurs and has expanded [25]. Growth of this perennial seagrass starts in April, peaks in late summer, and ends in late autumn (October/November) with seasonal senescence [26].

Field survey

Seagrass (*Zostera noltii*) patches occurred primarily on hummocks, whereas lugworms occurred mainly in gaps (distinct unvegetated areas) which occurred primarily in waterlogged hollows (see Fig. 1A). Since the edges of the patches often extended into the sides of the hollows, gaps were typically somewhat smaller than hollows. To assess the spatiotemporal dynamics of this mosaic, a field survey was conducted in two 100×50 m sites (situated 400 m from the highest shoreline and 350 m apart) in June, August and October 2009. In June, we first measured the relative height of one pair of hummocks and hollows in seven random locations (total n = 14), using a Trimble Spectra Precision LL500 Laser Level (Trimble, California, USA). Measurements were calibrated against two fixed reference metal poles of known height. In all three months, we visually assessed seagrass ground cover to the nearest 5% (a reasonable proxy for shoot density; see Fig. S1), and counted lugworm fecal casts (a good proxy for lugworm density, see Fig. S2) in 0.25 m^2 frames randomly placed in either of the two habitat types; (i) hummocks and (ii) hollows (n = 20 per habitat type and site, resulting in 240 observations in total). Third, average size of gaps in the seagrass beds were assessed by noting the position of gap edges along 50 m transects randomly placed parallel to the shore (n = 3 transects per site).

Statistical analyses. Hypotheses were tested using statistical modeling in the R environment [27]. A one-sided t-test was used to show that the difference in height between hummocks and hollows was greater than zero, after checking assumptions of normality. Effects of Patch type and Time (the three months) on Arc-sin-transformed seagrass cover, and effects of Time on gap width, were assessed using mixed linear models fitted using restricted maximum likelihood (REML) in the nlme package for R. Effects on lugworm densities, which were Poisson-distributed, were investigated using generalized linear mixed modeling with a Poisson distribution and a log-link function (using the *lme4* package). "Time" was treated as fixed (because the levels were chosen to capture seasonality), and "Site" (two levels) was included as random offset, since (i) site effects were not of specific interest but (ii) should be incorporated to reduce within-treatment and reduce the risk for type 1-errors associated with the repeated measures. Test assumptions were first assessed (by inspecting quantile-quantile plots, using the Shapiro-Wilk test for normality, and the Bartlett test for homoscedasticity), and if necessary, data was transformed (square root or log). The full models were fitted

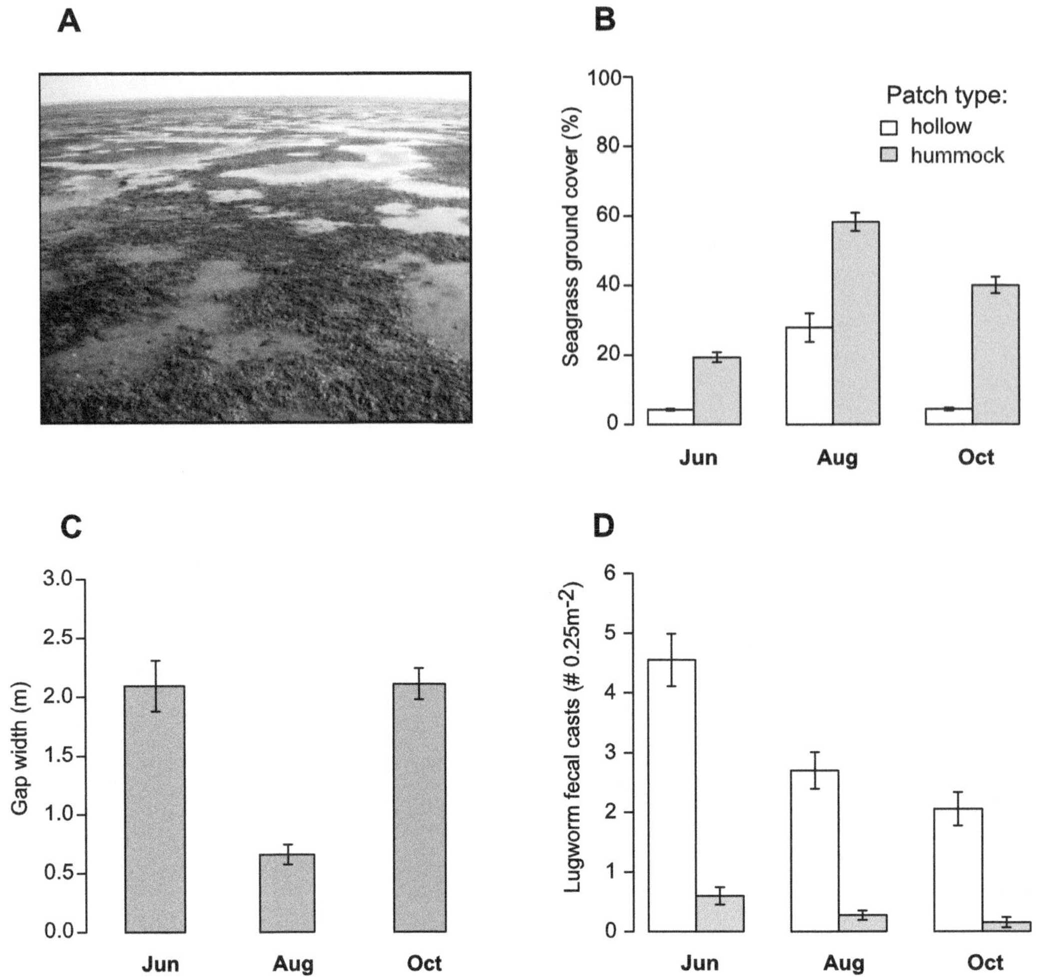

Figure 1. Field survey at the study sites. A) Photograph of site East in June, showing seagrass (*Zostera noltii* L.) patches growing on hummocks next to unvegetated hollows (Photo: Johan S Eklöf), (B) Seagrass ground cover (%) on hummocks and in hollows (n = 40 [20 per site]), (C) average gap width (in m, n = 6 [3 per site]), and (D) lugworm (*Arenicola marina* L.) fecal cast density (0.25 m^{-2}) on hummocks and in hollows (n = 40 [20 per site]), in June, August and October. Means ± 1S.E.

down to minimal adequate models using stepwise deletion of non-significant factors ($\alpha = 0.05$) followed by model fit comparisons using information theory (Aikaike's Information Criterion, or AIC scores).

Field experiment

A field experiment was conducted between June and November 2009 (the seagrass growth season), testing the relative and interactive effects of size of seagrass clearing (physical removal) and lugworm addition on ecosystem recovery. In each of the two sites (see Field survey), 24 quadratic plots were established within large (>5 m wide) seagrass patches. Plots were placed in a row parallel to the shore, >2 m apart and >2 m from natural gaps to avoid edge effects. Experimental quadratic clearings in plots were conducted in four sizes, mimicking natural disturbances (e.g. waterfowl grazing or uprooting caused by hydrodynamic stress): (i) "Control" (1 m^2 intact plots), (ii) "Small" (0.0625 m^2 clearing), (iii) "Medium" (0.25 m^2 clearing), and (iv) "Large" (1 m^2 clearing). To experimentally remove only seagrasses (and not the sediments they stabilized), plywood frames were hammered 20 cm deep along plot edges during low water. Next, the frame was filled with water and all seagrass shoots, roots and rhizomes were removed using a hand rake. After allowing suspended particles to settle, the water was slowly released and the frame gently removed. Lugworm (mean g. DW±SE = 0.43±0.05, n = 25) were added on half of the plots (lugworm addition) in densities corresponding to 32 ind. per m^{-2}. This relatively high density was intentional, as lugworm densities quickly self-regulate to local carrying capacity due to interspecific competition [28]. All treatment combinations were replicated three times per site.

Ecosystem engineering. First, we tested how Clearing size affected seagrass water flow attenuation (after three weeks, to minimize confounding effects of recolonization). This was assessed as % weight loss of plaster dissolution balls; a reliable method when – as here – used in similar water-flow environments [29]. Dissolution balls were molded of model plaster (Knauf Modelgips, Knauf B.V., Utrech, Netherlands) in plastic cups around 20 cm long galvanized steel nails, dried, weighed (mean g. DW±SE = 80.96±2.53, n = 24), and placed in the center of plots for three tidal cycles (40 hrs). No balls were placed in lugworm addition plots, to exclude potential disturbance from nails on added worms. The balls were then retrieved and dried until constant weight. Second, the effects of Clearing size and Lugworm addition on sediment erosion were assessed as the relative gap depth (the difference in height between the center and the mean of two sides of each plot after four and ten weeks), using the Trimble Laser Level; (see Methods: Field survey).

Lugworm density and seagrass recolonization rate. Lug worm fecal casts were counted in plots at the start of each month (six times) to assess if (i) the lugworm addition treatment remained over time (high vs. low densities), (ii) seagrass Clearing size, rather than seagrass removal *per se*, facilitated lugworm establishment (see Introduction), and (iii) seagrass presence explained low lugworm densities on natural hummocks (see Results, Field survey). To account for the differences in plot sizes, counts were standardized as no. of casts 0.25 m^{-2}; the area sampled in the Field survey. Seagrass recolonization occurred primarily as lateral rhizomal expansion from the edges into gaps (no seedlings were found). Therefore, *relative* rates of recolonization will decrease with clearing size even if the absolute rhizome expansion rate is constant, simply because the distance from the edge to the center increases. However, our "stress-induced shifts" model predicted that *absolute recovery rates* should decrease with increasing clearing size due to weaker engineering effects (see Introduction). We therefore estimated absolute recovery as the average rate at which the quadratic gaps contracted during the foregoing month (in mm day^{-1}). This resulted in five monthly estimations of daily gap contraction rates (from start of June until start of November). During days with low water and clear skies, plots were photographed (with a Konica Minolta Dimage ×50 at 5MP resolution) at a 90° angle from 1.5 m height (on the same dates as lugworm counts). The sizes of gaps were estimated to the closest 0.0001 m^2 using the freeware image analysis program ImageJ [30], by encircling the gap (between the bases of the most inward extending shoots) and measuring the size using a reference marker of known size. We then calculated the "gap contraction rate" (*gcr*, in mm day^{-1}) over the foregoing month as:

$$gcr = \frac{\sqrt{(x_{t1})} - \sqrt{(x_{t2})}}{2\Delta t}$$

where x_{t1} and x_{t2} are gap sizes at start and finish of that month, respectively (in mm^2), and Δt is the number of days between start (t1) and finish (t2). We also calculated the seagrass relative recovery at the start of each month as "percent of the gap size at start" (in June).

Data analyses. Treatment effects were explored using linear mixed effects modeling (see above) with the fixed factors Clearing size, Lugworm addition and Time (after the start of the experiment, where applicable), and their respective interactions, included in full models. "Plot" (n = 48) was included as a random factor, and effects of first order autoregressive and compound symmetry correlation structures were compared to account for effects of repeated measures [31]. "Site" was not included because (i) site differences were not of particular interest, (ii) the low number of levels makes it impossible to assess variation between sites [32], and (iii) potential effects of site differences on variation were accounted for by using individual plots as random offset.

Effects of Clearing size and Lugworm addition on flow-induced weight loss of dissolution balls were tested with a linear mixed effects model with "Site" (two levels) as a random factor (see Methods: Field survey). Effects of Clearing size, Lugworm addition and Time (repeated measure; June and August) on relative gap depth (caused by sediment erosion) were investigated using a mixed-effects model (see Methods: Field survey).

Fecal cast densities (a good proxy for lugworm density) were highly non-normal and zero-inflated. We therefore modeled effects of Clearing size, Lugworm addition, Site and Time using a generalized linear mixed model with a Poisson distribution, using a Bayesian approach. The initial density was modeled as a function of site, clearing size and lugworm addition, with all interactions. Subsequent densities were modeled with additional terms for months, and an autoregressive term for the previous lugworm abundance (i.e. the expected cast density), which varied by month. The time effect and autoregressive terms were given Normal priors with mean 0, and variance 10. The main effects, and first and second order interactions were modeled as random effects, with a Jeffreys prior on the variance (i.e. a uniform prior on the log variance): if interactions are small, this shrinks their effects towards zero [33]: the Jeffrey's prior gives more weight to smaller variances, and these in turn give more weight to small parameter estimates, i.e. estimates that are close to zero. Variances due to the effects were calculated from the distributions of the estimated parameters, rather than from the parameters of the model, as the low number of levels of the factors made the variance estimates uncertain. The fitting was carried out by MCMC in OpenBUGS3 1.2 [34]. Five chains were run, and after a burn-in of 150.000 iterations, a further 150.000 iterations were run, and the chains were thinned to every fifth iteration. Convergence and mixing were judged by eye and with plots of the modified Gelman-Rubin statistic [35].

Finally, effects on seagrass recovery (daily gap contraction rate) were evaluated by fitting a linear mixed effects model with "site" as a random offset, after excluding no clearing (Control) plots (which had no gaps, resulting in a mean and variance = 0).

Identifying factors(s) predicting the risk for sudden seagrass disturbance

To our surprise, gaps in three treatment combinations expanded on average instead of contracting during October; in Large clearings with and without lugworm addition, and Medium-sized clearings with lugworm addition (see Results). This effect was due to an additional disturbance on the seagrass that had recolonized plots subjected to these initial treatment combinations. Meanwhile, the recolonizing seagrass in other treatment plots were not impacted, and the gaps closed up completely. Even though this unanticipated pattern of additional disturbance was ultimately caused by an interaction between Clearing size and Lugworm addition (see Results), we wanted to in greater detail understand what factor(s) or conditions triggered this disturbance (e.g. the engineers or their effects), and thereby try to identify the additional disturbance. To identify *what local conditions* triggered the additional disturbance, we gathered plot-specific data on five potentially important variables serendipitously collected just before the additional disturbance (early October). The variables were (i) *seagrass gap size* and (ii) *lugworm fecal cast density* (reflecting the two treatments); (iii) *seagrass ground cover* (%) in recolonized areas, which may influence self-facilitation (cover estimated from the plot photographs using standard digitizing [see 36]), (iv) the *elevation* of the area surrounding each plot, (measured in August, see above), which may influence exposure, and (v) the *size of hollows* (water-logged low-tide depressions) encircling the gaps created by the seagrass clearing; a measure of the spatial extent of sediment erosion (estimated from plot photographs, see Methods; Field experiment) (For a summary of all data, see Fig. S3). These variables were then used as predictors in a multiple logistic regression model with "risk for disturbance" as response variable (where 0 = no gap expansion, and 1 = gap expansion, n = 48). "Site" (two levels) was included as a random offset to reduce the influence of natural variation (using the *lmer* function in the lme4 package for R). As "gap size" and "hollow size" were highly correlated (Pearson rho = 0.88), we removed "gap size" (regarded less important because the disturbance occurred outside the gaps but in the hollows). The minimal adequate model was identified as described above.

Results

Field survey

In June, hummocks were 5.81±0.5 cm (mean ± SE) higher than hollows (one-sided t-test; $t_6 = 12.41$, P<0.001). Seagrass ground cover (Fig. 1B) was at least three times higher on hummocks than in hollows (t = 14.68, P<0.001), increased fourfold between June and August both on hummocks and in hollows, and declined between August and October, but more so in hollows than on hummocks (Habitat×Month interaction; t = 3.7, P = 0.003). Due to the lateral expansion of seagrass into hollows, gaps in the seagrass beds shrunk in width from 2 to 0.6 m between June and August. However, between August and November, gaps expanded back to 2 m in size (t = 7.67, P<0.001). Finally, lugworm cast densities displayed an inverse spatial pattern compared to seagrass cover (Fig. 1D); they were on average eight times higher in hollows than on hummocks (Habitat effect: z = −13.4, P<0.001) and declined throughout the survey (linear Time effect: z = −6.78, P<0.001).

Experiment

Water flow attenuation (measured as % plaster dissolution) and sediment erosion (as relative gap height) indicated that seagrass ecosystem engineering strength in the center of plots depended on the *size* of seagrass clearing, rather than clearing *per se*. Relative weight loss of dissolution balls (Fig. 2) was higher in Large gaps than all other sizes, including no-clearing controls (t = 2.5, P = 0.02). Changes in relative gap depth over time (Fig. 3A) also depended on Clearing size (Clearing size×Time interaction, t = −8.01, P<0.001). The elevation of Control plots did not change between June and August, whereas gaps in clearing plots deepened – especially those in Medium and Large gaps. There was also a trend to an overall deepening of gaps in Lugworm addition plots (t = −1.8, P = 0.079).

Lugworm fecal cast densities were mainly affected by the main effects of Clearing size (explaining 55% of variation), Lugworm addition (20%) and Time (15%); cast densities increased with the size of initial clearing and with lugworm addition, but decreased

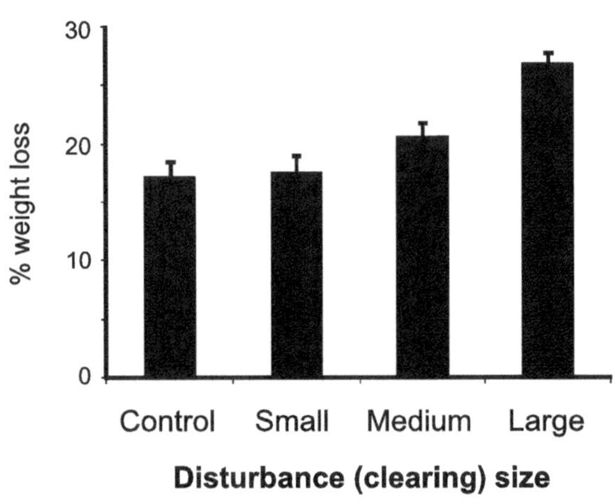

Figure 2. Effect of Clearing size (four levels) on seagrass water flow attenuation (% weight loss of plaster dissolution balls). Means ± 1S.E (n = 6).

over time in all treatments (Fig. 3B). The three-way interaction accounted for less than 1% of the total variation (posterior mode 0.5%, 95% highest posterior density interval: 0.03%–3.2%), and the two-way interactions accounted for about 10% of the variation between them (mostly split equally between the interactions involving Clearing size). More worms were found in the largest clearings, 3.7 times more than in the control plot (95% highest posterior density interval: 2.9–5.0). Site had no discernible effect.

The absolute rate of seagrass recovery into the gaps (Fig. 3C) depended on a complex interaction between Clearing size, Lugworm addition and Time (t = −2.72. P = 0.0073). This interaction largely consisted of three effects, all unexpected. First, seagrass recovery rate was between June and September either unaffected by Clearing size, or *higher* in Medium and Large than Small clearings (the opposite to the predicted effect, see hypothesis 1). This was explained by a much faster *relative* recovery in Small than Medium and Large plots in June and July (caused by their smaller size, see Fig. 3D), resulting in that recolonizing seagrass became space-limited in Small, but not in Medium and Large plots, during optimal growth conditions in late July and August (Fig. 3D). Second, lugworm addition did not slow down recovery; instead, lugworms appeared to be outcompeted by the seagrass (Fig. 3B). Third and final; during the last month of the experiment (October), the prevailing recovery trajectories changed drastically in ways that indirectly confirmed our third hypothesis. In Small plots, and in Medium plots without added lugworms, seagrass continued recovering and gaps contracted completely (disappeared). However, in those Medium plots where lugworms had been added five months earlier, the recolonizing seagrasses had disappeared in half of the plots. As a consequence, gaps in this treatment expanded (from an average of 0.0063 to 0.12 m^2), and absolute recovery rates shifted from positive to negative (Fig. 3C). In Large clearings, recolonizing seagrass were similarly lost in 11 of the 12 plots (92%), regardless of lugworm addition. Consequently, these gaps also expanded (from 0.12 to 0.51 m^2), and recovery rates shifted from positive to negative (Fig. 3C).

Factors predicting the risk for sudden seagrass disturbance

The risk for this additional disturbance on the recolonizing seagrass was predicted by hollow size (z = 3.0011, p = 0.0018, Fig. 4B), and was unaffected by seagrass cover, lugworm density and elevation (z = −0.59–0.34, p>0.55 for all factors). The threshold hollow size (where disturbance risk = 50%) was ~0.24 m^2 (see horizontal line in Fig. 4B). Based on the identification of hollow size as the sole predictor, we assessed how Clearing size and Lugworm addition in June had influenced this variable in early October, by fitting a linear mixed model with "Plot" as random factor, and excluding Control plots (which had no hollows, n = 36). The analysis showed that the size of the hollows was a simultaneous effect of Clearing size (t = 2.92, P = 0.00629) and *Arenicola* addition (t = 2.067, P = 0.046). Hollows in Large plots were 19 and 5 times larger than those in Small and Medium plots, respectively, and Lugworm addition increased hollow size by 45% across clearing sizes (Fig. 4C). Consequently, the risk for additional disturbance and expansion of gaps during October largely depended on the combined scale-dependent effects of seagrass removal and lugworm addition on surface sediment stability.

Discussion

Spatial mosaics of patches dominated by ecosystem engineers are common in ecosystems in general, and in inter- and sub-tidal

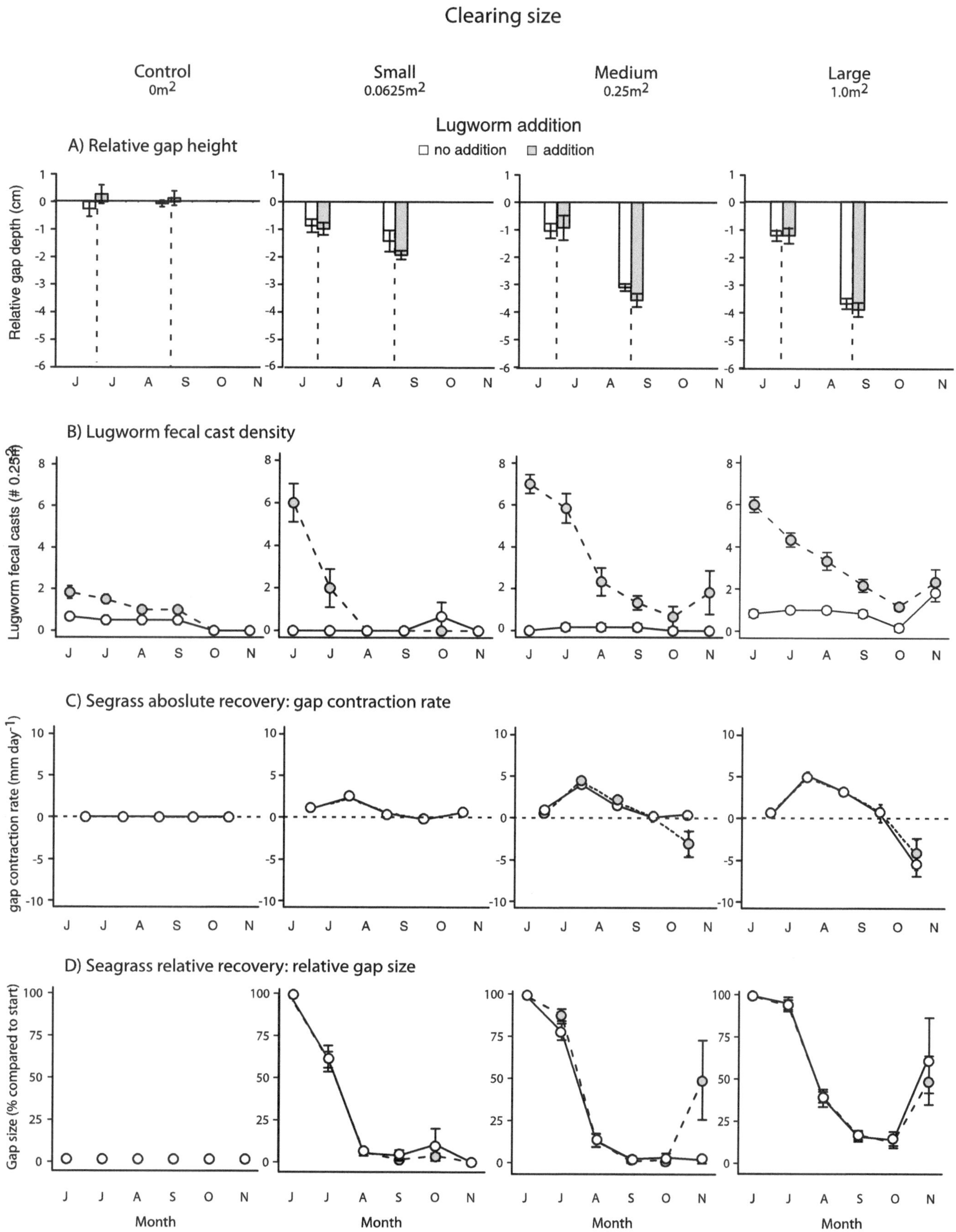

Figure 3. Effects of Clearing size (four levels), Lugworm addition (two levels, white and grey bars/points) and Time (months; Jun–Nov) on four variables reflecting recovery of seagrass ecosystems. Effects on (A) relative gap depth (mm), (B) lugworm (*Arenicola marina*) fecal cast density (0.25 m^{-2}), (C) absolute seagrass recovery rate (mm day^{-1}), and (D) relative seagrass gap size (% of that at start in June). Means ± 1S.E (n = 6).

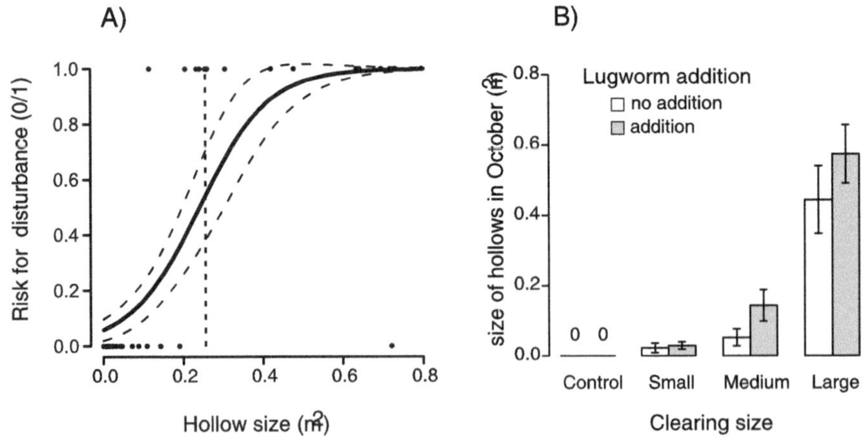

Figure 4. Clearing size and Lugworm addition influences the risk for additional disturbance by influencing hollow size. A) Effect of hollow size (m^2) in early October on the risk for disturbance on recolonizing seagrass during the following month (0–1). The lines show the predicted effect ± 95% CI (n = 48). The horizontal line depicts the threshold hollow size (0.24 m^2) where the risk is 50%, (B) Effect of Clearing size and Lugworm addition on October hollow size (m^2, n = 6). Means ± 1S.E.

marine systems in particular [22,37,38,39,40,41]. Such spatial patterning may be explained by local self-facilitation or by simple space preemption [24], but also by habitat-mediated competition where competitors exclude each other by exacerbating local abiotic stress on their competitors [17].

Here, we show that habitat-mediated self-facilitation and habitat-mediated competition may interactively result in scale-dependent ecosystem responses to disturbance. In the largest seagrass clearings (1 m^2), local environmental conditions shifted from sediment stabilizing to de-stabilizing over the course of the experiment regardless of lugworm addition, most likely because of weakened (i) water flow attenuation and (ii) sediment stabilization (the latter caused by a slower relative recovery rate). This essentially caused the formerly seagrass-dominated low-tide hummocks to switch into low-tide hollows during the autumn. However, this scale-dependent response was not due to simple slowing down of recovery due to lack of facilitation [e.g. 42] or space preemption by competitors [e.g. 8], but greatly increased risk for additional disturbance on recolonizing seagrass. Likewise, lugworm addition did not slow down seagrass recolonization, but lugworm surface sediment destabilization increased the size of hollows surrounding gaps. This was, in turn, the sole factor influencing the risk that recovering seagrass were disturbed again during autumn. As a consequence, the counteracting lugworm engineering decreased the threshold experimental clearing size associated with the additional disturbance from 1 to 0.25 m^2. Consequently, counteracting engineering reduced the threshold size of disturbance necessary to cause shifts in ecosystem as predicted, but via a mechanism that was largely unexpected.

The scale-dependent seagrass recolonization was – in contrast to our expectations – not caused by slower recolonization, but by increased risk for disturbance on recolonizing seagrasses during autumn. Since the study was not designed to identify the additional disturbance or the underlying cause(s) to it, we can only speculate on which was the disturbance. Visual observations in the field strongly suggested that waterfowl grazing – a well-known seasonal disturbance on *Z. noltii* in the Wadden Sea [26,43,44,45] – was the most likely candidate. Flocks of 150–200 Brent geese (*Branta bernicla*) and Wigeon (*Anas penelope*) were observed feeding daily on *Zostera noltii* from mid September to November (van der Heide et al., in preparation). Importantly, the grazing was spatially restricted to seagrasses in hollows, while seagrass-dominated hummocks covered 50% of the area (van der Heide et al. in preparation). Such hollow-specific waterfowl grazing has been observed elsewhere, and is explained by the low-tide standing water in hollows, which facilitates feeding while reducing ingestion of silt [46]. Moreover, geese typically cease feeding if food presence and/or accessibility is too low ("giving-up density"), which may happen when seagrass still remain [47]. Combined with our results, this suggests the disturbance on the recolonizing seagrass was waterfowl grazing in hollows of a size that harbored enough accessible food. Similar spatial thresholds in grazing have been demonstrated in terrestrial grasslands, and can sustain repeatedly grazed patches of short vegetation next to patches of high, ungrazed vegetation [48]. Until these suspected effects of waterfowl grazing have been thoroughly tested with experiments [see e.g. 43], we do not exclude the possibility that other disturbances (like uprooting due to increased hydrodynamics, see Fig. 2) contributed to or caused the seagrass loss. Regardless of which disturbance caused the seagrass loss, hollow size alone predicted the risk for sudden seagrass disturbance (Fig. 4A). Meanwhile, large enough hollows were caused by the combined effect of large disturbance and lugworm addition (Fig. 4B). Combined with the fact that disturbances on seagrass during autumn are already known to reduce local sediment stabilization in the end of the following growth season [43], this indicates the existence of a positive feedback between disturbance on seagrass and low sediment stability. If strong enough, such a feedback could trigger the formation of two alternative and potentially stable states; (i) undisturbed hummocks with high sediment stability, and (ii) repeatedly disturbed hollows with low sediment stability (Fig. 5). Shifts between the states should – based on our results – be mediated by factors influencing sediment stability; the size of disturbance and the bioturbation by lugworms. However, they may also be influenced by the overall risk for additional disturbance (e.g. geese density or hydrodynamic conditions), the erodability of sediments (e.g. silt content), and the strength of seagrass engineering (e.g. related to shoot density and height). The hypothesis of two alternative states is, moreover, supported by the pervasive spatial hummock-hollow mosaic in our study area (Fig. 1A–D), which is very similar to the bimodal state distribution typical for feedback-driven systems [see e.g. 17,22]. However, to prove that the states are truly *persistent* [49], it must be demonstrated that the risk for additional disturbance not only

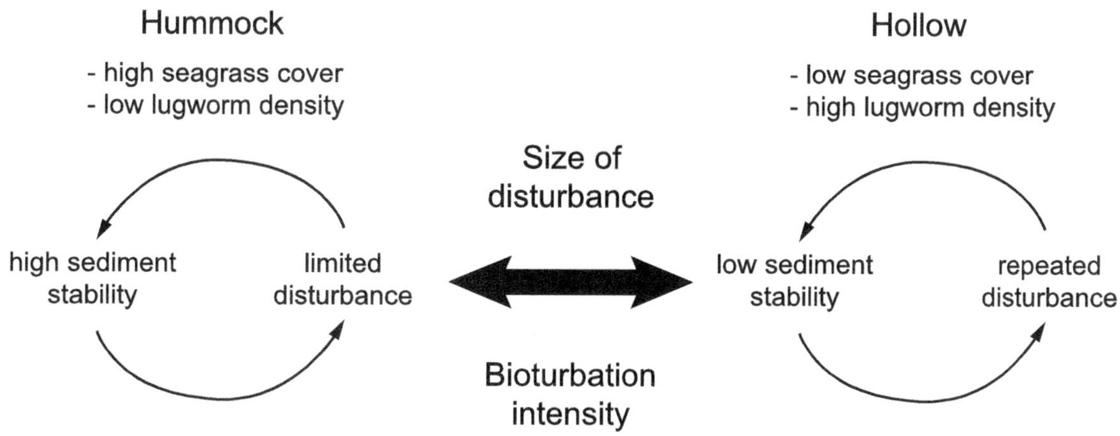

Figure 5. Conceptual model of how seagrass-dominated hummocks and lugworm-dominated hollows could constitute two alternative and potentially stable states in intertidal soft-bottom areas. Local seagrass cover and lugworm density interactively affects sediment net stability, which in turn determines the risk for disturbance on seagrass. The effects of disturbance feeds back positively on local sediment stabilization the following season, so that hollows disturbed one year has an increased risk of being disturbed again the following year. Shifts between the states should be mediated primarily by factors influencing local sediment stability; the size of disturbance on seagrass and lugworm bioturbation intensity, but also by risk for disturbance (e.g. geese grazing intensity or hydrodynamic stress), sediment relative erodability, and seagrass engineering strength.

depends on hollow size, but that disturbed hollows are much more prone to additional disturbance the following year.

Habitat-mediated competition can alone maintain spatial mosaics if counteracting engineering is strong enough to exclude competitors [11,20,22,50]. In our study area, lugworms were too weak of a competitor to exclude seagrasses. This unexpected result contrasts with documented seagrass exclusion by lugworms in nearby areas [24], and suggests the relative engineering strength of lugworms – and thereby the outcome of habitat-mediated competition – is conditional i.e. depends on other factors (e.g. sediment conditions and hydrodynamic conditions). This may explain the sometimes conflicting evidence for competition between bioturbators and intertidal plants around the world [22,37,38,39,40,41]. However, even though lugworms were more or less outcompeted by the end of the summer in our study, their destabilization of sediments increased the size of hollows, which in medium-sized clearings exacerbated the risk for additional disturbance. This demonstrates that bioturbation can exacerbate scale-dependent responses, in the same way as loss of sediment-stabilizing engineers has been shown to slow recovery [42]. It also emphasizes that such engineering effects may locally outlive engineers [13]. Moreover, lugworm densities in fact increased during the last month in plots where the recolonizing seagrasses were disturbed (see Fig. 3B, Medium and Large hollows during October). This suggests that removal of a superior competitor (seagrass) facilitated the starting point of population turnover; another prerequisite for persistence of an alternative state [49].

In conclusion, our study supports the notion of biota as an important factor for disturbance responses and as drivers of ecosystem heterogeneity [5,17,51]. The results illustrate that habitat-mediated self-facilitation, counteracting ecosystem engineering and physical disturbances should not be viewed in isolation, because they may interactively explain ecosystem trajectories following disturbance, and, over time, ecosystem spatial heterogeneity. Even though this study appears to be the first demonstrating the potential importance of such interactions, we believe they are a common in engineered systems [see e.g. 22]. Finally, from a local Wadden Sea perspective, the results strengthen the hypothesis [51] that recent shifts from a historical dominance of sediment-stabilizing engineering species like seagrass and reef-forming mussels, to current dominance by bioturbators such as lugworms is caused by species-specific disturbances (e.g. disease, eutrophication, overgrazing, harvest, etc.), but maintained by sediment-mediated positive feedback interactions.

Supporting Information

Figure S1 Relationship between *Zostera noltii* ground cover and seagrass shoot density, above-ground biomass and total biomass. To test how accurately estimations of *Zostera noltii* ground cover predicted *Z. noltii* shoot density and biomass, we in the field placed a 0.25 m^2 frame in random points ranging in seagrass cover from 5 to 90% ($n = 6$), photographed the frame (see the article for details), and excavated all seagrass material in the center of the frame using a 10.5 cm wide corer (to 20 cm depth). Seagrass ground cover in each frame was estimated with the same digital photography analysis as in the experiment (see article and Stewart et al. 2007). In the laboratory, we counted all shoots, separated and cleaned the above- and below-ground material in seawater, and dried both fractions until constant dry weight (at 60°C). Results from multiple regressions showed that estimated cover was a reasonable predictor of shoot density ($F_{1,5} = 14.4$, $P = 0.012$, $R^2 = 0.69$), above-ground biomass ($F_{1,5} = 21.3$, $P = 0.0056$, $R^2 = 0.77$) and total biomass ($F_{1,5} = 20.47$, $P = 0.0063$, $R^2 = 0.76$).

Figure S2 Relationship between *Arenicola marina* density and lugworm fecal cast density. Counts of *Arenicola marina* fecal casts is a commonly used and rapid method to assess worm densities (see e.g. van Weseenbeck et al. 2007 in article reference list). To ensure the reliability of the method in our study area, we compared counts of fecal casts in randomly placed 0.25 m^2 frames (both within and outside of *Zostera noltii* patches) with actual densities of worms in the sediment, by excavating all worms to 50 cm depth within the frame ($n = 10$). A linear regression analysis showed that the number of fecal casts accurately predicted worm densities ($F_{1,8} = 67.00$, $P < 0.0001$, $R^2 = 0.88$).

Figure S3 A–D. Five predictor variables used in multiple logistic regression analysis, as effects of Clearing and Lugworm addition (across two sites, means ± 1SE, n = 6).

Acknowledgments

We thank T. Bouma, M. van Katwijk, H. Olff, T. Piersma, and J. van de Koppel for discussions throughout the study, and M. Logan, P. Lavery, J. Stachowicz, P. Petraitis and two anonymous referees for advice that greatly improved the paper.

Author Contributions

Conceived and designed the experiments: JSE TvdH BKE. Performed the experiments: JSE TvdH SD EMvdZ BKE. Analyzed the data: JSE RO. Wrote the paper: JSE TvdH BKE.

References

1. Sousa WP (1984) The Role of Disturbance in Natural Communities. Annual Review of Ecology and Systematics 15: 353–391.
2. Holling CS (1973) Resilience and the stability of ecological systems. Annual Review of Ecology and Systematics 4: 1–23.
3. Folke C, Carpenter S, Walker B, Scheffer M, Elmqvist T, et al. (2004) Regime shifts, resilience, and biodiversity in ecosystem management. Annual Review of Ecology Evolution and Systematics 35: 557–581.
4. Clements FE (1916) Plant succession; an analysis of the development of vegetation. Washington: Carnegie Institution of Washington. 512 p.
5. Petraitis PS, Latham RE (1999) The importance of scale in testing the origins of alternative community states. Ecology 80: 429–442.
6. Wernberg T (2006) Scale of impact determines early post-disturbance assemblage structure in subtidal Fucus beds in the Baltic Sea (Bornholm, Denmark). European Journal of Phycology 41: 105–113.
7. Norkko A, Rosenberg R, Thrush SF, Whitlatch RB (2006) Scale- and intensity-dependent disturbance determines the magnitude of opportunistic response. Journal of Experimental Marine Biology and Ecology 330: 195–207.
8. Petraitis PS, Methratta ET, Rhile EC, Vidargas NA, Dudgeon SR (2009) Experimental confirmation of multiple community states in a marine ecosystem. Oecologia 161: 139–148.
9. Jones CG, Lawton JH, Shachak M (1994) Organisms as ecosystem engineers. Oikos 69: 373–386.
10. Jones CG, Lawton JH, Shachak M (1997) Positive and negative effects of organisms as physical ecosystem engineers. Ecology 78: 1946–1957.
11. Bouma TJ, Olenin S, Reise K, Ysebaert T (2009) Ecosystem engineering and biodiversity in coastal sediments: posing hypotheses. Helgoland Marine Research 63: 95–106.
12. van der Heide T, van Nes EH, Geerling GW, Smolders AJP, Bouma TJ, et al. (2007) Positive feedbacks in seagrass ecosystems: implications for success in conservation and restoration. Ecosystems 10: 1311–1322.
13. Hastings A, Byers JE, Crooks JA, Cuddington K, Jones CG, et al. (2007) Ecosystem engineering in space and time. Ecology Letters 10: 153–164.
14. Alvarado JL, Pinto R, Marquet P, Pacheco C, Guinez R, et al. (2001) Patch recolonization by the tunicate Pyura praeputialis in the rocky intertidal of the Bay of Antofagasta, Chile: evidence for self-facilitation mechanisms. Marine Ecology-Progress Series 224: 93–101.
15. Goldenheim W, Irving A, Bertness M (2008) Switching from negative to positive density-dependence among populations of a cobble beach plant. Oecologia 158: 473–483.
16. Bustamante-Sanchez MA, Armesto JJ, Halpern CB (2011) Biotic and abiotic controls on tree colonization in three early successional communities of Chiloe Island, Chile. Journal of Ecology 99: 288–299.
17. Rietkerk M, Van de Koppel J (2008) Regular pattern formation in real ecosystems. Trends in Ecology & Evolution 23: 169–175.
18. Maltese A, Cox E, Folkard AM, Ciraolo G, La Loggia G, et al. (2007) Laboratory measurements of flow and turbulence in discontinuous distributions of ligulate seagrass. Journal of Hydraulic Engineering-Asce 133: 750–760.
19. Folkard AM (2005) Hydrodynamics of model Posidonia oceanica patches in shallow water. Limnology and Oceanography 50: 1592–1600.
20. Reise K (2002) Sediment mediated species interactions in coastal waters. Journal of Sea Research 48: 127–141.
21. Rhoads DC, Young DK (1970) The influence of deposit-feeding organisms on sediment stability and community trophic structure. Journal of Marine Research 28: 150–178.
22. van Wesenbeeck BK, van de Koppel J, Herman PMJ, Bakker JP, Bouma TJ (2007) Biomechanical warfare in ecology; negative interactions between species by habitat modification. Oikos 116: 742–750.
23. Montserrat F, Suykerbuyk W, Al-Busaidi R, Bouma TJ, van der Wal D, et al. (2011) Effects of mud sedimentation on lugworm ecosystem engineering. Journal of Sea Research 65: 170–181.
24. Reise K (1983) Experimental removal of lugworms from marine sand affects small zoobenthos. Marine Biology 74: 327–332.
25. Philippart CJM, Dijkema KS (1995) Wax and Wane of *Zostera noltii* Hornem in the Dutch Wadden Sea. Aquatic Botany 49: 255–268.
26. Vermaat JE, Verhagen FCA (1996) Seasonal variation in the intertidal seagrass *Zostera noltii* Hornem: Coupling demographic and physiological patterns. Aquatic Botany 52: 259–281.
27. CRAN (2010) R: A Language and Environment for Statistical computing. Vienna, Austria: R Foundation for Statistical computing.
28. Flach EC (1992) Disturbance of Benthic Infauna by Sediment-Reworking Activities of the Lugworm Arenicola-Marina. Netherlands Journal of Sea Research 30: 81–89.
29. Porter ET, Sanford LP, Suttles SE (2000) Gypsum dissolution is not a universal integrator of 'water motion'. Limnology and Oceanography 45: 145–158.
30. Abramoff MD, Magelhaes PJ, Ram SJ (2004) Image Processing with ImageJ. Biophotonics International 11: 36–42.
31. Logan M (2010) Biostatistical design and analysis using R: a practical guide Wiley-Blackwell. 546 p.
32. Bolker BM, Brooks ME, Clark CJ, Geange SW, Poulsen JR, et al. (2009) Generalized linear mixed models: a practical guide for ecology and evolution. Trends in Ecology & Evolution 24: 127–135.
33. O'Hara RB, Sillanpää MS (2009) A Review of Bayesian Variable Selection Methods: What, How and Which. Bayesian Analysis 4: 85–118.
34. Thomas A, O'Hara RB, Ligges U, Sturtz S (2006) Making BUGS Open. R News 6: 12–17.
35. Brooks SP, Gelman A (1998) Alternative methods for monitoring convergence of iterative simulations. Journal of Computational and Graphical Statistics 7: 434–455.
36. Stewart AM, Edmisten KL, Wells R, Collins GD (2007) Measuring canopy coverage with digital imaging. Communications in Soil Science and Plant Analysis 38: 895–902.
37. Suchanec TH (1983) Control of seagrass communities and sediment distribution by *Callianassa* (Crustacea, Thalassinidea) bioturbation. Journal of Marine Research 41: 281–298.
38. Kuhnert J, Veit-Kohler G, Buntzow M, Volkenborn N (2010) Sediment-mediated effects of lugworms on intertidal meiofauna. Journal of Experimental Marine Biology and Ecology 387: 36–43.
39. Philippart CJM (1994) Interactions between *Arenicola marina* and *Zostera noltii* on a tidal flat in the Wadden Sea. Marine Ecology-Progress Series 111: 251–257.
40. Berkenbusch K, Rowden AA (2003) Ecosystem engineering - moving away from 'just-so' stories. New Zealand Journal of Ecology 27: 67–73.
41. Siebert T, Branch GM (2007) Influences of biological interactions on community structure within seagrass beds and sandprawn-dominated sandflats. Journal of Experimental Marine Biology and Ecology 340: 11–24.
42. Thrush SF, Whitlatch RB, Pridmore RD, Hewitt JE, Cummings VJ, et al. (1996) Scale-dependent recolonization: The role of sediment stability in a dynamic sandflat habitat. Ecology 77: 2472–2487.
43. Nacken N, Reise K (2000) Effects of herbivorous birds on intertidal seagrass beds in the northern Wadden Sea. Helgoland Marine Research 54: 87–94.
44. Ganter B (2000) Seagrass (Zostera spp.) as food for brent geese (Branta bernicla): an overview. Helgoland Marine Research 54: 63–70.
45. Jacobs RPWM, Den Hartog C, Braster BF, Carriere FC (1981) Grazing of seagrass *Zostera noltii* by birds at Terschelling (Dutch Wadden Sea). Aquatic Botany 10: 241–259.
46. Fox AD (1996) *Zostera* exploitation by Brent geese and wigeon on the Exe estuary, southern England. Bird Study 43: 257–268.
47. Tinkler E, Montgomery WI, Elwood RW (2009) Foraging ecology, fluctuating food availability and energetics of wintering brent geese. Journal of Zoology 278: 313–323.
48. Cromsigt J, Olff H (2008) Dynamics of grazing lawn formation: an experimental test of the role of scale-dependent processes. Oikos 117: 1444–1452.
49. Connell JH, Sousa WP (1983) On the evidence needed to judge ecological stability or persistence. American Naturalist 121: 789–824.
50. Woodin SA (1978) Refuges, disturbance, and community structure - a marine soft-bottom example. Ecology 59: 274–284.
51. Eriksson BK, van der Heide T, van de Koppel J, Piersma T, van der Veer HW, et al. (2010) Major changes in the ecology of the Wadden Sea: human impacts, ecosystem engineering and sediment dynamics. Ecosystems 13: 752–764.

Habitat-Mediated Variation in the Importance of Ecosystem Engineers for Secondary Cavity Nesters in a Nest Web

Hugo Robles[1*¤], **Kathy Martin**[1,2]

1 Department of Forest and Conservation Sciences, Centre for Applied Conservation Research, University of British Columbia, Vancouver, British Columbia, Canada, **2** Pacific Wildlife Research Centre, Environment Canada, Delta, British Columbia, Canada

Abstract

Through physical state changes in biotic or abiotic materials, ecosystem engineers modulate resource availability to other organisms and are major drivers of evolutionary and ecological dynamics. Understanding whether and how ecosystem engineers are interchangeable for resource users in different habitats is a largely neglected topic in ecosystem engineering research that can improve our understanding of the structure of communities. We addressed this issue in a cavity-nest web (1999–2011). In aspen groves, the presence of mountain bluebird (*Sialia currucoides*) and tree swallow (*Tachycineta bicolour*) nests was positively related to the density of cavities supplied by northern flickers (*Colaptes auratus*), which provided the most abundant cavities (1.61 cavities/ha). Flickers in aspen groves provided numerous nesting cavities to bluebirds (66%) and swallows (46%), despite previous research showing that flicker cavities are avoided by swallows. In continuous mixed forests, however, the presence of nesting swallows was mainly related to cavity density of red-naped sapsuckers (*Sphyrapicus nuchalis*), which provided the most abundant cavities (0.52 cavities/ha), and to cavity density of hairy woodpeckers (*Picoides villosus*), which provided few (0.14 cavities/ha) but high-quality cavities. Overall, sapsuckers and hairy woodpeckers provided 86% of nesting cavities to swallows in continuous forests. In contrast, the presence of nesting bluebirds in continuous forests was associated with the density of cavities supplied by all the ecosystem engineers. These results suggest that (i) habitat type may mediate the associations between ecosystem engineers and resource users, and (ii) different ecosystem engineers may be interchangeable for resource users depending on the quantity and quality of resources that each engineer supplies in each habitat type. We, therefore, urge the incorporation of the variation in the quantity and quality of resources provided by ecosystem engineers across habitats into models that assess community dynamics to improve our understanding of the importance of ecosystem engineers in shaping ecological communities.

Editor: Jane M Waterman, University of Manitoba, Canada

Funding: This research was funded by The Natural Sciences and Engineering Research Council of Canada (NSERC Special Strategic Grant to KM), Environment Canada, Sustainable Forest Management Network and Forest Renewal B.C. Tolko Industries Limited (Cariboo Woodlands). The funders had no role in study design, data collection and analysis, decision to publish, or preparation of the manuscript.

Competing Interests: This research was partly funded by Forest Renewal B.C. Tolko Industries Limited (Cariboo Woodlands). This does not alter the authors' adherence to all the PLOS ONE policies on sharing data and materials.

* E-mail: hugo.roblesdiez@uantwerpen.be

¤ Current address: Department of Animal Biology, Plant Biology and Ecology, Faculty of Sciences, University of A Coruña, A Coruña, Spain

Introduction

Ecosystem engineers are organisms that modulate resource availability to other organisms by maintaining or creating new habitat through physical state changes in biotic or abiotic components of the ecosystems [1]. Through this environmental modification, ecosystem engineers change the selective pressures to which other organisms are exposed (i.e. the "niche construction" process, see [2,3]). Consequently, ecosystem engineers are thought to be major drivers of evolutionary and ecological dynamics [3,4]. Given that niche construction via ecosystem engineering is a widespread phenomenon that structures ecological communities [3,5], a proper knowledge of the mechanisms underlying the associations between ecosystem engineers and resource users can improve our understanding of the structure of numerous ecological communities [6].

Several recent studies show the importance of ecosystem engineers in structuring ecological populations or communities in a wide variety of environments [7–16]. However, the variation in the influence of particular species of ecosystem engineers in different habitat types is poorly known, despite the recognized assumption that ecosystem engineering is a context-dependent process [1,5]. Because different habitats support communities that differ in the abundances of specific ecosystem engineers, the importance of each ecosystem engineer species for resource users may vary across habitats. Consequently, habitat may mediate the specific associations between ecosystem engineers and resource users. Some studies on burrowing mammals have found that the effects of ecosystem engineering vary spatially in extensive geographic ranges [12,17], but whether habitat types mediate the specific associations between ecosystem engineers and resource users after controlling for the confounding effect of the large spatial variation in ecosystem engineering remains unclear. Assessment of the specific associations between ecosystem engineers and resource users in different habitat types at local spatial scales is, however, important to shed light on whether such associations vary with habitat.

Different species of ecosystem engineer may provide resources of differing quality for a particular resource user [6,9]. Thus, ecosystem engineers can be considered as "suboptimal" or "preferred". A high abundance of suboptimal ecosystem engineers in habitats where the preferred ecosystem engineers are scarce may allow populations of resource users to persist. However, the influence of alternative species of ecosystem engineers on local populations of resource users in different habitats has received little investigation because most previous studies have assessed the influence of only single species of ecosystem engineer in one habitat type (but see [9]). A proper assessment of how different species of ecosystem engineers can supply resource users with alternative resources in different habitats can improve our understanding of how communities subjected to ecosystem engineering processes are structured.

Cavity-nesting birds depend on tree-cavities (i.e. the critical resource) for nesting in forest ecosystems [18,19] and are hierarchically structured in nest web communities according to their mode of acquiring cavities (e.g. [20,21]). While primary cavity nesters are able to build their own breeding cavities through excavation (e.g. [20]), secondary cavity-nesting birds require cavities provided by avian excavators or by fungal/insect decay (e.g. [20,22]). Fungal and insect infections induce cavity formation either directly by progressive heartwood decay or indirectly by providing suitable substrate for woodpecker excavation [23,24]. Therefore, avian excavators and rot fungi/insects function as key ecosystem engineers that facilitate the production of nest-sites for secondary cavity nesters [1].

In this study, we aimed (i) to assess whether habitat mediates the specific associations between ecosystem engineers and resource users in a cavity nest-web community, and (ii) to examine whether different species of ecosystem engineers provide resource users with alternative resources in different habitats. To assess the variation in the importance of the ecosystem engineers in relation to habitat, we investigated the relationships between the presence of nests of two secondary cavity nesters (the mountain bluebird *Sialia currucoides* and the tree swallow *Tachycineta bicolor*) in two different habitats (aspen groves and continuous mixed deciduous-coniferous forests), and the density of cavities supplied by four types of ecosystem engineers: three cavity excavating birds (northern flicker *Colaptes auratus*, red-naped sapsucker *Sphyrapicus nuchalis*, hairy woodpecker *Picoides villosus*) and rot fungi/insects.

Bluebirds are more likely to nest in cavities supplied by northern flickers, hairy woodpeckers and rot-fungi/insects than in cavities supplied by red-naped sapsuckers [6]. Swallows, however, are more likely to nest and to successfully produce fledglings in hairy woodpecker and sapsucker cavities than in flicker cavities [6]. We expected a higher density of flicker cavities relative to other cavity types in aspen groves, as flickers reach very high abundances in these patchy forests surrounded by extensive grasslands suitable for foraging [25]. We predicted that the high abundance of flicker cavities in aspen groves would lead to a high use of these cavities by the secondary cavity nesters (i.e. the resource users). This can be true particularly for bluebirds, but not so much for swallows, which avoid flicker cavities (see above). In continuous forests, however, we expected less difference in the density of cavities supplied by different ecosystem engineers compared to aspen groves. Consequently, we predicted a more similar contribution of the density of cavities provided by each ecosystem engineer on the presence of secondary cavity nesters in continuous forests.

Methods

Ethics statement

The surveys were conducted primarily on public lands. The Department of National Defence allowed us to conduct field research on two sites and lease holders allowed us access to three sites. All surveys were based on non-invasive observational procedures to minimize disturbance on birds. The research was approved by the Animal Care & Biosafety Committee of the University of British Columbia (permit numbers: A04-0101 to A07-130). All field activities were in agreement with federal and provincial legislation.

Study area and species

We collected the data in 35 sites (7–32 ha) within 50 km of Williams Lake (51° 52′ N, 122° 21′ W) in interior British Columbia, Canada. The study included two forest types: continuous mixed forests (27 sites) and aspen groves (8 sites). Continuous mixed deciduous-coniferous forest sites (≥80 years) were composed mainly of Lodgepole pine (*Pinus contorta* var. *latifolia*, 42% of trees), Douglas-fir (*Pseudotsuga menziesii*, 28%), hybrid white spruce (*Picea glauca x engelmannii*, 18%) and trembling aspen (*Populus tremuloides*, 12%), whereas black cottonwood (*Populus balsamifera*), alder (*Alnus* spp.), paper birch (*Betula papyrifera*), and willow (*Salix* spp.) were present in very low proportions [26]. Continuous mixed forests were occasionally disrupted by small grasslands, shallow ponds and selective harvesting. Aspen grove sites consisted of a few small clusters of trees (0.2–5 ha) surrounded by an extensive matrix of grasslands and shallow ponds. Aspen groves were composed principally of trembling aspen (54%) and, to a lesser extent, of lodgepole pine (38%) and Douglas-fir (8%) [26].

Mountain bluebirds and tree swallows are secondary cavity-nesting passerine birds. Because both insectivores require open areas for foraging [27,28], the extensive matrix of grasslands and shallow ponds surrounding aspen groves provides swallows and bluebirds with large areas of suitable foraging habitat. The availability of foraging habitat is more limited in continuous forests, where these birds forage over the interspersed grasslands, shallow ponds and clear-cuts within the forest. While populations of mountain bluebirds remain stable in North America [27], tree swallow populations have declined in recent decades in the areas of greatest swallow density within its range [28,29]. Populations of both secondary cavity nesters increased on our sites during the course of this study [30].

Field methods

We searched for nests of cavity excavators and secondary cavity nesters in May–July, 1995–2011 (average of 10.8±0.5 SE years of data per site). During nest surveys, which were conducted for an average of 6–7 observer-hours per sampling site per week, nest cavities were found by observing behavior of adults and listening for begging nestlings. We found most nests during the laying or early incubation stage. To inspect cavity contents, we used either a ladder, flashlights and mirrors, or a video camera system mounted in a pole (TreeTop Peeper; Sandpiper Technologies, Manteca, CA). We considered cavities with at least one egg or nestling to contain active nests.

Starting in 1999, we systematically monitored the cavities we found during routine searches in previous years until they were no longer available for nesting (details in [26]), which allowed us to calculate the density of cavities supplied by ecosystem engineers in a given site and year [6]. We excluded data from 1995 to 1998 from calculations of cavity densities because the survey effort differed from that of the later study years and the cavity formation agent (ecosystem engineer) of some cavities at the beginning of the study was unknown. We, therefore, calculated cavity density supplied by ecosystem engineers from 1999 to 2011 during which we knew the cavity history and the identity of the ecosystem engineer that formed the cavity [6].

Statistical analyses

Density of cavities supplied by ecosystem engineers. We used linear mixed-effects models (LMEs) to examine potential differences in the density of cavities (cavities ha^{-1}) supplied by ecosystem engineers between continuous forests and aspen groves. The response variable was the density of cavities supplied by the ecosystem engineers for each site-year, whereas habitat type (continuous forests vs. aspen groves) was fitted as a categorical fixed term. Site identity was fitted as a random term to control for multiple observations within the same sites across years.

We also assessed whether particular species of ecosystem engineers supplied higher densities of cavities than others in continuous forests and in aspen groves. For each habitat type, we performed a LME with the density of cavities as the response variable, the identity of each ecosystem engineer as a fixed term and site identity as a random term. All cavity densities in this paper were logarithmic-transformed. LMEs were conducted using the package "nlme" [31] in R 2.14.1 [32].

Presence of secondary cavity nesters in relation to the density of cavities supplied by ecosystem engineers. We used generalized linear mixed models (GLMMs) with binomial error distributions and logit link functions to assess the relationships between the presence of nesting bluebirds or swallows (presence of active nests in a given site and year = 1, no active nests = 0) and the density of cavities (cavities ha^{-1}) supplied by the ecosystem engineers in continuous mixed forests and aspen groves. Site identity was fitted as a random term. In addition, site size (ha) was fitted as a fixed term to control for its potential effect on the detection of nests. GLMMs were conducted using the package "lme4" [33] in R 2.14.1 [32].

We ran one set of models for each species of secondary cavity nester and each habitat type. Each model set was composed of 7 models. The first 4 models comprised the density of cavities supplied by each ecosystem engineer and site size as fixed terms. Another model included, in addition to site size, the density of cavities of the ecosystem engineers preferred by bluebirds (flickers, hairy woodpeckers and rot-fungi/insects; see above) or swallows (hairy woodpeckers and sapsuckers) for nesting. One model was composed of site size only (i.e. a null model). Finally, one model included site size and the density of cavities supplied by all the ecosystem engineers. This modeling approach allowed us to examine the importance of the ecosystem engineers by comparing models with the density of cavities supplied by each engineer species with models that included the preferred cavities for each secondary cavity nester as well as with null models and with models that included all cavities in the sites.

We used an information-theoretic approach [34] to assess the importance of the density of cavities facilitated by the ecosystem engineers on the presence of nesting swallows and bluebirds in a given site and year. We ranked the models according to Akaike's Information Criterion corrected for small sample sizes (AICc) and Akaike model weights [34,35]. Models with low AICc values are better supported by the data. Akaike model weights quantify the support of every model by the data, where higher weights indicate better explanatory power and the sum of all model weights is 1 [34].

Results

Density of cavities supplied by ecosystem engineers

Aspen groves had higher densities of cavities supplied by flickers (parameter estimate ± SE $= -2.617 \pm 0.372$, $t = -7.027$, $p < 0.001$) and rot-fungi/insects (parameter estimate ± SE $= -1.347 \pm 0.263$, $t = -5.122$, $p < 0.001$) than continuous forests (Fig. 1). In contrast, the densities of sapsucker (parameter estimate ± SE $= -0.503 \pm 0.500$, $t = -1.006$, $p = 0.324$) and hairy woodpecker (parameter estimate ± SE $= -0.179 \pm 0.453$, $t = -0.396$, $p = 0.697$) cavities did not differ significantly between both forest types (Fig. 1).

In aspen groves, flickers supplied the highest density of cavities, followed by sapsuckers, rot-fungi/insects and hairy woodpeckers (Table 1, Fig. 1). In continuous forests, sapsuckers provided significantly higher densities of cavities than all the other ecosystem engineers, and the densities of cavities supplied by hairy woodpeckers and flickers were significantly higher than those supplied by rot-fungi/insects (Table 1, Fig. 1).

Nesting cavities of secondary cavity nesters

Nesting cavities of bluebirds. In aspen groves, flickers supplied most nesting cavities to bluebirds, followed by rot-fungi/insects, sapsuckers and hairy woodpeckers (Fig. 2A). In continuous forests, bluebirds nested in flicker and hairy woodpecker cavities in similar proportions (Fig. 2A). Sapsucker cavities were used in lower proportions than flicker and hairy woodpecker cavities in continuous forests, whereas only a few cavities supplied by rot-fungi/insects were used by bluebirds for nesting (Fig. 2A).

Nesting cavities of swallows. In aspen groves, swallows mostly nested in flicker cavities and, to a lesser extent, in sapsucker cavities (Fig. 2B). Hairy woodpeckers and rot-fungi/insects provided few nesting cavities to swallows in aspen groves (Fig. 2B). In continuous forests, swallows mostly nested in sapsucker and hairy woodpecker cavities, whereas flickers and rot-fungi/insects provided swallows with few nesting cavities (Fig. 2B).

Presence of secondary cavity nesters in relation to the density of cavities supplied by ecosystem engineers

Presence of bluebird nests. In aspen groves, we found bluebird nests in 33 of 39 site-years (84.6%). Model selection of analyses that examined the presence of bluebird nests in aspen groves yielded two high ranked models that accounted for most weight in the candidate model set (Table 2). The best model included the density of flicker cavities as a predictor of the presence of bluebird nests, whereas the second-ranked model included the density of the preferred cavities (i.e. supplied by flickers, hairy woodpeckers and rot-fungi/insects). The presence of nesting bluebirds was positively related to the density of flicker cavities (parameter estimate ± SE $= 25.8 \pm 12.6$ $z = 2.053$, $p = 0.040$, Fig. 3) and to the density of the preferred cavities (parameter estimate ± SE $= 16.6 \pm 7.3$, $z = 2.277$, $p = 0.023$). However, among the preferred cavities, the models that included the density of cavities supplied by hairy woodpeckers and rot-fungi/insects had low model weights and AICc values similar to the null model (ΔAICc<3, see Table 2), which indicates low influence of these cavities on the presence of bluebird nests.

In continuous forests, we found bluebird nests in 64 of 276 site-years (23.2%). Model selection of analyses that examined the presence of bluebird nests in continuous forests yielded a single high ranked model that accounted for most weight in the candidate model set (Table 2). The best model included the density of cavities supplied by all the ecosystem engineers. The

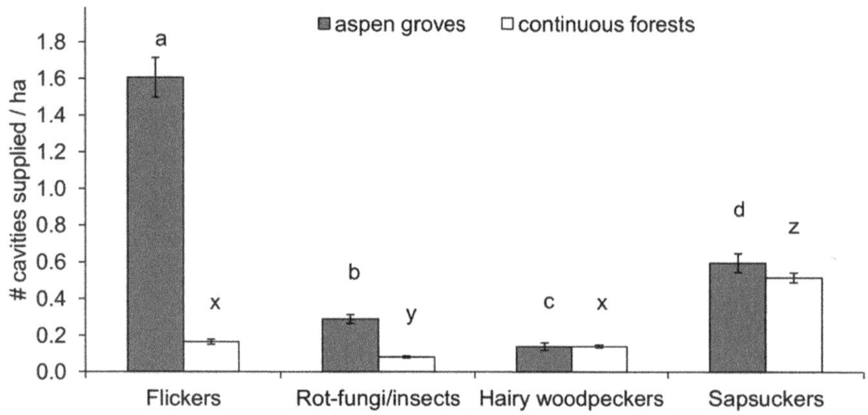

Figure 1. Average density of cavities supplied by the ecosystem engineers. Different letters above standard error bars indicate significant differences in cavity densities supplied by the ecosystem engineers for each habitat type (aspen grove or continuous forest; Tukey's post-hoc test for the models in table 1: all $p<0.001$).

presence of bluebird nests was most strongly and positively associated with the density of cavities supplied by all the ecosystem engineers (parameter estimate ± SE $=6.020\pm1.066$, $z=5.649$, $p<0.001$, Fig. 3).

Presence of swallow nests. In aspen groves, we found swallow nests in 32 of 39 site-years (82.1%). Model selection of analyses that examined the presence of swallow nests in aspen groves yielded two high ranked models that accounted for most weight in the candidate model set (Table 2). Similar to bluebird analyses, the best model included the density of flicker cavities as a predictor of swallow nest presence. The presence of nesting swallows in a given aspen grove and year was positively related to the density of flicker cavities (parameter estimate ± SE $=16.94\pm8.57$, $z=1.977$, $p=0.048$, Fig. 3). The second ranked model included the density of cavities supplied by all the ecosystem engineers (parameter estimate ± SE $=13.63\pm7.01$, $z=1.950$, $p=0.051$). However, the models composed of the density of cavities supplied by sapsuckers, hairy woodpeckers or rot-fungi/insects had low model weights and AICc values similar to the null model (ΔAICc<4, see Table 2), indicating low influence of these three ecosystem engineers on swallows nesting in aspen groves.

In continuous forests, we found swallow nests in 93 of 276 site-years (33.7%). Model selection of analyses that examined the presence of swallow nests in continuous forests yielded two high ranked models that accounted for most of the weight in the candidate model set (Table 2). The best model included the density of hairy woodpecker and sapsucker cavities as a predictor; that is, the cavities preferred by swallows for nesting. The presence of swallow nests was positively related to the density of these preferred cavities (parameter estimate ± SE $=4.024\pm0.817$, $z=4.927$, $p<0.001$, Fig. 3). The second best model included the density of cavities supplied by all the ecosystem engineers (parameter estimate ± SE $=3.671\pm0.736$, $z=4.989$, $p<0.001$). However, the models that included the density of cavities supplied by flickers and rot-fungi/insects had low model weights and AICc values similar to the null model (ΔAICc<5, see Table 2), which indicates low influence of these cavities on the presence of swallow nests in continuous forests.

Discussion

While niche construction via ecosystem engineering is considered to be a widespread phenomenon that structures ecological communities, the variation in the importance of ecosystem engineers across environmental gradients (e.g., across different habitats) is poorly understood [3,5]. Our results suggest that habitat type may mediate the inter-specific associations between ecosystem engineers and resource users. In aspen groves, the presence of bluebird and swallow nests was mainly and positively associated with the density of cavities supplied by flickers, which provided most nesting cavities to bluebirds and swallows (66% and

Table 1. Parameter estimates of models that examined the differences in the density of cavities supplied by the ecosystem engineers in aspen groves and in continuous forest sites.

Habitat	Parameters	Estimate ± SE	t	p
Aspen grove	Intercept	-2.507 ± 0.359	-6.984	<0.001
	Ecosystem engineer (Flickers)	2.853 ± 0.109	26.227	<0.001
	Ecosystem engineer (Rot-fungi/insects)	1.139 ± 0.086	10.363	<0.001
	Ecosystem engineer (Sapsuckers)	1.839 ± 0.109	16.904	<0.001
Continuous forest	Intercept	-2.194 ± 0.108	-20.389	<0.001
	Ecosystem engineer (Flickers)	-0.047 ± 0.081	-0.579	0.563
	Ecosystem engineer (Rot-fungi/insects)	-0.542 ± 0.104	-5.195	<0.001
	Ecosystem engineer (Sapsuckers)	1.160 ± 0.073	15.916	<0.001

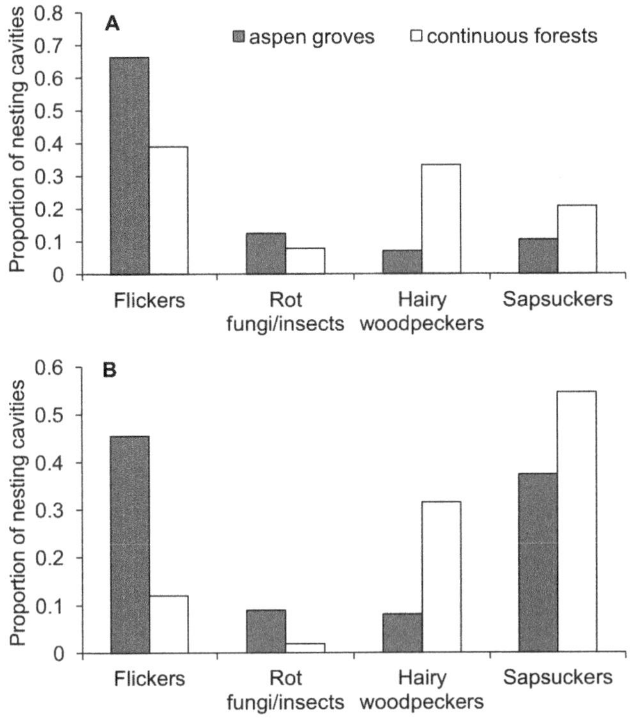

Figure 2. Proportion of nesting cavities supplied by the ecosystem engineers. The proportion of nests of bluebirds (A) and swallows (B) in cavities supplied by each ecosystem engineer are shown for aspen groves and continuous forests sites. Sample sizes: 229 and 77 nesting cavities of bluebirds in aspen groves and continuous forests, respectively; 222 and 108 nesting cavities of swallows.

45% of nesting cavities, respectively; Fig. 2). In continuous forests, however, the presence of bluebird nests was positively related to the density of cavities supplied by all ecosystem engineers. In addition, the presence of swallow nests in continuous forests was positively associated with the density of cavities supplied by sapsuckers and hairy woodpeckers, which provided 86% of nesting cavities used by swallows (Fig. 2).

To our knowledge, only one other study [9] has assessed the influence of habitat at local spatial scales to determine the variation in the associations between multiple ecosystem engineers and resource users in ecological communities. In their research, Machicote et al. [9] found that hairy armadillos (*Chaetophractus villosus*) produce suitable nest sites for burrowing owls (*Athene cunicularia*) in recently burned areas or areas maintained by anthropogenic disturbance, but in other habitat types plains vizcachas (*Lagostomus maximus*) play a major role as ecosystem engineers for these owls. These results suggest that habitat may mediate the associations between ecosystem engineers (i.e. vizcachas and armadillos) and resource users (i.e. owls) [9], which matches our results showing that habitat may mediate the inter-specific associations between cavity facilitators and secondary cavity users in a cavity-nest web community.

A better understanding of the context-dependent effects of ecosystem engineering can improve our knowledge of how ecological systems are structured [5]. Our results may help elucidate the mechanisms underlying the context-dependent variation in the importance of different ecosystem engineers. In our study system, both resource quantity and quality may explain the associations of resource users and ecosystem engineers in different habitats. In aspen groves, the presence of secondary cavity nesters was mainly related to the density of cavities supplied by flickers, which provided the most abundant cavities in this habitat type (Fig. 1). Flickers provide suitable nesting cavities to bluebirds, but swallows are less likely to occupy flicker cavities compared to cavities supplied by the other ecosystem engineers [6], suggesting that the high abundance (rather than the quality) of flicker cavities determines the high use of these cavities by swallows in aspen groves. Despite the relatively low abundance of sapsucker cavities compared to flicker cavities in aspen groves (Fig. 1), sapsuckers supplied 37% of the nesting cavities used by swallows (Fig. 2), possibly because these woodpeckers provide suitable cavities to swallows [6]. In continuous forests, despite the low abundance of hairy woodpecker cavities (Fig. 1), hairy woodpeckers provided approximately one third of the nesting cavities to both bluebirds and swallows (Fig. 2). The high use of hairy woodpecker cavities in continuous forests may be associated with the high quality of these cavities for both secondary cavity nesters, as cavities produced by hairy woodpeckers are the most likely to be occupied by swallows and bluebirds [6]. In addition, sapsuckers provided high numbers of nesting cavities to swallows in continuous forests (Fig. 2), perhaps because sapsuckers provided the most abundant cavities in this habitat type (Fig. 1). The small entrance size of many sapsucker cavities may denote that they are not all available for use by bluebirds [6], but even so, sapsuckers still provided 21% of the nesting cavities used by bluebirds in continuous forests. This may be due to the high cavity abundance that sapsuckers provided in this habitat type, increasing the probability that some sapsucker cavities were large enough to be accessible to bluebirds. These results suggest that the identification of the key ecosystem engineers for each resource user requires considering both the quantity and quality of the resources supplied by the ecosystem engineers in each habitat type.

Understanding how communities are structured requires a proper identification of the associations between ecosystem engineers and resource users that elucidate whether and when ecosystem engineers are functionally interchangeable [9]. The positive relationship between the presence of swallow nests and the density of flicker cavities in aspen groves suggests that nesting in suboptimal flicker cavities may be an acceptable alternative for swallows in this habitat type (flickers provided more nesting cavities to swallows in aspen groves than the other ecosystem engineers, Fig. 2). However, the exclusive use of flicker cavities might not guarantee the long-term persistence of swallow populations because swallows are less likely to produce fledglings in nesting cavities supplied by flickers than in those supplied by the other ecosystem engineers [6]. Our results also suggest that some ecosystem engineers may be functionally interchangeable for bluebirds, which used high numbers of flicker cavities in aspen groves but also high numbers of cavities supplied by hairy woodpeckers and sapsuckers in continuous forests (Fig. 2). Because the reproductive output of bluebirds nesting in cavities supplied by flickers, sapsuckers and hairy woodpeckers does not differ significantly [6], we found no evidence of fitness differences for bluebirds nesting in cavities supplied by these three different ecosystem engineers.

Cavities created by fungal or insect decay appear to be less important than those supplied by avian excavators, as avian excavators provide most nesting sites to secondary cavity nesters in North America (e.g. [21,36–40]). Our results also suggest a low influence of cavities created by fungal/insect decay on secondary cavity nesters, as the density of these cavities was not strongly correlated with the presence of cavities used by bluebirds and swallows in either habitat type. Moreover, both secondary cavity nesters used few cavities created by fungal/insect decay for nesting

Table 2. Model selection of analyses that examined the presence of bluebird or swallow nests in aspen groves and continuous forest sites.

Species	Habitat	Models	df	LogLik	AICc	ΔAICc	Weight
Bluebird	Aspen grove	Flickers	4	−6.03	21.24	0.00	0.67
		Flickers + hairy woodpeckers + rot-fungi/insects	4	−6.88	22.94	1.70	0.29
		All ecosystem engineers	4	−8.90	26.99	5.75	0.04
		Hairy woodpeckers	4	−12.09	33.36	12.12	0.00
		Null model	3	−14.70	36.09	14.85	0.00
		Sapsuckers	4	−13.73	36.64	15.40	0.00
		Rot-fungi/insects	4	−13.81	36.79	15.55	0.00
Bluebird	Continuous forest	All ecosystem engineers	4	−107.26	222.67	0.00	0.97
		Sapsuckers	4	−111.09	230.33	7.66	0.02
		Flickers + hairy woodpeckers + rot-fungi/insects	4	−112.43	233.01	10.34	0.01
		Flickers	4	−119.55	247.26	24.58	0.00
		Hairy woodpeckers	4	−119.72	247.59	24.91	0.00
		Null model	3	−129.24	264.56	41.89	0.00
		Rot-fungi/insects	4	−128.47	265.08	42.41	0.00
Swallow	Aspen grove	Flickers	4	−5.62	20.42	0.00	0.58
		All ecosystem engineers	4	−6.11	21.40	0.98	0.36
		Sapsuckers + hairy woodpeckers	4	−8.68	26.53	6.11	0.03
		Sapsuckers	4	−8.75	26.68	6.26	0.03
		Null model	3	−12.25	31.18	10.76	0.00
		Rot-fungi/insects	4	−11.02	31.21	10.79	0.00
		Hairy woodpeckers	4	−11.87	32.91	12.50	0.00
Swallow	Continuous forest	Sapsuckers + hairy woodpeckers	4	−150.78	309.70	0.00	0.46
		All ecosystem engineers	4	−150.82	309.80	0.10	0.44
		Sapsuckers	4	−152.25	312.65	2.95	0.10
		Hairy woodpeckers	4	−156.05	320.24	10.54	0.00
		Flickers	4	−160.26	328.66	18.96	0.00
		Null model	3	−164.22	334.53	24.82	0.00
		Rot-fungi/insects	4	−164.18	336.51	26.80	0.00

N = 39 and 276 site-years in aspen groves and continuous forests, respectively. The presence of nests was related to the density of cavities (logarithmic-transformed) supplied by the ecosystem engineers: northern flickers, red-naped sapsuckers, hairy woodpeckers and rot-fungi insects. Site size was fitted as a fixed term in all models to control for its potential effect on nest detection. LogLik: log-likelihood, AICc: AIC corrected for small sample size, ΔAICc: difference in AICc to the best model. Models of each model set for each bird species and habitat type are ranked according to their Akaike weight (Weight).

(2–12% of nesting cavities, Fig. 2). In addition, cavities created by fungal/insect decay are less likely to be used by swallows than hairy woodpecker cavities [6], which may explain why low-abundant hairy woodpecker cavities, but not cavities created by fungi/insects, were strongly associated with the presence of nesting swallows in continuous forests.

Bluebirds and swallows were less likely to nest in continuous forests than in aspen groves. One possibility is that bluebirds and swallows are less likely to occupy continuous forest sites because these sites have lower abundance of cavities supplied by the ecosystem engineers compared to aspen groves (Fig. 1). Another possibility is that aspen groves provide bluebirds and swallows with additional habitat resources. Supporting this hypothesis, cavities in aspen groves are more likely to be used by bluebirds and swallows after controlling for the cavity formation agent (i.e. ecosystem engineer identity) than those in continuous forests [6]. The high availability of open areas (grasslands and shallow ponds) surrounding aspen groves may provide swallows and bluebirds with suitable foraging habitat [27,28], whereas the availability of foraging habitat for these species is more limited in continuous forests. Other potential factors such as increased levels of predation or parasitism in continuous forests compared to aspen groves may also explain the lower presence of nesting bluebirds and swallows in continuous forests.

To summarize, by assessing the relationships between the presence of nests of two secondary cavity-nesting birds, and the density of cavities supplied by multiple ecosystem engineers in contrasting habitats, we show that ecosystem engineers may be interchangeable for resource users. The interchangeability of resources provided by different engineers was probably mediated by habitat type, as the two habitats supported different amounts of cavities supplied by each ecosystem engineer. Determining the generality of our results will require further evaluations of the associations between multiple ecosystem engineers and resource users in different habitats to assess the interchangeability of resources supplied by ecosystem engineers in different settings and communities. Such a study approach would also improve our understanding of the variation in ecosystem engineering effects

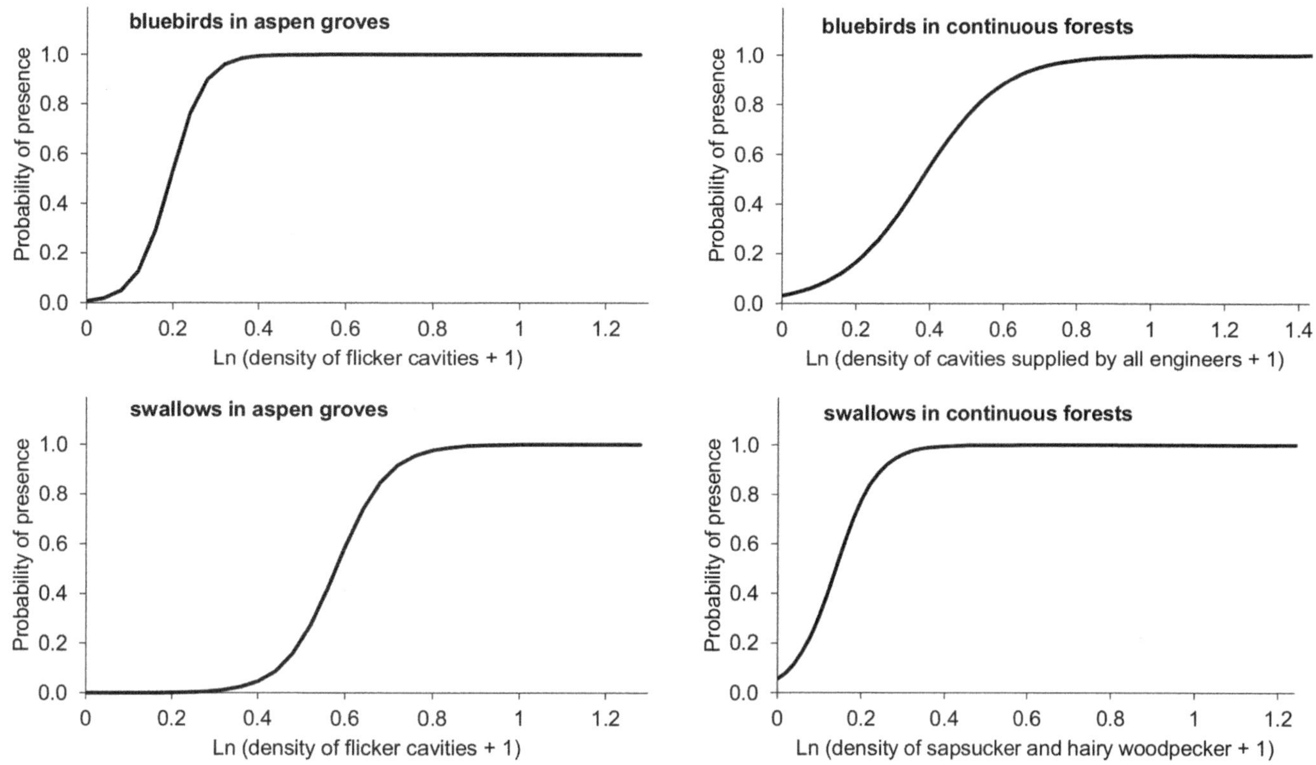

Figure 3. Probability of presence of bluebird and swallow nests in relation to cavity density supplied by the ecosystem engineers. The probability of presence was defined as the probability that a given aspen grove or continuous forest site had at least one cavity used by bluebirds or swallows for nesting in a given year. Presence probabilities are the predicted probabilities calculated from the best models that depicted the presence of nests in Table 2.

across environmental gradients, which in turn can improve the predictions on the general effects of ecosystem engineering on abiotic and biotic processes [5,41]. Overall, we recommend incorporating the variation in the quantity and quality of resources provided by ecosystem engineers across habitats into models that assess community structure and dynamics to provide a better understanding of the importance of ecosystem engineers in shaping ecological communities.

Acknowledgments

We are grateful to numerous field assistants and graduate students who contributed to data collection. We are also grateful to K. Cockle, M. Drever, A. Edworthy, A. Norris, C. Storey and other members of the K. Martin lab for their constructive feedback or assistance on the content of this manuscript. Three anonymous reviewers provided useful comments that improved a previous draft of this manuscript.

Author Contributions

Conceived and designed the experiments: KM. Performed the experiments: KM. Analyzed the data: HR. Contributed reagents/materials/analysis tools: HR. Wrote the paper: HR KM.

References

1. Jones CG, Lawton JH, Shachak M (1994) Organisms as ecosystem engineers. Oikos 69: 373–386.
2. Day RL, Laland KN, Odling-Smee FJ (2003) Rethinking adaptation: the niche-construction perspective. Perspect Biol Med 46: 80–95.
3. Boogert NJ, Paterson DM, Laland KN (2006) The implications of niche construction and ecosystem engineering for conservation biology. BioScience 56: 570–578.
4. Odling-Smee FJ, Laland KN, Feldman MW (2003) Niche construction: The neglected process in evolution. Princeton, New Jersey: Princeton University Press. 468.
5. Wright JP, Jones CG (2006) The concept of organisms as ecosystem engineers ten years on: progress, limitations, and challenges. BioScience 56: 203–209.
6. Robles H, Martin K (2013) Resource abundance and quality determine the inter-specific associations between ecosystem engineers and resource users in a cavity-nest web. PLoS ONE 8(9): e74694. doi: 10.1371/journal.pone.0074694.
7. Wright JP, Jones CG, Flecker AS (2002) An ecosystem engineer, the beaver, increases species richness at the landscape scale. Oecologia 132: 96–101.
8. Cardinale BJ, Gelmann ER, Palmer MA (2004) Net spinning caddisflies as stream ecosystem engineers: the influence of *Hydropsyche* on benthic substrate stability. Funct Ecol 18: 381–387.
9. Machicote M, Branch LC, Villareal D (2004) Burrowing owls and burrowing mammals: are ecosystem engineers interchangeable as facilitators? Oikos 106: 527–535.
10. Davidson AD, Lightfoot DC (2007) Interactive effects of keystone rodents on the structure of dessert grassland arthropod communities. Ecography 30: 515–525.
11. Pringle RM (2008) Elephants as agents of habitat creation for small vertebrates at the patch scale. Ecology 89: 26–33.
12. Cully JF, Collinge SK, VanNimwegen RE, Ray C, Johnson WC, et al. (2010) Spatial variation in keystone effects: small mammal diversity associated with black-tailed prairie dog colonies. Ecography 33: 667–677.
13. Gálvez-Bravo L, López-Pintor A, Rebollo S, Gómez-Sal A (2011) European rabbit (*Oryctolagus cuniculus*) engineering effects promote plant heterogeneity in Mediterranean dehesa pastures. J Arid Environ 75: 779–786.
14. Calderón-Cortés N, Quesada M, Escalera-Vázquez LH (2011) Insects as stem engineers: interactions mediated by the twig-girdler *Oncideres albomarginata chamela*

enhance arthropod diversity. PLoS ONE 6(4): e19083. doi:10.1271/journal.pone.0019083.
15. Joseph GS, Cumming GS, Cumming DHM, Mahlangu Z, Altwegg R, et al. (2011) Large termitaria act as refugia for all trees, deadwood and cavity-using birds in a miombo woodland. Landsc Ecol 26: 439–448.
16. Prugh LR, Brashares JS (2012) Partitioning the effects of an ecosystem engineer: kangaroo rats control community structure via multiple pathways. J Anim Ecol 81: 667–678.
17. Baker BW, Augustine DJ, Sedgwick JA, Lubow BC (2012) Ecosystem engineering varies spatially: a test of the vegetation modification paradigm for prairie dogs. Ecography 35: 1–10.
18. Aitken KEH, Martin K (2008) Resource selection plasticity and community responses to experimental reduction of a critical resource. Ecology 89: 971–980.
19. Robles H, Ciudad C, Matthysen E (2012) Responses to experimental reduction and increase of cavities by a secondary cavity-nesting bird community in cavity-rich Pyrenean oak forests. For Ecol Manage 277: 46–53.
20. Martin K, Aitken KEH, Wiebe KL (2004) Nest sites and nest webs for cavity-nesting communities in interior British Columbia, Canada: nest characteristics and niche partitioning. Condor 106: 5–19.
21. Blanc LA, Walters JR (2008) Cavity-nest webs in a longleaf pine ecosystem. Condor 110: 80–92.
22. Robles H, Ciudad C, Matthysen E (2011) Tree-cavity occurrence, cavity occupancy and reproductive performance of secondary cavity-nesting birds in oak forests: The role of traditional management practices. For Ecol Manage 261: 1428–1435.
23. Robles H, Ciudad C, Vera R, Olea PP, Purroy FJ, et al. (2007) Sylvopastoral management and conservation of the Middle Spotted Woodpecker at the south-western edge of its distribution range. For Ecol Manage 242: 343–352.
24. Blakely TJ, Jellyman PG, Holdaway RJ, Young L, Burrows B, et al. (2008) The abundance, distribution and structural characteristics of tree-holes in *Nothofagus* forest, New Zealand. Austral Ecol 33: 963–974.
25. Wiebe KL, Moore WS (2008) Northern Flicker (*Colaptes auratus*). In: Poole A, editor. The Birds of North America Online. Ithaca: Cornell Lab of Ornithology. Available: http://bna.birds.cornell.edu/bna/species/166a. Accessed 2012 Aug 22.
26. Edworthy AB, Wiebe KL, Martin K (2012) Survival analysis of a critical resource for cavity-nesting communities: patterns of tree cavity longevity. Ecol Appl 22: 1733–1742.
27. Power HW, Lombardo MP (1996) Mountain Bluebird (*Sialia currucoides*). In: Poole A, editor. The Birds of North America Online. Ithaca: Cornell Lab of Ornithology. Available: http://bna.birds.cornell.edu/bna/species/222. Accessed 2012 Aug 22.
28. Winkler DW, Hallinger KK, Ardia DR, Robertson RJ, Stutchbury BJ, et al. (2011) Tree Swallow (*Tachycineta bicolor*). In: Poole A, editor. The Birds of North America Online. Ithaca: Cornell Lab of Ornithology. Available: http://bna.birds.cornell.edu/bna/species/011. Accessed 2012 Aug 22.
29. Sauer JR, Hines JE, Fallon JE, Pardieck KL, Ziolkowski DJ Jr, et al. (2011) The North American Breeding Bird Survey, Results and Analysis 1966–2010. Version 12.07.2011 USGS Patuxent Wildlife Research Center, Laurel, MD. Available: http://www.mbr-pwrc.usgs.gov/bbs/. Accessed 2011 Dec 20.
30. Koch AJ, Martin K, Aitken KEH (2012) The relationship between introduced European Starlings and the reproductive activities of Mountain Bluebirds and Tree Swallows in British Columbia, Canada. Ibis 154: 590–600.
31. Pinheiro JC, Bates DM (2000) Mixed-Effects Models in S and S-PLUS. New York: Springer-Verlag. 528 p.
32. R Development Core Team (2011) R: A Language and Environment for Statistical Computing. Vienna: R Foundation for Statistical Computing. Available: http://www.r-project.org/. Accessed 2011 Dec 20.
33. Bates DM (2010) lme4: Mixed-effects modeling with R. Available: http://lme4.r-forge.r-project.org/book/.Accessed 2014 Jan 2.
34. Burnham KP, Anderson DR (2002) Model selection and multi-model inference: a practical information-theoretic approach. New York: Springer-Verlag. 496 p.
35. Johnson JB, Omland KS (2004) Model selection in ecology and evolution. Trends Ecol Evol 19: 101–108.
36. Stauffer DF, Best LB (1982) Nest-site selection by cavity-nesting birds in riparian habitats in Iowa. Wilson Bull 94: 329–337.
37. Raphael MG, White M (1984) Use of snags by cavity-nesting birds in Sierra Nevada. Wildlife Monographs 86: 3–66.
38. Waters JR (1988) Population and habitat characteristics of cavity-nesting birds in a California oak woodland (MS thesis). Arcata, CA: Humboldt State University. 200 p.
39. Aitken KEH, Martin K (2007) The importance of excavators in hole-nesting communities: availability and use of natural tree holes in old mixed forests of western Canada. J Ornithol 148: S425–S434.
40. Cockle KL, Martin K, Wesolowski T (2011) Woodpeckers, decay, and the future of cavity-nesting vertebrate communities worldwide. Front Ecol Environ 9: 377–382.
41. Gutierrez JL, Jones CG (2006). Physical ecosystem engineers as agents of biogeochemical heterogeneity. BioScience 56: 227–236.

18

Global Drivers and Tradeoffs of Three Urban Vegetation Ecosystem Services

Cynnamon Dobbs[1,2]*, Craig R. Nitschke[2], Dave Kendal[3]

1 School of Botany, The University of Melbourne, Melbourne, Australia, **2** School of Forest Science and Ecosystem, Melbourne School of Land and Environment, The University of Melbourne, Melbourne, Australia, **3** Australian Research Centre for Urban Ecology, Royal Botanic Gardens Melbourne, c/o School of Botany, The University of Melbourne, Melbourne, Australia

Abstract

Our world is increasingly urbanizing which is highlighting that sustainable cities are essential for maintaining human well-being. This research is one of the first attempts to globally synthesize the effects of urbanization on ecosystem services and how these relate to governance, social development and climate. Three urban vegetation ecosystem services (carbon storage, recreation potential and habitat potential) were quantified for a selection of a hundred cities. Estimates of ecosystem services were obtained from the analysis of satellite imagery and the use of well-known carbon and structural habitat models. We found relationships between ecosystem services, social development, climate and governance, however these varied according to the service studied. Recreation potential was positively related to democracy and negatively related to population. Carbon storage was weakly related to temperature and democracy, while habitat potential was negatively related to democracy. We found that cities under 1 million inhabitants tended to have higher levels of recreation potential than larger cities and that democratic countries have higher recreation potential, especially if located in a continental climate. Carbon storage was higher in full democracies, especially in a continental climate, while habitat potential tended to be higher in authoritarian and hybrid regimes. Similar to other regional or city studies we found that the combination of environment conditions, socioeconomics, demographics and politics determines the provision of ecosystem services. Results from this study showed the existence of environmental injustice in the developing world.

Editor: Zoe G. Davies, University of Kent, United Kingdom

Funding: These authors have no support or funding to report.

Competing Interests: The authors have declared no competing interests exist.

* Email: cdobbsbr@gmail.com

Introduction

Urban areas are dynamic and complex landscapes, where socio-ecological processes can deliver ecosystem services across multiple scales [1]. The ecosystem services concept provides a framework that integrates ecology with socioeconomics, creating a transdisciplinary approach for understanding the benefits that can be delivered by nature and the implications of these benefits on human wellbeing [2,3]. Population growth, consumption and governance can all influence the provision of ecosystem services which in turn affect human health, livelihood, culture and equity [4]. This concept is particularly relevant in urban systems where natural resources are under enormous pressure and where the demand for ecosystem services is increasing [5].

Cities differ in their governance, infrastructure, economy and social equity [6]. They also vary in their development, with some cities having high rates of urbanization and uncontrolled population growth while other cities are experiencing declines in population. The social, political and biophysical context of the city shapes how socio-ecological interactions affect the provision of environmental benefits [1]. Quantifying how urban ecosystem services are provided under these different socio-political-biophysical conditions provides a useful framework for understanding how socio-political-biophysical factors influence the provision of ecosystem services.

The structure and composition of urban vegetation influences the provision of ecosystem services. A number of regulating services (e.g. maintenance of air quality, climate regulation, maintenance of soil fertility), cultural services (e.g. aesthetics, sense of place and recreation) and supporting services (e.g. habitat for flora and fauna and space for reproduction) are linked to the patterns of urban vegetation [7–9]. The distribution of vegetation is a consequence of many factors including topography, climate, transportation infrastructure, plant dispersal mechanisms, real estate markets, planning, cultural practices and social preferences [1,9–12]. Kendal et al. [13] found that temperature influences the composition of cultivated trees in urban areas, while both education and income can influence local vegetation structure and composition [13–15]. The relationships between politics and urban vegetation have shown to be characterized by an inequitable distribution, often favouring urban elites over marginalized and deprived groups, either racial or socioeconomic [16–19]. However, a global analysis on how governance is related to the provision of ecosystem services is lacking, especially in relation to the national political context. The national political context influences the practice of participatory governance, local-level management and the prioritization of greening policies [20].

We know little about global patterns of ecosystem services, and research at larger scales has generally been restricted to a single country, region or rural landscape [21]. Urban ecosystem services research has focussed mainly on cities in the United States of America [9,12,22–24] with a few studies in other continents [25–29]. These studies mostly focus on the quantification of ecosystem services. We also know little about the tradeoffs and synergies that occur in the provision of ecosystem services [30], particularly in urban landscapes. When assessing services that represent different ecosystem functions, i.e. regulation, supporting and cultural [7], it is necessary to explore their synergies and tradeoffs [31]. Synergies occur when multiple services are simultaneously enhanced, while tradeoffs occur when one service is enhanced at the cost of reducing another [32]. For example, the provision of services such as recreation, spiritual enhancement and psychological benefits typically all increase when the amount and quality of green space available for urban dwellers increases [33–35]. In contrast, increasing tree cover in parks leads to increases in carbon storage and habitat provision, but could lead to a reduction in recreational services as the space available for sport fields decreases [36].

To our knowledge, no previous studies have explored the global drivers of urban ecosystem services. This research therefore represents one of the first attempts to quantify global urban ecosystem services and the existence of synergies and tradeoffs and their relation with development, climate and governance. The objectives of the study are 1) to test the effect of socio-political factors on the provision of ecosystem services to explore whether patterns previously found at local-levels scale up globally and 2) to explore whether common biophysical, demography and socioeconomic factors can explain the synergies and tradeoffs in ecosystem services. To achieve our objectives we quantified services that represent different ecosystem functions: carbon storage, recreation potential, and habitat provision. Carbon storage helps mitigate climate change at the global scale by offsetting the urban footprint [37], while at the regional and local scale it contributes to improving air quality [12,26]. Habitat provision in the urban landscape is strongly linked to biodiversity and to the well-being of urban inhabitants [23,38–43]. In comparison to carbon and biodiversity, recreation potential has a more local effect as it relates to the provision of space for leisure, contemplation and exercising which has been linked to improve public health [44].

Methods

Urban vegetation extraction

A sample of one hundred cities was selected to represent a diversity of biophysical, socioeconomic, demographic and cultural factors (Table 1; Figure S1; Table S1). Remotely sensed data were used to provide a standardized method to quantify ecosystem services and look for synergies and tradeoffs across a large number of cities. Cities were selected from a global pool where good quality satellite imagery (Landsat 5 TM) was available during the vegetation-growing season between years 2006 to 2011. Cities from tropical regions in Asia and Africa were not included because of cloud cover over the cities. Landsat images were of high resolution (30 m^2 multispectral pixels), which allowed for fine scale analysis. Landsat images are widely used in urban landscape studies [45,46].

Identifying city boundaries is a critical step in any analysis of urban landscapes, and one that is notoriously difficult at large scales [47,48]. The wide range of cities included in the study meant that standardised metadata (e.g. current administrative boundaries) were not available for all cities, therefore a method that could be applied to all cities was required. Following Schneider and Woodcock [47], the limits of a city were defined as the first area where less than 5% impermeable surface was present in a 200 m wide buffer located at the periphery of the urban area. To test the accuracy of this approach, the discrepancy between the administrative and calculated boundary was calculated for 30% of the cities in the study where administrative data was available. Discrepancy varied from a few square meters to 50 km^2 and was independent of the geographic location of the city, highlighting that that the error associated with our approach was randomly distributed.

The normalized difference vegetation index (NDVI) is an index of living green vegetation [49], and was calculated from the Landsat image of each city, using the red and infrared bands. An unsupervised classification of vegetation and impermeable surface was conducted; however, only vegetation (green cover) was retained for further analysis. To extract the classes representing vegetation the spectral value of 50 vegetation pixels by city were obtained from the NDVI image. The accuracy assessment was obtained using 80 random points within the vegetation class and cross referencing to Google Earth imagery. The Kappa coefficient for the vegetation classification was 0.8, while the user's accuracy corresponded to 75% and the producer's accuracy to 85% [50].

Quantification of ecosystem services

Recreation potential is derived from the vegetated areas that provide space for physical and psychological enjoyment. It includes vegetation in woodlands, grasslands and street trees that occurs in parks, smaller patches of vegetation and/or along streets. The recreation potential service was calculated as the amount of vegetated area per capita [23,51]. The area of vegetation from the NDVI imagery classification within our calculated urban boundary was divided by the population of the city, obtained from the United Nations global report on human settlements [6]. Due to the lack of data on census limits we assumed that the calculated urban boundary is consistent with the census data limits; while the accuracy of this approach may lead to over or underestimations, the error is randomly distributed and standardised across all cities.

Carbon storage in vegetation was calculated using an existing model based on Landsat derived NDVI [52]. This model was built for urban vegetation and has been validated with field data. The method has been previously used for assessing urban forest carbon offsets and quantifying carbon stock across an urban rural gradient in several cities of the United States [53,54]. The model is spatially explicit and calculates carbon storage per pixel (30 m^2) using the function:

$$Carbon(tonnes/pixel) = 0.10702e^{NDVI*0.0194}$$

Habitat potential is a function of vegetation structure at the landscape scale. Vegetation cover was obtained from the NDVI analysis. We recognize that different types of vegetation provide different degrees of habitat quality for floral and faunal guilds however this level of detail was not considered in the study. Areas of structural connectivity were identified using a Morphological Spatial Pattern Analysis (MSPA) available in the free software package GUIDOS (http://forest.jrc.ec.europa.eu/download/software/guidos/). MSPA has been used in studies focussed on assessing the connectivity of ecological habitats in both forested and urban landscapes [38,55,56]. It uses a land cover map of

Table 1. Socioeconomic, political and climatic characteristics for the 100 cities included in this study.

Climate (Köppen classification)	Population	Human Development Index (HDI)	Democracy Index (DI)
Tropical moist (11)	<1 million habitants (11)	Very high HDI (45)	Full democracies (38)
Dry climate (13)	1 to 2 million habitants (32)	High HDI (19)	Flawed democracies (32)
Moist mid latitude with mild winters (60)	2 to 6 million habitants (39)	Medium HDI (22)	Hybrid regimes (14)
Moist mid latitude with cold winters (16)	>6 million habitants (18)	Low HDI (14)	Authoritarian regimes (16)

vegetated/non vegetated areas to classify structural patterns following mathematical morphology methods [57].

The morphological segmentation of binary patterns obtained from an image with vegetated and non-vegetated pixels that produces seven categories according to their size, shape and connectivity.

Our measure of habitat potential is the proportion of area of vegetation larger than 1.44 ha which was classified as 'core' habitat [38]. 'Core' habitat areas are the pixels in patches where the distance to the non-vegetated area was greater than 60 m (2 Landsat pixels). Our definition of core areas include forest, woodland, shrubland and meadow areas and is consistent with broader ecological theory that shows that larger areas with relatively fewer edges are likely to support a wider of species (such as woodland birds [58]). However, we acknowledge that our measure based on NDVI is not a perfect measure of habitat potential, as it will include large areas of mown turf, which may have low ecological value, and exclude narrow linear corridors that may have high habitat potential. To achieve spatial concordance among the calculated ecosystem services across all cities values were standardised to the mean carbon storage per hectare, mean recreation potential per capita and habitat potential as the proportion of the total urban area covered by 'core' patches [59].

Socioeconomic and climatic factors

To explore the relationship between globally quantified ecosystem services with social and climatic factors we selected three social indicators: total population, the Human Development Index (HDI) and the Democracy Index (DI), and three climatic indicators: mean annual temperature, mean annual precipitation and the annual Heat Moisture Index (HMI) [60].

The Human Development Index (HDI) is a measure of development [61]. It is a composite index that combines indicators of life expectancy, educational attainment and income using a geometric mean [61]. The Democracy Index (DI) is an index that combines five metrics of governance: electoral process and pluralism, civil liberties, the functioning of the government, political participation and political culture [62] (Table S1). Values for mean annual temperature and precipitation were obtained from the World Meteorological Organization (http://www.wmo.int/pages/index_en.html) for the period between 1976 and 2005. The Heat Moisture Index (HMI) was calculated for each city following Wang et al. [60] as a ratio between mean annual temperature and mean annual rainfall. The HMI is a measure of the evaporative demand of the atmosphere and represents the aridity of the environment when the interaction between temperature and precipitation is considered.

Statistical analysis

The data were tested for normality using the Kolmogorov-Smirnov statistic. Recreation potential was log transformed to meet the analyses' assumptions of normality. Moran's I was calculated and used to assess if spatial autocorrelation exists within the analysed data [63].

A principal component analysis (PCA) was applied to the standardized data of each ecosystem service and social-climatic variables included in this study [63]. This multivariate data technique uses orthogonal transformation to group sets of correlated variables into principle components which are sets of linearly uncorrelated variables [64]. A Varimax rotation was used to improve result interpretation. The number of components was selected to provide the most interpretable solution with eigenvalues greater than one.

Following the PCA a Bayesian regression was used to assess the effect size of the socio-climatic variables on each of the ecosystem services considered in the study. This method uses a Markov Chain Monte Carlo technique to fit generalized linear models. Because of the lack of well established relationships between services and indicators, non-informative priors were used. The posterior distribution for each service model was simulated using a Markov chain Monte Carlo method. For each model we simulated 10000 iterations and a burn-in size of 2000, thinning the results by a factor of 1, reaching convergence [63]. For the socio-climatic parameters we report the 2.5% and the 97.5% credible interval of simulated posterior values, which represents the likely range of parameter values. We can be confident of a significant effect where the credible interval does not overlap with zero [65].

Once relations between ecosystem services were established, ANOVAs were used to test for differences between the categories of cities shown in Table 1, using Tukey's HSD multi-comparison test [63]. These are commonly used categories for classifying cities [6,61,62,66]. Synergies and tradeoffs were identified using pairwise Spearman correlations for all ecosystem services for all the cities together and separated by bio-socio-political relevant factors. Significance of the correlations was assessed at $p<0.001$, $p<0.05$ and $p<0.01$.

Results

Quantification of global ecosystem services

There was high variability in green cover. The city with the lowest amount of vegetation cover was Calcutta (0.4%), while Winnipeg had the highest (63.1%). The percent of green cover was normally distributed ($D=0.99$, $p<0.001$), with a mean of 32.6% (±12.1) with the majority of cities having between 20% and 40% green cover. Moran's I showed no indication of spatial autocorrelation ($Z=0.44$, $p=0.65$). The average recreation potential reached 8.9 m^2 per capita with the highest value (44 m^2) for Winnipeg and the minimum value for Istanbul (0.4 m^2). Carbon storage averaged 39 tonnes per ha, with the lowest value in Khartoum (0.2 tonnes/ha) and the maximum value in Paris (161 tonnes/ha). The average core habitat area provided within city habitat patches was 53%; the highest habitat potential was found for Montreal (98.3%) and the lowest for Bombay (2.4%). Results by city are detailed in Table S2.

Relationships between ecosystem services and urban climatic and development characteristics

The PCA was able to detect relationships between urban ecosystem services and development, political and climatic characteristics for the one hundred cities (Figure 1). A two-component solution provided the most interpretable result. Two components explained ~47% of the variance within the data (Table S3). Principal component 1 explained 31.6% of the variance in the data. The variables that loaded strongly (>0.4) on the first component were recreation potential, temperature, HDI and DI (Figure 1). Habitat provision, rainfall, HMI and population loaded strongly (>0.4) on the second component; this eigenvector explained 14.9% of the variance (Figure 1).

The PCA identified relationships between ecosystem services and some urban characteristics; however, to quantify the effect of these characteristics on services Bayesian regression was used to explore the size and direction of the effect for each urban characteristic (Figure 2; Table S3; Table S4). The association between carbon storage, HDI, DI and temperature, suggests that carbon storage tends to increase in wealthy, educated and democratic cities from cooler climates (Figure 1). The Bayesian analysis resulted in no significant effects detected for carbon storage (Figure 2); however, there were non-significant trends consistent with the PCA results suggesting a weak relationship with temperature and DI.

Recreation potential had a negative relationship with temperature, suggesting that cities from warmer climates tend to have a lower provision of this service. Higher recreation potential tends to occur in more democratic and more highly developed cities. Bayesian analysis showed that DI was the largest (positive) predictor of recreation potential, with population a significant (negative) predictor (Figure 2). There was also a non-significant trend suggesting that temperature may also be negatively related to recreation potential.

Habitat provision was positively related to HMI, suggesting that cooler and wetter cities have a lower provision of this service.

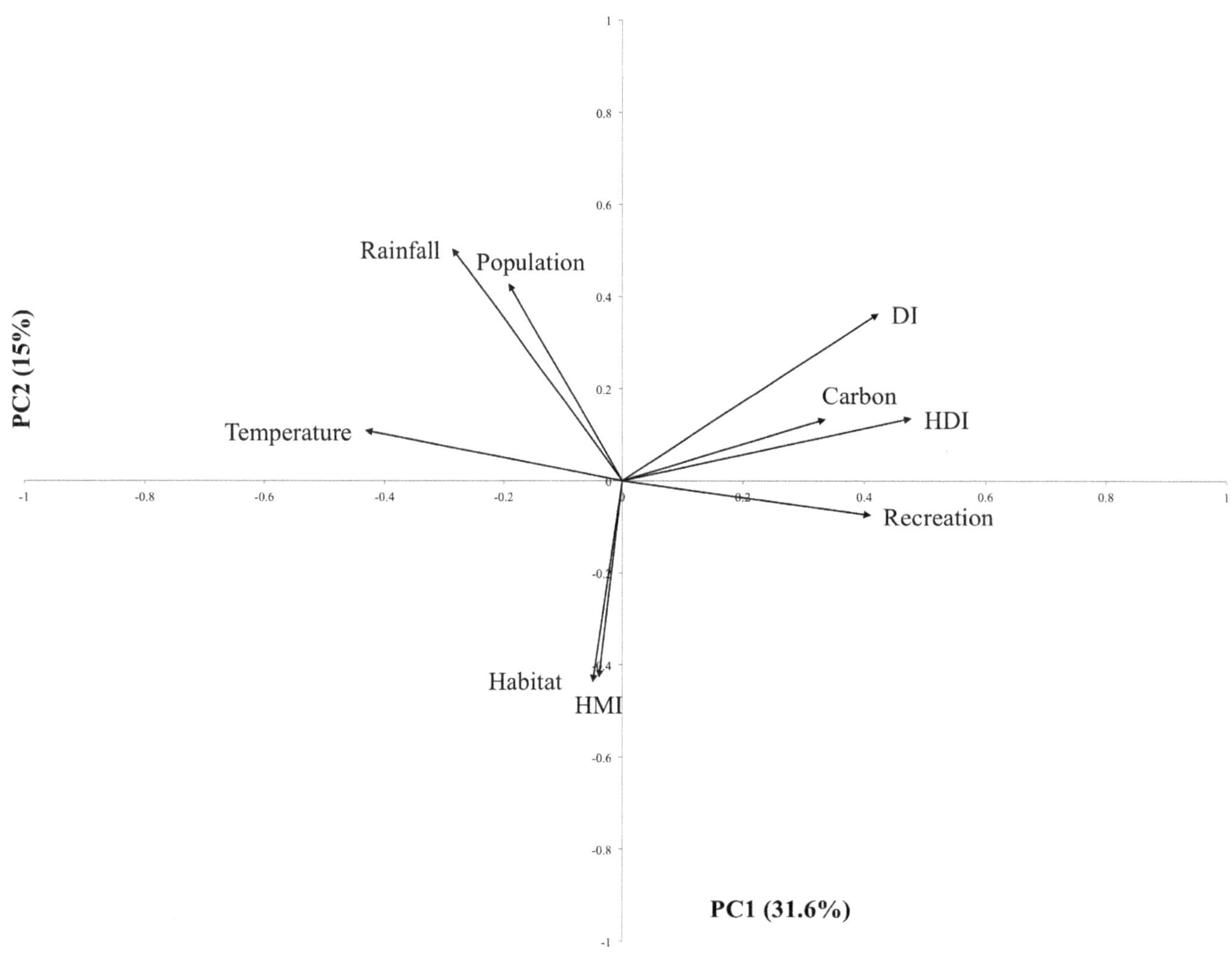

Figure 1. Ordination of the first and second standardized principal component for each ecosystem services and main drivers for 100 cities. The value of PC1 (Principal Component1) and PC2 (Principal Component 2) for the cities was standardised in order to more clearly show their location in the orthogonal space. The length of the arrow is an indication of the strength of the socio-political-climate variables and the ecosystem service in each PC. HMI: Heat Moisture Index, DI: Democracy Index, HDI: Human Development Index.

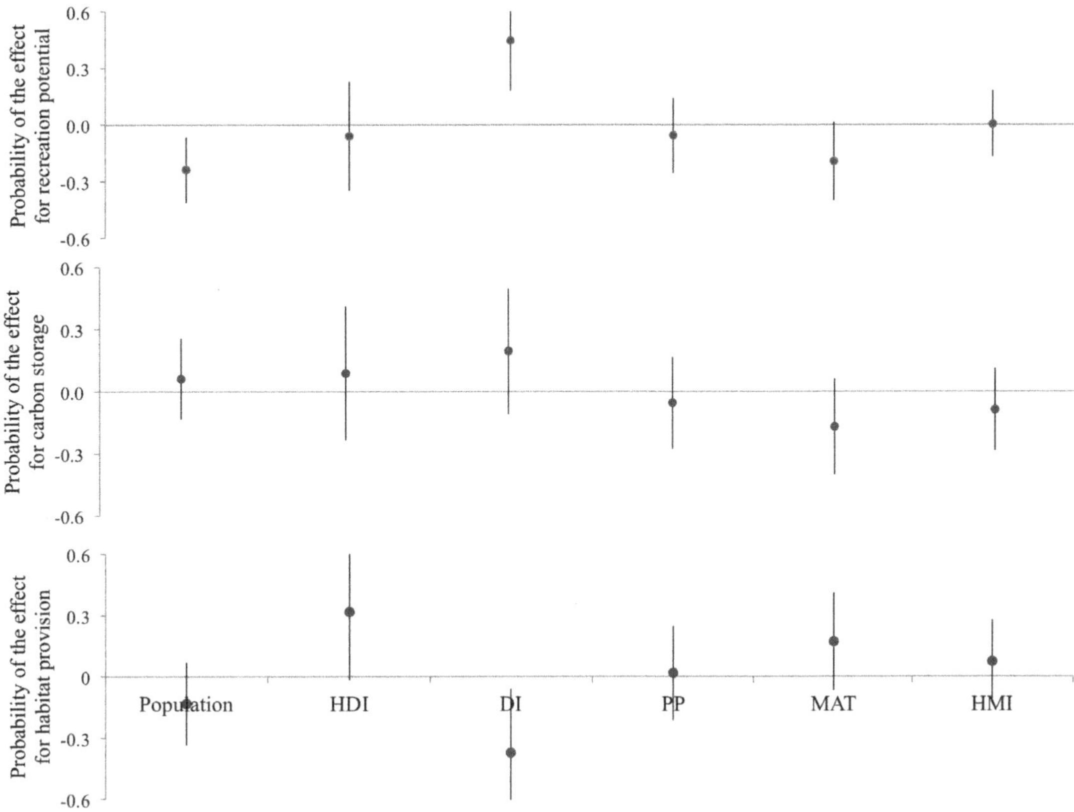

Figure 2. Bayesian models for three ecosystem services. Values overlapping zero imply a consistent effect of the bio-socio-political factor in the probability of having a positive or negative effect in the provision of ecosystem services. HDI: Human Development Index, DI: Democracy Index, PP: Annual precipitation, MAT: Mean Annual Temperature, HMI: Heat Moisture index.

Habitat provision had a negative relationship with population, suggesting that increases in population leads to fragmentation of urban vegetation (Figure 1). The Bayesian analysis showed that habitat potential was related (negatively) with DI (Figure 2), suggesting that more democratic cities tend to have lower habitat potential (more fragmented and less connected landscapes). A positive but non-significant trend was also detected between HDI and habitat potential.

The results from the ANOVAs and multiple comparison tests for the provision of ecosystem services by categories of city population, democratic regime and climate are summarised in Table 2. Carbon storage was higher in full than flawed democracies and authoritarian regimes. The analysis identified thresholds where the provision of ecosystem services declines. Cities with less than one million people tended to provide more recreation potential than larger cities, with megacities providing the least amount of recreational space. A higher provision of recreation service occurred in full democracies, and in continental cities over tropical and Mediterranean cities. Interestingly, authoritarian regimes tend to have a higher provision of habitat suggesting a higher degree of connectivity.

Synergies and tradeoffs of urban ecosystem services

There was a synergy between recreation potential and carbon storage across all cities (Table 3), which were positively and moderately correlated (Spearman's $\rho > 0.5$, $p < 0.001$). There were no significant correlations between habitat provision and recreation, or between carbon storage and habitat provision. Within city categories, the synergy between recreation and carbon is consistent for all population sizes with cities over 1 million people showing stronger correlations (Spearman's $\rho \approx 0.6$) than cities under 1 million (Spearman's $\rho \approx 0.45$). This synergy is also consistent for different political regimes except for full democracies. The synergy is strongest in Desert and Mediterranean cities and does not hold for continental and tropical cities. Synergies between recreation potential and habitat provision are only present in cities with under 1 million inhabitants and in democracies. A weak but significant synergy ($p < 0.01$) was found in Mediterranean cities between carbon storage and habitat provision (Table 3).

Discussion

Three ecosystem services were characterized for one hundred cities, revealing that the global distribution of ecosystem services are shaped by both development factors and climate. By integrating NDVI based land cover information with development indicators we demonstrated that anthropogenic variables do influence the provisioning of ecosystem services within urban systems [32,67,68]. Population and political factors more directly influenced recreation potential and habitat potential, while carbon storage was influenced by political regime and temperature. Similar to other urban studies, ecosystem services were influenced by both the natural environment (climate) and by demographics, socioeconomics and governance [14,69–71]. The congruence with local and regional studies demonstrates that these aforementioned relationships scale up to the global scale and highlight the need to

Table 2. Significant differences (ANOVA) in the provision of ecosystem services using categories from significantly influential urban characteristics.

		Recreation potential (m^2 per capita)	Carbon storage (kg/ha)	Habitat provision (%)
	r^2	*0.16*	*0.01*	*0.02*
Population	<1 million	16.5^a (2.6–44)	39.5^a (0.7–81)	45^a (5–74)
	1 to 2 million	8.6^b (1.7–31)	42.3^a (0.7–151)	54^a (4–93)
	2 to 6 million	7.5^b (0.6–33)	38.4^a (0.2–149)	56.2^a (4–98)
	>6 million	3.2^b (0.4–7.6)	27.7^a (5–161)	49.3^a (2–93)
	p-value	*0.0008*	*ns*	*ns*
Democracy Index	*r^2*	*0.29*	*0.14*	*0.04*
	Authoritarian	3.4^a (0.8–6.5)	21.3^b (0.2–59)	63.4^a (7–93)
	Hybrid	3.3^a (0.4–8)	32^{ab} (3.1–114)	52.6^b (4–78)
	Flawed democracy	5.6^a (0.7–31)	31.5^b (0.7–161)	52.9^b (2–90)
	Full democracy	13.9^b (1–44)	55.5^a (3.4–151)	48.9^b (4–78)
	p-value	*<0.0001*	*0.002*	*0.01*
Climate	*r^2*	*0.10*	*0.03*	*0.02*
	Tropical	3.3^a (0.6–8)	25.5^a (0.7–94.1)	58.2^a (2–88)
	Desert	8.5^{ab} (0.6–29)	31.7^a (0.2–84)	61.5^a (7–90)
	Mediterranean	7.4^a (0.4–33)	40.8^a (1.9–161)	51.6^a (4–93)
	Continental	13.7^b (3–44)	46.5^a (0.7–103)	49.2^a (4–98)
	p-value	*0.01*	*ns*	*ns*

Values label with the different letter a imply significant differences among categories of analysis for Tukey's HSD comparison test.

consider socio-political-environmental dimensions when developing urban areas to achieve sustainable development goals.

Our study found that recreational potential is lower in cities with more than 1 million inhabitants, which is consistent with trends observed in some European cities [51]. Recreation potential also decreased with changes in governance as cities in countries governed by non-democratic regimes had a lower provision of this service. It is well understood that developing countries have social inequalities caused by urbanization [72], and this study highlights the existence of environmental inequalities that may be exacerbate by urbanization. Of the 100 cities studied, only 26% of the cities had more green cover person than World Health Organization (WHO) recommendations of 9 m^2 per capita [73] and only 12% of the cities met the green space per capita of 20 m^2 per capita [74,75]. The only cities that reached the WHO recommendation were mid-size cities (1 to 6 million people) that are predominantly located in North America and South Africa, with the exception of

Table 3. Spearman correlations between different ecosystem services for all the cities and by categories of population, democracy and climate.

		Recreation vs. Carbon	Recreation vs. Habitat	Carbon vs. Habitat
All cities		0.53***	0.09	0.02
Population	Less 1 million	0.46*	0.47*	0.009
	1 to 2 million	0.64 ***	0.1	0.05
	2 to 6 million	0.59***	0.24	0.11
	more than 6 million	0.62**	−0.12	−0.14
Democracy	Authoritarian regimes	0.47**	0.07	−0.04
	Hybrid regimes	0.66**	−0.11	−0.02
	Flawed Democracy	0.63***	0.25*	0.17
	Full Democracy	0.12	0.36*	0.15
Climate	Tropical	0.18	0.3	−0.19
	Desert	0.64**	0.009	−0.1
	Mediterranean	0.58***	0.12	0.19*
	Continental	0.27	0.24	0.05

Fisher significant test: *p-value<0.01, **p-value<0.05, ***p-value<0.001.

Bismarck, Oklahoma City and Winnipeg all located in North America.

Cities with higher HDI and under democratic regimes were found to provide more green space for their inhabitants, which may be due to the increased demand for environmental quality by residents [76]. This group included cities in Canada, which have strong environmental and urban forestry programs and policies at both the national and local level. These policies promote an increase in urban green space and street tree plantings and the maintenance of conservation areas [77]. At the opposite end of the spectrum are cities with low HDI, where the main policies at the country and local level are mainly related to socioeconomics; in addition, many of these countries have poor institutional capacity and insufficient budgets to deliver environmental policies [78,79].

Provision of the carbon storage service was found to be strongly influence by climate, though HDI and DI (in interaction with HDI) were also important factors. Cities with the highest provision of this service were mostly from continental biomes (Frankfurt, Paris, Prague), while the lowest provisioning of this service in its majority occurred in cities located in tropical to desert biomes (Ulaanbaatar, Sana'a, Pyonyang). Certain cities however were outliers, interestingly Mendoza, Phoenix and Las Vegas have been able to increase carbon storage despite being located in desert biomes [80,81]. Humans have greatly increased the number of trees within cities such as Phoenix, U.S.A. and Mendoza, Argentina, despite climatic limitations, thereby increasing the provision of this service. In general, more affluent and democratic cities are more likely to have a greater biomass of vegetation that can sequester more carbon, either by maintaining larger patches of vegetation and, or in the case of many continental cities, larger tree populations in streets.

Habitat potential was mainly related to HDI and governance, which is consistent with other studies (e.g. Schwarz 2010; Huang et al. 2007). Cities governed by flawed democracies, authoritarian and hybrid regimes from low-income countries tend to be more compact, with vegetation restricted to peri-urban areas and consequently have low levels of fragmentation. This might reflect the effect of motorization with developed countries with full democracies as they tend to be sprawling cites characterized by high levels of vehicle ownership and the associated transportation infrastructure required to facilitate commuting by vehicles [71]. In addition, control over land ownership under socialist or communist regimes tend to result in cities that are less fragmented [71], therefore maximizing habitat potential. When a city is governed within a full democratic regime, habitat potential is likely to be reduced. This is particularly apparent in cities where urban development happened between the 18th and 19th century under European colonization and where high rates of sprawling and dispersed urbanization are still prevalent (e.g. U.S.A. and Australia; [71]).

The delivery of ecosystem services varies according to city context. In general, megacities provide low levels of ecosystem services; mid-size cities provided average levels of ecosystem services, while cities with less than 1 million inhabitants typically had higher levels of recreation potential. Fuller and Gaston [51] found the same trend in European cities which highlights that a common signal exists at the global scale. The level of democracy within the country the city is located in affected both recreation potential and carbon storage services positively but had a negative effect on habitat provision. Overall this suggests that inequalities in the provision of ecosystem services exists between less and more developed countries which further supports our thesis that environmental inequalities found within cities maybe relevant at the global scale [17,83].

The cities that have the highest values for the three services combined in this study were Prague, Paris and Frankfurt, old European cities of very high development level under full democracy. While the lowest values provided for the combined services were Ulaanbaatar, Buenos Aires and Tegucigalpa with medium level of development within flawed and hybrid democracies.

The analysis revealed that synergies between cultural and regulation functions exist as do tradeoffs with some supporting services. Synergies and tradeoffs of this nature have also been found for a variety of other land uses [29,31]. Relationships between ecosystem services were not linear and varied according to the combination of socioeconomic and political characteristics of the urban ecosystem which is consistent with findings of global coastal ecosystem services [84]. Unsurprisingly, there was a moderately strong synergy between recreation and carbon across most cities as both services are related to the amount of vegetation in cities. However, the synergistic relation between recreation and carbon is very weak for full democracies, which might be due to the existence of parks with relatively few trees, and a relatively larger population of trees located along streets. This is confirmed by the positive relation between recreation and habitat and the weak correlation with habitat and carbon. The relation between recreation and carbon is also weaker in tropical and continental cities. Tropical cities have smaller vegetated areas that are primarily covered by trees, while continental cities have larger vegetated areas cover by fewer trees. There was a substantial variability in the habitat and carbon data however and the power of the analysis may be improved by increasing the sampled size of cities within each climate classification.

The analysis of ecosystem service provision for a hundred cities has its limitations, which need to be acknowledged. City boundaries were difficult to obtain for a wide range of cities; therefore, errors may exist in the estimation of the services. However, for the relationships between democracy, development and climate our methodology allows for a robust assessment. A better estimation of recreation potential could have been obtained if we had further details on the structure and composition of these spaces along with the use of these spaces by people. The resolution of Landsat imagery however was not fine enough to achieve this, and as finer resolution imagery is difficult to source for a large proportion of cities included in this study this level of analysis could not be conducted. The estimation of carbon stored through a model that uses NDVI is commonly used in natural areas and some urban areas [52–54], where the outcomes are biased by model performance and the calculation of NDVI; we at least had control over biases and errors in the latter. The habitat potential estimation was not intended to assess the functionality of each vegetation patch and therefore is a coarse metric of habitat. We feel however that the results from this study provide a level of precision that is consistent between cities and our results are consistent with the findings of other urban studies conducted at finer scales.

Further research should include the addition of indicators that can represent cultural background and the legacy of historic development such as the effects of colonization, wars, ethnic diversity, industrialization, planning regulation and infrastructure development, among others [71]. A temporal analysis may also be able to shed light on different urban morphological trajectories and their relationships with ecosystem services. The inclusion of other ecosystem services would reveal more about how the context of the city affects the provision of ecosystem services; however, finding available information for a large range of cities is problematic. Existing standardised global datasets, such as the one developed in this study, are useful to explore the tradeoffs or synergies between services, along with finer scale indicators of biodiversity and other services.

Conclusion

Cities are areas of human agglomeration that depend on natural resources for the maintenance of human wellbeing. This study has identified that a relationship exists between the bio-socio-political context and the provision of ecosystem services. Cities in countries with democratic systems and more developed economies tend to provide more ecosystem services to their inhabitants; in theory, this should promote improved human wellbeing. This relation becomes more evident for the cultural and provisioning services included in this study, while regulating services such as carbon storage are primarily driven by biophysical conditions followed by social context. The context of the city also influences the synergies and tradeoffs between ecosystem services. This highlights that improvements in economic conditions may not maximise and can hinder the provision of ecosystem services. As a global city scale analysis, this study was able to identify the existence of environmental inequalities according to political, economic and demographic context, which suggests further research should explore these relations within and across cities. Understanding the synergies between services and social and environmental context should ameliorate the development of environmental and social inequalities that are typical of urbanization. The relationship between ecosystem services and bio-socio-political context provides a key understanding of the influential factors that urban planning and policy making impinge upon and thus provide insights for creating liveable, sustainable, and resilient cities globally.

Supporting Information

Figure S1 Map of studied cities.

Table S1 List of cities included in this study. Details on population, Human Development Index and Democracy Index by each city are provided.

Table S2 Three urban forest ecosystem services for one hundred cities included in the study.

Table S3 Scores for principal components and their respective eigenvalues.

Table S4 Mean, 2.5% and 97.5% confidence intervals for estimated probabilities of the effect of socioeconomic, political and climate variable using Bayesian regression on each studied ecosystem services.

Acknowledgments

The authors would like to thank Jorge Aubad for early revisions of this manuscript and the two anonymous reviewers that greatly improve this manuscript.

Author Contributions

Conceived and designed the experiments: CD CN DK. Performed the experiments: CD. Analyzed the data: CD. Contributed reagents/materials/analysis tools: CD CN DK. Contributed to the writing of the manuscript: CD CN DK.

References

1. Grimm NB, Faeth SH, Golubiewski NE, Redman CL, Wu J, et al. (2008) Global change and the ecology of cities. Science 319: 756–760.
2. Millenium Ecosystem Assessment (2003) Introduction and Conceptual Framework. Ecosystems and Human Well-being: A framework for Assessment. Island Press. 26–48.
3. Daily G (1997) What are ecosystem services? Nature's Services: Societal Dependance on Natural Ecosystems. Washington D.C: Island Press.
4. Millenium Ecosystem Assessment (2003) Ecosystems and human well being. Ecosystems and Human Well-being: A framework for Assessment. Washington D.C: Island Press. 85–106.
5. Bastian O, Haase D, Grunewald K (2011) Ecosystem properties, potentials and services – The EPPS conceptual framework and an urban application example. Ecol Indic 21: 7–16.
6. UN-Habitat (2011) Cities and Climate Change: Global report on human settlements. London, Washington D.C: EarthScan.
7. De Groot RS, Wilson MA, Boumans RM (2002) A typology for the classification, description and valuation of ecosystem functions, goods and services. Ecol Econ 41: 393–408.
8. Nowak DJ, Walton JT, Stevens JC, Crane DE, Hoehn RE (2008) A ground-based method of assessing urban forest structure and ecosystem services. Arboric Urban For 34: 347–358.
9. Dobbs C, Escobedo FJ, Zipperer WC (2011) A framework for developing urban forest ecosystem services and goods indicators. Landsc Urban Plan 99: 196–206.
10. Alberti M, Marzluff JM, Sculenberger E, Bradley G, Ryan C, et al. (2003) Integrating humans into ecology: Opportunities and challenges for studying urban ecosystems. Bioscience 53: 1169–1179.
11. Andersson E (2006) Urban Landscapes and Sustainable Cities. Ecol Soc 11. Available: http://www.ecologyandsociety.org/vol11/iss1/art34/.
12. Escobedo FJ, Kroeger T, Wagner JE (2011) Urban forests and pollution mitigation: analyzing ecosystem services and disservices. Environ Pollut 159: 2078–2087.
13. Escobedo FJ, Nowak DJ, Wagner J, De la Maza CL, Rodríguez M, et al. (2006) The socioeconomics and management of Santiago de Chile's public urban forests. Urban For Urban Green 4: 105–114.
14. Kinzig AP, Warren PS, Martin CA, Hope D, Katti M (2005) The Effects of Human Socioeconomic Status and Cultural Characteristics on Urban Patterns of Biodiversity. Ecol Soc 10.
15. Szantoi Z, Escobedo FJ, Wagner J, Rodriguez JM, Smith S (2012) Socioeconomic Factors and Urban Tree Cover Policies in a Subtropical Urban Forest. GIScience Remote Sens 49: 428–449.
16. Heynen NC, Lindsey G (2003) Correlates of urban forest canopy cover: implications for local public works. Public Work Manag Policy 8: 33–47.
17. Heynen NC, Perkins HA, Roy P (2006) The political ecology of uneven urban green space: The impact of political economy on race and ethnicity in producing environmental inequality in Milwaukee. Urban Aff Rev 42: 3–25.
18. Kitchen L (2013) Are Trees Always "Good"? Urban Political Ecology and Environmental Justice in the Valleys of South Wales. Int J Urban Reg Res 37: 1968–1983.
19. Perkins H, Heynen NC, Wilson J (2004) Inequitable access to urban reforestation: the impact of urban political economy on housing tenure and urban forests. Cities 21: 291–299.
20. Wilkinson C, Sendstad M, Parnell S, Schewenius M (2013) Urbanization, Biodiversity and Ecosystem Services: Challenges and Opportunities. Elmqvist T, Fragkias M, Goodness J, Güneralp B, Marcotullio PJ, et al., editors Dordrecht: Springer Netherlands.
21. Eigenbrod F, Armsworth PR, Anderson BJ, Heinemeyer A, Gillings S, et al. (2010) The impact of proxy-based methods on mapping the distribution of ecosystem services. J Appl Ecol 47: 377–385.
22. Nowak DJ, Crane DE (2002) Carbon storage and sequestration by urban trees in the United States. Environ Pollut 116: 381–389.
23. Tratalos JA, Fuller RA, Warren PH, Davies RG, Gaston KJ (2007) Urban form, biodiversity potential and ecosystem services. Landsc Urban Plan 83: 308–317.
24. Edmondson JL, Davies ZG, McCormack SA, Gaston KJ, Leake JR (2014) Land-cover effects on soil organic carbon stocks in a European city. Sci Total Environ 472: 444–453.
25. Tratalos JA, Fuller RA, Evans KL, Davies RG, Newson SE, et al. (2007) Bird densities are associated with household densities. Glob Chang Biol 13: 1685–1695.
26. Davies ZG, Edmondson JL, Heinemeyer A, Leake JR, Gaston KJ (2011) Mapping an urban ecosystem service: quantifying above-ground carbon storage at a city-wide scale. J Appl Ecol 48: 1125–1134.
27. Cilliers S, Cilliers J, Lubbe R, Siebert S (2012) Ecosystem services of urban green spaces in African countries-perspectives and challenges. Urban Ecosyst 16: 681–702.

28. Ernstson H, Sörlin S, Elmqvist T (2009) Social Movements and Ecosystem Services – the Role of Social Network Structure in Protecting and Managing Urban Green Areas in Stockholm. Ecol Soc 13.
29. Haase D, Schwarz N, Strohbach MW, Kroll F, Seppelt R (2012) Synergies, Trade-offs, and Losses of Ecosystem Services in Urban Regions: an Integrated Multiscale Framework Applied to the Leipzig-Halle Region, Germany. Ecol Soc 17.
30. Millenium Ecosystem Assessment (2003) Drivers of change in ecosystems and their services. Washington D.C: Island Press.
31. Raudsepp-Hearne C, Peterson GD, Bennett EM (2010) Ecosystem service bundles for analyzing tradeoffs in diverse landscapes. Proc Natl Acad Sci U S A 107: 5242–5247.
32. Bennett EM, Peterson GD, Gordon LJ (2009) Understanding relationships among multiple ecosystem services. Ecol Lett 12: 1394–1404.
33. Fuller RA, Irvine KN, Devine-wright P, Warren PH, Gaston KJ (2007) Psychological benefits of greenspace increase with biodiversity. Biol Lett 3: 390–394.
34. Tzoulas K, Korpela K, Venn S, Yli-pelkonen V, Ka A, et al. (2007) Promoting ecosystem and human health in urban areas using Green Infrastructure: A literature review. Landsc Urban Plan 81: 167–178.
35. Jim CY, Chen WY (2009) Ecosystem services and valuation of urban forests in China. Cities 26: 187–194.
36. Bjerke T, Østdahl T, Thrane C, Strumse E (2006) Vegetation density of urban parks and perceived appropriateness for recreation. Urban For Urban Green 5: 35–44.
37. McDonald RI (2008) Global urbanization: can ecologists identify a sustainable way forward? Front Ecol Environ 6: 99–104.
38. Tannier C, Foltête J, Girardet X (2012) Assessing the capacity of different urban forms to preserve the connectivity of ecological habitats. Landsc Urban Plan 105: 128–139.
39. Aronson MFJ, La Sorte FA, Nilon CH, Katti M, Goddard MA, et al. (2014) A global analysis of the impacts of urbanization on bird and plant diversity reveals key anthropogenic drivers. Proc Biol Sci 281: 20133330.
40. Luck GW, Davidson P, Boxall D, Smallbone L (2011) Relations between Urban Bird and Plant Communities and Human Well-Being and Connection to Nature. Conserv Biol 25: 816–826.
41. Mayer FS, Ã CMF (2005) The connectedness to nature scale: A measure of individuals ' feeling in community with nature. J Environ Psychol 24: 503–515.
42. Keniger LE, Gaston KJ, Irvine KN, Fuller RA (2013) What are the benefits of interacting with nature? Int J Environ Res Public Health 10: 913–935.
43. Zhang JW, Howell RT, Iyer R (2014) Engagement with natural beauty moderates the positive relation between connectedness with nature and psychological well-being. J Environ Psychol 38: 55–63.
44. Pataki DE, Alig RJ, Fung AS, Golubiewski NE, Kennedy CA, et al. (2006) Urban ecosystems and the North American carbon cycle. Glob Chang Biol 12: 1–11.
45. Van de Voorde T, Vlaeminck J, Canters F (2008) Comparing Different Approaches for Mapping Urban Vegetation Cover from Landsat ETM+ Data: A Case Study on Brussels. Sensors 8: 3880–3902.
46. Zhu Z, Woodcock CE, Rogan J, Kellndorfer J (2011) Assessment of spectral, polarimetric, temporal, and spatial dimensions for urban and peri-urban land cover classification using Landsat and SAR data. Remote Sens Environ 117: 72–82.
47. Schneider A, Woodcock CE (2008) Compact, Dispersed, Fragmented, Extensive? A Comparison of Urban Growth in Twenty-five Global Cities using Remotely Sensed Data, Pattern Metrics and Census Information. Urban Stud 45: 659–692.
48. Potere D, Schneider A, Angel S, Civco L (2009) Mapping urban areas on a global scale: which of the eight maps now available is more accurate? Int J Remote Sens 30: 37–41.
49. Tucker CJ (1979) Red and photographic infrared linear combinations for monitoring vegetation. Remote Sens Environ 8: 127–150.
50. Congalton RG, Green K (2009) Assessing the Accuracy of Remotely Sensed Data: Principles and Practices. 2nd Editio. Boca Raton, FL: CRC/Taylor & Francis.
51. Fuller RA, Gaston KJ (2009) The scaling of green space coverage in European cities. Biol Lett 5: 352–355.
52. Myeong S, Nowak DJ, Duggin MJ (2006) A temporal analysis of urban forest carbon storage using remote sensing. Remote Sens Environ 101: 277–282.
53. Poudyal NC, Siry JP, Bowker JM (2010) Urban forests' potential to supply marketable carbon emission offsets: A survey of municipal governments in the United States. For Policy Econ 12: 432–438.
54. Hutyra LR, Yoon B, Alberti M (2011) Terrestrial carbon stocks across a gradient of urbanization: a study of the Seattle, WA region. Glob Chang Biol 17: 783–797.
55. Saura S, Vogt P, Velázquez J, Hernando A, Tejera R (2011) Key structural forest connectors can be identified by combining landscape spatial pattern and network analyses. For Ecol Manage 262: 150–160.
56. Wickham JD, Riitters KH, Wade TG, Vogt P (2010) A national assessment of green infrastructure and change for the conterminous United States using morphological image processing. Landsc Urban Plan 94: 186–195.
57. Vogt P, Riitters KH, Estreguil C, Kozak J, Wade TG, et al. (2006) Mapping Spatial Patterns with Morphological Image Processing. Landsc Ecol 22: 171–177.
58. Begon M, Townsend C, Harper J (2006) Ecology: from individuals to ecosystems. 4th Editio. Blackwell Publishing Ltd.
59. Naidoo R, Balmford a, Costanza R, Fisher B, Green RE, et al. (2008) Global mapping of ecosystem services and conservation priorities. Proc Natl Acad Sci U S A 105: 9495–9500.
60. Wang T, Hamann A, Yanchuk A, O'Neill GA, Aitken SN (2006) Use of response functions in selecting lodgepole pine populations for future climates. Glob Chang Biol 12: 2404–2416.
61. UNDP (2011) Human Development Report 2011 A Better Future for All. New York.
62. Economist Intelligence Unit (2011) Democracy index 2011: Democracy under stress. Available: www.eiu.com.
63. Fortin M-J, James PM a., MacKenzie A, Melles SJ, Rayfield B (2012) Spatial statistics, spatial regression, and graph theory in ecology. Spat Stat 1: 100–109.
64. Demšar U, Harris P, Brunsdon C, Fotheringham a S, McLoone S (2013) Principal Component Analysis on Spatial Data: An Overview. Ann Assoc Am Geogr 103: 106–128.
65. McMahon SM, Diez JM (2007) Scales of association: hierarchical linear models and the measurement of ecological systems. Ecol Lett 10: 437–452.
66. Bowler DE, Buyung-ali L, Knight TM, Pullin AS (2010) Urban greening to cool towns and cities: A systematic review of the empirical evidence. Landsc Urban Plan 97: 147–155.
67. Ellis EC, Ramankutty N (2008) Putting people in the map: anthropogenic biomes of the world. Front Ecol Environ 6: 439–447.
68. Mikkelson GM, Gonzalez A, Peterson GD (2007) Economic inequality predicts biodiversity loss. PLoS One 2: e444.
69. Angel S, Sheppard SC, Civco DL, Buckley R, Chabaeva A, et al. (2005) The Dynamics of Global Urban Expansion. Washington D.C: Transport and Urban Development Department, The World Bank.
70. Hope D, Gries C, Zhu W, Fagan WF, Redman CL, et al. (2003) Socioeconomics drive urban plant diversity. P Natl Acad Sci-Biol 100: 8788–8792.
71. Huang J, Lu XX, Sellers JM (2007) A global comparative analysis of urban form: Applying spatial metrics and remote sensing. Landsc Urban Plan 82: 184–197.
72. United Nations (2011) Are we building competitive and liveable cities? Guidelines for developing eco-efficient and socially inclusive infrastructure. United Nations.
73. Kuchelmeister G (1998) Urban forestry in the Asia-Pacific Region: Status and prospects. Rome.
74. Sukopp H, Numata M, Huber A (1995) Urban Ecology as the Basis of Urban Planning. The Hague: SPB Academic Pub.
75. Wang X-J (2009) Analysis of problems in urban green space system planning in China. J For Res 20: 79–82.
76. Farzin YH, Bond CA (2006) Democracy and environmental quality. J Dev Econ 81: 213–235.
77. Conway TM, Urbani L (2007) Variations in municipal urban forestry policies: A case study of Toronto, Canada. Urban For Urban Green 6: 181–192.
78. Camara de Comercio de Bogota (2009) Capítulo 3. Gestión del Espacio Público de Bogotá. Observatorio de la Gestion Urbana. Camara de Comercio de Bogota. 37–60.
79. Singh VS, Pandey DN, Chaudry P (2010) Urban forest and open green spaces: lessons for Jaipur, Rajasthan, India. Rajasthan, India.
80. Strohbach MW, Arnold E, Haase D (2012) The carbon footprint of urban green space-A life cycle approach. Landsc Urban Plan 104: 220–229.
81. Imhoff ML, Bounoua L, Defries R, Lawrence T, Stutzer D, et al. (2004) The consequences of urban land transformation on net primary productivity in the United States. Remote Sensing of. Remote Sens Environ 89: 434–443.
82. Schwarz N (2010) Urban form revisited-Selecting indicators for characterising European cities. Landsc Urban Plan 96: 29–47.
83. Pedlowski MA, Da Silva VAC, Adell JJC, Heynen NC (2002) Urban forest and environmental inequality in Campos dos Goytacazes, Rio de Janeiro, Brazil. Urban Ecosyst 6: 9–20.
84. Koch EW, Barbier EB, Silliman BR, Reed DJ, Perillo GME, et al. (2009) Non-linearity in ecosystem services: temporal and spatial variability in coastal protection. Front Ecol Environ 7: 29–37.

Total Nitrogen Concentrations in Surface Water of Typical Agro- and Forest Ecosystems in China, 2004-2009

Zhiwei Xu, Xinyu Zhang*, Juan Xie, Guofu Yuan, Xinzhai Tang, Xiaomin Sun, Guirui Yu

Key Laboratory of Ecosystem Network Observation and Modeling, Institute of Geographic Sciences and Natural Resources Research, Chinese Academy of Sciences, Beijing, China

Abstract

We assessed the total nitrogen (N) concentrations of 28 still surface water (lake and pond), and 42 flowing surface water (river), monitoring sites under 29 typical terrestrial ecosystems of the Chinese Ecosystem Research Network (CERN) using monitoring data collected between 2004 and 2009. The results showed that the median total N concentrations of still surface water were significantly higher in the agro- (1.5 $mg \cdot L^{-1}$) and oasis agro- ecosystems (1.8 $mg \cdot L^{-1}$) than in the forest ecosystems (1.0 $mg \cdot L^{-1}$). This was also the case for flowing surface water, with total N concentrations of 2.4 $mg \cdot L^{-1}$, 1.8 $mg \cdot L^{-1}$ and 0.5 $mg \cdot L^{-1}$ for the agro-, oasis agro- and forest ecosystems, respectively. In addition, more than 50% of the samples in agro- and oasis agro- ecosystems were seriously polluted (>1.0 $mg \cdot L^{-1}$) by N. Spatial analysis showed that the total N concentrations in northern and northwestern regions were higher than those in the southern region for both still and flowing surface waters under agro- and oasis agro- ecosystems, with more than 50% of samples exceeding 1.0 $mg \cdot L^{-1}$ (the Class III limit of the Chinese National Quality Standards for Surface Waters) in surface water in the northern region. Nitrogen pollution in agro- ecosystems is mainly due to fertilizer applications, while the combination of fertilizer and irrigation exacerbates nitrogen pollution in oasis agro- ecosystems.

Editor: Manuel Reigosa, University of Vigo, Spain

Funding: This research was funded by the Key Direction in Knowledge Innovation Program of the Chinese Academy of Sciences (KZCX2-EW-310) and the National Natural Science Foundation of China (No. 41171153). The authors declare that no additional external funding was received for this study. The funders had no role in study design, data collection and analysis, decision to publish, or preparation of the manuscript.

Competing Interests: The authors have declared that no competing interests exist.

* E-mail: zhangxy@igsnrr.ac.cn

Introduction

Nitrogen (N) pollution, leading to eutrophication of inland waters, has resulted in an increase in global algal biomass and photosynthesis, such that primary production is approximately 60% higher than expected background levels in lakes [1], streams and rivers [2]. As a major contributor to eutrophication of water bodies, non-point losses of N (e.g. in runoff) have received particular attention. N pollution causes water eutrophication, which disrupts ecology and causes, among other problems, toxic algal blooms, loss of oxygen, loss of biodiversity (including species that are important for commerce and recreation) [3–5]. Eutrophication can also seriously affect our ability to use water for drinking, industry, agriculture, recreation, and other purposes.

Total N concentrations in surface water have been classified in China and other countries to control water eutrophication and improve water quality. To identify at-risk surface water bodies and protect them from eutrophication, the US EPA developed guidelines, which state that N concentrations should not exceed 0.3 $mg \cdot L^{-1}$ in streams and rivers or 0.1 $mg \cdot L^{-1}$ in lakes and reservoirs [6]. Water has been divided into 14 distinct aggregate nutrient ecoregions in the U.S. according to total P and total N concentrations, chlorophyll *a* and turbidity. In China surface water has been divided into five categories according to the Chinese National Quality Standards for Surface Water. Water categorized as class I to III can be used as drinking water, while class IV and V water is only suitable for industrial and agricultural uses. The total N values for categories I – V are < 0.2 $mg \cdot L^{-1}$, 0.2 – 0.5 $mg \cdot L^{-1}$, 0.5 – 1.0 $mg \cdot L^{-1}$, 1.0 – 1.5 $mg \cdot L^{-1}$ and 1.5 – 2.0 $mg \cdot L^{-1}$, respectively [7].

In China, stream total N concentrations have tended to increase since the 1980s as a consequence of demographic, industrial and agricultural development [8,9]. Rivers and lakes in the Taihu Lake region are polluted to varying degrees by N [10], with 80% of samples having concentrations exceeding 1.0 $mg \cdot L^{-1}$, meaning that the water is only suitable for industrial, agricultural and landscape uses according to the Chinese National Quality Standards for Surface Water [11]. Cai et al. [12] reported that eutrophication of lakes is serious in southern China, but that it is worse across a large part of northern China. Rural rivers in eastern China are severely polluted [13]. As a general rule, water pollution in China tends to intensify from tributaries to the main stems of river systems, from urban to rural areas, from surface water to groundwater and from the regional to the basin scale [12].

Natural and anthropogenic sources both contribute to surface water N inputs [14,15]. Now that point-source pollution has been controlled effectively, non-point source pollution, especially that which results from agricultural land management, has become the main influence on surface water and is an important environmental issue worldwide [16,17]. Crop production is by far the largest cause of human alteration of the global N cycle. Global industrial N fixation for fertilizers has increased rapidly to 80×10^{6} $Mg \cdot year^{-1}$ [18]. In the United States (US) and Europe, only 18% of the N input in fertilizer is removed from farms in produce, leaving behind, on average, 174 kg·ha^{-1}·year^{-1} of surplus N [19,20]. This surplus may accumulate in soils, from where it may be either eroded and transported to surface water, leached to surface and ground water, or lost to the atmosphere [18]. Inorganic fertilizers now contribute 80 Tg N year^{-1} to the environment, while 32 – 45 Tg year^{-1} of inorganic fertilizers find their way into freshwater (from leaching and erosion) [21].

Fertilizer consumption has grown rapidly in China since 1978, and China recently became the biggest producer and consumer of N fertilizer in the world. The average annual application rate of N in China gradually increased to 130 kg·ha^{-1} in 1985, then rapidly increased to 236 kg·ha^{-1} in 1995 and then to 262 kg·ha^{-1} in 2001 [22]. However, less than half of the fertilizer N applied in China is taken up by crops [23]. The losses of total N through runoff and leaching were 5.6 – 9.0 kg·ha^{-1} from paddy fields and 12.5 – 19.5 kg·ha^{-1} from upland fields, accounting for 2.6 – 4.2% and 7.6 – 12.3% of the total applied N [24]. It is estimated that as much as 0.644 Tg of total N is leached, accounting for 2.8% of China's total N application [24].

Because of the increasing incidence of surface water N pollution, many countries are beginning to be concerned about sustainable water management. The European Union has implemented the Water Framework Directive, which aims to recover good status in water resources by 2015 [25]. Canada (both at the national and provincial levels) and the US have begun to take an effective and integrated approach to land-use management with respect to protection of drinking water sources. On a national level, Canada has established aboriginal water systems, while at the provincial level; British Columbia has established a water policy, similar to that developed by the US Environmental Protection Agency [26]. The Chinese government has published many water environmental protection measures such as the Water Pollution Prevention Program of Yangtze River basin (2011–2015) and other measures for important river basins (Yellow River, Huaihe River, Weihe River, Liaohe River, Haihe River, Songhuajiang River, Chaohu and Dianchi) [27]. Until now however, there have been few national scale assessments of N concentrations in surface water in China.

Using the Chinese Ecosystem Research Network (CERN) as the monitoring framework, total N concentrations at 28 still, and 45 flowing, surface water monitoring sites of 28 CERN monitoring stations were assessed from 2004 to 2009. The aims of this study were: (1) to assess surface water total N pollution under agro-, oasis agro- and forest ecosystems and (2) to identify sites vulnerable to N pollution in agro- and oasis agro- ecosystems. Findings from this study will provide key information that will be useful in controlling N pollution in surface water in China.

Materials and Methods

2.1 Ethics statement

The authors declare that no specific permits were required for the described field studies. The authors also declare that no specific permissions were required for these locations/activities. Locations are not privately-owned or protected in any way. We confirm that this study did not involve endangered or protected species and no protected species were sampled during the monitoring campaign.

2.2 Monitoring sites

Surface water monitoring of the CERN focuses on assessing the effects of typical terrestrial ecosystems, i.e. agro-, oasis agro- and forest ecosystems, on water quality. Twenty-eight monitoring stations of the CERN were chosen to represent typical agricultural, oasis agricultural and forest ecosystems, geology, soils, and land use types found across a wide range of climatic zones in China (Figure 1). The annual average rainfall of the monitoring ecosystems (80°43′39″-133°18′03″E, 18°13′01″-47°27′15″N) varied from approximately 35 mm (Cele) to 1956 mm (Dinghushan), while the annual average temperature ranged from 1.5°C (Hailun) to 21.8°C (Xishuangbanna).

The 12 agro- ecosystems were located in (1) humid and subhumid regions in the temperate zone of northeastern China (Hailun, Shenyang) and the warm temperate zone of northern China (Luancheng, Yucheng, Fengqiu), and (2) the Loess Plateau (Ansai and Changwu), (3) humid areas in the sub-tropical zone in the Yangtze River Delta (Changshu) and southern China (Huanjiang, Qianyanzhou, Yanting, Taoyuan,Yingtan), respectively (Figure 1). Based on geographical and climatic zones, we designated Hailun, Shenyang, Luancheng, Yucheng, Fengqiu, Ansai and Changwu agro- ecosystems as the northern group, while the Changshu, Huanjiang, Qianyanzhou, Yanting, Taoyuan and Yingtan agro- ecosystems were designated as the southern group. Land use in these agro- ecosystems was mainly crop growing, including wheat, maize, soybean, rice and cotton (Table 1).

Seven oasis agro- ecosystems (Akesu, Cele, Eerduosi, Fukang, Linze, Naiman, Shapotou) were located in the warm temperate zone of northwest and northern China in arid and semi-arid areas. The oasis agro- ecosystem is unique and found only in arid agricultural areas. Unlike agro- ecosystems in other areas represented in this study, crop cultivation depends heavily on irrigation. Land use in the oasis agro- areas was mainly cotton, maize and wheat growing (Table 1). Therefore, we classified these ecosystems as an independent group and called it oasis agro-ecosystem. We designated these seven oasis agro- ecosystems as the northwestern group.

Total N concentrations in surface water in forest ecosystems were assessed. The forest ecosystems were located along the north-south transect of eastern China (Figure 1), and were representative of native and secondary forests and were free from human fertilization and irrigation activities.

2.3 Monitoring and analysis methods

CERN surface water quality samples were collected according to the Water Monitoring Protocol of the Chinese Ecosystem Research Network [28]. 530 samples were collected from the still surface water sites, and 703 samples were collected from the 45 flowing surface water monitoring sites. The monitoring frequency ranged from 2 to 12 times per year, with sampling distributed evenly through the wet and dry seasons. The maximum sampling frequency was monthly (Ansai, Fengqiu, and Changshu), and the minimum sampling frequency was twice a year (for most of the other monitoring ecosystems), in both the dry and wet seasons. Water samples were analyzed at the Chinese Science Academy's laboratory following standard protocols and methods [28]. Total N was determined by spectrophotometry after potassium persulfate digestion. Information about crop rotation, soil type and fertiliser application rate was recorded for each sampling event at each monitoring station.

2.4 Statistical analysis

Data were analyzed with Matlab 7.11.0 (Massachusetts, USA). A Lilliefors test was conducted to test the normality of the data. Data were not normally distributed, so the nonparametric Kruskal-Wallis test was used to test for differences between the median total N concentrations for data grouped by ecosystem type (for agro-, oasis agro- and forest ecosystems) and geographical region (southern, northern and northwestern) for agro- and oasis

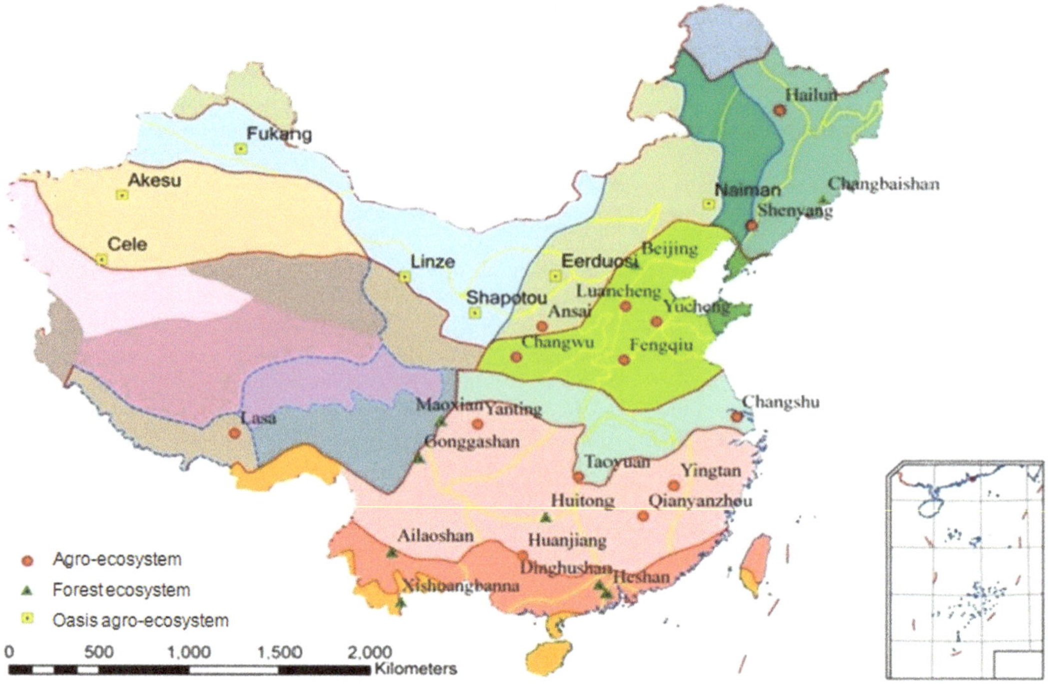

Figure 1. Distribution map of total nitrogen monitoring sites in agro- , oasis agro- and forest ecosystems of the Chinese Ecosystem Research Network (CERN).

agro- ecosystems. Where there were differences between data groups, the Kruskal-Wallis test was combined with a multi-comparison method in Matlab R 2010b to determine which groups were different ($p < 0.05$). We used linear regression to test for relationships between total N concentrations and soil N application rates using SPSS 19.0 for Windows. We used $p < 0.05$ as the significance level. We used Origin 8.0 software for box plots.

1.0 mg·L^{-1} (Class III limit of the Chinese National Quality Standards for Surface Waters) was used as the guideline to assess the exceedance frequency of total N concentrations at the monitoring sites in agro-, oasis agro- and forest ecosystems. Sites with high exceedance frequencies were identified as total N vulnerable zones.

Results

3.1 Total N concentrations under different ecosystems

Total N concentrations in the agro- and oasis agro- ecosystems were significantly higher than in the forest ecosystems. The typical total N concentrations (10th and 90th percentiles in box plots) in the agro-, oasis agro- and forest ecosystems ranged between 0.4 – 8.7 mg·L^{-1}, 0.7 – 15.2 mg·L^{-1}, and 0.2 – 6.6 mg·L^{-1}, respectively for still surface water (Figure 2(A)), and 0.4 – 10.9 mg·L^{-1}, 0.6 – 11.6 mg·L^{-1}, and 0.2 – 3.3 mg·L^{-1}, respectively for flowing surface water (Figure 2(B)). The median total N concentrations of still and flowing surface water in forest ecosystems were 1.1 mg·L^{-1} and 0.5 mg·L^{-1}, respectively. These concentrations were significantly lower than those of still (1.5 mg·L^{-1} and 1.8 mg·L^{-1}, respectively) and flowing surface water (2.4 mg·L^{-1} and 1.8 mg·L^{-1}, respectively) in the agro- and oasis agro-ecosystems ($p < 0.05$).

The surface water in agro- and oasis agro- ecosystems was seriously polluted by N. The median concentrations of total N under the agro- and oasis agro- ecosystems exceeded 1.0 mg·L^{-1} [7] (Figure 2), indicating that more than 50% of surface water samples were heavily polluted by N in these ecosystems.

The total N concentrations in surface water were higher in Beijing, Huitong, Heshan and Dinghushan forest ecosystems than the other forest ecosystems. The exceedance frequencies (> 1.0 mg·L^{-1}) of flowing surface water total N concentrations in Huitong and Dinghushan were about 100%, which indicates that surface water in some forest ecosystems is eutrophic. The water quality standard was not exceeded for either still or surface water in Gongga, Ailao and Xishuangbanna (Table 2).

3.2 Total N concentrations of agro- and oasis agro- ecosystems in different regions

The surface water total N concentrations in northern and northwestern regions in China were higher than those in the southern region. The typical total N concentrations of still surface water in northern, southern and northwestern regions ranged from 0.6 – 4.3 mg·L^{-1}, 0.3 – 3.5 mg·L^{-1}, 0.6 – 15.6 mg·L^{-1} (Figure 3(A)) and 0.9 – 39.1 mg·L^{-1}, 0.3 – 4.4 mg·L^{-1}, 0.8 – 12.2 mg·L^{-1} for flowing surface water (Figure 3(B)). There were no significant differences between total N concentrations in still surface water for the different regions. However, the total N concentrations in flowing surface water in the northern region (3.8 mg·L^{-1}) were significantly higher than those in the southern (1.2 mg·L^{-1}) and northwestern regions (1.8 mg·L^{-1}) ($p < 0.05$).

Surface water in northern, southern and northwestern regions showed varying levels of N pollution, with median total N concentrations of still and flowing surface water all exceeding 1.0 mg·L^{-1} [7] (Figure 3). Results indicate that more than 50% of surface water samples were heavily polluted by N in northern agro- ecosystems, especially flowing surface water samples. The median total N concentrations for surface water in the southern and northwestern regions exceeded 1.0 mg·L^{-1} [7]. About 50% of

Table 1. TN concentrations in still and flowing surface water under agro- and oasis agro- ecosystems.

Spatial Regions	Station	Mean Precipitation (mm)	Soil type	Land use	N application rate ($kg \cdot ha^{-1} year^{-1}$)	Still surface water				Flowing surface water			
						n	Median	Max	>1.0 $mg \cdot L^{-1}$ frequency(%)	n	Median	Max	>1.0 $mg \cdot L^{-1}$ frequency(%)
North	Ansai	500	Loessial Soil	Soybean-Millet	120	—	—	—	—	56(1)	2.8	13.8	96
	Changwu	584	Loessial Soil	Maize-Wheat	345	24(1)	3.0	9.7	100	17(2)	1.6	3.9	65
	Fengqiu	597	Fluvo-aquic Soil	Maize- Wheat	345	—	—	—	—	57(1)	3.3	39.1	96
	Hailun	500-600	Black Soil	Wheat-Maizerotation	120	—	—	—	—	19(1)	57.0	84.8	100
	Shenyang	650-700	Aquic brown Soil	Maize	75	13(1)	0.8	9.4	23	11(1)	0.3	0.4	0
	Yucheng	582	Fluvo-aquic Soil	Maize-Wheat	510	—	—	—	—	75(2)	5.4	27.8	91
South	Changshu	1038	Red Soil	Paddy-Wheat	466	46(2)	1.9	7.9	80	96(1)	1.8	12.3	75
	Huanjing	1389	Calcareous soil	Maize-Soybean	—	5(1)	0.6	9.0	20	10(2)	0.3	10.3	20
	Qianyanhzou	1542	Red Soil	Paddy-Paddy	320	19(2)	1.0	7.3	53	9(1)	0.9	2.0	44
	Yanting	826	Purple Soil	Maize -Wheat	300	32(1)	1.7	4.5	84	—	—	—	—
	Taoyuan	1450	Red Soil	Paddy-Paddy	270	9(2)	1.0	4.4	44	20(2)	0.9	8.0	40
	Yingtan	1785	Red Soil	Peanut	150	44(2)	0.6	3.0	9	43(6)	0.8	34.8	35
Northwest	Akesu	45.7	Aeolian sandy soil	Cotton	160	15(2)	0.9	1.6	40	7(1)	0.8	1.8	43
	Cele	35	Aeolian sandy soil	Cotton-Maize rotation	468	18(1)	1.0	3.2	39	19(1)	2.8	13.4	84
	Eerduosi	348.3	Aeolian sandy soil	Cotton-Maize rotation	—	4(1)	1.0	1.2	50	21(1)	2.2	4.3	95
	Fukang	164	Aeolian sandy soil	Cotton-Maize rotation	275	9(1)	1.5	17.4	89	8(1)	1.6	91.0	63
	Linze	117	Aeolian sandy soil	Wheat–Maize rotation	122	8(2)	23.3	69.5	100	10(1)	12.2	72.0	100
	Naiman	340-450	Aeolian sandy soil	Wheat-Maize rotation	207	8(1)	1.4	8.3	75	14(2)	0.9	1.3	36
	Shapotou	180-220	Aeolian sandy soil	Wheat-Maize rotation	256	10(1)	11.5	43.0	100	—	—	—	—

Note: The "n" values represent the number of sampling sites and the monitoring sites (within brackets). "-"s illustrate that no detection data were available.

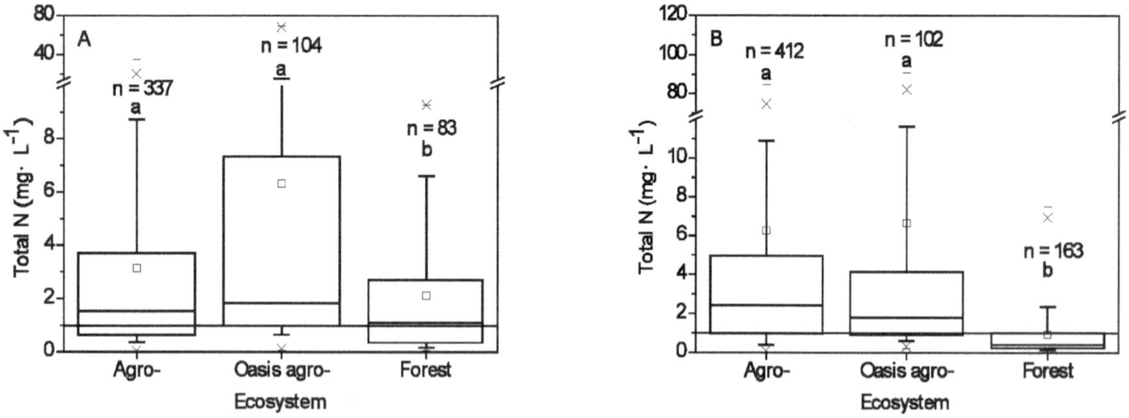

Figure 2. Total nitrogen concentrations in still surface water (A) and flowing surface water (B) of the agro-, oasis agro- and forest ecosystems between 2004 and 2009. Box plots illustrate the 25th, 50th, and 75th percentiles, the whiskers indicate the 10th, 90th percentiles, the "-"s indicate the maximum and minimum percentiles, the "×"s indicate the 1st and 99th percentiles, the "□"s indicate the mean values. Box plots labeled with different letters (a, b) indicate that differences in median values among the three ecosystem types are significant at $p < 0.05$. "—" illustrate the Class III guideline of 1.0 mg·L^{-1} for total nitrogen according to the national quality standards for surface waters of China.

surface water samples were seriously polluted by N in northwestern oasis agro- ecosystems.

The total N concentrations varied by station. The flowing surface water total N concentration was highest in the Hailun agro- ecosystem located in northern China, with a median value of 57.0 mg·L^{-1} (Table 1). The flowing surface water total N concentrations in the northern agro- ecosystems, except Shenyang, all exceeded 1.0 mg·L^{-1}. Exceedance frequencies at Ansai, Fengqiu, Hailun and Yucheng were 90 – 100%. In addition, the still surface water at Changwu was heavily polluted by N and had an exceedance frequency of 100%.

In southern agro- ecosystems, the two highest still and flowing surface water total N concentrations occurred in Changshu, the median values of which were 1.9 mg·L^{-1} (still) and 1.8 mg·L^{-1} (flowing). At this station about 80% of total N concentrations in still surface water and 75% of total N concentrations for flowing surface water exceeded 1.0 mg·L^{-1} (Table 1). The still and flowing surface water total N concentrations of other stations (Huanjiang, Qianyanzhou, Taoyuan, Yingtan) were about 1.0 mg·L^{-1}, and the exceedance frequencies varied from 20% to 50%.

The total N concentrations in the northwestern oasis agro-ecosystems were highest in Linze and Shapotou. The still and flowing surface water median total N concentrations were 23.3 mg·L^{-1} and 12.2 mg·L^{-1}, respectively, in Linze, while for still surface water in Shapotou, the median total N concentration was 11.5 mg·L^{-1}. The total N concentrations all exceeded 1.0 mg·L^{-1} in these two stations, indicating that the surface water was heavily polluted by N (Table 1).

3.3 Correlations between total N concentrations and N fertilizer application rates in agro- and oasis agro- ecosystems

Correlation analysis between the surface water total N concentrations and N application rates showed that the flowing surface water total N concentrations and N application rates were significantly correlated ($r^2 = 0.415$, $p = 0.009$) (Figure 4). Results indicated that total N concentrations tended to increase as N application rates increased. While there was a similar trend for still surface water and N application rates, the correlation was not significant ($r^2 = 0.225$, $p = 0.107$).

Discussion

4.1 Surface water nitrogen pollution in national and international

The median surface water total N concentrations under agro- and oasis agro- ecosystems of the CERN were lower than those in the Tai Lake region (6.4 mg·L^{-1}) [29], but similar to those in surface water in northern China [30,31]. Compared with studies in other countries, the total N concentrations for this study were higher than those of surface water for a Japanese agro- ecosystem [32], but similar to the total N concentrations of the Calapooia River Basin in a Western Oregon agro- ecosystem, where they ranged from 0.5 to 43 mg·L^{-1}[33]. Generally, total N concentrations in northern and northwestern regions were higher than in the southern region of China. A spatial assessment of lakes in China from 1990 to 2010 showed that total N concentrations decreased with rising latitude, but were not related to longitude [12]. Different farming practices, irrigation practices and crop rotations may influence N leaching [34]. For instance, total N losses due to leaching were 9.875 kg·ha^{-1} in the wheat growing season, but were only 1.868 kg N·ha^{-1} in the rice growing season [29]. Li et al [35]. However, reported that nitrate-N concentrations in surface water from rice growing land were significantly higher than those in corn land. In addition, the dilution effect of precipitation could decrease the total N concentrations in the southern region, as precipitation is much higher than in the northern region.

The total N concentrations in surface water under the forest ecosystems were lower than in agro- and oasis agro- ecosystems. This result is similar to what was observed for surface water total phosphorus concentrations in the same study area [36]. Larned et al. [37] also indicated that dissolved N concentrations in pastoral and urban ecosystems were 2 – 7 times higher than in native and plantation forest ecosystems. Total N concentrations were higher in Chinese typical forest ecosystems than in surface water in Japanese forest ecosystems, where they ranged from 0.01 to 1.3, with a median value of 0.14 mg·L^{-1} [38]. In the US, total N concentrations in surface water in northern Californian national forests ranged from 0.03 to 0.1 mg·L^{-1} [39]; these concentrations are also lower than those in China.

Table 2. TN concentrations and background information of still and flowing surface water under the 9 forest ecosystems.

Ecotype	Station	Altitude (m)	Mean Precipitation (mm)	Soil type	Vegetation	Still surface water				Flowing surface water			
						n	Median	Max	>1.0 mg·L^{-1} frequency (%)	n	Median	Max	>1.0 mg·L^{-1} frequency (%)
Humid, Sub- humid Areas in Temperate Zone	Changbai	740	695	Dark Brown soil	Broad-leaved korean pine forest	—	—	—	—	9(1)	1.0	1.4	44
Humid, Sub-humid Areas in Warm Temperate Zone	Beijing	1248	612	Mountain brown soils	Man-made Pinus tabulaeformis Forests	—	—	—	—	9(1)	1.6	5.7	56
Humid Areas in North Sub- tropical Zone	Maoxian	1826	825	Cinnamon soil	Warm temperate coniferous forest	—	—	—	—	8(1)	0.7	1.0	13
	Gongga	2950	1974	Podzolie brown taiga soils	Sulbalpine dark coniferous forest	—	—	—	—	54(7)	0.3	0.8	0
	Huitong	541	1079	Yellow soil	Broad-leaved Trees Mixed Forests	11(1)	1.5	3.5	100	13(1)	1.8	3.7	92
Humid Areas in South Sub- tropical Zone	Ailao	2481	1880	Yellow Brown soil	Subtropical mid mountain humid evergreen broadleaved forest	18(1)	0.3	0.9	0	19(1)	0.2	0.5	0
	Heshan	90	1927	Ferrisols	Acacia mangium pure forests	12(1)	0.8	3.3	42	19(1)	1.6	7.5	63
	Dinghu	90	1700	Lateritic Red Soil	Broad-leaved Trees Mixed Forests	30(3)	5.2	9.3	100	10(1)	1.3	7.0	70
Humid Areas in tropical Zone	Banna	560	1539	Red soil	Tropical seasonal rain forest	12(1)	0.3	0.5	0	23(1)	0.2	0.9	0

Note: The "n" values represent the number of sampling sites and the monitoring sites (within brackets). "-"s illustrate that no detection data were available.

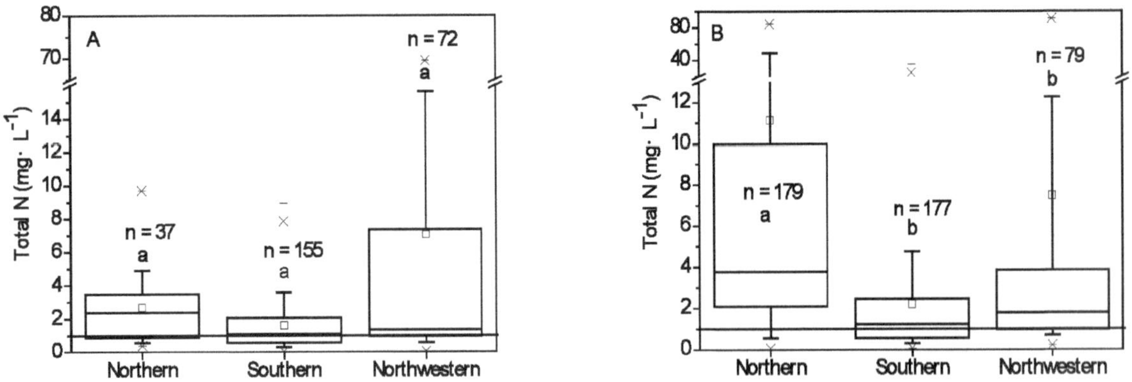

Figure 3. Total nitrogen concentrations of still surface water (A) and flowing surface water (B) under agro- and oasis agro-ecosystems in northern, southern and northwestern areas. The illustrations about Box plots are the same as those in figure 2.

Increasing N concentrations have been observed in many watersheds in the US, Europe, New Zealand and Canada because of land use change, atmospheric deposition, fertilizer application and burning fossil fuels [40–43]. Total N at the Konza Prairie Biological Station (Konza), located in the Flint Hills region of the Great Plains in Kansas, increased from 0.4 mg·L^{-1} to 1.2 mg·L^{-1} due to land use change [44]. There are concerns about N exports from streams in the US midwest because of excessive nutrient enrichment and eutrophication [45,46]. In agro- ecosystems in Baltimore, the annual total N exports were about 30 kg N ha^{-1} during 2005 [47]. The annual total N export rates from the world's rivers in different geographical areas between 76 °N and 43 °S ranged from 1 – 20 630 kg N ha^{-1} year^{-1} [48].

4.2 Factors influencing TN in surface water

Climate, hydrology, soil properties, geomorphology, topography, soil cover and land use are the main factors that influence nutrients in stream water [49-51]. As we were restricted by the sampling frequency, we did not explore relationships between mean annual precipitation and total N in surface waters. However, a significant inverse relationship between total N concentrations and precipitation was reported by a study which examined seasonal variation of total N in the Beijing urban ecosystem [52].

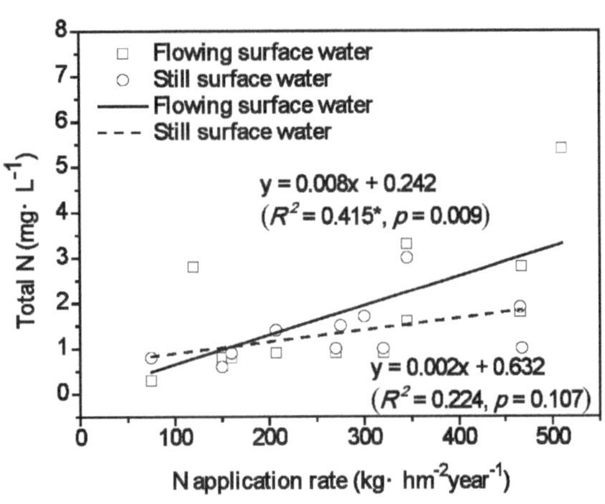

Figure 4. Correlations of total nitrogen concentrations and soil N fertilizer application rates in agro- and oasis agro- ecosystems ($p < 0.05$, n = 15). The total nitrogen concentrations in this figure were the median values of agro- and oasis agro- ecosystems.

Land use change can influence N concentrations. The N concentrations in rural residential areas and cultivated land were much higher than those in grass and forest ecosystems [53,54]. Kvítek et al. [55] reported that, as ploughed land in the catchment increased, nitrate contamination of surface water also increased. Studies of N pollution in the US and the Netherlands showed that 60% – 80% of N came from agricultural non-point sources [56–58]. Alvarez-Cobelas [48] also reported that exports of total N were four times higher from catchments dominated by crops than from forested catchments.

Nitrogen fertilizer, as a non-point source pollutant, is a critical source of pollution in agro- and oasis agro- ecosystems. Surface water N pollution in agro- ecosystems was closely related to agricultural activity. China surpassed the US and the European Union in its production and use of N fertilizers in approximately 2000. However, less than half of the fertilizer N applied in China was taken up by crops [23]. The rest was largely lost to the environment in gaseous (NH_3, NO, N_2O and N_2) or dissolved (NH_4^+ and NO_3^-) forms [59,60]. Leaching is the most important pathway for water eutrophication and poor N fertilizer management could lead to N leaching into the drainage water through run off and drainage [61,62]. In spite of considerable controls on point source pollution, water quality standards have not reached the criterion of the US and European countries due to the increasing contribution of non-point pollution [63]. In general, the waters under the agro- ecosystems have been widely polluted by total N, while only been polluted by total P in a few agro- ecosystem sites [36]. Therefore, the surface water eutrophication under the agro-ecosystems was P limited, which indicated that N fertilizer should generally be reduced and P fertilizer should only be controlled in the P polluted areas of the agro-ecosystems.

Soil textures differ according to location; therefore, losses of N fertilizer through leaching, drainage and runoff are also variable [64]. The relatively high total N concentrations in surface water in northwestern regions may be attributed to soil types. Studies in France have shown that leached N varied from 31 mg·L^{-1} in deep loamy soils to 92 mg·L^{-1} in shallow sandy soils [34]. Soils in the northwestern region of China are mainly aeolian sandy soils, and have low water-holding capacity, meaning that N fertilizer is likely to be lost from soil to surface water. Combined with frequent irrigation, serious soil losses in this region may contribute to the high total N concentrations [65].

Atmospheric N deposition may be an important source of surface water N [66]. While its effects may be weaker than other factors [69], it may still aggravate water eutrophication and affect ecosystem stability [67,68]. The 9 forest ecosystems in our study showed eutrophication tendencies, especially the Beijing, Huitong, Heshan and Dinghushan forest ecosystems. The reactive nitrogen (N_r) species emitted during fossil fuel combustion, have resulted in some of the most pronounced air pollution ever recorded in China, and the increased N_r emissions may have influenced atmospheric N deposition near the study areas. Research has shown that N deposition in Dinghushan and Beijing exceeds 30 kg·ha^{-1} year^{-1}. These rates are higher than those recorded for the other forest ecosystems in China [70–72]. Therefore, the higher atmospheric N deposition may contribute to the higher total N concentrations in the Beijing and Dinghu forest ecosystems.

Conclusions

The total N concentrations of surface water under agro- and oasis agro- ecosystems were much higher than those under forest ecosystems. About 50% of median total N concentrations exceeded 1.0 mg·L^{-1}. There was an obvious spatial pattern in surface water total N concentrations, with much higher total N concentrations in the northern region of China, especially in flowing surface water. The surface water of some agro- ecosystems (Ansai, Changwu, Fengqiu, Hailun) and oasis agro- ecosystems (Cele, Linze, Eerduosi) were severely polluted by N with exceedance (> 1.0 mg·L^{-1}) frequencies greater than 50%. Fertilizer applications were the main source of N pollution in the flowing surface water in agro- and oasis agro- ecosystems.

Acknowledgments

We gratefully acknowledge the many Chinese Ecosystem Research Network (CERN) personnel, who designed the monitoring networks and collected and analyzed the samples used in this study.

Author Contributions

Conceived and designed the experiments: ZX XZ XS GY. Performed the experiments: ZX XZ XS. Analyzed the data: ZX XZ JX. Contributed reagents/materials/analysis tools: JX. Wrote the paper: ZX XZ JX GY XT XS GY.

References

1. Lewis WM Jr (2011) Global primary production of lakes: 19th Baldi Memorial Lecture. Inland Waters 1(1): 1–28.
2. OECD (1982) Eutrophication of Waters-monitoring, Assessment and Control; Paris, France: Organization for Economic Co-operation and Development.
3. Tian YH, Yin B, Yang LZ, Yin SX, Zhu ZL (2007) Nitrogen runoff and leaching losses during rice-wheat rotations in Taihu Lake region, China. Pedosphere 17(4): 445–456.
4. Lassaletta L, García-Gómez H, Gimeno BS, Rovira JV (2009) Agriculture-induced increase in nitrate concentrations in stream waters of a large Mediterranean catchment over 25 years (1981–2005). Science of the Total Environment 407(23): 6034–6043.
5. Mian IA, Begum SM, Ridealg M, McClean CJ, Cresser MS (2010) Spatial and temporal trends in nitrate concentration in the river Derwent, North Yorkshire and its need for NVZ status. Science of the Total Environment 408(4):702–712.
6. US EPA (2002) National Recommended Water Quality Criteria: 2002. Office of Water, EPA-822-R-02-047, U.S. Environmental Protection Agency, Washington DC. http://www.epa.gov/waterscience/standards/wqcriteria.html.
7. China Ministry of Environmental Protection, China General Administration of Quality Supervision and Quarantine (2002) Environmental Quality Standards for Surface Water. GB3838-2002.
8. Yang AL, Zhu YM (1999) Study on nonpoint source pollution in surface water environment. Advance of Environmental Sciences 7(5):60–67.
9. Bao QS, Wang HD (1996) Perspective on water environment and nonpoint source pollution in China. Geographical Sciences 16(1): 66–71.
10. Xu H, Liu ZP, Jiao JG, Yang LZ (2008) Nitrogen pollution status of various types of passing-by water bodies in upper reaches of Taihu Lake. Chinese Journal of Ecology, 27(1): 43–49.
11. Liu ZD, Yu XG, Wang ZX (2003) The current water pollution of Taihu drainage basin and the new management proposals. Journal of Natural Resources 16(4): 467–474.
12. Cai LY, Li Y, Zheng ZH (2010) Temporal and spatial distribution of nitrogen and phosphorus of lake systems in China and their impact on eutrophication. Earth and Environment 38 (2):235–241.
13. Wang LX, Lü JL, Zhuang SY, Hu ZY (2009) Characteristics of phosphorus adsorption on paddy soil and river sediment in east China. Soils, 41(3):402–407.
14. He B, Oki K, Wang Y, Oki T (2009b) Using remotely sensed imagery to estimate potential annual pollutant loads in river basins. Water Science and Technology 60 (8): 2009–2015.
15. He B, Oki T, Kanae S, Mouri G, Kodama K, et al. (2009a) Integrated biogeochemical modelling of nitrogen load from anthropogenic and natural sources in Japan. Ecological Modelling 220(18): 2325–2334.
16. Galloway JN, Cowling EB (2002) Reactive nitrogen and the world: 200 years of change. Ambio 31(2): 64–71.
17. Galloway JN, Schlesinger WH, Levy II H, Michaels A, Schnoor JL (1995) Nitrogen fixation: anthropogenic enhancement-environmental response. Global Biogeochemical Cycles 9(2): 235–252.
18. Vitousek PM, Aber JD, Howarth RW, Likens GE, Matson PA, et al. (1997) Human alteration of the global nitrogen cycle: Sources and consequences. Ecological Applications 7(3):737–750.
19. Isermann K (1991) Share of agriculture in nitrogen and phosphorus emissions into the surface waters of Western Europe against the background of their eutrophication. Fertilizer Research 26(1–3):253–269.
20. National Research Council (1993b) Soil and water quality: an agenda for agriculture. National Academy Press, Washington, D.C., USA.
21. Smil V (1999) Nitrogen in crop production: An account of global flows. Global Biogeochemistry Cycles 13(2): 647–662.
22. Zhang D, Li G, Yang YS, Zhang X, Guo H (2009) Bio-geological processes of nitrogen transport and transformation in the aeration zone and aquifer. Hydrological Sciences Journal 54(2): 316–326.
23. Zhang FS, Wang YQ, Zhang WF, Cui ZL, Ma WQ, et al. (2008) Nutrient use efficiencies of major cereal crops in China and measures for improvement. Acta Pedologica Sinica 45(5):915–924.
24. Hu YT, Liao Qian JH, Wang SW, Yan XY (2011) Statistical analysis and estimation of N leaching from agricultural fields in China. Soils 43 (1): 19–25.
25. Letcher RA, Giupponi C (2005) Policies and tools for sustainable water management in the European Union. Environmental Modeling & Software 20(2):93–98.
26. Davies JM, Mazumder A (2003) Health and environmental policy issues in Canada: the role of watershed management in sustaining clean drinking water quality at surface sources. Journal of Environmental Management 68(3): 273–286.
27. The Central People's Government of the People's Republic of China (2012) Water pollution prevention program of Yangtze River basin (2011-2015). Gazette of the State Council of the People's Republic of China 11:51–60.
28. Yuan GF, Tang DY, Sun XM (2007) Water Monitoring Protocol of Chinese Ecosystem Research Network. Beijing: China Environmental Science Press.
29. Xie YX, Xiong ZQ, Xing GX, Sun GQ, Zhu ZL (2007) Assessment of nitrogen pollutant sources in surface waters of Taihu Lake region. Pedosphere 17(2): 200–208.
30. Chen LD, Peng HJ, Fu BJ, Qiu J, Zhang SR (2005) Seasonal variation of nitrogen-concentration in the surface water and its relationship with land use in a catchment of northern China. Journal of Environmental Science 17(2): 224–231.
31. Meng LY, Shi MW, Yan XX, Lu H, Zhu XL (2009) Evaluation of Nonpoint Source Pollution on Nitrogen and Phosphorus of Surface Water of Hutuo River Valley under Background State. Geography and Geo-information Science 25(3): 95–98.
32. Mouri G, Takizawa S, Oki T (2011) Spatial and temporal variation in nutrient parameters in stream water in a rural-urban catchment, Shikoku, Japan: Effects of land cover and human impact. Journal of Environmental Management 92(7):1837–1848.
33. Mueller-Warrant GW, Griffith SM, Whittaker GW, Banowetz GM, Pfender WF, et al. (2012) Impact of land use patterns and agricultural practices on water quality in the Calapooia River Basin of western Oregon. Journal of Soil and Water Conservation 67 (3): 183–201.
34. Beaudoin N, Saad JK, Laethem CV, Machet JM, Maucorps J, et al. (2005) Nitrate leaching in intensive agriculture in Northern France: Effect of farming practices, soils and crop rotations. Agriculture, Ecosystems and Environment 111(1–4): 292–310.
35. Li J, Lu WX, Zeng XK, Yuan JH, Yu FR (2010) Analysis of spatial-temporal distributions of nitrate-N concentration in Shitoukoumen catchment in northeast China. Environmental Monitoring and Assessment 169(1–4): 335–345.
36. Xie J, Zhang XY, Xu ZW, Sun XM, Ballantine D (2013) Total Phosphorus concentrations in surface water of Typical Chinese Ecosystems, 2004–2010. Frontiers of Environmental Science & Engineering, DOI: 10.1007/s11783-013-0601-5.
37. Larned ST, Scarsbrook MR, Snelder TH, Norton NJ, Biggs BJF (2004) Water quality in low-elevation streams and rivers of New Zealand: recent state and

trends in contrasting land-cover classes. New Zealand Journal of Marine and Freshwater Research 38(2): 347–366.
38. Zhang Z, Fukushima T, Shi PJ, Tao FL, Onda Y, et al. (2008) Baseflow concentrations of nitrogen and phosphorus in forested headwaters in Japan. Science of the Total Environment 402(1): 113–122.
39. Roche LM, Kromschroeder L, Atwill ER, Dahlgren RA, Tate KW (2013) Water quality conditions associated with cattle grazing and recreation on national forest lands. PLoS ONE 8(6): e68127.
40. Stoddard JL (1994) Long-term changes in watershed retention of nitrogen: Its causes and aquatic consequences. Environmental Chemistry of Lakes and Reservoirs 237: 223–284.
41. Boesch DF, Brinsfield RB, Magnien RE (2001) Chesapeake Bay eutrophication: Scientific understanding, ecosystem restoration, and challenges for agriculture. Journal of Environment Quality 30(2): 303–320.
42. Dayyani S, Prasher SO, Madani A, Madramootoo CA (2012) Impact of climate change on the hydrology and nitrogen pollution in a tile-drained agricultural watershed in eastern Canada. Transactions of the Asabe 55(2): 389–401.
43. Duggan IC, Collier KJ, Champion PD, Croker GF, Davies-Colley RJ, et al. (2002) Ecoregional differences in macrophyte and macroinvertebrate communities between Westland and Waikato: are all New Zealand lowland streams the same? New Zealand Journal of Marine and Freshwater Research 36(4): 831–845.
44. Kemp MJ, Dodds WK (2001) Spatial and temporal patterns of nitrogen concentrations in pristine and agriculturally-influenced prairie streams. Biogeochemistry 53(2): 125–141.
45. Dodds WK, Welch EB (2000) Establishing nutrient criteria in streams. Journal of the North American Benthological Society 19(1): 186–196.
46. US Environmental Protection Agency (USEPA) (2000) US Environmental Protection Agency (USEPA) Nutrient criteria technical guidance manual: rivers and streams, EPA-822-B-00-002. US Government Printing Office, Washington, DC.
47. Kaushal SS, Groffman PM, Band LE, Elliott EM, Shields CA, et al. (2011) Tracking nonpoint source nitrogen pollution in human-impacted watersheds. Environmental Science & Technology 45(19): 8225–8232.
48. Alvarez-Cobelas M, Angeler DG, Sánchez-Carrillo S (2008) Export of nitrogen from catchments: A worldwide analysis. Environmental Pollution 156(2): 261–269.
49. Ometto JPHB, Martinelli LA, Ballester MV, Gessner A, Krusche AV, et al. (2000) Effects of land use on water chemistry and macroinvertebrates in two streams of Piracicaba river basin, southeast Brazil. Freshwater Biology 44(2): 327–337.
50. McKee LJ, Eyre BD, Hossain S, Pepperell PR (2001) Influence of climate, geology and humans on spatial and temporal nutrient geochemistry in the subtropical Richmond River catchment, Australia. Marine and Freshwater Research 52(2): 235–248.
51. Likens GE, Buso, DC (2006) Variation in streamwater chemistry throughout the Hubbard Brook Valley. Biogeochemistry 78(1): 1–30.
52. Ren YF, Xu ZW, Zhang XY, Wang XK, Sun XM, et al. (2013) Nitrogen pollution and source identification of urban ecosystem surface water in Beijing. Frontiers of Environmental Science & Engineering, DOI: 10.1007/s11783-012-0474-z.
53. Zhu B, Wang T, Xu TP, Kuang FH, Luo ZX, et al. (2006) Nonpoint-source nitrogen movement and its environmental effects in a small watershed in hilly area of purple soil. Journal of Mountain Science 24(5):601–606.
54. Zhang XY, Xu ZW, Sun XM, Dong WY, Ballantine DJ (2013) Nitrate in shallow groundwater in typical agricultural and forest ecosystems in China, 2004–2010. Journal of Environmental Sciences 25(5): 1007–1014.
55. Kvítek T, Žlábek P, Bystřický V, Fučík P, Lexa M, et al. (2009) Changes of nitrate concentrations in surface waters influenced by land use in the crystalline complex of the Czech Republic. Physics and Chemistry of the Earth 34(8–9):541–551.
56. Boer PM (1996) Nutrient Emissions from agriculture in the Netherlands: causes and remedies. Water Science Technology 33(4–5): 183–189.
57. Kronvang B, Graesbøll P, Larsen SE, Svendsen LM, Andersen HE (1996) Diffuse nutrient losses in Denmark. Water Science technology 33(4–5):81–88.
58. Shepard R (2000) Nitrogen and phosphorus management on Wisconsin farms: Lessons learned for agricultural water quality programs. Journal of Soil and Watershed Conservation 55(1): 63–68.
59. Zhu ZL, Chen DL (2002). Nitrogen fertilizer use in China: contributions to food production, impacts on the environment and best management strategies. Nutrient Cycling in Agroecosystems 63(2–3): 117–127.
60. Ju XT, Xing GX, Chen XP, Zhang SL, Zhang LJ, et al. (2009). Reducing environmental risk by improving N management in intensive Chinese agricultural systems. Proceedings of the National Academy of Sciences of the United States of America 106(9): 3041–3046.
61. Barton L, Colmer TD (2006) Irrigation and fertilizer strategies for minimising nitrogen leaching from turfgrass. Agricultural Water Management 80(1–3):160–175.
62. Cang HJ, Xu LF, Li Z, Ren H (2004) Nitrogen losses from farmland and agricultural nonpoint source pollution. Tropical Geography, 24: 332–336.
63. Novotny V (2005) Diffuse pollution from agriculture in the world. In: Proceedings from the International Workshop on Where Do Fertilizers Go? Belgirate, Italy.
64. Nie SW, Gao W, Chen YQ, Sui P, Eneji AE (2009) Review of current status and research approaches to nitrogen pollution in farmlands. Agricultural Sciences in China 8(7):843–849.
65. Zhao SZ (2003) Ecological environmental problems resulting from exploitation of water resources in Northwest China. Northwestern Geology 36(3):92–96.
66. Krusche AV, de Camargo PB, Cerri CE, Ballester MV, Lara LBLS, et al. (2003). Acid rain and nitrogen deposition in a sub-tropical watershed (Piracicaba): ecosystem consequences. Environmental Pollution 121(3): 389–399.
67. Schindler DW (1988) Effects of acid rain on freshwater ecosystems. Science 239(4836): 149–157.
68. Bergström A, Jansson M (2006) Atmospheric nitrogen deposition has caused nitrogen enrichment and eutrophication of lakes in the northern hemisphere. Global Change Biology 12(4): 635–643.
69. Gorham E (1998). Acid deposition and its ecological effects: a brief history of research. Environmental Science & Policy 1(3): 153–166.
70. Huang ZL, Ding MM, Zhang ZP, Yi WM (1994) The hydrological processes and nitrogen dynamics in a monsoon evergreen broad-leafed forest of Dinghushan. Acta Phtoecologica Sinica 18(2): 194–199.
71. Zhou GY, Yan JH (2001) The influence of region atmospheric precipitation characteristics and its element inputs on the existence and development of Dinghushan forest ecosystems. Acta Ecologica Sinica 21(12): 2002– 2012.
72. Zhang Y, Liu XJ, Zhang FS, Ju XT, Zhou GY, et al. (2006) Spatial and temporal variation of atmospheric nitrogen deposition in North China Plain. Acta Ecologica Sinica 26(6): 1633–1638.

A Mixed Methods Approach to Exploring the Relationship between Norway Rat (*Rattus norvegicus*) Abundance and Features of the Urban Environment in an Inner-City Neighborhood of Vancouver, Canada

Chelsea G. Himsworth[1,2]*, Kirbee L. Parsons[1], Alice Y. T. Feng[1], Thomas Kerr[3,4], Claire M. Jardine[5], David M. Patrick[1]

1 School of Population and Public Health, University of British Columbia, Vancouver, British Columbia, Canada, **2** Animal Health Centre, British Columbia Ministry of Agriculture, Abbotsford, British Columbia, Canada, **3** British Columbia Centre for Excellence in HIV/AIDS, St. Paul's Hospital, Vancouver, British Columbia, Canada, **4** Department of Medicine, University of British Columbia, Vancouver, British Columbia, Canada, **5** Department of Pathobiology, University of Guelph, Guelph, Ontario, Canada

Abstract

Urban rats (*Rattus* spp.) are among the most ubiquitous pest species in the world. Previous research has shown that rat abundance is largely determined by features of the environment; however, the specific urban environmental factors that influence rat population density within cities have yet to be clearly identified. Additionally, there are no well described tools or methodologies for conducting an in-depth evaluation of the relationship between urban rat abundance and the environment. In this study, we developed a systematic environmental observation tool using methods borrowed from the field of systematic social observation. This tool, which employed a combination of quantitative and qualitative methodologies, was then used to identify environmental factors associated with the relative abundance of Norway rats (*Rattus norvegicus*) in an inner-city neighborhood of Vancouver, Canada. Using a multivariate zero-inflated negative binomial model, we found that a variety of factors, including specific land use, building condition, and amount of refuse, were related to rat presence and abundance. Qualitative data largely supported and further clarified observed statistical relationships, but also identified conflicting and unique situations not easily captured through quantitative methods. Overall, the tool helped us to better understand the relationship between features of the urban environment and relative rat abundance within our study area and may useful for studying environmental determinants of zoonotic disease prevalence/distribution among urban rat populations in the future.

Editor: Georges Chapouthier, Université Pierre et Marie Curie, France

Funding: This work was supported by the Canadian Institutes of Health Research [MOP – 119530 and CGV – 104833]. The funders had no role in study design, data collection and analysis, decision to publish, or preparation of the manuscript.

Competing Interests: The authors have declared that no competing interests exist.

* E-mail: Chelsea.Himsworth@gov.bc.ca

Introduction

Norway rats (*Rattus norvegicus*) are among the most widespread rodent species, inhabiting every continent except for Antarctica [1]. The ubiquity of this species is largely attributable to their remarkable ability to quickly and successfully adapt to new environments and resources [1]. Norway rats are best suited to close cohabitation with people, and are considered a true commensal species as few populations are found in a truly sylvatic state [1].

Their propensity to exploit human settlements has caused Norway rats to become a particularly ubiquitous and problematic pest in cities around the world. Within urban settings, Norway rats can damage property, contaminate food, and act as a source of infectious disease for people [1,2]. For these reasons there is considerable interest in identifying and understanding the factors that promote or deter rat infestations in urban centers.

Previous studies have indicated that the environment is probably the single most important determinant of rat density and distribution across the urban landscape [1,3]. More specifically, it is clear that rats require suitable harborage and food sources in order to establish an infestation [1,3]. However, the adaptability of Norway rats means that it is often difficult to identify what specific features of the urban environment combine to create optimal rat habitat.

Researchers have employed a variety of methods in an effort to identify environmental features that promote or deter rat infestations. For example, a study in Buenos Aires, Argentina, compared the relative abundance of Norway rats in several locations demonstrating varying degrees of human modification [4]. They found that Norway rats were more common in shantytowns and parklands compared to industrial-residential neighborhoods and natural preserves [4]. Another study in the same city found that Norway rats were more abundant in shantytowns compared to parklands [5]. In Salzburg, Austria, Norway rats were trapped in a variety of urban locations with distinct environmental features [6]. This study found that Norway rats were most common in habitats with water, vegetation, natural soil, and organic waste, while rats were rarely found in heavily built-up areas and areas frequently utilized by people [6].

The main drawback of the aforementioned approaches is that they are capable of capturing only 'high-level' environmental impacts on rat populations (i.e., differences in rat abundance between different ecosystems). Rat population density and distribution, however, may vary considerably even over very short geographic distances within a single urban ecosystem [7,8], suggesting that minute differences in the micro-environment of a city block could determine its capacity to support rat populations. These micro-environmental data, however, are relatively difficult to collect in a systematic fashion, and is seldom captured by cursory observation or pre-established databases.

A similar problem has been encountered in the social and health sciences, where researchers have attempted to identify links between an individual's health, welfare, behaviour, etc., and the features of the neighborhood in which they reside. In order to overcome this problem, researchers have developed the science of systematic social observation (SSO) [9,10]. The tools employed by SSO require the researcher to closely observe and quantify multiple different aspects of a city block, resulting in the creation of a comprehensive dataset that captures the complexity of the block environment [9,10]. These tools have revolutionized our understanding of impact of the urban environment on the people who reside within it.

Given that the urban environment also influences rat populations, it seems possible that a similar tool could be developed and used to predict rat abundance. There are several studies that have used a similar approach to study rats. However the tools and methodologies employed by these studies had several limitations including the following: capturing a limited number of environmental features [11], being specific to rural [12] or developing areas [13,14], and being difficult to reproduce [13,14].

The goals of this study were: 1) to develop a systematic environmental observational tool that captures and quantifies features of a developed urban microenvironment that could influence rat populations; 2) to develop an approach to using that tool that includes analysis of both quantitative and qualitative data; and 3) to use that tool/approach to identify environmental factors associated with relative rat abundance within an inner-city neighborhood of Vancouver, Canada.

Methods

This study was a part of a larger project (the Vancouver Rat Project, www.vancouverratproject.com) the goal of which is to characterize the health risks that rats pose to people through the transmission of infectious diseases. More specifically, the project aims to understand the ecology of zoonoses in rats and people by understanding the dynamic interplay among rats, people, pathogens, and the urban environment.

Ethics Statement

This study was approved by the University of British Columbia's Animal Care Committee (A11-0087) and adhered to national guidelines set out by the Canadian Council on Animal Care (www.ccac.ca), including those pertaining to animal user training, euthanasia, protocol review, and wildlife (http://www.ccac.ca/en_/standards/guidelines). This study did not involve any endangered or protected species.

Study Area

The study area was composed of 43 contiguous city blocks (0.82 km^2) which included the core of Vancouver's Downtown Eastside (49°15″0′N/123°8″0′W). The Downtown Eastside (DTES) is an impoverished inner-city neighborhood characterized by a large number of single-room occupancy hotels, high rates of homelessness, and a large open air illicit drug scene [15]. It was selected as a priority for study within the city because residents and community groups identified rat infestations as a significant but neglected issue in the area. Also included in the study were several blocks from the more affluent, adjacent Gastown district. All trapping took place on public property only and no specific permissions were required.

Development of an Environmental Observation Tool

The environmental observation tool (Appendix S1) was modeled after the SSO tool developed by Parsons et al. (2010). We utilized scales, items, and constructs from SSO tools [10,16–18] that measured factors with the potential to influence rat populations, as well as from prior studies of rat abundance [11,13,14]. We also developed new items based on a review of the literature regarding rat ecology [1] and our field observations. The final tool included 58 items covering 6 major domains: (1) land use characteristics, (2) property condition, (3) green space characteristics, (4) alley surface characteristics, (5) presence of waste, and (6) alley use (i.e., for loitering and transportation). Additionally, there was a free form section at the end of the tool where observers could make notes about the block under evaluation. For this free form section, particular attention was paid to identifying and describing features that were perceived to be associated with rat presence/abundance, as well as unusual or unique features of the block under study.

In order to facilitate the use of the tool in this study, and in the future, each domain and item was accompanied by detailed instructions regarding its purpose and/or use. Additionally, a supplementary document was produced with photographic examples of different items and ratings (see Appendix S2).

Tool Implementation

Two observers were trained in the use of the tool, and these raters completed their observations together between 8:00 h and 16:00 h, with final ratings for each item being arrived at via consensus. To complete their observations, observers evaluated all aspects of the block including block faces (i.e., the area of the block facing a street), alleyways, and aerial photographs.

Determining Rat Abundance

This study was part of a larger project characterizing the ecology of zoonotic pathogens in rat populations for which whole rat carcasses were required for testing. For this reason a trapping-removal method (vs. a non-lethal method, such as mark-recapture) was required. It should be noted, however, that systematic trapping-removal is considered to be among the most accurate methods for enumerating rats, and has been used frequently in the past to determine rat abundance in urban centers [7,19–21].

Each city block randomly assigned to a 3-week trapping period over the course of 1 year in order to capture seasonal variations in rat abundance within the study as a whole. Of the 43 blocks included in the study, 9 were trapped in the fall, 10 in the winter, 14 in the spring and 12 in the summer.

Within each block, approximately 20 Tomahawk Rigid Traps for rats (Tomahawk Live Trap, Hazlelhurst, USA) per block were set out along each side of the back alley that bisected the block.

Traps were evenly spaced where possible, but had to be placed on outdoor public property in a location where they did not obstruct traffic and could be secured to prevent theft. Traps were pre-baited (filled with bait but fixed open) for one week to acclimatize rats to trapping equipment and bait, followed by two weeks of active trapping. Baits used included peanut butter, bacon fat, flour, and oats. Traps were set at 16:00 h and checked the following morning at 8:00 h.

Trapped rats were anesthetized with isoflurane in a rodent Inhalation Narcosis Chamber (Harvard Apparatus, Holliston, USA) prior to pentobarbital euthanasia via intracardiac injection. Rats were identified to species by external morphology [22].

Dependent Variable

Of the 673 rats trapped over the course of the study, 665 (98.8%) were Norway rats (*Rattus norvegicus*) and 8 (1.2%) were black rats (*Rattus rattus*). Given the relatively small number of black rats trapped, and the fact that the ecology of black rats has been shown to be significantly different than Norway rats [1], black rats were excluded from the analysis.

The unit of analysis was the city block ($n = 43$). The dependent variable was trap success rate in each block, with trap success rate being equal to the number of rats trapped in a block over the entire trapping period divided by the total trap effort over that period. The total trap effort (# traps set x # nights) was adjusted according to the method described by Nelson and Clark [23]. This method involves subtracting half a unit from the total trap effort for each trap spring for any cause (e.g., trapping of a rat, trapping of a non-target species, tripping of a trap). This adjustment was considered particularly important because of the marked variation in rat abundance among blocks, and because of the frequency with which traps were sprung by non-target species and by members of the public, both of which can significantly bias trap success indices if they are not taken into account [23].

Given that each block under study had roughly the same geometric surface area (1.2 ha) and was trapped in the same manner, and given that rat populations are thought to be relatively isolated to the level of a city block with minimal movement of animals among blocks [1], it is appropriate to use the city block as the unit of analysis and the number of rats trapped offset by trap effort as a proxy for relative population density/rat abundance.

To confirm the absence of geographic autocorrelation of rat abundance among the blocks, a georeferenced map of block outlines within the study area was imported into ArcGIS 10.0 (ESRI, Redlands, USA). Global Moran's I was used to test the null hypothesis that trap success (# rat trapped/trap effort) within a block was randomly distributed (i.e., no spatial autocorrelation) using inverse distance for conceptualization of spatial relationships and a Euclidean distance method.

Predictor Variables

A total of 59 potential predictor variables were included for initial consideration. Each of the 58 items on the environmental observational tool was used to create 1 predictor variable (Table S1 – note that variables were numbered according to the item numbering in the observation tool). The season in which the block was trapped (fall = September – November, winter = December – February, spring = March – May, summer = June – August) was also included as a predictor.

Seven variables (variable nos. 1, 2, 5, 13, 25, 31, and 41) were eliminated initially because they captured the same and/or less information as subsequent variables. For example, variable 13 is a nominal-categorical variable that characterizes the predominant housing type in a block, with 8 categories that each correspond to a class of housing. However, variable nos. 14–19 are ordinal variables that capture both the presence of each class of housing in a block, as well as the proportion of each block occupied by each class of housing.

Next, variables representing different aspects of the same concept were consolidated into a single new composite variable representing the weighted average of the underlying component variables. For example, variable nos. 26–30 all represent building condition (the underlying concept of interest), and pertain to the proportion of each block occupied by buildings in 'extremely poor', 'poor', 'fair', 'good', and 'excellent' condition. These variables were combined into a single composite variable, as were variable nos. 32–36 and 42–44.

Predictor variables where $> 95\%$ of observations had the same score (variable nos. 8, 12, 18, and 54) were deemed non-discriminatory and removed. Variable no. 46 (number of rat holes) was also removed because it was deemed to be a sign of rat abundance rather than a cause.

A total of 37 predictor variables were considered for statistical modeling.

Model Building

The goal of the model-building protocol was to identify the most parsimonious set of predictor variables that best explained the outcome, and all predictor variables were given equal consideration going into the model building procedure (i.e., we had no a priori hypotheses regarding the relative importance of specific predictors). Similarly, we had no a priori hypothesis regarding interactions among predictors, and, given the large number of potential predictors ($n = 37$) relative to the number of observations ($n = 43$), no interaction terms were included for consideration.

A zero-inflated negative binomial (ZINB) model was used to model the relationship between the predictors and outcome in order to account for excess zeros (no rats were trapped in 11 of 43 blocks) and overdispersion in the outcome variable. This ZINB model combined a binary model with a logit link to model the odds of not trapping any rats (zero-inflation model), and a negative binomial model with a log link to model the number of rats trapped offset by the natural logarithm of the trap effort (count model). The unit of analysis was the city block ($n = 43$).

The relationship between each predictor and the outcome was first modeled in a bivariate manner, including the predictor in the binary and count portions of the model independently (Table S2). The majority of the predictor variables captured the proportion of the block occupied by an environmental factor of interest, and were therefore modeled in a linear manner. For several ordinal variables (variable nos. 48, 49, 57, and 58) where the spacing between categories was more subjective (e.g., 'none' vs. 'some' vs. 'a little' vs. 'a lot'), bivariate models treating the variable in a linear and categorical manner were generated and compared using the Bayesian Information Criterion (BIC) [24]. For all 4 predictors, the linear form of the variable was superior to the categorical form in both the binary and count portions of the model, therefore these variables were treated as numeric variables. Season was treated as a categorical variable.

In order to narrow down the pool of potential predictors for inclusion in a multivariate model, a program was created in R (R Development Core Team, Vienna, Austria) that performed automated combined forward and backward selection on each part of the model independently using Akaike's Information Criterion (AIC) to select the best potential group of predictors for further consideration [24]. Briefly, in order to select the binary model, the negative binomial model was set at 1 (offset by the natural logarithm of the trap effort). The null binary model (a binary model with no predictors) was used as the starting point, and predictors were added sequentially based on their associated AIC (i.e., the predictor that produced the lowest AIC for the binary model was added first). After each addition, the model was checked to determine if removal of one or more predictors included in the model affected the AIC. If removal of the predictor(s) decreased the AIC for the binary model, then the predictor(s) was eliminated. If removal increased the AIC, then the predictor(s) was retained. This process was repeated until addition or elimination of any subsequent predictors could not further decrease the binary model AIC. The selection procedure described above was then repeated for the count model holding the binary model at 1. All 37 predictor variables, including season, were allowed to compete equally for both components of the model, with the stepwise selection procedure ensuring that only those variables that improved model fit (according to the AIC) were retained. This process should control for confounding among predictors because it will select the variables with the strongest relationship to the outcome in the presence of all other predictor variables.

Variables identified through automated selection for the binary and count components of the ZINB model were then combined in a single model and manual forward and backward selection was performed using the BIC to maximize model fit and parsimony (as the BIC penalizes model complexity more heavily than the AIC) [24]. The final model was that with the lowest BIC.

To investigate collinearity among explanatory variables in the final model, Spearman's rank correlation (ρ) was used to examine bivariate relationships between each of the included predictors [24]. Any highly correlated variables ($\rho > 0.8$) were modeled separately, with only the variable making the most significant contribution to the model (based on the BIC) being retained [24]. Finally, the variables included in the final model were used to build a negative binomial (NB) model and the Vuong test was used to compare the NB and ZINB models with the null hypothesis that the two models were equivalent and the alternative hypothesis that the ZINB model was superior.

Qualitative Analysis

Thematic and narrative qualitative analysis [25] was used in order to: 1) provide insight into the results of the quantitative analysis, 2) capture information not captured by the quantitative aspects of the tool, 3) provide increased resolution for environmental data by allowing the observer to note special features within the block (vs. the quantitative aspects of the tool, which pertained to the block as a whole), and 5) aid in triangulation for the purposes of identifying the environmental characteristics most significantly associated with rat abundance.

Data included in the qualitative analysis included written notes recorded in the environmental observational tool, as well as field notes recorded during a two-day long debrief at the end of the project. This debriefing involved re-visiting each block included in the study area and discussing observer opinions of the block environment and its relationship to rat abundance.

Handwritten notes were transcribed into Microsoft Excel (Microsoft, USA) and labeled with block number and block trap success. Emergent thematic analyses of early observations were discussed among the two observers and served to inform the focus of subsequent observations, as well as ongoing analyses. The coding framework employed a priori codes derived from the quantitative observational tool, as well as emergent codes based upon ongoing observational work. All qualitative data were reviewed, and text segments related to each individual code were categorized/classified (to identify commonalities among blocks and special features within a block). Subsequent coding passes were used to refine and expand code categories and to identify instances of negative evidence.

Results

Quantitative Findings

There was no significant autocorrelation of rat abundance among blocks ($p = 0.14$) based on Global Moran's I, supporting our assertion that the city blocks could be treated as independent units. Bivariate relationships between the outcome and each of the 37 potential predictors are presented in Table S1. The final ZINB model included one variable in the zero-inflation portion of the model and six variables in the negative binomial portion of the model (Table 1). None of the variables included in the final model were highly correlated ($\rho < 0.45$ for all comparisons) and the Vuong test indicated that the ZINB model was superior to the NB model ($P = 0.03$).

The final model indicated that as the proportion of a block occupied by institutional parcels (i.e., those dedicated to not-for-profit services, such as churches, outreach centres, etc.) increased, the odds of not trapping any rats decreased (Table 1). As the proportion of a block occupied by single-family homes increased, the trap success rate decreased. In contrast, as the proportion of the block occupied by housing over commercial and by abandoned parcels (i.e., those with structures that were previously used but are now vacant and not undergoing construction or demolition) increased, the trap success rate increased. As general building condition increased, trap success rate decreased, whereas trap success rate increased as the as the amount of garbage/trash/junk/litter in the alley increased. Finally, as the use of an alley as a transportation corridor increased, trap success rate decreased.

Qualitative Findings

Relationship between rat abundance and land use. Observers often reported trapping rats around abandoned buildings, demolished buildings, and vacant lots. In one block, a large parcel filled with debris from a demolished building was observed to be heavily infested with rats.

Blocks with low trap success tended to be described as being primarily residential (particularly single family homes) or primarily industrial (including industrial sites producing food), while blocks with high trap success tended to be described as being more mixed-use. Observers noted that there was no clear association between the presence of facilities dealing with food (e.g., restaurants, groceries, etc.) and rat presence/abundance.

Relationship between rat abundance and socioeconomic status of a block. Observers reported that rats were much more abundant in the impoverished DTES neighborhood than the adjacent more affluent Gastown neighborhood. Observers also reported that blocks composed primarily of single-family homes appeared to be much more affluent than other residential areas, and that many of the low-rise apartments and housing over commercial was low income housing and single-room occupancy hotels (SROs). Low-income apartments, SROs, and outreach centres for impoverished community members were more commonly reported in blocks with high trap success. Also, within a block, observers often reported trapping rats near SROs.

Relationship between rat abundance and property/building condition. Buildings in disrepair were much more commonly reported in blocks with high trap success vs. blocks with low trap success. Also, within a block, observers often reported trapping rats around buildings in disrepair. Conversely, buildings and properties were much more commonly reported to be in good to excellent condition in blocks with low trap success. However,

Table 1. Relationship between features of the urban environment and relative abundance of Norway rats (*Rattus norvegicus*).

Variable #	Variable description	β	SE	P-Value	exp (β)	95% CI
Count Model (negative binomial with log link)						
10	Proportion of block occupied by **abandoned** parcels	0.70	0.32	0.027	2.02	1.08–3.77
14	Proportion of block occupied by **single family houses**	−0.67	0.25	0.007	0.51	0.31–0.83
19	Proportion of block occupied by **housing over commercial**	0.65	0.21	0.002	1.91	1.27–2.87
c26t30	General **building condition**	−2.01	0.58	0.001	0.13	0.04–0.42
48	Amount of **garbage/trash/junk/litter**	0.83	0.28	0.003	2.29	1.33–3.93
58	Amount of **transport**	−0.99	0.24	< 0.001	0.37	0.23–0.59
Zero-Inflation Model (binomial with logit link)						
5	Proportion of block occupied by institutional parcels	−3.42	1.44	0.012	0.03	0.002–0.55

Zero-inflated negative binomial model for the number of rats trapped in a city block offset by the natural logarithm of the trap effort in that block.

observers reported trapping rats around buildings in good repair and vice versa. Observers noted that all aspects of a block had to be assessed to adequately evaluate repair. In some blocks, the street face appeared to be in good repair while the alley face was in very poor repair.

Relationship between rat abundance and green space. Well maintained green space and food gardens were more often reported in blocks with low trap success. Unkempt green space was reported in blocks with high and low trap success. However, within a block, observers often reported trapping rats near unkempt green space.

Relationship between rat abundance and waste. Observers reported debris, strewn garbage, and overflowing garbage bins much more commonly in blocks with high trap success. Within a block, observers also reported trapping rats in close proximity to areas with accumulated refuse and around dumpsters. Blocks with low trap success were most commonly described as clean with very little garbage. However, observers also noted that some blocks with high trap success were relatively clean. Observers commented that rats were not caught around organic waste bins or compost piles.

Relationship between rat abundance and condition of the alley surface. Blocks with high trap success tended to be described as having pavement in disrepair, cracked pavement, unpaved areas, and rat holes, while blocks with low trap success tended to be described as being well paved and having pavement in good repair. Within an alley, observers reported trapping rats around areas where the pavement was in disrepair and/or deeply cracked, around unpaved areas, and around rat holes, regardless of the condition of the rest of the alley. However, some blocks with high trap success were reported to have good alley surface condition and vice versa.

Sources of harborage for rats. Observers suggested that most rats likely live in buildings or in underground burrows (rat holes were commonly seen in blocks with high trap success). However, observers also reported trapping rats near areas with stored wood pallets, plastic barrels, and other containers. In four blocks, observers reported frequently catching rats near rat corridors (i.e., access points, such as gaps between buildings, which connected the alley face to deeper aspects of the block).

Effect of human activity on rat presence. Heavy use of alleyway by people (particularly people loitering) was reported in blocks with high and low trap success, but was more commonly reported in blocks with high trap success.

Observers noted that the alleyways more heavily used for loitering tended to be dirtier (i.e., have more accumulated refuse).

Discussion

Using quantitative and qualitative analysis, we were able to identify a variety of environmental factors associated with rat abundance. We also found that once observers were comfortable with the use of the tool, its implementation was relatively efficient and straight forward.

As has been found in previous studies using SSO, the combination of quantitative and qualitative analysis was particularly beneficial. Not only did the qualitative data support and elucidate the quantitative results, these data also provided additional insight into environmental variations within a block, and drew our attention to factors not well captured by the quantitative aspect of the tool (e.g., the apparent association between rat abundance and poverty), as well as inconsistencies in the relationship between certain environmental factors and relative rat abundance.

This study was able to identify a number of factors associated with relative rat abundance that have not been studied and/or well characterized previously (e.g., specific land use, residential density, and building disrepair). Several of our findings, however, are in agreement with previous research on the relationship between environmental factors and rat abundance. For example, studies have been able to show that the presence of exposed garbage is a strong predictor/promoter of rat infestations [6,13]. Our study also found that rats were relatively less common in blocks where the land use was predominantly industrial and single-family residential, and this is consistent with the findings of Cavia et al. (2009), who also found that Norway rats were uncommon in industrial and residential neighborhoods.

Other studies have also indicated that factors associated with poverty can promote rat infestations, likely due to a combination of factors including infrastructural disrepair and decreased environmental hygiene, which combine to create ideal urban rat habitat [3,5]. Poverty might also impair an individual's or community's ability to prevent or eliminate infestations (e.g., through neighborhood rehabilitation or employing pest control professionals). In our study, there were several variables that, on first glance, are difficult to intuitively link to rat abundance, including proportion of block occupied by institutional parcels and housing over commercial. However, upon further examination, and in light of the qualitative data, it became clear that much of the housing over commercial was low-income housing, including single room occupancy hotel rooms (SROs). Similarly, institutional parcels in the area are primarily composed of outreach centers for impoverished members of the community. Observers specifically noted that rats were often trapped around SROs and that blocks with high trap success tended to have more SROs, low-income housing, and outreach centers. Observers also noted that far more rats were trapped in the impoverished DTES compared to the adjacent, more affluent Gastown district, and that blocks composed primarily of single-family residences were also more affluent than the rest of the DTES. Factors potentially associated with poverty were associated with trap success independent of variables accounting for infrastructural disrepair and accumulation of waste. This suggests that there are likely one or more environmental variables associated with poverty that were not accounted for in this study.

It is interesting to note that season did not appear to have a significant relationship with rat presence or abundance. Indeed, there has been considerable debate in the literature regarding the impact of season on rats in urban ecosystems [1]. The absence of seasonal variation in urban rat population size might not be surprising given that urban ecosystems, being largely under anthropogenic control, would seem to be less prone to seasonal fluctuation in resource availability (i.e., the presence of suitable food and harborage) compared to more sylvatic settings. However, given that this study only included one year's worth of data, the true impact of season could not be definitively determined [26].

There were a number of findings in our study that contrast with those found in past research on the subject. For example, green space has been reported to be a strong promoter of rat infestations [4,6]. In particular, a study by Traweger et al. (2008) found that rats were most abundant in more natural areas with abundant vegetation, and that trap success was very low in areas with little vegetation, as well as heavily built-up areas, and areas frequented by humans. Our study found that there was no relationship between amount of green space and rat abundance, that rats were very common in built up areas, and that human activity (in the form of loitering) appeared to have a negligible to positive relationship with rat abundance. Increased use of an alley as a transport corridor did appear to have a negative association with the number of rats trapped. However, it is difficult to determine if this because rats avoid areas of increased activity or if alleys used for transport tended to be in better condition or in areas with other features that deter rat infestations. Although Traweger et al. (2008) speculate that natural areas are essential to providing appropriate rat habitat, the results of our study indicate that Norway rats can thrive even in the most urbanized of settings.

Interestingly, the qualitative data also indicated the presence of conflicts within our own study. For example, although strewn garbage seemed to be strongly associated with rat abundance in general, there were some alleys and/or alley segments where rats were caught that appeared relatively clean. Similarly, although many rats were caught in alleys with unpaved areas, cracked pavement, and/or rat holes, some alleys with robust rat populations appeared to have relatively good pavement condition. These conflicts, in combination with the variety of environmental factors that were associated with trap success in our study, suggest that the relationship between rat abundance and the environment is a function of a complex interaction of factors. In other words, there is not one single environmental factor that is a necessary determinant of rat abundance. Rather, a variety of different factors can combine to create suitable rat habitat.

Despite the success of the methodology, this study had several limitations. The relatively low number of observations ($n = 43$) limited the number of variables that could be included in the model. Future studies should seek to increase power (i.e., the number of blocks included) in order to better investigate the full suite of variables that might predict rat abundance. With regard to our trapping methodology, only rats accessing the back alley were accounted for in our measure of rat abundance. Although this 'alley trapping' approach is consistent with previous trapping studies of urban rats [7,8], it is possible that some rats could reside entirely indoors and/or not access the alley. This may particularly be the case for black rats, as it has been suggested that these species prefer to reside in human structures [1]. Indeed, in this study, it could not be determined if the low number of black rats was a result of the trapping methodology, or whether this species is truly rare within this urban area. Ideally, future studies should account for all indoor and outdoor rat populations within a block. However, as this would require intensive trapping on private property, it may not be a feasible proposition. Finally, it should be noted that while many factors have a clear direct relationship with rat abundance (e.g., presence of exposed garbage); other factors (e.g., land use) likely represent some underlying and unmeasured factor(s) that influence rat populations. Although these environmental 'proxies' may be valuable for predicting rat abundance, they are less useful for developing intervention strategies. After all, it is not reasonable to change the land use in a block to prevent rat infestations. It may therefore be useful for future studies to focus on and dissect the relationship between rat abundance and factors like land use and poverty.

Overall, this tool proved to be an efficient and effective way to examine the effect of urban environmental factors on relative rat abundance. In addition to their effects of rat abundance, environmental factors are also known to influence pathogen prevalence in rat populations [2]. In the future, we hope to determine whether this tool could also be used to identify features of the urban environment associated with the prevalence of zoonotic microbes in rats.

Supporting Information

Appendix S1 A systematic environmental observation tool to predict relative rat abundance in urban centres.

Appendix S2 Guidance document for the systematic environmental observation tool.

Table S1 Quantitative predictors generated from the environmental observation tool.

Table S2 Bivariate relationships between features of the urban environment and relative abundance of Norway rats (*Rattus norvegicus*). Zero-inflated negative binomial model for the number of rats trapped in a city block offset by the natural logarithm of the trap effort in that block.

Acknowledgments

The field work was made possible by the assistance of the City of Vancouver (Murray Wightman and Stuart McMillan), the Urban Health Research Initiative, the Vancouver Injection Drug Users Study, and the Vancouver Area Network of Drug Users. We would like to thank Rick White of the University of British Columbia Statistical Consulting and Research Laboratory for his help with the statistical analysis.

Author Contributions

Conceived and designed the experiments: CGH KLP AYTF. Performed the experiments: CGH KLP AYTF. Analyzed the data: CGH TK CMJ DMP. Wrote the paper: CGH KLP AYTF TK CMJ DMP.

References

1. Feng AYT, Himsworth C (2013) The secret life of the city rat: A review of the ecology of urban norway and black rats (*Rattus norvegicus* and *Rattus rattus*). Urban Ecosyst 17: 149–162.
2. Himsworth CG, Parsons KL, Jardine C, Patrick DM (2013) Rats, cities, people, and pathogens: A systematic review and narrative synthesis of literature regarding the epidemiology of rat-associated zoonoses in urban centers. Vector Borne Zoonotic Dis 13: 349–359.
3. Davis DE (1953) The characteristics of rat populations. Q Rev Biol 28: 373–401.
4. Cavia R, Ruben Cueto G, Virginia Suarez O (2009) Changes in rodent communities according to the landscape structure in an urban ecosystem. Landscape Urban Plann 90: 11–19.
5. Vadell MV, Cavia R, Suárez OV (2010) Abundance, age structure and reproductive patterns of *Rattus norvegicus* and *Mus musculus* in two areas of the city of Buenos Aires. Int J Pest Manage 56: 327–336.
6. Traweger D, Travnitzky R, Moser C, Walzer C, Bernatzky G (2006) Habitat preferences and distribution of the brown rat (*Rattus norvegicus* berk.) in the city of Salzburg (Austria): Implications for an urban rat management. J Pest Sci 79: 113–125.
7. Easterbrook JD, Shields T, Klein SL, Glass GE (2005) Norway rat population in Baltimore, Maryland, 2004. Vector Borne Zoonotic Dis 5: 296–299.
8. Glass GE, Gardner-Santana L, Holt RD, Chen J, Shields TM, et al. (2009) Trophic garnishes: Cat-rat interactions in an urban environment. PLoS ONE: e5794.
9. Raudenbush SW, Sampson RJ (1999) Ecometrics: Toward a science of assessing ecological settings with application to the systematic social observation of neighborhoods. Sociol Methodol 29: 1–41.
10. Parsons JA, Singh G, Scott AN, Nisenbaum R, Balasubramaniam P, et al. (2010) Standardized observation of neighbourhood disorder: Does it work in Canada? Int J Health Geogr 9: 6-072X-9-6.
11. Figgs LW (2011) Assessing the association between brown rat (*Rattus norvegicus*) activity and food and harbourage characteristics in a residential community of a medium-sized United States city. Int J Pest Manage 57: 217–222.
12. Yin JX, Geater A, Chongsuvivatwong V, Dong XQ, Du CH, et al. (2008) Predictors for presence and abundance of small mammals in households of villages endemic for commensal rodent plague in Yunnan province, China. BMC Ecology 8: 18.
13. Promkerd P, Khoprasert Y, Virathavone P, Thoummabouth M, Sirisak O, et al. (2008) Factors explaining the abundance of rodents in the city of Luang Prabang, Lao PDR, as revealed by field and household surveys. Integr Zool 3: 11–20.
14. Frantz SC, Comings JP (1976) Evaluation of urban rodent infestation - an approach in Nepal. Proceedings of the 7th Vertebrate Pest Conference. Paper 19.
15. Smith HA (2000) Where worlds collide: Social polarization at the community level in Vancouver's Gastown/Downtown Eastside. Available: https://circle.ubc.ca/handle/2429/10914. Accessed 31 January 2014.
16. Caughy MO, O'Campo PJ, Patterson J (2001) A brief observational measure for urban neighborhoods. Health Place 7: 225–236.
17. Weich S, Burton E, Blancard M, Prince M, Sproston K, et al. (2001) Measuring the build environment: Validity of a site survey instrument for use in urban settings. Health Place 7: 283–292.
18. RAND Corperation (2011) The Los Angeles family and neighborhood survey. Available: http://lasurvey.rand.org/. Accessed 8 September 2013.
19. Emlen J, Stokes A, Davis D (1949) Methods for estimating populations of brown rats in urban habitats. Ecology 30: 430–442.
20. Villafane IE, Cavia R, Vadell MV, Suarez OV, Busch M (2013) Differences in population parameters of *Rattus norvegicus* in urban and rural habitats of central Argentina. Mammalia 77: 187–193.
21. Leslie PH, Davis DHS (1939) An attempt to determine the absolute number of rats on a given area. J Anim Ecol 8: 94–113.
22. Nagorsen DW (2002) An identification manual to the small mammals of British Columbia. Victoria: Province of British Columbia.
23. Nelson L, Clark FW (1973) Correction for sprung traps in catch/effort calculations of trapping results. J Mammal 54: 295–298.
24. Dohoo I, Martin W, Stryhn H (2007) Veterinary epidemiologic research. Charlottetown: AVC Inc.
25. Bricki N, Green J (2007) A guide to using qualitative research methodology. New York: Medicins Sans Frontieres.
26. Krebs C (1999) Current paradigms of rodent population dynamics - What are we missing? In: Singleton GR, Hinds LA, Leirs H, Zhang Z, editors. Ecologically-Based Management of Rodent Pests. Canberra: ACIAR. pp. 33–48.

21

Demographic Outcomes and Ecosystem Implications of Giant Tortoise Reintroduction to Española Island, Galapagos

James P. Gibbs[1]*, Elizabeth A. Hunter[2], Kevin T. Shoemaker[3], Washington H. Tapia[4], Linda J. Cayot[5]

1 Department of Environmental and Forest Biology, State University of New York College of Environmental Science and Forestry, Syracuse, New York, United States of America, **2** Warnell School of Forestry and Natural Resources, University of Georgia, Athens, Georgia, United States of America, **3** Department of Ecology and Evolution, Stony Brook University, Stony Brook, New York, United States of America, **4** Department of Applied Research, Galapagos National Park Service, Puerto Ayora, Galapagos, Ecuador, **5** Galapagos Conservancy, Fairfax, Virginia, United States of America

Abstract

Restoration of extirpated species via captive breeding has typically relied on population viability as the primary criterion for evaluating success. This criterion is inadequate when species reintroduction is undertaken to restore ecological functions and interactions. Herein we report on the demographic and ecological outcomes of a five-decade-long population restoration program for a critically endangered species of "ecosystem engineer": the endemic Española giant Galapagos tortoise (*Chelonoidis hoodensis*). Our analysis of complementary datasets on tortoise demography and movement, tortoise-plant interactions and Española Island's vegetation history indicated that the repatriated tortoise population is secure from a strictly demographic perspective: about half of tortoises released on the island since 1975 were still alive in 2007, *in situ* reproduction is now significant, and future extinction risk is low with or without continued repatriation. Declining survival rates, somatic growth rates, and body condition of repatriates suggests, however, that resources for continued population growth are increasingly limited. Soil stable carbon isotope analyses indicated a pronounced shift toward woody plants in the recent history of the island's plant community, likely a legacy of changes in competitive relations between woody and herbaceous plants induced by now-eradicated feral goats and prolonged absence of tortoises. Woody plants are of concern because they block tortoise movement and hinder recruitment of cactus–a critical resource for tortoises. Tortoises restrict themselves to remnant cactus patches and areas of low woody plant density in the center of the island despite an apparent capacity to colonize a far greater range, likely because of a lack of cactus elsewhere on the island. We conclude that ecosystem-level criteria for success of species reintroduction efforts take much longer to achieve than population-level criteria; moreover, reinstatement of endangered species as fully functioning ecosystem engineers may often require large-scale habitat restoration efforts in concert with population restoration.

Editor: Cheryl A. Nickerson, The Biodesign Institute, Arizona State University, United States of America

Funding: The 2010 field surveys of tortoise-plant interactions as well as this synthesis of the accumulated information was funded by the Galapagos Conservancy in collaboration with the Galapagos National Park Service, the Prometeo Program of Ecuador's National Secretariat for Higher Education, Science, Technology and Innovation, and U.S. National Science Foundation grant DEB 1258062. The funders had no role in study design, data collection and analysis, decision to publish, or preparation of the manuscript.

Competing Interests: The authors have declared that no competing interests exist.

* Email: jpgibbs@esf.edu

Introduction

Restoration of extirpated species and populations via captive breeding and reintroduction has become a widespread practice in biodiversity conservation [1–2]. As such projects proliferate there is growing need to define and document reintroduction success [2–3]. Documentation of a viable, self-sustaining population is essential for declaring success of any reintroduction project [4]. Yet reintroduction efforts are often motivated not just by population viability considerations but also by broader concerns about restoring ecological functions and interactions, thereby highlighting the need to document the role of reintroduced species as agents of ecosystem change [4].

Species reintroduction can generate profound and complex ecosystem changes, especially if the focal species can be characterized as an "ecosystem engineer" [5–7]. For example, carnivore reintroduction to Yellowstone National Park in the western United States likely triggered a trophic cascade whereby predation pressure on large herbivores has led to the restoration of plant communities [6],[8]. Yet ecosystem-level effects of species reintroduction may manifest only after the focal species crosses a threshold of effective density far higher than needed to ensure population persistence alone [9–10] and thereby result in separate time frames and planning horizons for achieving population-level versus ecosystem-level objectives of species reintroduction programs.

Despite the promise of species reintroduction as a tool for restoring populations and in some cases ecosystems, prolonged absence of certain species can result in habitat degradation and thereby lead to failed reintroduction efforts despite repeated attempts [1],[4],[11–12]. In some cases, management interventions may be required to render release sites hospitable to a focal species before the species can assume its former ecological role [13]. For example, invasive species removal may facilitate successful repatriation in cases where a species' former ecological role has been supplanted by an invasive competitor [14]. For these reasons assessments of the success and feasibility of reintroduction projects must carefully consider ecological changes wrought in the absence of the focal species.

Reintroduction of tortoise populations is increasingly proposed not only to conserve biodiversity [15] but also as tools for ecosystem restoration [16–18]. Giant tortoises, once widespread on all continents except Antarctica, are ecosystem engineers that manipulate the distribution and abundance of other organisms through direct effects of herbivory, disturbance and seed dispersal on plant communities and subsequent indirect impacts on animal communities [14],[19–20]. The effects of giant tortoises on terrestrial ecosystems of oceanic islands (to which giant tortoises are currently restricted) are potentially on par with those of megaherbivores as drivers of savanna structure and function, a system for which the role of ecosystem engineers as agents of ecosystem change is far better understood [21–22]. Consequently, giant tortoises have been proposed as generic tools for island ecosystem restoration [17].

Herein we examine the demographic and ecological outcomes of the repatriation of Galapagos giant tortoises (*Chelonoidis hoodensis*) to Española Island (Fig. 1). *C. hoodensis* is a CITES Appendix I and IUCN Red List "critically endangered" species [23] that once numbered in the thousands of individuals but dropped to just 15 individuals by 1960 due to historical exploitation of tortoises throughout the archipelago [24]. Between 1963 and 1974, all survivors (12 females and 3 males) were brought into captivity and to date >1500 of their captive-raised offspring have since been released onto the island [25]. Successful reproduction of repatriated *C. hoodensis* was first documented in 1990 [24] suggesting the repatriated population might become self-sustaining.

Concerns remain, however, about current and future habitat quality. Specifically, the island's dense and extensive woody plant cover is hypothesized to inhibit tortoise movement (thereby excluding tortoises from colonizing new areas) and to outcompete grasses and juvenile cactus that are critical to tortoises as forage [16]. The present-day dominance of woody plants on Española Island may be a legacy of both the absence of giant tortoises and the presence of large numbers of feral goats on the island for ¾ of a century (1905 to 1978), which might have shifted competitive relationships between woody and herbaceous plants by altering soil water availability in ways that favor woody plants [26–28]. Furthermore, sudden eradication of introduced livestock (in this case, goats) in the absence of native herbivores (in this case, giant tortoises) may have exacerbated woody plant incursion [28]. A particular concern relates to tortoise interactions with a large-seeded, arboreal prickly pear cactus (*Opuntia megasperma*). The cactus is restricted to just three islands and is listed by the IUCN as a "vulnerable" species due to extensive population reduction caused by feral goats. The cactus serves as a critical source of food, water, and shade for tortoises [29]. Although tortoises have been observed to injure and crush young *O. megasperma* individuals, in aggregate tortoise impacts on cactus populations, including seed dispersal [20], appear to be marginally positive on Española Island [16].

We integrated complementary datasets on tortoise demography, tortoise-plant interactions and vegetation history on Española Island to comprehensively assess the reintroduced tortoise population's trajectory in the context of past, current and future ecological conditions. We first examined a 32-year-long tortoise mark-recapture dataset for evidence that the population is self-sustaining. We then assessed the relationship between tortoise density and habitat features (including cactus and woody plant density) in the context of historical changes in woody plant cover on the island. Last, we simulated how different management scenarios, including cessation of the repatriation program and propagation of cactus and control of woody plants, might affect the long-term stability of the tortoise population. We demonstrate that the first step of restoring a viable population of giant tortoises has been achieved; however, significant habitat restoration challenges remain before the role of tortoises as self-sustaining, island-wide ecosystem engineers is fully realized.

Materials and Methods

Study area and the tortoise monitoring program

Española Island (−1° 38′ N; −89° 67′ W) occupies the southeastern sector of the Galapagos Archipelago and is small (60.48 km^2, <1% of the archipelago), low elevation (max. 206 m), flat, and arid (annual precipitation varies between 10–600 mm with average annual temperatures of 23.8°C). Dominant herbaceous plants include *Galactia striata* and *G. tenuiflora, Phaseolus mollis*, *Rhynchosia minima* and *Paspalum* spp. Woody plants, primarily *Cordia lutea* and *Prosopis juliflora*, occur at high densities throughout the island (Fig. 2). The arboreal *Opuntia* cactus is almost entirely restricted to the island's central region [30], a 1250-hectare area that encompasses the three sites used for releasing repatriates from the *ex situ* breeding program: El Caco, Las Tunas, and Gardner Bay (Fig. 2a). Releases began with 17 captive-bred tortoises in 1975 and continued with 2- to 5-year-old tortoises introduced every year from 1975–1994 and sporadically thereafter from 1997–2007 for a total of 1482 tortoises repatriated [29].

Since the reintroduction program was initiated all repatriated tortoises as well as individuals encountered in the field during surveys have been given unique identification markings. Up to 1991, a unique combination of notches (made with machete or metal file) was carved onto the tortoises' marginal scutes. Thereafter a duplicate marking system was introduced consisting of numbers branded on the rear costal scutes combined with insertion of passive integrated transponder (PIT) tags (AVID Identification Systems Inc., Norco, California, USA) into the left rear leg of each tortoise. Locations of individual tortoises have been registered since 2002 with hand-held Global Positioning System (GPS) devices, thereby enabling tracking movements of marked individuals between survey bouts during later phases of the tortoise monitoring program. Most (96%) tortoises captured were weighed with a calibrated spring scale. Curved carapace length, measured as the distance from the notch at the anterior of the carapace over the tops of the vertebral scutes to the tip of the supracaudal scute, was used as the measure of tortoise size. Sex and age were assessed on the basis of external morphology. Native-hatched tortoises (offspring of repatriates), first encountered in significant numbers after 2000, were identified on the basis of lack of notches or PIT tags and the distinct appearance of their scutes: more compressed growth segments, lighter coloration, and more scute abrasion given size.

Figure 1. Adult male Galapagos giant tortoise (*Chelonoidis hoodensis*) resting beneath an adult arboreal prickly pear cactus (*Opuntia megasperma*) surrounding by woody plants (mostly muyuyo or *Cordia lutea*), Española Island, May 2010 (image: J. P. Gibbs).

Fieldwork was carried out on an endangered species and on public land in strict accordance with permit conditions of the Galapagos National Park Directorate and following the protocol approved by State University of New York, College of Environmental Science and Forestry's Institutional Animal Care & Use Committee (Protocol Number: 121201).

Tortoise-plant interactions

Because of the hypothesized importance of woody plants as a potential factor structuring tortoise-plant interactions and thereby affecting tortoise population recovery, we quantified the extent of woody vegetation on Española Island over both space and time. Spatial extent was determined by delineating woody vegetation from cloud-free Quickbird satellite imagery (December 29, 2006, 0.6 meters resolution) using a supervised classification calibrated with woody plant density estimates from field plots (see below, IDRISI Andes 5.0). We aggregated these data to obtain a map of percent woody cover by summing the number of 0.6×0.6 m cells classified as woody in each 24 m cell. Temporal trend in woody plant cover, emphasizing possible changes associated with goat impacts and tortoise near-absence over the last 100 years, was inferred from soil carbon isotope analysis [31]. To do so, we first determined diagnostic $\delta^{13}C$ signatures of the island's dominant woody plants (*Bursera*, *Cordia*, *Croton*, *Parkinsonia*, *Prosopis*) and herbaceous plants (*Cyperus* and *Paspalum*) as well as cactus (*Opuntia*) ($n = 10$ of each species). We then sampled soil at 10 cm increments through the soil horizon from soil pits dug at Gardner Bay and in the Caco area. All analyses were made at the Environmental Stable Isotope Laboratory at SUNY-ESF in Syracuse, New York. Accuracy and precision of stable isotope measurements were verified using National Institute of Standards and Technology (NIST) reference materials (M. A. Teece, pers. comm.). Radiocarbon dating of two soil samples at two points in the soil profile of the Caco pit was performed by Beta Analytic Inc., Miami, Florida, to enable calibration of soil depth and soil sample age. We tested trends in the dominance of woody plants in the soil record using linear models of $\delta^{13}C$ by depth for two time periods (recent history versus earlier, see Results) using a general linear mixed model with soil pit treated as a random effect.

To assess specific hypotheses about tortoise-plant interactions we performed two studies. The first was re-censusing cactus in June 2014 in a 60 ha macroplot in the Tunas tortoise release area first censused in May 2004 [16] to examine cactus population trends in an area where tortoises have long been established at high density. The second involved measuring tortoise and cactus abundance as well as woody vegetation extent on 96 field plots within the central "tortoise zone" in May-June 2010. The sampling frame was established by buffering by 250 m all tortoise occurrences recorded during previous surveys. We then randomly placed one plot within every 250×250 m grid cell. Each plot comprised five 20-m-radius circular sub-plots with a central sub-plot and four sub-plots directly adjacent at the cardinal directions. Within each sub-plot we recorded total numbers of tortoises, tortoise droppings, juvenile cactus (no bark on trunk), subadult cactus (some bark on trunk), and adult cactus (trunk entirely bark-covered). Density of woody plants was estimated using the point-quarter method [32] for two basal diameter classes: ≥ 5 cm (mature plants) and≤ 0.5 cm (recruits). All measurements were subsequently averaged at the plot-level for analysis of interactions among tortoises, cactus, and woody vegetation via binary classification trees [33]. We also modeled tortoise density as a function of cactus and woody plant density, as well as distance from release site, using a linear negative binomial regression model. The model was run in WinBUGS, using 30,000 MCMC iterations (treating the first 10% as a burn-in and capturing every 10th observation to reduce serial autocorrelation).

After 30 years of repatriation, tortoises remain restricted to the central part of the island (Fig. 2a); therefore, we evaluated whether tortoises have the intrinsic capacity to move further afield if suitable habitat were available. To do so, observed annual displacement rates (km per year) of individuals from 2000 to

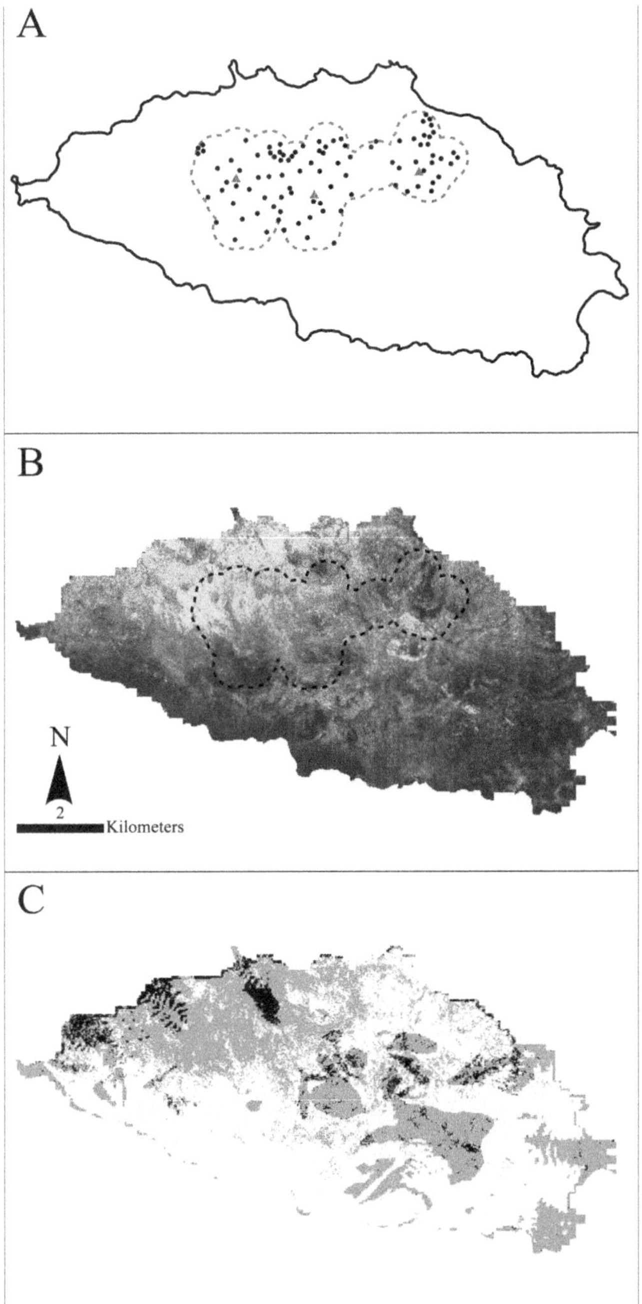

Figure 2. Study areas on Española Island. (A): gray dashed line is "tortoise zone" or maximum extent of tortoises observed on the island since the repatriation program was initiated; black circles are 2010 survey plot locations for tortoises, cactus, and woody plants; gray triangles are tortoise introduction sites (West: Tunas, central: Cacos, East: Gardner). Woody plant cover on Española Island (B; darker shades indicate higher percent woody plant cover), classified using cloud-free Quickbird satellite imagery (December 29, 2006, 0.6 meters resolution) into areas with and without woody vegetation using a supervised classification calibrated with plot woody plant density estimates (IDRISI Andes 5.0) (percent cover is the percent vegetated cells per 24 m×24 m area). Potentially highly suitable cactus restoration areas (C) with low slopes (<2.06 degrees, bottom quartile of available slopes) and low woody plant cover (<52% cover woody plants, bottom quartile of available woody plant cover); black areas have both low slopes and low woody cover, gray areas have either low slope or low woody cover.

2003 were used as a proxy for intrinsic capacity for range expansion. We tested for differences among age/sex classes (juvenile and adult, female and male) in mean displacement rates with a linear mixed effects model and then examined whether the observed expansion in range area (computed as a minimum convex polygon, or MCP, over all known occurrence locations) differed from a null expectation associated with random movement. We generated the null distribution for range expansion by simulating 500 replicate movement scenarios in which observed starting locations were randomly permuted (thereby de-linking starting locations from observed movement distances and directions). Finally, we tested the hypothesis that tortoises near the edge of the tortoise zone would be more likely to move away from the current range than tortoises situated far from the edge by regressing the direction of tortoise movement relative to the nearest edge with the distance of the starting location from the nearest edge. All analyses were run in the R statistical computing language (raster package and adehabitatHR package) [35–36].

Last, to estimate carrying capacity (K) for tortoises of the central zone they now occupy (Fig. 2a), we used the regression model of tortoise density (described previously) to compute the expected tortoise density at release sites, extrapolating the resulting estimate to the rest of the tortoise zone (assuming tortoises had reached carrying capacity at distance = 0 from release sites and that mean cactus and woody plant densities computed from plot-level data pertained to the entire tortoise zone). In addition, we used the same model to evaluate island-wide carrying capacity under three potential restoration scenarios: (1) a cactus regeneration scenario in which the cactus density of the tortoise zone is restored to the entire island, (2) a woody plant removal scenario in which woody plant density is reduced by half across the entire island, and (3) a combined cactus regeneration/woody plant removal scenario. The potential effect of these restoration scenarios was expressed as the percent increase in the tortoise carrying capacity over the current estimated carrying capacity.

Tortoise demography

We assessed the demographic outcome of the tortoise repatriation program in two ways. The first was direct estimation of tortoise population size in 2010 (after 34 years of repatriations) from mean tortoise densities measured on the 96 field plots in the tortoise zone (see above). The second involved estimation of abundance, vital rates, growth rates and body condition based on a capture history database assembled from surveys between 1975 and 2007 (recapture data were available every year except 1995, 1996 and 1998), during which an average of 120 tortoises was recaptured per year (range: 16 to 377). Trends in somatic growth (cm/year) and body condition (deviations from expected mass-length relationships) were assessed using generalized additive mixed models (GAMMs, mgcv [34]) that included the categorical covariates sex/age (juvenile, adult, male, female or unknown) and individual (as a random effect in order to account for repeated measures [37]) as well as year (1975–2007) and body size (curved carapace length). Growth was determined from change in curved carapace length between subsequent measurements of tortoises (2.5% percent of observations were removed from both extremes of the growth rate estimates distribution as well as recaptures >3 years apart to guard against anomalies caused by errors in field recording or data entry).

We also extracted annual abundance and survival rates from the capture history database using a Bayesian open-population model (Cormack-Jolly-Seber analogue) in conjunction with the known numbers of tortoises released each year. We estimated survival (φ) separately for juveniles (<8 y.o.) and subadults/adults ($\geq$8 y.o.).

To accommodate temporal dynamics in survival rate (e.g., environmental stochasticity, trends in survival, density dependence), survival rates were estimated separately for each 4-year time period (subadults/adults), 2-year time periods (juveniles), and 1-year time period (year-of-release). Mean capture probability was estimated separately for juveniles/subadults (<15 y.o.) and reproductive adults (≥15 y.o.) with inter-annual variance in capture probability modeled as a logit-normal random effect. Because our modeling framework explicitly tracked the status of each released tortoise (i.e., alive or dead) each year, we were able to estimate annual abundance of released tortoises (excluding native-born individuals) at two distinct release areas (Tunas and Caco versus Gardner Bay) as the sum of living individuals occupying each of these release areas (Fig. 2a) (due to proximity and lack of barriers to movement of tortoises between the release sites of El Caco and Las Tunas these were treated as a single release site for analysis).

We estimated population parameters with Markov-Chain Monte Carlo (MCMC) in WinBUGS (version 1.4) via the R2WinBUGS package in R [34]. Uninformative uniform prior probability distributions were assigned to all parameters and the initial 5,000 MCMC samples were discarded as a burn-in. We performed a further 10,000 MCMC iterations and saved every 10th iteration to reduce serial autocorrelation among samples [38]. Convergence of the Markov chains to the stationary posterior distribution was tested with the Gelman–Rubin diagnostic [38]. We summarized posterior distributions for all parameters with the mean of all MCMC samples as a point estimate and the 2.5 and 97.5 percentiles of the MCMC samples as a 95% credible interval [38]. A detailed description of our demographic and population projection model is provided in Appendix S1 with accompanying R and WinBUGS code in Appendix S2.

We projected over the next 100 years the abundance of giant tortoises on Española Island on the basis of estimates of survival and temporal variation in survival derived from the capture-recapture analysis, carrying capacity estimates from tortoise-habitat relationships (see above), and fecundity estimates extrapolated from observations of native-born tortoises (see Appendix S3). We used a female-only model, because females are generally presumed to be the limiting sex for most turtles and tortoises [39]. Survival rates of tortoises aged 1–2 were not estimable (all tortoises were released at age 3 and above) and were assigned a range of values typical of long-lived turtle species [40]. Initial female abundance was approximated using the estimated total number of released individuals that were alive in 2007 (derived from capture-recapture analyses, and ignoring native-born individuals). We used a "ceiling" type density dependence model in which the maximum population growth rate (R_{max}) was defined by the vital rates estimated for this study (a reasonable assumption given the population was likely below carrying capacity for much of the study period). All parameter values used in our population models, and the information sources for these parameter values, are presented in Table 1.

Total fertility rate (total number of yearlings produced per reproductive adult) was computed as the product of four parameters: mean proportion of females in the population, mean number of eggs produced per female, mean hatching success rate, and mean neonate (hatchling) survival. We used outside sources of information to estimate the first two fertility components (Table 1) and an approximate-likelihood technique to estimate the remaining fertility parameters (see Appendix S1 for details): on the basis of the increase in the proportion of native-born individuals observed over time we concluded that the product of egg viability and hatchling survival on Española Island was between 0.023 (equivalent to both parameters set at 0.15) and 0.066 (equivalent to both parameters set at 0.26). Recruitment was modeled as a Poisson process [38]. Lacking information on the covariance structure of the four components of fertility, we conservatively assumed that all four parameters were perfectly correlated, which may underestimate population stability. All population projections consisted of 100 separate population simulations for 150 years each starting in the year 2011. For all simulations we recorded total abundance of tortoises and total abundance of adults (the ecological effects of tortoises are largely driven by adults; see below). Last, we estimated extirpation probability as the proportion of simulation runs reaching the quasi-extinction threshold (75 adults or fewer at any point during the simulation) after 150 years. All population projection models were coded in the R statistical computing language, and the code is provided in Appendix S3.

Results

Spatial and temporal pattern in woody plant extent on Española Island

Extent of woody plants on Española Island showed dramatic variation across the island in 2006 (Fig. 2b). Regression tree analysis indicated that the fraction of a given 24 m×24 m area on the island dominated by woody plants was highest (0.96) within 1.1 km of the coast; farther from the coast, woody cover was more extensive on lower slopes (<3.2 degrees, 0.86) than steeper ones and lower (0.58) in the tortoise zone than outside (0.68). In terms of temporal variation in woody cover, extent of woody plants apparently increased strongly and recently in terms of the history of the island's plant community (Fig. 3). Radiocarbon dating from the Caco soil pit samples at 40 and 60 cm yielded dates of 510+/−40 BP and 1100+/−40 BP, respectively; thus our 0.8–1.1 m soil pits captured >1,000 years of vegetation history on the island. Stable carbon isotope signatures ($\delta^{13}C$ values) were highly distinct between herbaceous and woody plants sampled: all woody plants were consistently more negative (−26 to −30‰) and all herbaceous plants consistently less negative (−13 to −16‰; Fig. 3) with cactus (−15‰) grouping among the complement of herbaceous plants sampled. In recent history (soil depths of ≦40 cm or within 510 YBP), $\delta^{13}C$ values have become more negative with time ($t = 5.49$, $p<0.001$) indicating a shift towards greater dominance by woody plants (Fig. 3); prior to 510 YBP, $\delta^{13}C$ values indicated no consistent trend ($t = 0.38$, $p = 0.71$).

Tortoise-habitat relationships

In 2010 tortoise density was highest (1.8 tortoises/ha) in locations where adult cactus density was highest (>4.0 cactus/ha) (Fig. 4a). Where adult cactus density was lower (<4.0 cactus/ha), tortoises occurred at much lower densities (0.66/ha) and only if large diameter woody plant density was also low (<10.7 stems/100 m^2). No tortoises occurred on plots with low adult cactus densities when combined with high densities of large diameter woody plants. Tortoise dropping density (reflecting tortoises in transit) was greatest (15.2/ha) where woody plant density was lowest (Fig. 4b) yet was still high (13.0/ha) in areas of high woody plant extent if adult cactus were present; otherwise there were dramatically fewer droppings (4.1/ha) suggesting that tortoises avoid woody plant areas particularly if no cactus is present.

Tortoise density in the tortoise zone declined with distance from reintroduction release points (Tunas, Caco and Gardner), increased with adult cactus density, and decreased with woody plant density. The credible interval of the effect of distance from introduction points did not overlap with zero (95% C.I.: −5.31 to −2.42), but it did for the cactus (95% C.I.: −0.14 to 2.33) and

Table 1. Summary of input parameters used to simulate the population dynamics of Española giant tortoises during and after repatriation to Española Island, Galapagos.

Parameter	Lower-Upper Bound	Source
Age at first reproduction	17–17	Age of oldest repatriated tortoises at year of first appearance of native-born individuals on the island [24]
Proportion of population female	0.3–0.7	Not known; wide interval centered on 0.5
Carrying capacity	424–3921	Estimated from tortoise density relationships to distance from introduction sites, and cactus and woody plant density (see Methods)
Fecundity	4–10	Mean annual eggs produced per female [68]
Annual survival, ages 1–3	0.6–0.9	Not known; estimated using rates typical for turtles [40] with wide interval to accommodate uncertainty
Hatchling survival (ages 0–1 years)	0.15–0.26	Estimated via simulation to obtain values of a fraction of native-born animals similar to that observed in the field
Mean hatching success rate	0.15–0.26	Estimated via simulation to obtain values of a fraction of native-born animals similar to observed in the field

Parameters estimated from the 32-year capture-recapture dataset include juvenile and adult survival, temporal variation in survival, and survival at year-of-release (see text).

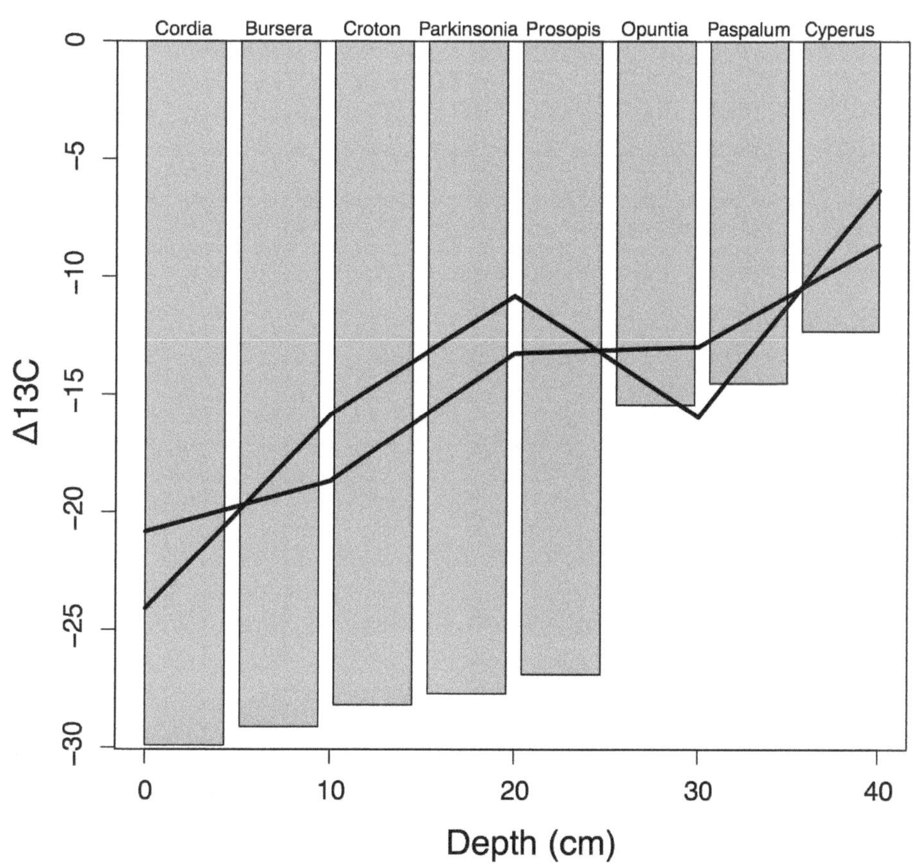

Figure 3. Carbon isotope signatures ($\delta^{13}C$ values) of soils at increasingly greater depths in two soil pits (Gardner and El Caco) on Española Island, Galapagos. Radiocarbon dating indicated that depths of 40 cm are approximately 500 years BP; therefore, samples below the 40 cm depth are grouped to show pre-disturbance averages. Points are $\delta^{13}C$ values of the island's dominant woody plants (*Bursera, Cordia, Croton, Parkinsonia, Prosopis*) and herbaceous plants (*Cyperus* and *Paspalum*) as well as cactus (*Opuntia*) (averaged across 10 samples of each, with standard errors for all means too miniscule to be evident in this graphic).

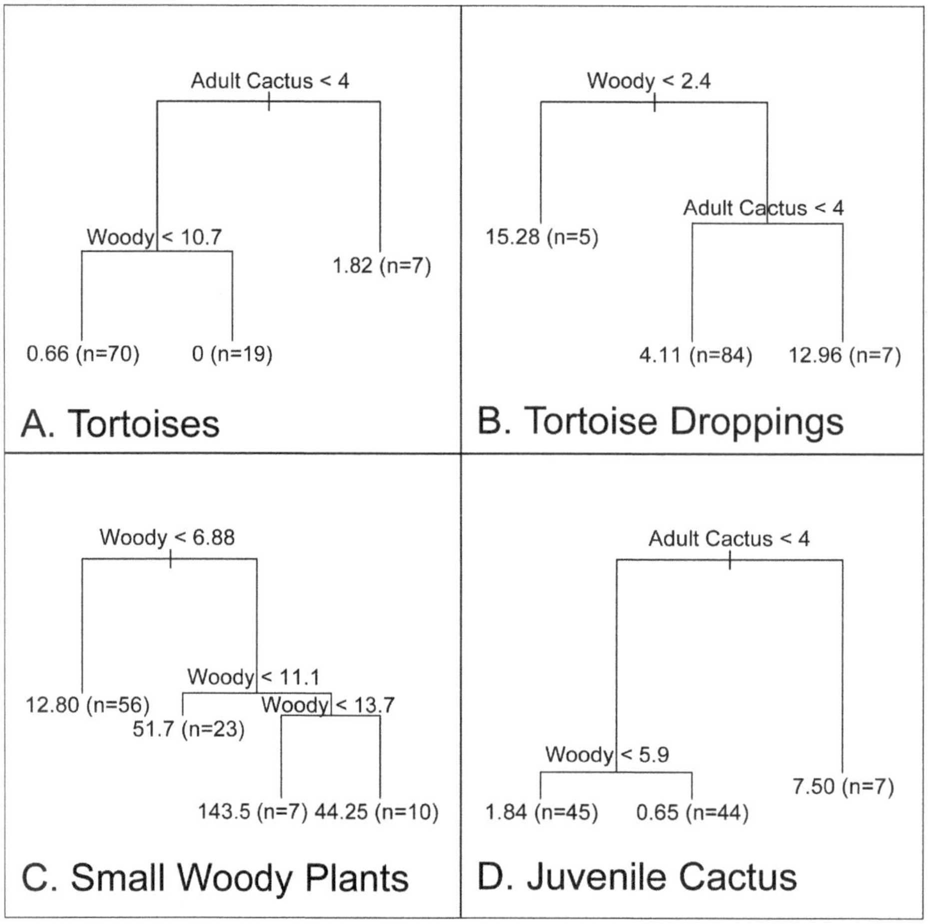

Figure 4. Relationships derived from binary classification trees between *density of tortoises* (A) and *tortoise droppings* (B) with thresholds of adult cactus density and large diameter woody plant density; between *small diameter woody plants* (C) with thresholds of large diameter woody plant density; and between *density of juvenile cactus* (D) with thresholds of large diameter woody plant density and adult cactus density. Density of tortoises, tortoise droppings, and cactus are expressed as number/hectare; woody plants as stems/100 m^2.

woody plant (95% C.I.: −2.45 to 2.04) effects. Using these relationships, we estimated a carrying capacity of 2081 tortoises (95% C.I.: 424 to 3921) within the tortoise zone. Under the three potential restoration scenarios, the change in the tortoise carrying capacity over the current carrying capacity was estimated to be a 7% increase under the cactus regeneration scenario, a 52% increase under the woody plant removal scenario, and a 62% increase under the cactus regeneration and woody plant removal scenario.

The area observed to be used by tortoises (MCP) expanded little during the period for which GPS locations were available (2000 and after; Fig. 5a). During this same period random movements by tortoises would have generated a substantial expansion of the currently occupied core area (Fig. 5b); in fact, the observed final MCP was entirely outside the statistical distribution of plausible final MCPs produced by random movement (Fig. 5c). This disparity suggests that tortoise movements were not random with respect to the edge of the observed core area, which was also confirmed by our final movement analysis: tortoises near to the edges of the core area were much more likely to move away from the current range edge than tortoises situated more centrally within the core area (Fig. 5d).

For cactus populations in the tortoise zone, we estimated densities of 1.7 juvenile cactus/ha, 0.4 subadult cactus/ha, and 1.0 adult cactus/ha for a total of 2142 juvenile cacti (95% C.I.: 1303.8–2980.3), 457.5 subadult cacti (201.7–713.2) and 1310.2 adult cacti (627.4–1992.9). Density of juvenile cactus was highest (7.5/ha) where adult cactus densities were >4 individuals/ha, next highest yet dramatically lower (1.8/ha) where adult cactus densities were <4 individuals/ha and density of large diameter woody plants was <5.9 stems/100 m^2, and virtually absent (0.65/ha) where adult cactus reached densities of <4 individuals/ha and density of large diameter woody plants was >5.9 stems/stems/100 m^2 (Fig. 4d). Density of small diameter woody plants (woody plant regeneration) was extremely high (143 stems/100 m^2) at intermediate densities of large diameter woody plants (11–13 stems/100 m^2) and not associated with tortoise density (Fig. 4c).

Over a 10-year period in the Tunas tortoise release zone macroplot [16], total individual cacti censused increased from 315 to 396 individuals (+26%): stage structure (adults versus juveniles and subadults) did not vary between census periods ($G_{adj} = 2.38$, df = 1, P = 0.12) despite juveniles and subadults increasing in abundance (194 to 266 individual juveniles and subadults, +37%) more rapidly than adult cacti (121 to 130 adults, or +7%).

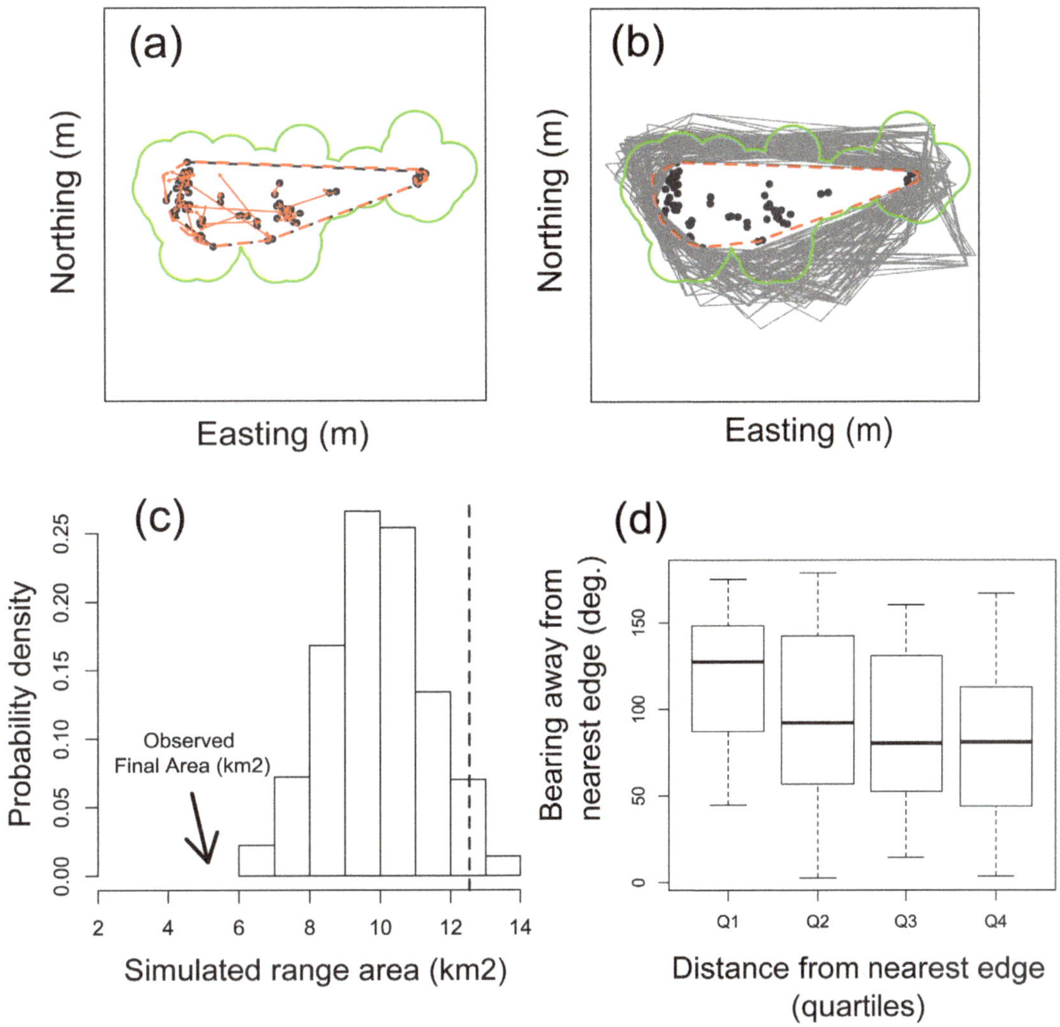

Figure 5. Movement patterns of giant tortoises reintroduced to Española Island, Galapagos. Closed circles in (A) and (B) represent the initial locations of all tortoises captured from 2000 to 2003 that were later recaptured. Arrows in (A) represent observed movement distances and directions. Minimum convex polygons (MCP) enclose the initial locations (grey) and initial plus final locations (red; differences represent the observed range expansion resulting from the observed movements). The grey polygons in (B) depict hypothetical MCPs (n = 500) for which initial locations have been decoupled from observed movement distances and directions. The observed final range area was smaller than all resulting hypothetical MCPs (C). In fact, those tortoises nearest to the range margins exhibited a strong tendency to move away from the nearest edge (D).

Demographic reconstruction of the reintroduction program and future projections

Our recapture history comprised 1459 individuals from among cohorts initially released in El Caco (*n* = 780), Las Tunas (*n* = 370), and Gardner Bay (*n* = 309). Individuals were captured on average 2.2 times each (range: 1 to 21). Annual somatic growth rates (cm per year in curved carapace length) averaged 2.99 cm±2.19 cm (SD), with 35.6% of deviance explained by sex and age, carapace length, and year (Table 2): juveniles grew an average of 4.46 cm/year (±1.18 SD, n = 175), unknowns (intermediate between juveniles and adults) 3.23 cm/year (±2.19 SD, n = 361), males 2.25 cm/year (±2.27, n = 278) and females 2.31 cm/year (±2.01, n = 212). Our suite of independent variables accounted for virtually all variation (98%) in tortoise body mass: after controlling for the inherent size-mass relationship, males and unknowns were relatively lighter than females, and tortoise condition declined with time (Table 2).

Annual probability of capture for juvenile giant tortoises averaged 0.38 (CI 0.17–0.58), and was slightly lower for adult giant tortoises (Fig. S1-1 in Appendix S1). Mean annual survival was high for both juvenile (mean = 0.958, CI 0.86–0.99) and adult giant tortoises (mean = 0.977, CI 0.93–0.99), with mean juvenile survival approximately 1% lower than adult survival (Fig. S1-1 in Appendix S1). Adult survival rate varied temporally from 0.94 to 0.99, juvenile survival rate varied temporally from 0.88 to 0.99, and survival at year-of-release varied from 0.29 to 0.98, with generally lower survival rates after 1994 (Fig. 6a). Maximum population growth rate (R_{max}, or the discrete population growth rate excluding density dependent effects) was estimated at 1.01 on the basis of our vital rate (survival and fecundity) estimates (popbio package in R [34]). With only 6 years of data for which to estimate survival and abundance at the "Gardner" release area, credible intervals for all survival terms were much more diffuse (Fig. 6b) and most survival information was derived from the much larger and longer-term records from the Tunas and Caco release site. We estimate that more than half of tortoises released to the island since 1975 were still alive in 2007 (Fig. 6a). Notably, the independent estimate of tortoise population size in 2010 derived from plot-

Table 2. Summary of generalized additive mixed models fitted to giant tortoises (n = 909) on Española Island, Galapagos between 1975 and 2007 to explore relationships between (A) somatic growth rates (cm/year) and age/sex while controlling for time (year) and individual length (carapace length) (35.6% of deviance explained) and (B) body mass and age/sex while controlling for time (year) and individual length (carapace length) (97.8% of deviance explained).

(A) Somatic growth							
Parametric coefficients					**Nonlinear effects (nonparametric)**		
Parameter	**Estimate**	**Standard Error**	*t*	*P*	**df**	*F*	*P*
Constant	27.665	18.323	1.510	0.131			
Sex/Age (Female vs. Juvenile)	0.649	0.177	3.662	0.0003			
Sex/Age (Female vs. Male)	0.527	0.145	3.626	0.0003			
Sex/Age (Female vs. Unknown)	−0.209	0.172	−1.211	0.226			
Year	−0.0122	0.00918	−1.330	0.183			
Carapace length					5.466	49.550	0.001
(B) Mass-length							
Parameter	**Estimate**	**Standard Error**	*t*	*P*	**df**	*F*	*P*
Constant	173144.0	16114.663	10.745	0.001			
Sex/Age (Female vs. Juvenile)	29.967	162.393	0.184	0.853			
Sex/Age (Female vs. Male)	−447.683	144.858	−3.0912	0.00202			
Sex/Age (Female vs. Unknown)	−1190.300	175.237	−6.793	0.001			
Month							
Year	−78.221	8.073	−9.689	0.001			
Carapace length					8.992	8413	0.001

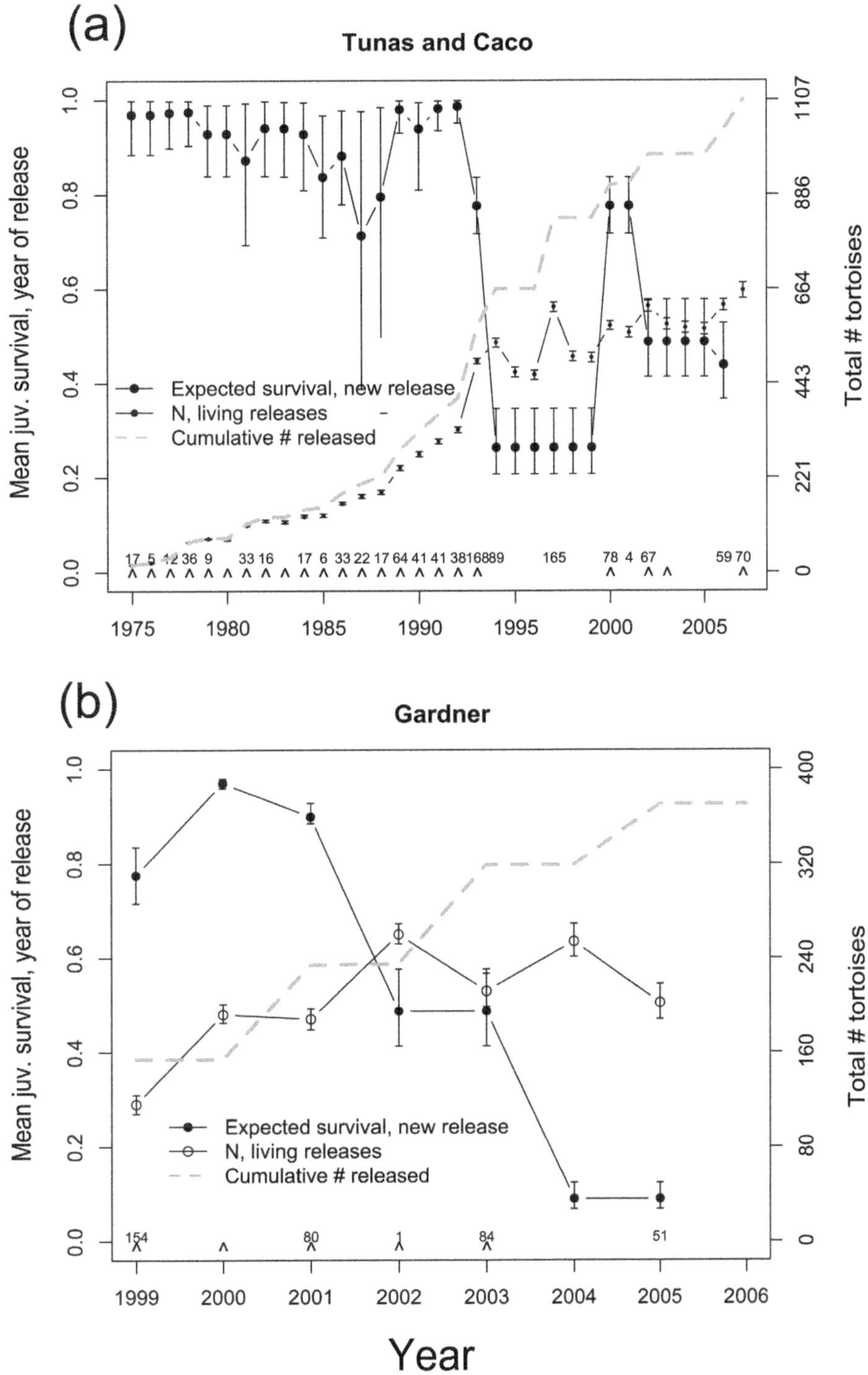

Figure 6. Estimates derived from a 32-year capture-recapture data set of abundance and vital rates of giant tortoises reintroduced to Española Island, Galapagos depicted by release sites (see Fig. 1): Tunas and Caco combined (A) and Gardner (B). Closed circles depict expected survival rates for newly released juveniles, the dashed line cumulative number of tortoises released, and open circles the estimated number of released individuals remaining. Release years are denoted with carets located above the x axis, and the numbers above the carets indicate the total number of tortoises released during each repatriation event.

based counts of tortoises for the Tunas and Caco area of 0.613 tortoises/ha (or total population of 769.5 tortoises, 439.7–1099.2, 95% confidence limits) corresponded well to the population size estimate derived from the mark-recapture analysis: 864 tortoises surviving in 2007 (Fig. 6a & b).

Population projections built from vital rates estimated from the mark-recapture data suggest that the benefits of continuing tortoise repatriation efforts to Española in terms of population viability would be very low. Continued repatriation for 25 more years, termination of repatriation, and termination of repatriation coupled with one-time removal of 50 adults (for possible translocation to other islands) yielded nearly equal and negligible extinction risk estimates over the 100-year time frame (Figs. 7a–c). In terms of restoring the role of giant tortoises as ecosystem engineers, however, the scenarios' outcomes contrasted in a substantive manner: carrying capacity was projected to be reached under the continued repatriation scenario by 2075 versus by 2100 for repatriation termination and by 2115 for removal of 50 adults in 2012 with termination of repatriation. This pattern was even more pronounced when adult (female) abundance was examined in isolation: adult abundance plateaus at ca. 1000 adult female tortoises (ca. 2000 total adults) by the year 2045 under the continued repatriation scenario (Fig. 7d), whereas adult abundance only reached half this level by the end of the simulation period under scenarios in which repatriation is terminated (Fig. 7e, f).

Discussion

The science of species reintroduction has typically focused on the initial steps of a reintroduction project: how to physically release the animals [41], how to improve genetic mixing of the small initial population [42], how suitable receiving habitat is for a release [43], and how to prevent excessive dispersal away from the intended release site [44–45]. Although these considerations are important for initial re-establishment of any population, how to generate wide-scale re-delivery of ecosystem-level processes via the introduced species is more complicated and rarely examined. In the case of the giant tortoise repatriation on Española Island in Galapagos, we determined that the re-introduced tortoise population was likely now self-sustaining over the long-term, but realization of ecological goals for the tortoise re-introduction program (notably, full occupation of the island by tortoises and the restoration of the island's once-dominant herbaceous and graminoid vegetation) will require considerably more time and perhaps further intervention in the form of habitat management.

More specifically, our analysis indicated that the repatriated Española tortoise population is secure from a strictly demographic perspective: extinction risk was very low (<1%) with or without continued repatriation, and population growth rates were weakly positive. We are confident in our population reconstruction in part because the estimate of tortoise population size derived from plot-based counts of tortoises yielded an estimate of population size that corresponded to the final abundance estimate derived in a

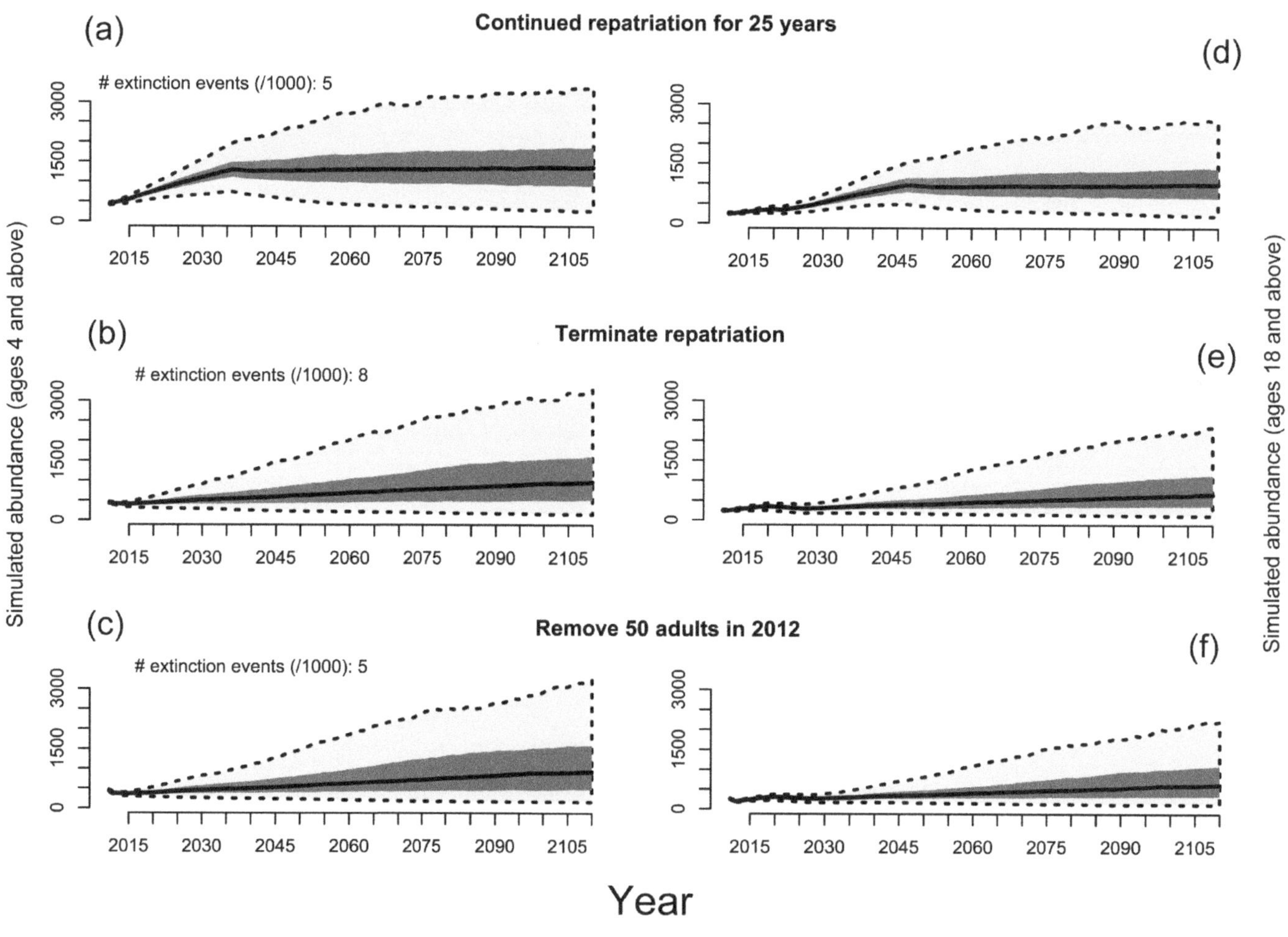

Figure 7. Summary of projected 100-year population dynamics for the repatriated giant tortoise population on Española Island, Galapagos, for (A, B, C) all female tortoises ≥4 years old, and (D, E, F) only adult female tortoises ≥18 years. Vital rates for the projection models were developed largely from a 32-year-capture-recapture database and carrying capacity (K) was determined using tortoise densities measured at field plots and modeled as a function of vegetation characteristics and distance from release sites. We modeled three management scenarios: (A, D) continued release of captive-raised tortoises for 25 years, (C, E) termination of tortoise repatriation, and (C, F) removal of 50 reproductive adult females at the beginning of the simulation (e.g., to accelerate repatriation efforts on other islands).

completely independent manner from analysis of capture-recapture data. This analysis was made possible by a rigorous monitoring program conducted in parallel with and since the inception of the reintroduction effort 32 years ago, and because ancillary sources of information were available (e.g., plot-based analyses of vegetation characteristics and tortoise densities) from which we could estimate carrying capacity.

Our demographic analyses for the Española population imply that approximately half of tortoises released on the island since 1975 were still alive in 2007, and that estimated fecundity rates were generally sufficient to offset mortality, resulting in modestly positive population growth even if the repatriation program were to be terminated (Fig. 6). Similar, high intrinsic population growth rates were reported for Aldabran giant tortoises, whose population rebounded from very low numbers in the early 20th century (when they were in the hundreds or low thousands [46]) to over 100,000 tortoises by the 1970s after cessation of severe harvesting pressures, which implies a discrete annual population growth of at least 1.07 [46–47]. The Aldabra example and our case study of Española tortoises together illustrate that giant tortoise conservation efforts starting from very few individuals have the potential for relatively rapid success if the deterministic factors that caused population decline originally are ameliorated.

The apparent decline in the survival rate of release cohorts we observed since ca. 1994 (Fig. 6a), along with decreasing growth rates and body condition, may be evidence, however, that resources on Española island are becoming limiting to continued population growth. In other words, the population may now be at or near carrying capacity of the area occupied. Two caveats may apply to this conclusion. First, declines in apparent survival could be due to emigration of young tortoises from the core area that is the focus on tortoise monitoring, perhaps motivated by density-dependent food competition among tortoises in a growing population. This said, recent surveys outside the tortoise zone [48] have revealed very little sign (droppings, bedding sites, trails) of tortoises elsewhere on the island. Second, the reduction we observed in growth rates and condition over time could be simple artifacts of ontogenetic changes in carapace length in an aging population given that this is a prominently saddle-backed species whose distinct morphology is manifested after sexual maturity [49]. However, the magnitude of relative change of carapace length (the measure of size used in this study) versus other measures of body size is indistinguishable in male and very modest in female saddle-backed tortoises between younger and older individuals [50]).

Population growth is generally not limited by environmental constraints during early phases of reintroduction efforts because densities are below carrying capacity [51–52], with population limiting factors not manifesting their effects until population densities increase. With high estimated adult, subadult and juvenile annual survival rates of ca. 98% throughout the reintroduction effort on Española Island, and in the absence of opportunity for migration (as seems to be the case for this island population where suitable habitat is limited to a small sector of the island), low survival of hatchlings and very young individuals is likely to limit population growth as the population approaches carrying capacity. This was evidenced by the generally low estimated survival ($\leq$50%) of tortoises released since 1994, after which population growth rates tended to be reduced (Fig. 6a). While juvenile survival rates estimated for giant tortoises in this study are not atypical for turtles and tortoise populations experiencing stable growth [40], higher juvenile survival rates are likely required to boost population growth rates [53].

We estimated a low probability of extinction for the Española tortoise population under all of the future management scenarios we assessed. We have used an admittedly short time-frame (100 years) for considering extinction risk for such a long-lived animal (generation time is estimated at ca. 50 to 60 years on the basis of our vital rate estimates) yet modeling time frames beyond 100 years are of little relevance to managers and conservation planners [54]. Clearly, projecting the success of reintroduction efforts can be difficult for long-lived species such as tortoises, whose reproduction can be delayed for a decade or more and for which high adult annual survival often obscures critical population dynamics [53]. For example, few reintroduction projects for long-lived species like tortoises and turtles have monitored populations long enough to estimate adult survival (with some exceptions [41]), let alone to estimate carrying capacity and to project density-dependent declines in population growth rates, yet such estimates are critical for confirming reintroduction success and estimating the long-term ecological effects of reintroduction efforts.

Despite evident achievement of population viability through this repatriation program, significant constraints evidently remain on the tortoise population's ability to achieve high densities and extensive spatial distribution. More specifically, the tortoises' preferred habitat, areas with high adult cactus density and low woody plant density, is currently limited across the island, and consequently the tortoises' movements are restricted to a small portion of the island despite an intrinsic movement capacity to expand to a far larger area than the species currently occupies. Our plot-based analysis of tortoise-cactus-woody-plant interactions supports the general hypothesis that woody plants have negative impacts on tortoises. This likely occurs because woody plants block tortoise movement thereby increasing travel costs and restricting access to habitat. Woody plants also reduce cactus recruitment, likely through light and water competition and perhaps by mediating tortoise movement thereby limiting dispersal and germination of cactus seed [16].

The current predominance of woody plants appears to be anomalous in the context of the history of vegetation on the island given that temporal trends in stable carbon isotopes of soils indicated a pronounced shift toward woody plants in the recent history of Española Island's plant community. This inference relied on the clear separation we observed in $\delta^{13}C$ values for woody and herbaceous plants but also relies on a reasonable but impossible-to-validate assumption that the plant species sampled in the present also were those that dominated in the past. The cause of the recent shift towards woody plants cannot be determined from data available to us because in addition to the herbivory by goats and centuries-long effective absence of tortoises, climate change may have altered plant communities and the $\delta^{13}C$ values expressed in the soil profile. This said, other studies in the Galapagos have indicated a change in the historical plant community composition with the reduction in giant tortoise populations and invasion of exotic herbivores [55]. Moreover in other semi-arid ecosystems, including ones that share some of the woody plant genera that dominate Española Island, the loss of a native herbivore and the introduction of an invasive herbivore has resulted in woody plant encroachment to an extent that the ecosystem cannot evidently recover without intervention [28]. Given the current widespread dominance of woody plants on Española Island, it is possible that the system has already reached a new, stable and woody-plant-dominated state that may require extensive woody plant removal or other habitat management to transition to a more mixed woody plant-cactus-grassland state [56].

The status of cactus on the island also remains of particular concern, not just because the species itself is of conservation concern due to its highly restricted distribution and reduced abundance due to historical goat impacts, but also because tortoises are heavily dependent on it. There are currently only an estimated 1310 adult cacti in the zone occupied by tortoises on Española Island, or just 1–2 cactus per tortoise. Moreover, our density estimates for cactus on Española Island of 1.7 juvenile cactus/ha, 0.4 subadult cactus/ha, and 1.0 adult cactus/ha (3.1 total stems/ha) are anomalously low for Galapagos in comparison to, for example, those on other comparable arid islands (Pinzón and Santa Fe Islands), which have stem densities of about 300–440/ha, that is, two orders of magnitude higher [57]. Our surveys also revealed a stage structure among cacti that is dramatically skewed to adults, thereby indicating a constraint on recruitment of young cacti into the adult population. Cactus populations elsewhere in Galapagos show a typical juvenile-biased size distribution [58], with a clear dominance of younger age classes typical of non-declining plant populations.

The tortoise re-introduction does appear to have a net positive impact on the cactus population, itself of conservation concern due to destruction by goats now removed. In this study we observed that the cactus population is increasing in one of the primary tortoise release sites, Las Tunas, confirming the suggestion [16] that tortoises have a modest yet positive impact on cactus populations. However, the population growth rate of adult cacti observed (about 7% over the last decade) is considerably less than the per decade rate at which the tortoise population has grown over the last four decades (about 200 tortoises or 33% population growth per decade). Notably tortoise population growth is comparable to the growth rate of subadult and juvenile cacti observed herein, or 37%. Only adult cactus substantially produce resources needed by tortoises (fruits, pads, shade) and the disparity of population growth rates of tortoises versus adult cacti indicate that ratios of tortoises to adult cacti will increase with time; however, as long as adult cactus survival remains high we predict a slowly improving status for the cactus population resulting from interactions with reintroduced tortoises, which foster recruitment of young cacti that will eventually transition to adult status producing resources (pads, fruits and shade) of use to tortoises.

Given that tortoises are now restricted to a small portion of Española Island and that tortoise habitat has likely degraded, it comes as little surprise that carrying capacity estimated for tortoises on Española appears to be quite low: approximately 2100 tortoises for the 1250 ha "tortoise zone" or 1.7 tortoises/hectare. This is striking given that tortoise densities on other islands in the archipelago have been reported as high as 8 tortoises/hectare [59]. Therefore, continued reintroductions of juvenile tortoises to the island would seem ill-advised. Currently juveniles have low survival rates and are therefore recruited infrequently into the population and likely play only a small role in sustaining it.

Instead of further repatriations, a more useful approach to improve the long-term outlook for this species to expand its ecological role on the island would be to improve habitat in such a way that the tortoises, now with significant numbers of older repatriates evidently reproducing, can colonize more of the island. More specifically, woody plant density is evidently a limiting factor to population growth given that a 50% reduction in woody plant density was projected to correspond to a 52% increase in the tortoise carrying capacity. Achieving this scale of restoration via manual removal of woody plants (the only currently available option) on half the island would, however, be extraordinarily expensive and logistically difficult given the remoteness of the island. Prescribed burns are not currently used as a habitat management tool in Galapagos given the unclear role of natural fires as an historical agent of natural disturbance in the archipelago [60], thus mechanical or chemical means may be required to reduce woody plant extent to levels more consistent with the recent history of the plant community on the island (Fig. 3).

Given the depleted and evidently declining status of tree cactus on the island combined with its importance to tortoises, a special focus on cactus management may be warranted given the near absence of cactus on much of the uncolonized parts of the island. Large-scale reseeding of cactus, now feasible given the availability of new technologies (e.g., [61]), could increase cactus recruitment, growth and survival dramatically. Given light competition between cactus and woody plants, cactus out-planting would best coincide with woody plant control (see Fig. 2c for potential cactus restoration areas with low slopes and low woody plant cover). Because tortoises are suspected to not only disperse cactus seeds and thereby increase recruitment but also mechanically flatten small cactus and thereby reduce cactus recruitment [19], a phased approach involving clearing of woody plants in new zones of the island unoccupied by tortoises and distant from occupied areas coupled with cactus out-plantings would be most effective. Timing of tortoise colonization could be controlled by establishing corridors to cactus groves in areas cleared of woody plants once sufficient number of out-planted cactus had grown to "tortoise-proof" size (subadult and older stages), that is, a decade or more after planting [62].

The reintroduction of an ecosystem engineer can have indirect benefits for other species of conservation concern [6],[8]. We have demonstrated that in the case of the Española tortoise in addition to recruitment benefits conferred to the endangered Opuntia cactus that is a primary tortoise food source, woody plant reduction could also aid the Waved Albatross (*Phoebastria irrorata*), a critically endangered seabird that effectively only nests on Española Island and that is likely limited by access to open-ground take-off and landing sites [48],[63]. Tortoises may be able to reduce or at least control woody plant density near albatross sites, but the current tortoise zone only marginally overlaps with that of the Waved Albatross, which are restricted to the southern part of the island [64]. Española tortoises and Waved Albatross may have a long co-evolutionary relationship: the absence of actual nest-building by albatross and their habit of rolling their eggs may be a response to the threat of trampling by tortoises [65]. The potential cross-species benefits of this tortoise reintroduction supports the more general claim that re-establishment of ecosystem engineers should be prioritized in conservation and ecological restoration [66–67]. For these benefits to be realized in the case of Española Island, however, we conclude that reinstatement of Española giant tortoises as fully functioning ecosystem engineers and not simply a viable population will likely require large-scale habitat restoration efforts in concert with further population restoration.

Supporting Information

Appendix S1 Contains a detailed description of demographic model and population projection model used.

Appendix S2 Contains R and WinBUGS code for estimating survival from 32 years of tortoise capture recapture data gathered on Española island.

Appendix S3 Contains R code for projecting population dynamics for the *in situ* population of Española giant tortoises.

Acknowledgments

Mark Teece kindly oversaw the stable carbon isotope analyses as part of a larger archipelago-wide analysis and he and Steven Blake provided many useful comments on the manuscript. The mark-recapture data set upon which many components of this study were based was generated over 32 years during many arduous site visits by many park guards of the Galapagos National Park Service, biologists from the Charles Darwin Research Station (now Charles Darwin Foundation) and cooperating scientists. The 2010 field surveys of tortoise-plant interactions as well as this synthesis of the accumulated information was made possible by support from the Galapagos Conservancy in collaboration with the Galapagos National Park Service, the Prometeo Program of Ecuador's National Secretariat for Higher Education, Science, Technology and Innovation, and U.S. National Science Foundation grant DEB 1258062.

Author Contributions

Conceived and designed the experiments: JG WT LC. Performed the experiments: JG WT LC. Analyzed the data: JG EH KS. Contributed reagents/materials/analysis tools: EH KS JG. Contributed to the writing of the manuscript: JG EH KS WT LC.

References

1. Griffith B, Scott JM, Carpenter JW, Reed C (1989) Translocation as a species conservation tool: status and strategy. Science 245: 477–480.
2. Seddon PJ, Armstrong DP, Maloney RF (2007) Developing the science of reintroduction biology. Conservation Biology 21: 303–312.
3. Dodd CK, Seigel RA (1991) Relocation, repatriation, and translocation of amphibians and reptiles: are they conservation strategies that work? Herpetologica 47: 336–350.
4. Armstrong DP, Seddon PJ (2008) Directions in reintroduction biology. Trends in Ecology and Evolution 23: 20–25.
5. Wilmers CC, Crabtree RL, Smith DW, Murphy KM, Getz WM (2003) Trophic facilitation by introduced top predators: grey wolf subsidies to scavengers in Yellowstone National Park. Journal of Animal Ecology 72: 909–916.
6. Beyer HL, Merrill EH, Varley N, Boyce MS (2007) Willow on Yellowstone's northern range: evidence for a trophic cascade? Ecological Applications 17: 1563–1571.
7. Johnson BE, Cushman JH (2007) Influence of a large herbivore reintroduction on plant invasions and community composition in a California grassland. Conservation Biology 21: 515–526.
8. Ripple WJ, Beschta RL (2012) Trophic cascades in Yellowstone: the first 15 years after wolf reintroduction. Biological Conservation 145: 205–213.
9. Hunter EA, Gibbs JP (2013) Densities of ecological replacement herbivores required to restore plant communities: a case study of giant tortoises on Pinta Island, Galapagos. Restoration Ecology: 10.1111/rec.12055.
10. Ripple WJ, Beschta RL (2007) Restoring Yellowstone's aspen with wolves. Biological Conservation 138: 514–519.
11. Davidson AD, Friggens MT, Shoemaker KT, Hayes CL, Erz J, et al. (2014) Population dynamics of reintroduced Gunnison's prairie dogs in the southern portion of their range. In press. Journal of Wildlife Management.
12. Byers JE, Cuddington K, Jones CG, Talley TS, Hastings A, et al. (2006). Using ecosystem engineers to restore ecological systems. Trends in Ecology and Evolution 21: 493–500.
13. Field KJ, Tracy CR, Medica PA, Marlow RW, Corn PS (2007) Return to the wild: translocation as a tool in conservation of the Desert Tortoise (*Gopherus agassizii*). Biological Conservation 136: 232–245.
14. Hamann O (1993) On vegetation recovery, goats and giant tortoises on Pinta Island, Galapagos, Ecuador. Biodiversity and Conservation 2: 138–151.
15. Truett J, Phillips M (2009) Beyond historic baselines: restoring Bolson tortoises to Pleistocene range. Ecological Restoration 27: 144–151.
16. Gibbs JP, Márquez C, Sterling EJ (2008) The role of endangered species reintroduction in ecosystem restoration: tortoise-cactus interactions on Española Island, Galapagos. Restoration Ecology 16: 88–93.
17. Hansen DM, Donlan CJ, Griffiths CJ, Campbell KJ (2010) Ecological history and latent conservation potential: large and giant tortoises as a model for taxon substitutions. Ecography 33: 272–284.
18. Griffiths CJ, Hansen DM, Jones CG, Zuël N, Harris S (2011) Resurrecting extinct interactions with extant substitutes. Current Biology 21: 762–765.
19. Gibbs JP, Sterling EJ, Zabala FJ (2010) Giant tortoises as ecological engineers: a long-term quasi-experiment in the Galapagos Islands. Biotropica 42: 208–214.
20. Blake S, Wikelski M, Cabrera F, Guezou A, Silva M, et al. (2012) Seed dispersal by Galapagos tortoises. Journal of Biogeography 82: 310–321.
21. Asner GP, Levick SR (2012) Landscape-scale effects of herbivores on treefall in African savannas. Ecology Letters 15: 1211–1217.
22. Cromsigt JPGM, Te Beest M (2014) Restoration of a megaherbivore–landscape-level impacts of white rhinoceros in Kruger National Park, South Africa. Journal of Ecology 102: 1365–2745.
23. Tortoise & Freshwater Turtle Specialist Group (1996) *Chelonoidis nigra* ssp. *hoodensis*. In: IUCN 2013. IUCN Red List of Threatened Species. Version 2013.2. <www.iucnredlist.org>. Downloaded on 28 January 2014.
24. Márquez C, Morillo G, Cayot LJ (1991) A 25-year management program pays off: repatriated tortoises on Española reproduce. Noticias de Galapagos 50: 17–18.
25. Cayot LJ (2008) The restoration of giant tortoise and land iguana populations in Galapagos. Galapagos Research 65: 39–43.
26. Walker BH, Ludwig D, Holling CS, Peterman RM (1981) Stability of semi-arid savanna grazing systems. Journal of Ecology 69: 473–498.
27. Archer S (1995) Tree-grass dynamics in a *Prosopis*-thornscrub savanna parkland: reconstructing the past and predicting the future. Ecoscience 2: 83–99.
28. Browning DM, Archer SR (2011) Protection from livestock fails to deter shrub proliferation in a desert landscape with a history of heavy grazing. Ecological Applications 21: 1629–1642.
29. Cayot LJ (1987) Ecology of giant tortoises (Geochelone elephantopus) in the Galapagos Islands. Syracuse University, Syracuse, New York.
30. Márquez C, Vargas H, Snell H, Mauchamp A, Gibbs J, et al. (2003) ¿Por qué tan pocas Opuntia en la Isla Española-Galapagos? Ecología Aplicada 2: 21–29.
31. Bai E, Boutton TW, Liu F, Wu XB, Archer SR (2012) Spatial patterns of soil $\delta^{13}C$ reveal grassland-to-woodland successional processes. Organic Geochemistry 42: 1512–1518.
32. Mueller-Dombois D, Ellenberg H (1974) Aims and methods of vegetation ecology. New York: Wiley.
33. De'ath G, Fabricius KE (2000) Classification and regression trees: a powerful yet simple technique for ecological data analysis. Ecology 81: 3178–3192.
34. R Development Core Team (2012) R: A language and environment for statistical computing. R Foundation for Statistical Computing, Vienna, Austria. ISBN 3-900051-07-0. Available: http://www.R-project.org/.
35. Hijmans RJ (2013) raster: Geographic data analysis and modeling. R package version 2.1-49. http://CRAN.R-project.org/package=raster.
36. Calenge C (2006) The package adehabitat for the R software: a tool for the analysis of space and habitat use by animals. Ecological Modelling 197: 516–519.
37. Crawley MJ (2007) The R Book. John Wiley & Sons, West Sussex, England.
38. Bolker BM (2008) Ecological models and data in R. Princeton, NJ: Princeton University Press.
39. Pearse D, Avise JC (2001) Turtle mating systems: behavior, sperm storage, and genetic paternity. Journal of Heredity 92: 206–211.
40. Pike DA, Pizzatto L, Pike BA, Shine R (2008) Estimating survival rates of uncatchable animals: the myth of high juvenile mortality in reptiles. Ecology 89: 607–611.
41. Tuberville TD, Norton TM, Todd BD, Spratt JS (2008) Long-term apparent survival of translocated gopher tortoises: A comparison of newly released and previously established animals. Biological Conservation 141: 2690–2697.
42. Milinkovitch MC, Monteyne D, Gibbs JP, Fritts TH, Tapia W, et al. (2004) Genetic analysis of a successful repatriation programme: giant Galapagos tortoises. Proceedings of the Royal Society of London 271: 341–345.
43. Schadt S, Revilla E, Wiegand T, Knauer F, Kaczensky P, et al. (2002) Assessing the suitability of central European landscapes for the reintroduction of Eurasian lynx. Journal of Applied Ecology 39: 189–203.
44. Reinert HK, Rupert R (1999) Impacts of translocation on behavior and survival of timber rattlesnakes, *Crotalus horridus*. Journal of Herpetology 33: 45–61.
45. Yott A, Rosatte R, Schaefer JA, Hamr J, Fryxell J (2010) Movement and spread of a founding population of reintroduced elk (*Cervus elaphus*) in Ontario, Canada. Restoration Ecology 19: 70–77.
46. Bourn D, Gibson C, Augeri D, Wilson CJ, Church J, et al. (1999) The rise and fall of the Aldabran giant tortoise population. Proceedings of the Royal Society B: Biological Sciences 266: 1091–1100.
47. Swingland I, Coe M (1979) The natural regulation of giant tortoise populations on Aldabra Atoll: Recruitment. Philosophical Transactions of the Royal Society of London Series B, Biological Sciences 286: 177–188.
48. Gibbs JP, Woltz HW (2010) A pilot survey of the central colony of the waved albatross *Phoebastria irrorata* on Española Island. Galapagos Research 67: 18–21.
49. Chiari Y, Claude J (2011) Study of the carapace shape and growth in two Galapagos tortoise lineages. Journal of Morphology 272: 379–386.
50. Chiari Y, Hyseni C, Fritts TH, Glaberman S, Marquez C, et al. (2009) Morphometrics parallel genetics in a newly discovered and endangered taxon of Galapagos tortoise. PLoS One 4: e6272.
51. Nicoll MAC, Jones CG, Norris K (2003) Declining survival rates in a reintroduced population of the Mauritius kestrel: evidence for non-linear density

dependence and environmental stochasticity. Journal of Animal Ecology 72: 917–926.
52. Armstrong DP, Davidson RS, Perrott JK, Roygard J, Buchanan L (2004) Density-dependent population growth in a reintroduced population of North Island saddlebacks. Journal of Animal Ecology 74: 160–170.
53. Congdon JD, Dunham AE, van Loben Sels RC (1993) Delayed sexual maturity and demographics of Blanding's turtles (*Emydoidea blandingii*): implications for conservation and management of long-lived organisms. Conservation Biology 7: 826–833.
54. Shoemaker KT, Breisch AR, Jaycox JW, Gibbs JP (2014) Disambiguating the minimum viable population concept: Response to Reed and McCoy. In Press. Conservation Biology.
55. Froyd CA, Coffey EED, van der Knaap WO, van Leeuwen JFN, Tye A, et al. (2014) The ecological consequences of megafaunal loss: giant tortoises and wetland biodiversity. Ecology Letters 17: 144–154.
56. Scheffer M, Carpenter S, Foley JA, Folke C, Walker B (2001) Catastrophic shifts in ecosystems. Nature 413: 591–596.
57. Racine CH, Downhower JF (1974) Vegetative and reproductive strategies of *Opuntia* (Cactaceae) in the Galapagos Islands. Biotropica 6: 175–186.
58. Hicks DJ, Mauchamp A (2000) Population structure and growth patterns of *Opuntia echios* var. *gigantea* along an elevational gradient in the Galapagos Islands. Biotropica 32: 235–243.
59. Márquez C, Wiedenfeld D, Snell H, Fritts T, MacFarland C, et al. (2004) Estado actual de las poblaciones de tortugas terrestres gigantes (*Geochelone* spp., Chelonia: Testudinae) en las Islas Galapagos. Ecologia Aplicada 3: 98–111.
60. Trueman M, Hobbs RJ, Van Niel K (2013). Interdisciplinary historical vegetation mapping for ecological restoration in Galapagos. Landscape Ecology 28: 519–532.
61. Liu M, Zhiyang L, Ren W (2014) Research on the effect of waterboxx technology on *Haloxylon ammodendron* afforestation in arid and semiarid areas. Advanced Science, Engineering and Medicine 6: 236–239.
62. Hicks DJ, Mauchamp A (1996) Evolution and conservation biology of the Galapagos opuntias (Cactaceae). Haseltonia 4: 89–102.
63. Douglas H (1998) Changes in the distribution and abundance of waved albatrosses at Isla Española, Galapagos Islands, Ecuador. Condor 100: 737–740.
64. Anderson DJ, Huyvaert KP, Apanius V, Townsend H, Gillikin CL, et al. (2002) Population size and trends of the waved albatross *Phoebastria irrorata*. Marine Ornithology 30: 63–69.
65. Anderson DJ (2008) Waved Albatross. In T. De Roy, M. Jones, and J. Fitter, editors. Albatross: their world, their ways. Richmond Hill, Ontario, Canada: Firefly Books Ltd.
66. Soulé ME, Estes JA, Berger J, Del Rio CM (2003) Ecological effectiveness: conservation goals for interactive species. Conservation Biology 17: 1238–1250.
67. Hobbs RJ, Cramer VA (2008) Restoration ecology: interventionist approaches for restoring and maintaining ecosystem function in the face of rapid environmental change. Annual Review of Environment and Resources 33: 39–61.
68. Márquez C, Cayot LJ, Rea S (1999) La Crianza de tortugas gigantes en cautiverio: Un manual operativo. Quito, Ecuador: A & B Editores.

PERMISSIONS

All chapters in this book were first published in PLOS ONE, by The Public Library of Science; hereby published with permission under the Creative Commons Attribution License or equivalent. Every chapter published in this book has been scrutinized by our experts. Their significance has been extensively debated. The topics covered herein carry significant findings which will fuel the growth of the discipline. They may even be implemented as practical applications or may be referred to as a beginning point for another development.

The contributors of this book come from diverse backgrounds, making this book a truly international effort. This book will bring forth new frontiers with its revolutionizing research information and detailed analysis of the nascent developments around the world.

We would like to thank all the contributing authors for lending their expertise to make the book truly unique. They have played a crucial role in the development of this book. Without their invaluable contributions this book wouldn't have been possible. They have made vital efforts to compile up to date information on the varied aspects of this subject to make this book a valuable addition to the collection of many professionals and students.

This book was conceptualized with the vision of imparting up-to-date information and advanced data in this field. To ensure the same, a matchless editorial board was set up. Every individual on the board went through rigorous rounds of assessment to prove their worth. After which they invested a large part of their time researching and compiling the most relevant data for our readers.

The editorial board has been involved in producing this book since its inception. They have spent rigorous hours researching and exploring the diverse topics which have resulted in the successful publishing of this book. They have passed on their knowledge of decades through this book. To expedite this challenging task, the publisher supported the team at every step. A small team of assistant editors was also appointed to further simplify the editing procedure and attain best results for the readers.

Apart from the editorial board, the designing team has also invested a significant amount of their time in understanding the subject and creating the most relevant covers. They scrutinized every image to scout for the most suitable representation of the subject and create an appropriate cover for the book.

The publishing team has been an ardent support to the editorial, designing and production team. Their endless efforts to recruit the best for this project, has resulted in the accomplishment of this book. They are a veteran in the field of academics and their pool of knowledge is as vast as their experience in printing. Their expertise and guidance has proved useful at every step. Their uncompromising quality standards have made this book an exceptional effort. Their encouragement from time to time has been an inspiration for everyone.

The publisher and the editorial board hope that this book will prove to be a valuable piece of knowledge for researchers, students, practitioners and scholars across the globe.

LIST OF CONTRIBUTORS

Maria Włodarska-Kowalczuk, Emilia Jankowska, Lech Kotwicki and Piotr Balazy
Department of Marine Ecology, Institute of Oceanology Polish Academy of Sciences, Sopot, Poland

Francesca Rossi
Laboratoire Ecologie des Systèmes marins côtiers, Université Montpellier 2, Montpellier, France

Britta Gribsholt
Department of Ecosystems, Royal Netherlands Institute for Sea Research, Yerseke, the Netherlands

Frederic Gazeau
Centre National de la Recherche Scientifique-Institut National des Sciences de l'Univers, Laboratoire d' Oceanographie de Villefranche, Villefranche-sur-mer, France
Université Pierre et Marie Curie, Observatoire Oce´anologique de Villefranche, Villefranche-sur-mer, France

Valentina Di Santo
Department of Biology, Boston University, Boston, Massachusetts, United States of America

Jack J. Middelburg
Department of Ecosystems, Royal Netherlands Institute for Sea Research, Yerseke, the Netherlands
Department of Earth Sciences - Geochemistry, Faculty of Geosciences, Utrecht University, Utrecht, The Netherlands

Johan Erlandsson
ARONIA Coastal Zone Research Team, Åbo Akademi University/Novia University of Applied Sciences, Ekenäs, Finland

Christopher D. McQuaid
Coastal Research Group, Department of Zoology and Entomology, Rhodes University, Grahamstown, South Africa

Martin Sköld
Department of Mathematics, Stockholm University, Stockholm, Sweden

Ricard V. Solé
ICREA-Complex Systems Lab, Universitat Pompeu Fabra, Barcelona, Spain
Santa Fe Institute, Santa Fe, New Mexico, United States of America
Institut de Biologia Evolutiva, UPF-CSIC, Barcelona, Spain
European Centre for Living Technology, C Foscari University of Venice, Venice, Italy

Sergi Valverde
ICREA-Complex Systems Lab, Universitat Pompeu Fabra, Barcelona, Spain
Institut de Biologia Evolutiva, UPF-CSIC, Barcelona, Spain
European Centre for Living Technology, C Foscari University of Venice, Venice, Italy

Tyler B. Muhly
Faculty of Environmental Design, University of Calgary, Calgary, Alberta, Canada
Ecosystem Management, Alberta Innovates – Technology Futures, Vegreville, Alberta, Canada

Mark Hebblewhite
Department of Ecosystem and Conservation Sciences, University of Montana, Missoula, Montana, United States of America

Dale Paton and Marco Musiani
Faculty of Environmental Design, University of Calgary, Calgary, Alberta, Canada

Justin A. Pitt and Mark S. Boyce
Department of Biological Sciences, University of Alberta, Edmonton, Alberta, Canada

Tjisse van der Heide and Han Olff
Community and Conservation Ecology Group, Centre for Ecological and Evolutionary Studies (CEES), Groningen University, Haren, The Netherlands

Egbert H. van Nes
Aquatic Ecology and Water Quality Management Group, Department of Environmental Sciences, Wageningen University, Wageningen, The Netherlands

Marieke M. van Katwijk
Department of Environmental Science, Institute for Wetland and Water Research, Faculty of Science, Radboud University Nijmegen, Nijmegen, The Netherlands

Alfons J. P. Smolders
Department of Aquatic Ecology and Environmental Biology, Institute for Water and Wetland Research, Faculty of Science, Radboud University Nijmegen, Nijmegen, The Netherlands

Shuqing Zhao, Chao Zhu, Decheng Zhou and Dian Huang
College of Urban and Environmental Sciences, and Key Laboratory for Earth Surface Processes of the Ministry of Education, Peking University, Beijing, China

Jeremy Werner
U.S. Geological Survey (USGS) Earth Resources Observation and Science (EROS) Center, Sioux Falls, South Dakota, United States of America

Erin K. Cameron, Heather C. Proctor and Erin M. Bayne
Department of Biological Sciences, University of Alberta, Edmonton, Alberta, Canada

Xiao-dong Zhang
Coastal Ecosystems Research Station of the Yangtze River Estuary, Ministry of Education Key Laboratory for Biodiversity Science and Ecological Engineering, Institute of Biodiversity Science, Fudan University, Shanghai, China
Key Laboratory of Wetland Services and Restoration, Institute of Wetland Research, Chinese Academy of Forestry, Beijing, China

Xin Jia
School of Soil and Water Conservation, Beijing Forestry University, Beijing, China

Yang-yun Chen, Jun-jiong Shao, Xin-ru Wu, Lei Shang and Bo Li
Coastal Ecosystems Research Station of the Yangtze River Estuary, Ministry of Education Key Laboratory for Biodiversity Science and Ecological Engineering, Institute of Biodiversity Science, Fudan University, Shanghai, China

Tjisse van der Heide, Han Olff and Ellen J. Weerman
Community and Conservation Ecology Group, Centre for Ecological and Evolutionary Studies (CEES), University of Groningen, Groningen, The Netherlands

Egbert H. van Nes
Aquatic Ecology and Water Quality Management Group, Department of Environmental Sciences, Wageningen University, Wageningen, The Netherlands

Johan S. Eklöf
Marine Benthic Ecology and Evolution, Centre for Ecological and Evolutionary Studies (CEES), University of Groningen, Groningen, The Netherland
Department of Biology and Environmental Sciences - Kristineberg, University of Gothenburg, Fiskebäckskil, Sweden
Department of Systems Ecology, Stockholm University, Stockholm, Sweden

Els M. van der Zee
Department of Marine Ecology, NIOZ Royal Netherlands Institute for Sea Research, Den Burg, Texel, The Netherlands
Animal Ecology Group, Centre for Ecological and Evolutionary Studies (CEES), University of Groningen, Groningen, The Netherlands

Serena Donadi and Britas Klemens Eriksson
Marine Benthic Ecology and Evolution, Centre for Ecological and Evolutionary Studies (CEES), University of Groningen, Groningen, The Netherlands

Jill L. Edmondson, Odhran S. O'Sullivan, Jonathan Potter, Nicola McHugh and Jonathan R. Leake
Department of Animal and Plant Sciences, University of Sheffield, Sheffield, United Kingdom

Richard Inger and Kevin J. Gaston
Environment and Sustainability Institute, University of Exeter, Penryn, Cornwall, United Kingdom

Lee A. Vierling
Department of Forest, Rangeland, and Fire Sciences, McCall Outdoor Science School, University of Idaho, Moscow, Idaho, United States of America

Kerri T. Vierling
Department of Fish and Wildlife Sciences, University of Idaho, Moscow, Idaho, United States of America

Patrick Adam
Environmental Science, University of Idaho, Moscow, Idaho, United States of America

Andrew T. Hudak
Rocky Mountain Research Station, US Forest Service, Moscow, Idaho, United States of America

Theresa Sinicrope Talley and Kim-Chi Nguyen
Scripps Institution of Oceanography, University of California San Diego, La Jolla, California, United States of America

Anthony Nguyen
Department of Biology, University of San Diego, San Diego, California, United States of America

Nancy Calderó n-Cortés and Mauricio Quesada
Centro de Investigaciones en Ecosistemas, Universidad Nacional Autónoma de México, Morelia, Michoacán, México

Luis H. Escalera-Vázquez
Departamento de Zoología, Instituto de Biología, Universidad Nacional Autónoma de México, México D.F., México

Carl Van Colen and Magda Vincx
Ghent University, Department of Biology, Marine Biology Research Group, Ghent, Belgium

Simon F. Thrush
National Institute of Water and Atmospheric Research, Hamilton, New Zealand
University of Auckland, School of Environment, Auckland, New Zealand

Tom Ysebaert
Netherlands Institute for Sea Research (NIOZ-Yerseke), Yerseke, The Netherlands

Johan S. Eklöf, Serena Donadi and Britas Klemens Eriksson
Department of Marine Benthic Ecology and Evolution, Center for Ecology & Evolutionary Studies, Groningen University, Groningen, The Netherlands

Tjisse van der Heide
Community and Conservation Ecology Group, Center for Ecology & Evolutionary Studies, Groningen University, Groningen, The Netherlands

Els M. van der Zee
Animal Ecology Group, Center for Ecology & Evolutionary Studies, Groningen University, Groningen, The Netherlands
Department of Marine Ecology, NIOZ Royal Netherlands Institute for Sea Research, Den Burg,Texel, The Netherlands

Robert O'Hara
Biodiversity and Climate Research Center, Frankfurt, Germany

Hugo Robles
Department of Forest and Conservation Sciences, Centre for Applied Conservation Research, University of British Columbia, Vancouver, British Columbia, Canada

Kathy Martin
Department of Forest and Conservation Sciences, Centre for Applied Conservation Research, University of British Columbia, Vancouver, British Columbia, Canada
Pacific Wildlife Research Centre, Environment Canada, Delta, British Columbia, Canada

Cynnamon Dobbs
School of Botany, The University of Melbourne, Melbourne, Australia
School of Forest Science and Ecosystem, Melbourne School of Land and Environment, The University of Melbourne, Melbourne, Australia

Craig R. Nitschke
School of Forest Science and Ecosystem, Melbourne School of Land and Environment, The University of Melbourne, Melbourne, Australia

Dave Kendal
Australian Research Centre for Urban Ecology, Royal Botanic Gardens Melbourne, c/o School of Botany, The University of Melbourne, Melbourne, Australia

Zhiwei Xu, Xinyu Zhang, Juan Xie, Guofu Yuan, Xinzhai Tang, Xiaomin Sun and Guirui Yu
Key Laboratory of Ecosystem Network Observation and Modeling, Institute of Geographic Sciences and Natural Resources Research, Chinese Academy of Sciences, Beijing, China

Kirbee L. Parsons, Alice Y. T. Feng and David M. Patrick
School of Population and Public Health, University of British Columbia, Vancouver, British Columbia, Canada

Chelsea G. Himsworth
School of Population and Public Health, University of British Columbia, Vancouver, British Columbia, Canada
Animal Health Centre, British Columbia Ministry of Agriculture, Abbotsford, British Columbia, Canada

Thomas Kerr
British Columbia Centre for Excellence in HIV/AIDS, St. Paul's Hospital, Vancouver, British Columbia, Canada

Department of Medicine, University of British Columbia, Vancouver, British Columbia, Canada

Claire M. Jardine
Department of Pathobiology, University of Guelph, Guelph, Ontario, Canada

James P. Gibbs
Department of Environmental and Forest Biology, State University of New York College of Environmental Science and Forestry, Syracuse, New York, United States of America

Elizabeth A. Hunter
Warnell School of Forestry and Natural Resources, University of Georgia, Athens, Georgia, United States of America

Kevin T. Shoemaker
Department of Ecology and Evolution, Stony Brook University, Stony Brook, New York, United States of America

Washington H. Tapia
Department of Applied Research, Galapagos National Park Service, Puerto Ayora, Galapagos, Ecuador

Linda J. Cayot
Galapagos Conservancy, Fairfax, Virginia, United States of America

Index